Current Natural Sciences

Max CERF

Optimization Techniques I

Continuous Optimization

Science Press | edp sciences

EDP Sciences – ISBN(print): 978-2-7598-3162-3 – ISBN(ebook): 978-2-7598-3164-7
DOI: 10.1051/978-2-7598-3162-3

I would like to express my sincere thanks to the people who helped me to produce this book, especially:

France Citrini *for having accepted the publication of the book by EDP Sciences;*

Sophie Hosotte *and* ***Magalie Jolivet*** *for their assistance in the editorial process;*

Thomas Haberkorn *for his verification of scientific content, and for the countless numerical tricks he shared with me;*

Emmanuel Trélat *for his preface, proofreading and presentation advices, and also for our long-standing collaboration and the expertise I have acquired owing to him;*

Claude Blouvac*† my high school teacher who set me on the path of mathematics in 1982 and who would have been very happy to see this book.*

Preface

Algorithms are omnipresent in our modern world. They show us the optimal routes, prevent and assess risks, provide forecasts, anticipate or assist our decisions. They have become an essential part of our daily lives. These algorithms are mostly based on optimization processes and consist in minimizing or maximizing a criterion under certain constraints, thus indicating feasible and intelligent solutions, allowing us to plan a process to be carried out in the best possible way.

There are many different optimization methods, based on various heuristics, sometimes simple and intuitive, sometimes more elaborate and requiring fine mathematical developments.

It is through this jungle of algorithms, resulting from decades of research and development, that Max Cerf guides us with all his expertise, his intuition and his pragmatism.

Max Cerf has been a Senior engineer at ArianeGroup for 30 years. As a recognized specialist in trajectography, he designs and optimizes spacecraft trajectories under multiple constraints. He has thus acquired and developed a comprehensive knowledge of the best algorithms in continuous optimization, discrete optimization (on graphs) and optimal control. He is also an exceptional teacher, with a real talent for explaining complicated concepts in a clear and intuitive way.

With this double book, he offers an invaluable guide to the non-specialist reader who wishes to understand or solve optimization problems in the most efficient way possible.

Emmanuel Trélat, Sorbonne University, Paris

Introduction

Mathematical optimization has undergone a continuous development since the Second World War and the advent of the first computers. Facing the variety of available algorithms and software, it is sometimes difficult for a non-specialist to find his way around. The aim of this two-volume book is to provide an overview of the field. It is intended for students, teachers, engineers and researchers who wish to acquire a general knowledge of mathematical optimization techniques.

Optimization aims to control the inputs (variables) of a system, process or model to obtain the desired outputs (constraints) at the best cost. Depending on the nature of the inputs to be controlled, a distinction is made between continuous, discrete or functional optimization.

Continuous optimization (Chapters 1 to 5 of Volume I) deals with real variable problems. Chapter 1 presents the optimality conditions for differentiable functions and the numerical calculation of derivatives. Chapters 2, 3 and 4 give an overview of gradient-free, unconstrained and constrained optimization methods. Chapter 5 is devoted to continuous linear programming, with the simplex and interior point methods.

Discrete optimization (Chapters 1 and 2 of Volume II) deals with integer variable problems. Chapter 1 deals with mixed variable linear programming by cutting or tree methods. Chapter 2 presents an overview of combinatorial problems, their modelling by graphs and specific algorithms for path, flow or assignment problems.

Functional optimization (Chapters 3 to 5 of Volume II) deals with infinite dimension problems. The input to be controlled is a function and no longer a finite number of variables. Chapter 3 introduces the notions of functional and calculus of variations. Chapter 4 presents optimal control problems and optimality conditions. Chapter 5 is devoted to numerical methods (integration of differential equations, direct and indirect methods).

In each chapter, the theoretical developments and the demonstrations are limited to the essentials. The algorithms are accompanied by detailed examples to facilitate their understanding.

Table of contents

1. Continuous optimization

Continuous optimization also called parametric optimization deals with real variable problems. This chapter presents the main theoretical notions forming the basis of numerical algorithms.

Section 1 introduces the standard formulation of a constrained optimization problem. After a reminder about functions of several variables, the optimization problem is tackled through the notions of level lines and directions of descent. This geometrical approach illustrates the constraint effect on the solution.

Section 2 is devoted to the calculation of derivatives by finite differences. This method is the only one that can be used in most practical applications. Its efficiency depends on the increment setting which is discussed in detail.

Section 3 presents the reduction methods. The objective is to use the constraints to eliminate part of the variables so that the problem can be locally reduced to an unconstrained one. The reduction methods are first detailed for linear constraints and then generalized to nonlinear constraints.

Section 4 formulates global optimality conditions based on the existence of a saddle point of the Lagrangian. These conditions are applicable especially to linear programming, but they are much more difficult to apply in the general case of nonlinear problems.

Section 5 formulates the local optimality conditions of Karush, Kuhn and Tucker. These conditions, which form the basis of continuous optimization and numerical algorithms, lead to a system of nonlinear equations and inequalities. The focus is on their geometrical interpretation in connection with level lines.

1.1 Formulation

This section introduces the basic notions of continuous optimization. After some reminders on functions of several variables, the focus is set on the geometric interpretation of level lines and directions of descent. The characteristics of the solution and the effect of the constraints are then discussed.

1.1.1 Standard form

The standard form of a continuous optimization problem is as follows.

$$\min_{x\in\mathbb{R}^n} f(x) \;\text{ s.t. }\; \begin{cases} c_E(x) = 0 \\ c_I(x) \le 0 \end{cases} \tag{1.1}$$

("s.t." is the abbreviation for "subject to").

The vector $x \in \mathbb{R}^n$ represents the n variables or parameters of the problem.

$$x = \begin{pmatrix} x_1 \\ \vdots \\ x_n \end{pmatrix} \in \mathbb{R}^n \tag{1.2}$$

The function to be minimized $f:\mathbb{R}^n \to \mathbb{R}$ is called either the cost function or the objective function or the criterion.
The function $c_E:\mathbb{R}^n \to \mathbb{R}^p$ represents a vector of p equality constraints.

$$c_E(x) = \begin{pmatrix} c_{E1}(x) \\ \vdots \\ c_{Ep}(x) \end{pmatrix} \quad \text{with} \quad c_{Ej} : \mathbb{R}^n \to \mathbb{R} \tag{1.3}$$

The function $c_I:\mathbb{R}^n \to \mathbb{R}^q$ represents a vector of q inequality constraints.

$$c_I(x) = \begin{pmatrix} c_{I1}(x) \\ \vdots \\ c_{Iq}(x) \end{pmatrix} \quad \text{with} \quad c_{Ij} : \mathbb{R}^n \to \mathbb{R} \tag{1.4}$$

The constraints c_E and c_I are grouped in the vector c of dimension $m = p + q$.

$$c(x) = \begin{pmatrix} c_E(x) \\ c_I(x) \end{pmatrix} \in \mathbb{R}^m \tag{1.5}$$

Any optimization problem can be put into the standard form (1.1) owing to the following transformations:

- maximization/minimization: $\max\limits_x f(x) \;\Leftrightarrow\; \min\limits_x -f(x)$;
- superior/inferior constraint: $c(x) \geq 0 \;\Leftrightarrow\; -c(x) \leq 0$.

An inequality constraint can be transformed into an equality constraint by introducing a positive slack variable. The positivity constraint on the slack variable is simpler than a general nonlinear inequality.

- inequality/equality constraint: $c(x) \leq 0 \;\Leftrightarrow\; \begin{cases} c(x) + y = 0 \\ y \geq 0 \end{cases}$.

Some algorithms assume moreover positive variables. This is done by replacing a variable (of any sign) with the difference of two positive variables.

- switching to positive variables: $x = x^+ - x^-$ with $\begin{cases} x^+ \geq 0 \\ x^- \geq 0 \end{cases}$.

1.1.2 Function of several variables

The functions f and c are assumed to be "sufficiently differentiable".
The gradient of the function f is the vector of its first partial derivatives.

$$\nabla f = \begin{pmatrix} \dfrac{\partial f}{\partial x_1} \\ \vdots \\ \dfrac{\partial f}{\partial x_n} \end{pmatrix} \tag{1.6}$$

By convention the notation ∇f will represent a <u>column vector</u> of $\mathbb{R}^n$.

The constraint gradient matrix has the constraint gradients as columns.
The Jacobian matrix J of the constraints is the transpose of the gradient matrix.

$$\nabla c = \left(\nabla c_1, \cdots, \nabla c_m\right) = \begin{pmatrix} \dfrac{\partial c_1}{\partial x_1} & \cdots & \dfrac{\partial c_m}{\partial x_1} \\ \vdots & \ddots & \vdots \\ \dfrac{\partial c_1}{\partial x_n} & \cdots & \dfrac{\partial c_m}{\partial x_n} \end{pmatrix} = J^T \tag{1.7}$$

The Hessian of the function f is the matrix of its second partial derivatives. This matrix of $\mathbb{R}^{n\times n}$ is symmetric.

$$\nabla^2 f = \begin{pmatrix} \frac{\partial^2 f}{\partial^2 x_1} & \cdots & \frac{\partial^2 f}{\partial x_1 \partial x_n} \\ \vdots & \ddots & \vdots \\ \frac{\partial^2 f}{\partial x_n \partial x_1} & \cdots & \frac{\partial^2 f}{\partial^2 x_n} \end{pmatrix} \tag{1.8}$$

The Taylor expansion of a function of one variable in the point x_0 is

$$f(x) = f(x_0) + h\frac{\partial f}{\partial x}(x_0) + \cdots + \frac{h^k}{k!}\frac{\partial^k f}{\partial x^k}(x_0) + \cdots \tag{1.9}$$

This formula gives the value of f in $x_0 + h$, where h is the increment.

For a function of several variables, Taylor's formula is generalized by expanding with respect to each variable and grouping terms of the same order. The increment h is then a <u>vector</u> of $\mathbb{R}^n$ with components $(h_1, \cdots, h_n)$.

$$f(x) = f(x_0) + D_h(x_0) + \cdots + \frac{1}{k!}D_h^k f(x_0) + \cdots \tag{1.10}$$

with the notation: $D_h^k f = \left(\sum_{i=1}^{n} h_i \frac{\partial}{\partial x_i}\right)^k f$, where k is the order of derivation.

The first two terms are expressed in terms of the gradient and the Hessian.

$$f(x_0 + h) = f(x_0) + \nabla f(x_0)^T h + \frac{1}{2}h^T \nabla^2 f(x_0) h + \cdots \tag{1.11}$$

<u>Remark on notations</u>

The notation x_0 refers here to a vector of $\mathbb{R}^n$. In some formulas, the notation x_k may designate the k-th component of the vector x. In case of ambiguity, the meaning will be clarified.

In the case of a quadratic function defined by a symmetric matrix $Q \in \mathbb{R}^{n\times n}$ and a vector $c \in \mathbb{R}^n$, the gradient and the Hessian are given by

$$f(x) = \frac{1}{2}x^T Q x + c^T x \quad \Rightarrow \quad \begin{cases} \nabla f = Qx + c \\ \nabla^2 f = Q \end{cases} \tag{1.12}$$

1.1.3 Level lines

A level line of a function is the set of points in which the function takes the same value. The level line noted L_0 passing through the point $x_0 \in \mathbb{R}^n$ is defined by

$$L_0 = \{x \in \mathbb{R}^n \ / \ f(x) = f(x_0)\} \tag{1.13}$$

A level line in $\mathbb{R}^n$ is a hypersurface (space of dimension n – 1). To simplify the text the term "level line" will be used everywhere.

The drawing of level lines in dimension 2 helps understanding the main concepts of optimization. This plot is comparable to a geographical map on which the topography is materialized by iso-altitude curves. The minima and maxima correspond respectively to valleys and peaks surrounded by level lines. Tightened level lines indicate a "steep" slope. The following example shows the level lines of the so-called Rosenbrock function.

Example 1-1: Rosenbrock function

This function is defined by: $f(x_1,x_2)=100\left(x_2-x_1^2\right)^2+\left(1-x_1\right)^2$.

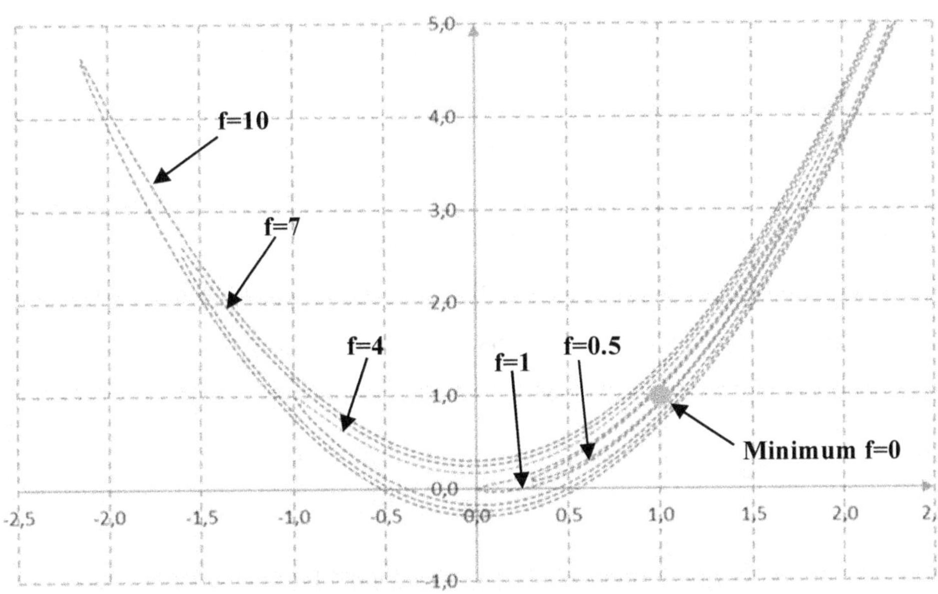

Figure 1-1: Level lines of the Rosenbrock function.

This function exhibits a long narrow "valley" around its minimum in (1 ; 1).
The Rosenbrock function is sometimes called "banana function" because of the shape of its level lines.
Figure 1-1 shows the lines of respective levels 0 ; 0.5 ; 1 ; 4 ; 7 ; 10.

1.1.4 Direction of descent

A small move d from the point x_0 produces a variation of the function f. The value of f in the new point $x = x_0 + d$ is given by the Taylor expansion to the first order. Noting $g_0 = \nabla f(x_0)$ the gradient of f in the point x_0, we have

$$f(x_0 + d) = f(x_0) + g_0^T d + \cdots \tag{1.14}$$

The variation of f depends on the scalar product $g_0^T d$ and therefore on the angle between the move direction d and the gradient g_0:

- a move along the gradient $(g_0^T d > 0)$ produces an increase of f.
  The direction d is an ascent direction;
- a move opposite to the gradient $(g_0^T d < 0)$ produces a decrease of f.
  The direction d is a descent direction;
- a move orthogonal to the gradient $(g_0^T d = 0)$ does not change f.
  The direction d is tangent to the level line.

The gradient has therefore the following geometrical property.

Property 1-1 : Direction of the gradient vector

The gradient at a point is orthogonal to the level line passing through that point.

It also represents the direction of steepest ascent at this point.

Figure 1-2 shows the level line passing in x_0, whose equation is $f(x) = f(x_0)$, and the minimum located in x*. This level line separates the space into two domains:

- inwards are the points x such that $f(x) < f(x_0)$ (shaded area);
- outwards are the points x such that $f(x) > f(x_0)$.

The hyperplane tangent to the level line in x_0 determines the directions of ascent on the one hand and the directions of descent on the other. Among the three directions of descent plotted, it can be seen that the direction d_2 is the one that passes closest to the minimum x* while the direction d_1 is farther away.

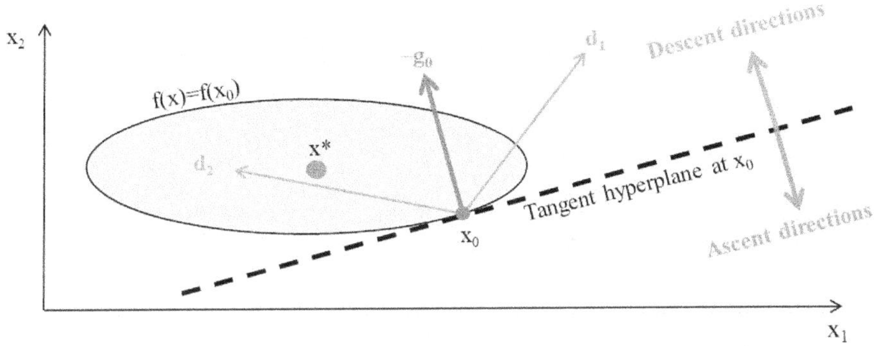

Figure 1-2: Ascent/descent directions and level line.

1.1.5 Directional variation

The variation of f along a given unit vector d is expressed by the function φ.

$$\varphi(s) \underset{def}{=} f(x_0 + sd) \tag{1.15}$$

This "directional function" depends only on the real variable s which represents the step length along the unit vector $d \in \mathbb{R}^n$ (direction of move).

The second order Taylor expansion of the function φ in $s = 0$ is

$$\varphi(s) = \varphi(0) + s\varphi'(0) + \frac{1}{2}s^2\varphi''(0) + \cdots \tag{1.16}$$

Let us identify this expansion with that of f noting $g_0 = \nabla f(x_0)$ and $H_0 = \nabla^2 f(x_0)$

$$f(x_0 + sd) = f(x_0) + sg_0^T d + \frac{1}{2}s^2 d^T H_0 d + \cdots \tag{1.17}$$

This yields the derivatives of φ depending on the gradient and Hessian of f in x_0.

$$\begin{cases} \varphi'(0) = g_0^T d \\ \varphi''(0) = d^T H_0 d \end{cases} \tag{1.18}$$

The first derivative of φ is called the **directional derivative** of f along d. It is positive for an ascent direction, negative for a descent direction.

The second derivative of φ is called the curvature of f along d. A strong curvature indicates a fast variation of f and tight level lines. The value of the curvature depends on the eigenvalues of the Hessian H_0.

Let us recall some useful definitions and properties of eigenvalues:

- a matrix $A \in \mathbb{R}^{n\times n}$ has the eigenvalue $\sigma \in \mathbb{R}$ and the associated eigenvector $u \in \mathbb{R}^n$ (non null) if $Au = \sigma u$;
- a real symmetric matrix $A \in \mathbb{R}^{n\times n}$ has n real eigenvalues and it admits an orthonormal basis of eigenvectors;
- a matrix $A \in \mathbb{R}^{n\times n}$ is positive semidefinite if $\forall u \in \mathbb{R}^n$, $u^T A u \geq 0$.
A positive semidefinite symmetric matrix has positive or zero eigenvalues.
We will note in abbreviated form: $A \geq 0$;
- a matrix $A \in \mathbb{R}^{n\times n}$ is positive definite if $\forall u \in \mathbb{R}^n$, $u \neq 0$, $u^T A u > 0$.
A positive definite symmetric matrix has strictly positive eigenvalues.
We will note in abbreviated form: $A > 0$.

The symmetric matrix H_0 has n real eigenvalues ordered by increasing absolute values $|\sigma_1| \leq |\sigma_2| \leq \cdots \leq |\sigma_n|$ and it admits an orthonormal basis of associated eigenvectors $(u_1, u_2, \cdots, u_n)$ satisfying

$$H_0 u_k = \sigma_k u_k \quad \text{with} \quad u_i^T u_j = \begin{cases} 0 & \text{if } i \neq j \\ 1 & \text{if } i = j \end{cases} \tag{1.19}$$

By decomposing the unit direction d on this orthonormal basis

$$d = \alpha_1 u_1 + \cdots + \alpha_n u_n \tag{1.20}$$

then replacing in (1.18), we obtain the value of the second derivative of φ.

$$\varphi''(0) = \alpha_1^2 \sigma_1 + \cdots + \alpha_n^2 \sigma_n \tag{1.21}$$

Since d is a unit vector $(\alpha_1^2 + \cdots + \alpha_n^2 = 1)$ and the eigenvalues are in increasing order, we deduce that the curvature of f is comprised between $|\sigma_1|$ and $|\sigma_n|$.

The ratio $\kappa = \frac{|\sigma_n|}{|\sigma_1|}$ from the largest to the smallest eigenvalue is called the **conditioning** of the matrix H_0. A conditioning close to 1 indicates that the function varies similarly in all directions, resulting in almost circular level lines. A large conditioning results in level lines "flattened" along the directions (eigenvectors) associated to the largest eigenvalues.

Example 1-2 : Conditioning of the Rosenbrock function

Let us consider the Rosenbrock function: $f(x_1,x_2)=100\left(x_2-x_1^2\right)^2+\left(1-x_1\right)^2$.

Its Hessian is: $H(x_1,x_2)=\begin{pmatrix} -400(x_2-3x_1^2)+2 & -400x_1 \\ -400x_1 & 200 \end{pmatrix}$.

The Hessian in the point $x^*(1\,;1)$ is: $H^*=\begin{pmatrix} 802 & -400 \\ -400 & 200 \end{pmatrix}$.

The eigenvalues are the solutions of the characteristic equation.

$$\det(H^*-\sigma I)=\det\begin{pmatrix} 802-\sigma & -400 \\ -400 & 200-\sigma \end{pmatrix}=0 \quad \rightarrow \quad \begin{cases} \sigma_1 = 1001{,}6 \\ \sigma_2 = 0{,}3994 \end{cases}$$

The large conditioning ($\kappa = 2508$) results in very elongated level lines around the minimum x^* as shown in figure 1-1.

Figure 1-3a shows three directions of descent d_0,d_1,d_2 in a point x_0 and figure 1-3b shows their respective directional functions $\varphi_0,\varphi_1,\varphi_2$.

The direction d_0 opposite to the gradient gives the steepest decrease at the starting point $(s=0)$, but it does not lead to the lowest minimum. The direction d_2 passes closer to the minimum x^* and the associated directional function φ_2 has a lower minimum. The different behaviors among the directions come from the Hessian conditioning in x^*. A large conditioning leads to quite different locations (value of s) and levels of the minimum (value of f) depending on the chosen direction.

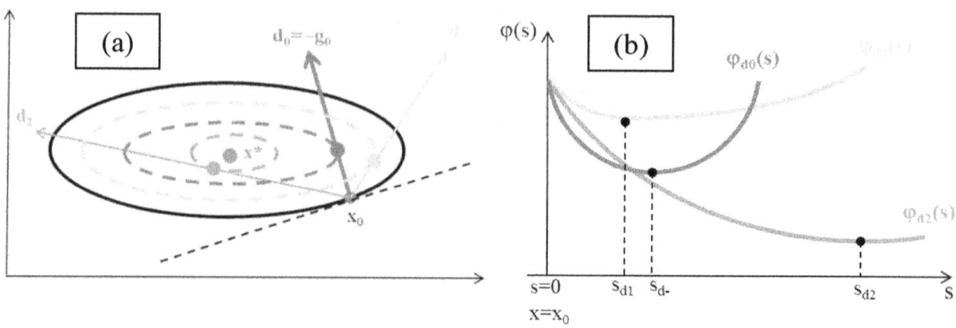

Figure 1-3: Descent directions (a) and directional variation (b).

1.1.6 Solution

Solving a constrained optimization problem involves the notions of feasible domain, global or local minimum and active constraints.

Feasible domain

The feasible domain denoted X_{feas} is the set of points $x \in \mathbb{R}^n$ satisfying the constraints of the optimization problem (1.1).

$$x \in X_{feas} \Leftrightarrow \begin{cases} c_E(x) = 0 \\ c_I(x) \le 0 \end{cases} \tag{1.22}$$

A feasible point lies therefore:
- on a zero-level line for an equality constraint $c_E(x)=0$;
- on the negative side of a zero-level line for an inequality constraint $c_I(x) \le 0$.

Figure 1-4 shows a problem with one equality constraint $c_1(x)=0$ and two inequality constraints $c_2(x) \le 0$, $c_3(x) \le 0$. The zero-level line is drawn for each constraint. The shaded areas show the forbidden domains for the inequality constraints. The overall feasible domain X_{feas} is the intersection of the feasible domains of each constraint. It reduces to a curve arc.

Depending on the constraints the feasible domain may be connected or not, convex or not. Figure 1-5 illustrates the notions of connectivity (one-piece domain) and convexity (domain without holes). Non-connectivity is usually due to inequality constraints. A non-connected feasible domain greatly complicates the resolution of the optimization problem.

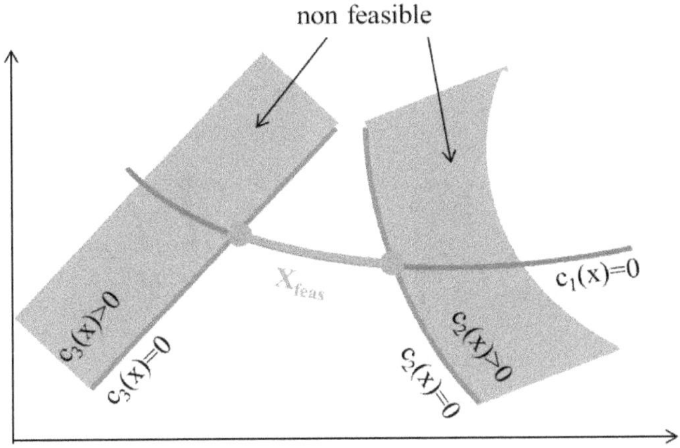

Figure 1-4: Feasible domain and level lines.

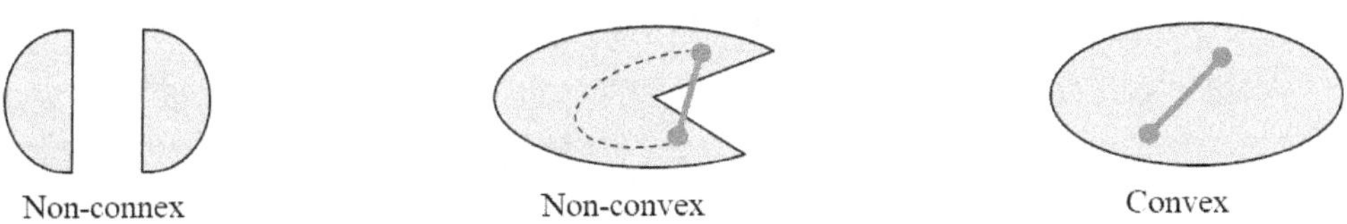

Figure 1-5: Possible shapes of the feasible domain.

Global and local minimum

A point x* is a **global minimum of** the optimization problem (1.1) if there exists no feasible point better than x*.

$$\forall x \in X_{feas}\ ,\ f(x^*) \le f(x) \tag{1.23}$$

A point x* is a **local minimum of** the optimization problem (1.1) if there exists no feasible point better than x* in the neighborhood of x*.

$$\exists \varepsilon > 0\ /\ \forall x \in X_{feas}\ ,\ \|x - x^*\| < \varepsilon\ \Rightarrow\ f(x^*) \le f(x) \tag{1.24}$$

A global or local minimum can be unique (strict minimum) or not.

Figure 1-6 illustrates these definitions for a function of one variable.

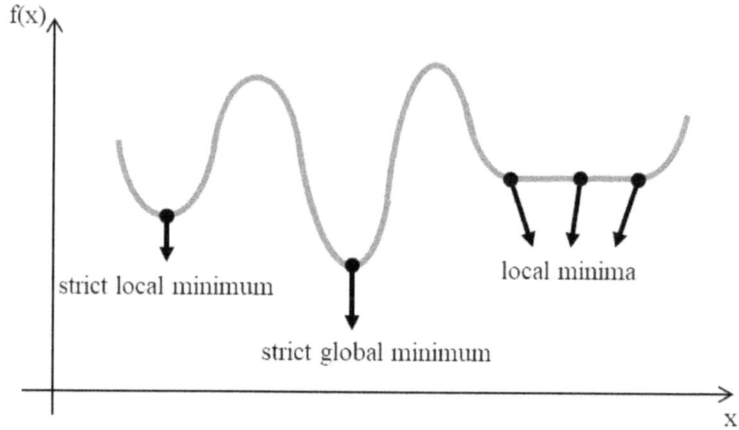

Figure 1-6: Global and local minimum.

For an unconstrained problem $(X_{feas} = \mathbb{R}^n)$, the local minimum conditions in x* are derived from the Taylor expansion to order 2 (1.11).

$$\begin{aligned} &\forall d \in \mathbb{R}^n \,,\ f(x^*) \le f(x^*+d) \\ &\quad \text{with } f(x^*+d) = f(x^*) + \nabla f(x^*)^T d + \frac{1}{2} d^T \nabla^2 f(x^*) d + \cdots \end{aligned} \tag{1.25}$$

This inequality must be satisfied for any move $d \in \mathbb{R}^n$. This is possible only if the gradient $\nabla f(x^*)$ is zero (otherwise opposite moves $\pm d$ would make the function either increase or decrease) and if the Hessian $\nabla^2 f(x^*)$ is a positive semidefinite matrix $(\forall d \,,\, d^T \nabla^2 f(x^*) d \ge 0)$. We can thus state the necessary/sufficient conditions for a local minimum without constraints.

Theorem 1-2: Necessary conditions for an unconstrained local minimum

$$x^* \text{ local minimum of } f \;\Rightarrow\; \begin{cases} \nabla f(x^*) &= 0 \\ \nabla^2 f(x^*) &\ge 0 \end{cases}$$

Theorem 1-3: Sufficient conditions for an unconstrained local minimum

$$\begin{cases} \nabla f(x^*) &= 0 \\ \nabla^2 f(x^*) &> 0 \end{cases} \;\Rightarrow\; x^* \text{ (strict) local minimum of } f$$

Active constraints

An inequality constraint c_I is said to be **active** in x if $c_I(x)=0$, and **inactive** in x if $c_I(x)<0$.

If an inequality constraint c_I is inactive in a local minimum x*, it remains negative in a neighborhood of x* (the function c_I is assumed to be continuous) and it does not locally restrict the feasible domain. The local minimum x* defined by (1.24) is in this case unsensitive to the constraint c_I .

This is no longer true if the constraint is active, because the point x* lies then on the boundary of the feasible domain and a part of its neighborhood is no longer feasible.

Figure 1-7 shows in grey the feasible domain associated with the inequality constraint $c_I(x)\le 0$:

- in figure 1-7a, the point x* is within the feasible domain.

  The constraint is inactive and all points in the neighborhood (dotted circle) are feasible;

- in figure 1-7b, the point x* is on the boundary of the feasible domain.

  The constraint is active and only a part of the neighborhood (dotted arc) is feasible. If the constraint were ignored, there could be a minimum better than x* in the region outside the feasible domain.

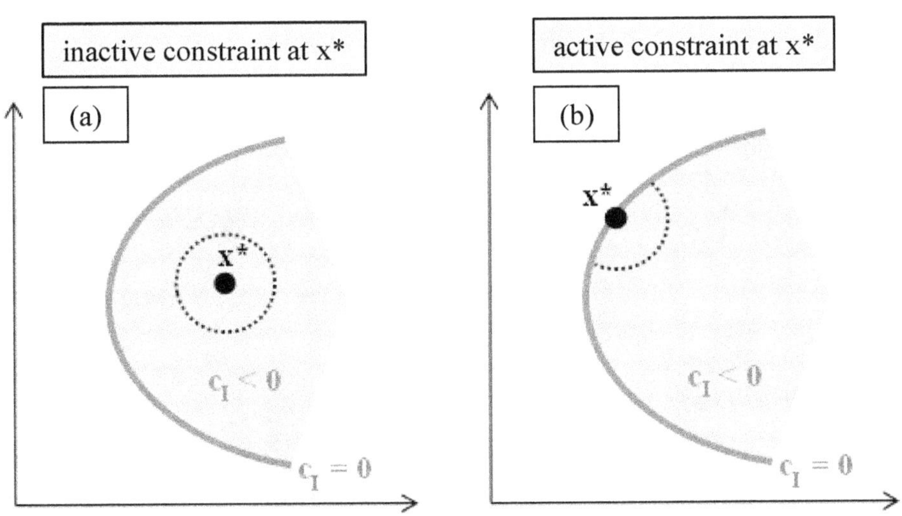

Figure 1-7: Inactive (a) or active inequality (b) constraint.

Since the solution x* is unsensitive to the inactive constraints, we can reformulate the optimization problem (1.1) by removing these constraints, and transforming the active inequality constraints into equality constraints.

$$\min_{x\in\mathbb{R}^n} f(x) \text{ s.t. } \begin{cases} c_E(x)=0 \\ c_I(x)\le 0 \end{cases} \Leftrightarrow \min_{x\in\mathbb{R}^n} f(x) \text{ s.t. } \begin{cases} c_E(x) = 0 \\ c_I^{act}(x)=0 \end{cases} \quad (1.26)$$

These two problems have the same solution, but the second one, which has only equality constraints, is much easier to solve.

Note

This observation may seem useless, since to eliminate the inactive constraints in x, one must already know the solution x*. A heuristic method to "guess" the active constraints consists in first solving the problem without the inequality constraints, then introducing as equality constraints those that would not be satisfied. This approach does not offer any guarantee, but it can be effective on problems whose features are already mastered.*

Example 1-3: Progressive activation of the inequality constraints

Let us consider the following problem: $\min_{x_1,x_2} x_1^2+x_2^2 \text{ s.t. } \begin{cases} x_1+x_2\ge 2 \\ x_1\ge 1 \end{cases}$.

The solution without the inequality constraints would be: $x_1=x_2=0$.

It does not satisfy any of the constraints. We therefore resume the resolution by activating the two unsatisfied constraints, which are now treated as equalities.

$\min_{x_1,x_2} x_1^2+x_2^2 \text{ s.t. } \begin{cases} x_1+x_2=2 \\ x_1=1 \end{cases}$ which gives the correct solution: $x_1=x_2=1$.

1.2 Numerical derivatives

In most applications, the cost function and the constraints are evaluated by numerical simulations and the analytical expression of their derivatives is not known. This section presents the evaluation of derivatives by finite differences and the methods to reduce the numerical errors.

1.2.1 First derivatives

The gradient of the function $f(x_1,\cdots,x_n)$ is the vector of its partial derivatives.

The simplest method for estimating the partial derivative $\frac{\partial f}{\partial x_i}$ in the point $a \in \mathbb{R}^n$ of components $(a_1,\cdots,a_n)$ is to apply an increment h_i on the i-th component and to use the **simple finite difference** formula.

$$\frac{\partial f}{\partial x_i}(a_1,\cdots,a_i,\cdots,a_n) \approx \frac{f(a_1,\cdots,a_i+h_i,\cdots,a_n)-f(a_1,\cdots,a_i,\cdots,a_n)}{h_i} \quad (1.27)$$

Writing the Taylor expansion (1.9) of f in a_i+h_i, we observe that formula (1.27) truncates all the terms from h_i^2 in the derivative estimate.

This formula therefore has a truncation error of order h_i^2.

The increment h_i can be positive or negative and it is not necessarily the same for all partial derivatives. Its setting is discussed in section 1.2.3.

The estimation of the gradient by this method requires the evaluation of f in the point a and n evaluations in the points a_i+h_i, giving a total of n + 1 evaluations.

The **centered finite difference** formula is very similar.

$$\frac{\partial f}{\partial x_i}(a_1,\cdots,a_i,\cdots,a_n) \approx \frac{f(a_1,\cdots,a_i+h_i,\cdots,a_n)-f(a_1,\cdots,a_i-h_i,\cdots,a_n)}{2h_i} \quad (1.28)$$

Writing the Taylor expansion (1.9) of f in a_i+h_i and in a_i-h_i, we observe that the terms in h_i^2 cancel out. This formula has thus a truncation error of order h_i^3 instead of h_i^2 for the simple finite difference. This gain in precision is obtained at the expense of two evaluations of the function f for each partial derivative, giving a total of 2n + 1 evaluations.

1.2.2 Second derivatives

The Hessian of the function $f(x_1,\cdots,x_n)$ is the matrix of its second partial derivatives. The second partial derivative $\frac{\partial^2 f}{\partial x_i \partial x_j}$ can be estimated by applying again the finite difference formula on the partial derivative evaluated by (1.27).

$$\frac{\partial^2 f}{\partial x_i \partial x_j} \approx \frac{f_{ij} - f_i - f_j + f}{h_i h_j} \quad \text{with} \quad \begin{cases} f_{ij} = f(a_1, \cdots, a_i + h_i, \cdots, a_j + h_j, \cdots, a_n) \\ f_i = f(a_1, \cdots, a_i + h_i, \cdots, a_j, \quad \cdots, a_n) \\ f_j = f(a_1, \cdots, a_i, \quad \cdots, a_j + h_j, \cdots, a_n) \\ f = f(a_1, \cdots, a_i, \quad \cdots, a_j, \quad \cdots, a_n) \end{cases} \tag{1.29}$$

Writing the Taylor expansions (1.9) of the functions f_{ij}, f_i and f_j, we obtain a truncation error of order $h_i h_j$. The estimation of the Hessian by (1.29) requires $n(n-1)/2$ evaluations to obtain the values f_{ij}. The values f_i and f_j have already been calculated for the gradient. This formula is seldom used because of its computational cost. Quasi-Newton methods are preferred (section 3.2).

1.2.3 Increment setting

A computer stores real variables with a finite number of significant digits. The machine precision is the smallest real ε_m such that $1+\varepsilon_m \neq 1$ ($\varepsilon_m = 10^{-16}$ for a double precision calculation). The rounding error on the real a is $\delta a = \varepsilon_m a$.
The evaluation of a function $f(a)$ is subject to a rounding error $\delta f = \varepsilon_f f(a)$. The relative error ε_f is usually much higher than ε_m since the function may result of a complex numerical simulation. This rounding error affects the estimate of the derivative. To evaluate the error on the derivative, let us note respectively f_{true} and f_{calc} the exact and calculated values of f.

$$\begin{cases} f_{calc}(a) \quad = f_{true}(a) \quad \pm \delta f \\ f_{calc}(a+h) = f_{true}(a+h) \pm \delta f \end{cases} \tag{1.30}$$

The derivative of f is estimated by simple finite difference (1.27).

$$f'_{calc}(a) = \frac{f_{calc}(a+h) - f_{calc}(a)}{h} = \frac{f_{true}(a+h) - f_{true}(a) \pm 2\delta f}{h} \tag{1.31}$$

Writing the Taylor expansion of f to order 2

$$f_{true}(a+h) = f_{true}(a) + h f'_{true}(a) + \frac{h^2}{2} f''_{true}(a) + o(h^2) \tag{1.32}$$

and replacing in (1.31), we obtain

$$f'_{calc}(a) = f'_{true}(a) + \frac{h}{2} f''_{true}(a) \pm \frac{2\delta f}{h} + o(h) \tag{1.33}$$

The maximum error $\delta f'$ on the derivative estimate is obtained by summing the absolute values of each term and replacing $\delta f = \varepsilon_f f(a)$.

$$\delta f' = \frac{h}{2}\left|f''_{true}(a)\right| + \frac{2}{h}\varepsilon_f \left|f_{true}(a)\right| \tag{1.34}$$

The term in h is the **truncation error** of the Taylor expansion. The term in 1/h is the **rounding error** due to the finite precision calculation. Decreasing the increment h reduces the truncation error, but increases the rounding error.

The optimal value of the increment h is the one that minimizes the error $\delta f'$.

$$\min_h \delta f' \quad \rightarrow h_{opt} = 2\sqrt{\varepsilon_f \left|\frac{f_{true}(a)}{f''_{true}(a)}\right|} \tag{1.35}$$

Assuming arbitrarily that the (unknown) second derivative is close to 1, the optimal increment would be

$$h_{opt} = \sqrt{\varepsilon_f \left|f_{true}(a)\right|} \tag{1.36}$$

Furthermore, the increment for calculating $f(a+h)$ must be greater than $\varepsilon_m a$ (otherwise $a+h=a$). If the value of a is large and/or the value of $f(a)$ is small, the optimal increment (1.36) may not satisfy the condition $h_{opt} > \varepsilon_m a$.

This issue is avoided by **scaling**. Scaling consists in making affine variable changes to reduce them to quantities of the order of unity.

A variable x between the bounds x_{min} and x_{max} is replaced by the variable

$$\tilde{x} = \frac{x - x_{min}}{x_{max} - x_{min}} \quad \rightarrow 0 \le \tilde{x} \le 1 \tag{1.37}$$

A function f whose magnitude is f_0 is replaced by the function

$$\tilde{f}(x) = \frac{f(x)}{f_0} \quad \rightarrow \tilde{f}(x) \approx 1 \tag{1.38}$$

The order of magnitude f_0 can be either the value of f in an arbitrary point or an estimate of the minimum.

After scaling, the optimal increment (1.36) is simply expressed as

$$\tilde{h}_{opt} = \sqrt{\varepsilon_f} \tag{1.39}$$

where the relative error ε_f on f is greater than the machine precision ε_m.

The increment can thus be set by the following rule of thumb.

Property 1-4: Derivative by finite difference

To obtain an accurate derivative, **scale the problem** and choose an increment greater than the **root of the machine precision.**

The error on the derivative (1.34) is then of the order of $\sqrt{\varepsilon_f}$, which means that a finite difference derivative is about half as accurate as the function in terms of significant digits.

Example 1-4: Optimal increment for finite differences

We seek to estimate by finite difference the derivative of $f(x)=x^2$ in $a=1$. The increment is varied from 10^{-1} to 10^{-15}. The difference between the numerical and the exact derivative $f'(1)=2$ is plotted with a logarithmic scale in figure 1-8. It can be seen that the best increment is about 10^{-8} and it yields an accuracy about 10^{-8} on the derivative, corresponding to the root of the machine precision of 10^{-16}.

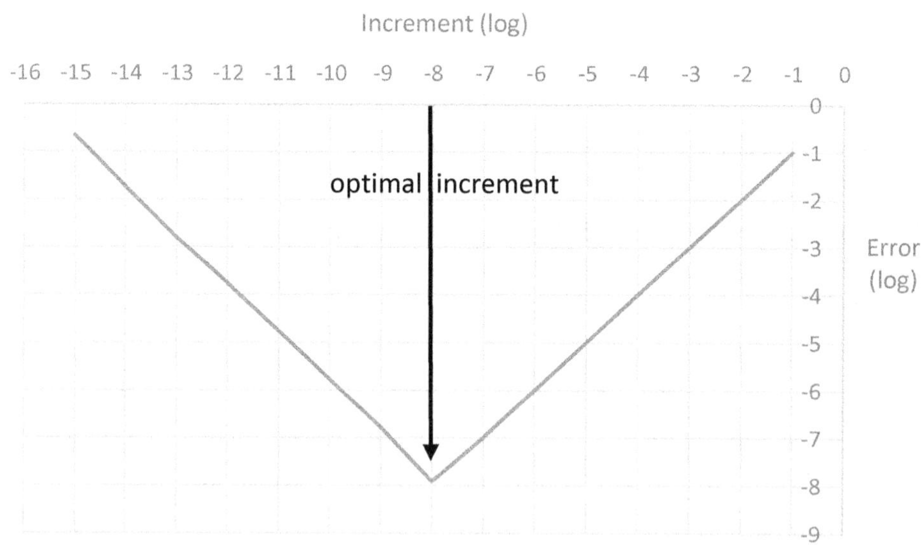

Figure 1-8: Error on the derivative depending on the increment.

1.2.4 Complex derivative

By applying a complex increment ih (instead of a real increment h), the Taylor expansion to order 3 takes the form

$$f_{true}(a+ih) = f_{true}(a) + ihf'_{true}(a) - \frac{h^2}{2} f''_{true}(a) - \frac{ih^3}{6} f'''_{true}(a) + o(h^3) \tag{1.40}$$

The function f and the variable a are still reals. The calculated value $f_{calc}(a+ih)$ is affected by rounding errors δf_r on the real part and δf_i on the imaginary part.

$$f_{calc}(a+ih) = f_{true}(a+ih) + \delta f_r + i\delta f_i \tag{1.41}$$

These rounding errors are proportional to the first real term and to the first imaginary term of the expansion (1.40) respectively

$$\begin{cases} \delta f_r = \varepsilon_f f_{true}(a) \\ \delta f_i = \varepsilon_f h f'_{true}(a) \end{cases} \tag{1.42}$$

The derivative is estimated by the formula

$$f'_{calc}(a+ih) = \frac{\mathrm{Im}\left[f_{calc}(a+ih)\right]}{h} \tag{1.43}$$

Taking the imaginary part of (1.41), and using (1.40) and (1.42), we have

$$\begin{aligned} \mathrm{Im}\left[f_{calc}(a+ih)\right] &= \mathrm{Im}\left[f_{true}(a+ih)\right] + \delta f_i \\ &= (1+\varepsilon_f)h f'_{true}(a) - \frac{h^3}{6} f'''_{true}(a) + o(h^4) \end{aligned} \tag{1.44}$$

then replacing in (1.43), we obtain

$$f'_{calc}(a) = (1+\varepsilon_f) f'_{true}(a) - \frac{h^2}{6} f''_{true}(a) + o(h^3) \tag{1.45}$$

The maximum error on the derivative estimate is

$$\delta f' = \varepsilon_f \left| f'_{true}(a) \right| + \frac{h^2}{6} \left| f''_{true}(a) \right| \tag{1.46}$$

The truncation error is in h^2 and the rounding error does not depend on h. It is thus possible to choose an arbitrarily small increment and obtain the derivative with a relative error ε_f of the same order of magnitude as on the function.

In practice, taking $h \approx \sqrt{\varepsilon_f}$ is sufficient to make the truncation error smaller than the rounding error. This complex derivation technique is at the origin of automatic code differentiation methods.

Example 1-5: Finite difference with complex increment

Let us estimate by complex finite difference the derivative of $f(x) = x^4$ in $a = 2$.

Figure 1-9 compares derivatives with real and complex increments varying from 10^{-1} to 10^{-15}. The best increment is 10^{-8} in both cases, but the complex derivative does not suffer from rounding errors and it is close to machine accuracy.

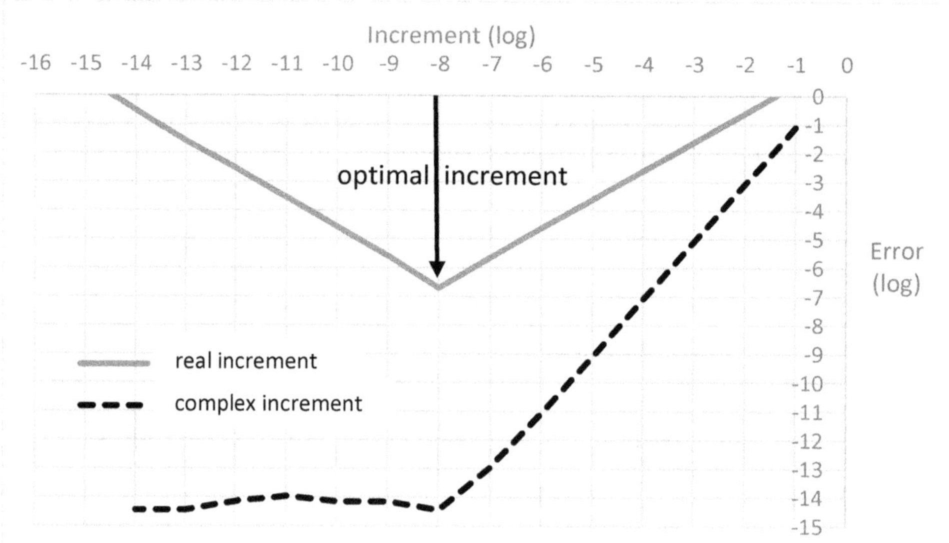

Figure 1-9: Error on the derivative depending on the increment.

The complex derivative formula (1.43) uses only one evaluation of the function and it allows the derivative to be estimated with very good accuracy. But, it requires that all variables in the simulator calculating the function f are defined as complexes, which may be an obstacle to its practical use.

1.2.5 Derivatives by extrapolation

The Richardson extrapolation method aims to estimate the limit $\lim_{h \to 0} A(h)$ of a function $A(h)$ which is not defined in $h = 0$. The principle is to evaluate the function in several points in the vicinity of 0, then combine them to eliminate the successive terms of the Taylor expansion. This expansion is of the form

$$A(h) = \alpha_0 + \alpha_1 h + \alpha_2 h^2 + \cdots + \alpha_n h^n + o(h^n) \tag{1.47}$$

The coefficients $\alpha_0, \alpha_1, \cdots, \alpha_n$ are unknown. The limit we are trying to estimate corresponds to the coefficient $\alpha_0 = \lim_{h \to 0} A(h)$.

Richardson algorithm

The function is evaluated in m + 1 points $h_0, h_1, \cdots, h_m$. These points are of the form $h_k = r^k h$, where h is an arbitrary increment and r is a real between 0 and 1.

The m + 1 values of A are called rank 0 values and they are noted $A_{k,0} \underset{\text{def}}{=} A(h_k)$.

The Taylor expansion in these points $(h_k)_{k=0 \text{ to } m}$ gives the system

$$\begin{cases} A_{0,0} = A(r^0 h) = \alpha_0 + \alpha_1 h + \alpha_2 h^2 + \cdots + \alpha_n h^n + o(h^n) \\ A_{1,0} = A(r^1 h) = \alpha_0 + \alpha_1 r h + \alpha_2 r^2 h^2 + \cdots + \alpha_n r^n h^n + o(h^n) \\ A_{2,0} = A(r^2 h) = \alpha_0 + \alpha_1 r^2 h + \alpha_2 r^4 h^2 + \cdots + \alpha_n r^{2n} h^n + o(h^n) \\ \vdots \\ A_{m,0} = A(r^m h) = \alpha_0 + \alpha_1 r^m h + \alpha_2 r^{2m} h^2 + \cdots + \alpha_n r^{mn} h^n + o(h^n) \end{cases} \tag{1.48}$$

To eliminate the terms in h, the rank 0 values are combined by the formula

$$A_{k,1} \underset{\text{def}}{=} \frac{A_{k,0} - r A_{k-1,0}}{1 - r} \quad \text{for } k = 1 \text{ to } m \tag{1.49}$$

This gives the m rank 1 values with coefficients noted $\beta_2, \cdots, \beta_n$.

$$\begin{cases} A_{1,1} = \alpha_0 + \beta_2 h^2 + \cdots + \beta_n h^n + o(h^n) \\ A_{2,1} = \alpha_0 + \beta_2 r^2 h^2 + \cdots + \beta_n r^n h^n + o(h^n) \\ A_{3,1} = \alpha_0 + \beta_2 r^4 h^2 + \cdots + \beta_n r^{2n} h^n + o(h^n) \\ \vdots \\ A_{m,1} = \alpha_0 + \beta_2 r^{2m} h^2 + \cdots + \beta_n r^{mn} h^n + o(h^n) \end{cases} \tag{1.50}$$

In the same way, the terms in h^2 are eliminated → m−1 rank 2 values $A_{k,2}$

then the terms in h^3 → m−2 rank 3 values $A_{k,3}$

....

until the terms in h^m → 1 rank m value $A_{m,m}$

The formula for passing from rank j−1 to rank j is

$$A_{k,j} \underset{def}{=} \frac{A_{k,j-1} - r^j A_{k-1,j-1}}{1-r^j} \quad \text{for } k = j \text{ to } m \tag{1.51}$$

In practice, the successive calculations are arranged in columns.

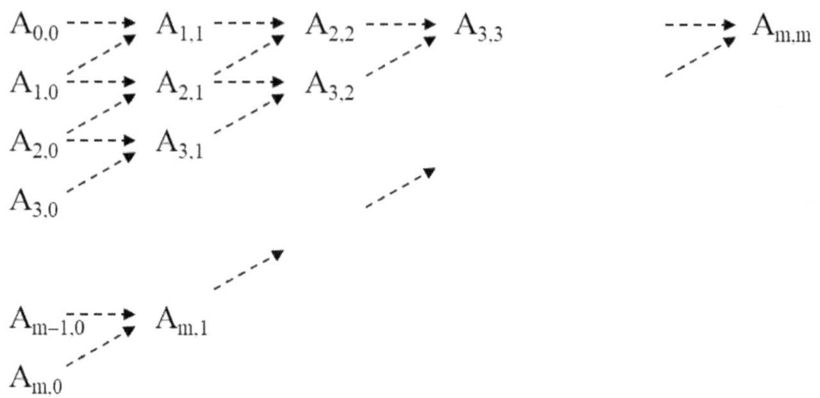

Figure 1-10: Calculation of the successive ranks of the Richardson algorithm.

Each of the (m−j+1) values of rank j is an approximation of α_0 to the order h^j.

The last value $A_{m,m}$ (of rank m) is an approximation of α_0 to the order h^m .

$$A_{m,m} = \alpha_0 + O\left(r^{m(m+1)} h^{m+1}\right) \tag{1.52}$$

The truncation error in (1.52) limits the order m of extrapolation. It is indeed unnecessary for this error to become smaller than the machine accuracy ε_m (assuming the function is scaled so that α_0 is about unity).

The increment h and the order m must thus satisfy the inequality

$$r^{m(m+1)} h^m > \varepsilon_m \tag{1.53}$$

This formula suggests that it is preferable to choose an increment h large enough in order to maximize the order m of extrapolation.

With the standard values $r = 0.5$ and $\varepsilon_m = 10^{-16}$, an increment $h = 0.1$ allows the order $m = 6$ to be reached, needing to perform 7 evaluations of the function.

Application to the estimation of derivatives

The Richardson method is applied to estimate the first and second derivatives of the function f in point a.
For that purpose, the functions $A(h)$ and $B(h)$ are defined by

$$\begin{cases} A(h) = \dfrac{f(a+rh)-f(a+h)}{(r-1)h} \\ B(h) = 2\dfrac{f(a+rh)-rf(a+h)+(r-1)f(a)}{r(r-1)h^2} \end{cases} \tag{1.54}$$

Their Taylor expansions in $h=0$ are

$$\begin{cases} A(h) = f'(a) + \dfrac{1}{2}\dfrac{r^2-1}{r-1}hf''(a) + \cdots + \dfrac{1}{n!}\dfrac{r^n-1}{r-1}h^{n-1}f^{(n)}(a) + \cdots \\ B(h) = f''(a) + \dfrac{1}{3}\dfrac{r^2-1}{r-1}hf'''(a) + \cdots + \dfrac{2}{n!}\dfrac{r^{n-1}-1}{r-1}h^{n-2}f^{(n)}(a) + \cdots \end{cases} \tag{1.55}$$

Their respective limits when $h \to 0$ are the derivatives $f'(a)$ and $f''(a)$ that we are trying to estimate. The Richardson algorithm described above is applied by choosing an increment h, a ratio r (generally 0.5) and an extrapolation order m.

The function f is first evaluated in m + 1 points $(a+r^k h)_{k=0 \text{ to } m}$.

$$f_k = f(a+r^k h) \quad \text{for } k=0 \text{ to } m \tag{1.56}$$

The rank 0 values of the functions $A(h)$ and $B(h)$ are then calculated by

$$\begin{cases} A_{k,0} = \dfrac{f_{k+1}-f_k}{(r-1)r^k h} \\ B_{k,0} = 2\dfrac{f_{k+1}-rf_k+(r-1)f(a)}{(r-1)r^{2k+1}h^2} \end{cases} \tag{1.57}$$

Richardson formula (1.51) applied to functions A and B provides m-order estimates of $f'(a)$ and $f''(a)$.

Example 1-6: Numerical derivatives by extrapolation

Let us estimate by extrapolation the first and second derivatives of $f(x) = x^2 + \dfrac{1}{x}$ in $a=1$. The exact values are: $f'(1) = 1$ and $f''(1) = 4$.

The initial increment is set to h = 0.1 and the extrapolation order is set to m = 6. The values of f are first calculated for increments from h to $h/2^6$.

h_k	$a+h_k$	$f(a+h)_k$
0.10000000	1.10000000	2.11909091
0.05000000	1.05000000	2.05488095
0.02500000	1.02500000	2.02623476
0.01250000	1.01250000	2.01281057
0.00625000	1.00625000	2.00632788
0.00312500	1.00312500	2.00314450
0.00156250	1.00156250	2.00156738

Table 1-1: Function values for extrapolation to order 6.

We then calculate the rank 0 values of the function A (1.57), and extrapolate to rank 5 by (1.51). Each column corresponds to a rank and produces an estimate of the derivative. We observe that each rank gains 2 correct digits and that the last rank estimate yields the first derivative with 10 correct digits.

A_0	A_1	A_2	A_3	A_4	A_5
1.2841991342	1.0074965685	1.0001968322	1.0000025180	1.0000000159	1.0000000000
1.1458478513	1.0020217662	1.0000268073	1.0000001723	1.0000000005	
1.0739348088	1.0005255470	1.0000035017	1.0000000113		
1.0372301779	1.0001340130	1.0000004476			
1.0186820955	1.0000338389				
1.0093579672					

$f'(1) = 1$

Table 1-2: Extrapolation to order 6 of the first derivative.

Similarly the values of f allow us to calculate the rank 0 values of function B (1.57), and then extrapolate to rank 5 by the formula (1.51). The second derivative is also estimated with 10 correct significant digits.

B_0	B_1	B_2	B_3	B_4	B_5
3.7316017316	3.9850068631	3.9996063356	3.9999949640	3.9999999683	4.0000000000
3.8583042973	3.9959564675	3.9999463855	3.9999996555	3.9999999990	
3.9271303824	3.9989489060	3.9999929968	3.9999999775		
3.9630396442	3.9997319741	3.9999991049			
3.9813858091	3.9999323222				
3.9906590657					

$f''(1) = 4$

Table 1-3: Extrapolation to order 6 of the second derivative.

1.3 Problem reduction

The constraints limit the domain of the optimization variables. They can be taken into account either by reduction, or by multipliers or by penalization. The multiplier and penalization methods are discussed in sections 1.5 and 4.2 respectively. This section presents the reduction methods whose interest is both practical (for some of the algorithms presented in chapter 4) and theoretical (to prove the optimality conditions). The reduction is first developed in the case of linear constraints and then generalized to the case of nonlinear constraints.

1.3.1 Linear reduction

Consider an optimization problem with linear equality constraints.

$$\min_{x \in \mathbb{R}^n} f(x) \text{ s.t. } Ax = b \ , \ A \in \mathbb{R}^{m \times n} \ , b \in \mathbb{R}^m \tag{1.58}$$

The constraints form a linear system with m equations and n unknowns.

$$Ax = b \Leftrightarrow \begin{cases} A_{1,1}x_1 + A_{1,2}x_2 + \cdots + A_{1,n}x_n = b_1 \\ A_{2,1}x_1 + A_{2,2}x_2 + \cdots + A_{2,n}x_n = b_2 \\ \cdots\cdots\cdots\cdots\cdots\cdots\cdots\cdots\cdots\cdots \\ A_{m,1}x_1 + A_{m,2}x_2 + \cdots + A_{m,n}x_n = b_m \end{cases} \tag{1.59}$$

This system must admit at least one solution for the optimization problem (1.58) to make sense.

This is achieved if $m \le n$ (less equations than unknowns) and if the matrix A has full rank equal to m (independent rows).

The reduction technique consists in using the m equations (1.59) to express m unknowns in terms of the remaining n–m, and replacing them in the function f.

The reduction uses any basis of $\mathbb{R}^n$ composed of n independent vectors. The first m vectors form a matrix $Y \in \mathbb{R}^{n \times m}$ with n rows and m columns, the last n–m vectors form a matrix $Z \in \mathbb{R}^{n \times (n-m)}$ with n rows and n–m columns.

$$(Y \mid Z) = \begin{matrix} & \begin{matrix} 1 & \cdots & m & 1 & \cdots & n-m \end{matrix} \\ \begin{matrix} 1 \\ \vdots \\ n \end{matrix} & \left(\begin{array}{ccc|ccc} \times & \cdots & \times & \times & \cdots & \times \\ \vdots & \vdots & \vdots & \vdots & \vdots & \vdots \\ \times & \cdots & \times & \times & \cdots & \times \end{array} \right) \end{matrix} \tag{1.60}$$

The components of a vector $p \in \mathbb{R}^n$ in this basis are denoted p_Y and p_Z.

p_Y and p_Z represent line vectors with m and n−m components respectively.

$$p = Yp_Y + Zp_Z \tag{1.61}$$

Suppose we start from an initial point x_0 and look for a new point of the form $x = x_0 + p$, where the vector p represents the move from x_0.

The initial point is not necessarily feasible: $Ax_0 = b_0 \neq b$.

For the new point to be feasible, the move p must satisfy

$$A(x_0 + p) = b \;\Rightarrow\; Ap = b - b_0 \;\Rightarrow\; AYp_Y + AZp_z = b - b_0 \tag{1.62}$$

The m×m matrix AY is invertible, because A and Y are of rank m.

This allows expressing the components p_Y depending on the components p_Z

$$p_Y = (AY)^{-1}(b - b_0 - AZp_Z) \tag{1.63}$$

and the expression for the <u>feasible</u> move p

$$p = Y(AY)^{-1}(b - b_0) + \left[I - Y(AY)^{-1}A\right]Zp_z \tag{1.64}$$

The optimization problem (1.58) now depends only on the variables p_Z.

$$\min_{p_Z \in \mathbb{R}^{n-m}} f\left(x_0 + Y(AY)^{-1}(b - b_0) + \left[I - Y(AY)^{-1}A\right]Zp_z\right) \tag{1.65}$$

The n−m components p_Z are called **independent** (or free) **variables**, the m components p_Y are deduced by (1.63) and are called **dependent variables**. If the initial point x_0 is feasible $(Ax_0 = b_0 = b)$, the feasible move (1.64) simplifies to

$$p = \left[I - Y(AY)^{-1}A\right]Zp_z \tag{1.66}$$

The reduction allows us to pass from a problem with n unknowns and m linear constraints to a problem with n−m unknowns without constraints. The reduced problem (1.65) is equivalent to the original problem, but it is much simpler to solve (fewer variables, no constraints). The reduction basis (1.60) is chosen arbitrarily and it can simply be the canonical basis of $\mathbb{R}^n$. In practice, it is preferable to construct this basis from the matrix A in order to exploit the structure of the constraints.

Basis formed from columns of A

The basis can be defined by selecting m independent columns of the matrix A. These columns form a sub-matrix B of dimension m×m, the remaining columns form a sub-matrix N of dimension m×(n−m). Assuming that the selected columns are the m first ones (this amounts to swapping the variables), the matrix A and the move p are decomposed into

$$A = {}_{m}\begin{pmatrix} \overset{m}{B} & \overset{n-m}{N} \end{pmatrix} \quad , \quad p = \begin{pmatrix} p_B \\ p_N \end{pmatrix} \begin{matrix} \to m \\ \to n-m \end{matrix} \tag{1.67}$$

With this decomposition, the feasible move condition (1.62) gives

$$\begin{aligned} Ap = b - b_0 \quad &\Rightarrow \quad Bp_B + Np_N = b - b_0 \\ &\Rightarrow \quad p_B = B^{-1}(b - b_0) - B^{-1}Np_N \end{aligned} \tag{1.68}$$

The matrices Y, Z and the components p_Y , p_Z are then defined by

$$\begin{aligned} &Y = \begin{pmatrix} \overset{m}{B^{-1}} \\ 0 \end{pmatrix} \begin{matrix} m \\ n-m \end{matrix} \quad , \quad Z = \begin{pmatrix} \overset{n-m}{-B^{-1}N} \\ I \end{pmatrix} \begin{matrix} m \\ n-m \end{matrix} \\ &p_Y = b - b_0 \quad , \quad p_Z = p_N \end{aligned} \tag{1.69}$$

Indeed, the move (1.61) with these matrices gives the form (1.67).

$$p = Yp_Y + Zp_Z = \begin{pmatrix} B^{-1}(b - b_0) - B^{-1}Np_N \\ p_N \end{pmatrix} = \begin{pmatrix} p_B \\ p_N \end{pmatrix} \tag{1.70}$$

This basis choice is the simplest as it is made directly on the matrix A.

The disadvantage is that the matrix B to be inverted may turn out to be ill-conditioned. To choose the "best" matrix B, it would be necessary to examine all possible combinations of m columns of A, which is unfeasible in practice.

Figure 1-11 illustrates this conditioning problem with a constraint in $\mathbb{R}^2$ of the form: $a_1x_1 + a_2x_2 = b$. The constraint matrix is: $A = \begin{pmatrix} a_1 & a_2 \end{pmatrix}$ and the basis is formed by the first column of the matrix A.

The independent move p_N is along the component x_2 and the dependent move p_B along x_1 is defined by (1.68), which in this simple case gives

$$\begin{cases} B = (a_1) \\ N = (a_2) \end{cases} \quad \Rightarrow \quad p_B = -\frac{a_2}{a_1} p_N \tag{1.71}$$

If the coefficient a_1 is small (straight line almost parallel to the x_1 axis), the move p_B is much larger than p_N and it is subject to numerical inaccuracies. This is due to the ill-conditioned matrix $B = (a_1)$ (eigenvalue close to zero). It is better in this case to select the second column of A to form the basis.

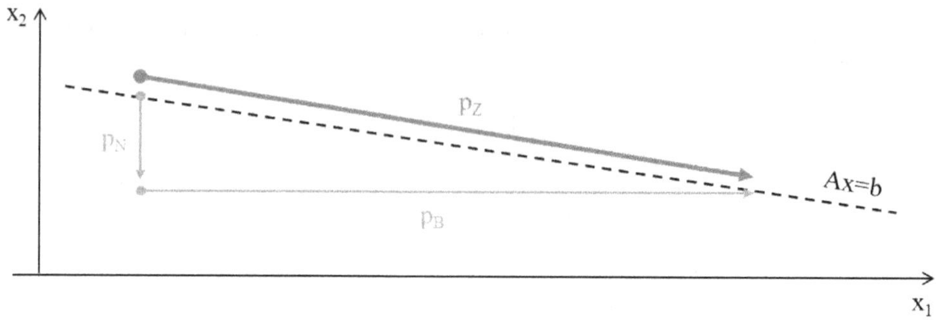

Figure 1-11: Geometric interpretation of the reduction.

The conditioning issue is avoided if the basis is a direction vector of the straight line such as $Z = \begin{pmatrix} a_2 \\ -a_1 \end{pmatrix}$. The move $p = Zp_Z$ is thus directly feasible and it does not lead to numerical inaccuracies even if a_1 is small. This procedure of using the kernel of the matrix A is detailed below.

<u>Basis formed from the kernel of A</u>

The kernel (or null space) of the matrix A is the set of vectors $z \in \mathbb{R}^n$ such that $Az = 0$. This system with more unknowns (n) than equations (m) defines a space of dimension n−m. The basis of $\mathbb{R}^n$ (1.60) is composed of n−m independent kernel vectors (Z-matrix) and m independent off-kernel vectors (Y-matrix). The out-of-kernel vectors can be for example the columns of the matrix A^T, as shown by the following decomposition theorem.

Theorem 1-5: Decomposition on the kernel and the image

Any vector $x \in \mathbb{R}^n$ can be decomposed in a unique way as $x = A^T y + z$ with vectors $y \in \mathbb{R}^m, z \in \mathbb{R}^n$, where z belongs to the kernel of A $(Az = 0)$.

The matrix A $(m \times n\ ,\ m \le n)$ is assumed to be of full rank. This theorem expresses that the kernel of A and the image of A^T are supplementary in $\mathbb{R}^n$.

Demonstration of the existence

If we give ourselves $z \in \mathbb{R}^n$ belonging to the kernel of A, then we can find y by: $Ax = AA^T y + Az = AA^T y \Rightarrow y = (AA^T)^{-1} Ax$, because AA^T is a $(m \times m)$ matrix of full rank.

Demonstration of the uniqueness

Assuming $x = A^T y + z = A^T y' + z'$, with z and z' belonging to the kernel of A, then $Ax = AA^T y = AA^T y' \Rightarrow AA^T(y - y') = 0 \Rightarrow y - y' = 0$, because AA^T is of full rank and then $A^T y + z = A^T y' + z' \Rightarrow z = z'$.

With such a (Y,Z) basis, the feasible move condition (1.62) gives

$$Ap = b - b_0 \Rightarrow AYp_Y = b - b_0 \quad \text{because} \quad AZ = 0 \tag{1.72}$$

This choice has the advantage of making the components p_Y and p_Z independent. In particular, if the initial point x_0 is feasible, then $p_Y = 0$.

Among all kernel bases, it is preferable to choose an orthogonal basis. For this purpose, the matrix A^T $(n \times m,\ m \le n)$ is factorized in a QR form by Householder or Givens method with an orthogonal matrix Q $(QQ^T = I)$.

$$A^T = QR = {}_n\begin{pmatrix} \overset{m}{Q_1} & \overset{n-m}{Q_2} \end{pmatrix} \begin{pmatrix} \overset{m}{R_1} \\ 0 \end{pmatrix} \begin{matrix} m \\ n-m \end{matrix} \tag{1.73}$$

Q_1 is orthogonal $n \times m$, Q_2 is orthogonal $n \times (n - m)$ and R_1 is upper triangular $m \times m$. The kernel basis Z and the non-kernel basis Y are defined by

$$\begin{cases} Z = Q_2 & \rightarrow \ AZ = R_1^T Q_1^T Q_2 = 0 \\ Y = Q_1 & \rightarrow \ AY = R_1^T Q_1^T Q_1 = R_1^T \end{cases} \tag{1.74}$$

The matrix $AY = R_1^T$ must be inverted to calculate the move (1.64). This matrix has the same conditioning as A.

$$\text{cond}(A) = \text{cond}(QR)^T = \text{cond}(Q_1 R_1)^T = \text{cond}(R_1^T) = \text{cond}(AY) \tag{1.75}$$

because the matrix Q_1 is orthogonal.

One can show that the conditioning of AY is at least that of A. The choice of an orthogonal basis from the kernel thus combines the decoupling of the components p_Y and p_Z (1.72) and the minimization of the numerical inversion errors.

Reduced problem

By choosing as matrix Z a basis of the kernel of A, and assuming that the initial point x_0 is feasible $(b = b_0)$, the reduced problem (1.65) simplifies to

$$\min_{p_Z \in \mathbb{R}^{n-m}} f(x_0 + p) = f(x_0 + Zp_Z) \underset{\text{def}}{=} f_r(p_Z) \tag{1.76}$$

The reduced function noted f_r then depends only on the $n-m$ variables p_Z.

The gradient and the Hessian of f_r are called **reduced gradient** and **reduced Hessian**. Their expressions are obtained from (1.76) by derivatives composition

$$\begin{cases} g_r = Z^T g \\ H_r = Z^T HZ \end{cases} \tag{1.77}$$

with the notations $\begin{cases} g_r = \nabla f_r \\ g = \nabla f \end{cases}$ and $\begin{cases} H_r = \nabla^2 f_r \\ H = \nabla^2 f \end{cases}$.

The optimality conditions of the unconstrained problem (1.76) are

$$\begin{cases} g_r = Z^T g = 0 \\ H_r = Z^T HZ \geq 0 \end{cases} \tag{1.78}$$

Case of a linear problem

If the cost function is linear: $f(x) = c^T x$, the reduced gradient is: $g_r = Z^T c$.

Consider a basis of the kernel formed by columns of A as in (1.67). The matrix A and the vector c are split into basic and non-basic components.

$$A = {}_m\begin{pmatrix} \overset{m}{B} & \overset{n-m}{N} \end{pmatrix} \quad , \quad c = \begin{pmatrix} c_B \\ c_N \end{pmatrix}\begin{matrix} m \\ n-m \end{matrix} \tag{1.79}$$

The matrix Z is then given by (1.69) and the reduced gradient becomes

$$g_r = c_N - (B^{-1}N)^T c_B \tag{1.80}$$

For linear programming (chapter 5), the components of c are called the costs and the components of the reduced gradient g_r are called the **reduced costs**.

1.3.2 Generalized reduction

The reduction technique can be applied to nonlinear constraints. A natural idea is the **elimination** method which consists in using constraints to express part of the variables (called dependent) in terms of the others (called independent or free). This reduces to a reduced problem without constraints.

Example 1-7: Box problem

We wish to build a cylindrical box having a given volume while minimizing the surface area.

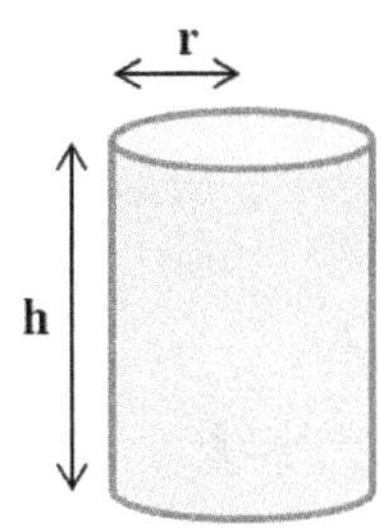

The variables are the radius r and the height h.
The cost function is the area to minimize: $S = 2\pi r^2 + 2\pi rh$.
The constraint is the volume equal to V_0: $V = \pi r^2 h = V_0$.

The problem is formulated by dividing S by 2π and noting $V_0 = 2\pi v_0$.

$$\min_{h,r} f(h,r) = r^2 + rh \quad \text{s.t.} \quad c(h,r) = r^2 h - 2v_0 = 0$$

The constraint allows eliminating the variable h: $h = \frac{2v_0}{r^2}$.

Substituting in the cost function, we obtain

$$\min_{r}\left(r^2 + \frac{2v_0}{r}\right) \Rightarrow r = v_0^{\frac{1}{3}} \Rightarrow h = 2v_0^{\frac{1}{3}} \Rightarrow f = 3v_0^{\frac{2}{3}}$$

The elimination technique requires some precautions in the case of nonlinear constraints. Indeed, the direct elimination of variables can lead to a false result as shown in the following example.

Example 1-8: Risks associated to the direct elimination method

Consider the problem: $\min_{x_1,x_2} x_1^2 + x_2^2 \quad \text{s.t.} \quad x_1^2 + 4x_2^2 = 1$.

The figure on the next page shows the 0-level line of the constraint (solid line) and two level lines of the function (dashed lines).

The constrained minimum is at

$x_1 = 0\ ,\ x_2 = \pm\frac{1}{2}\ .$

Let us eliminate x_1 from the constraint.

$x_1^2 = 1 - 4x_2^2$

The function to be minimized now depends only on x_2 : $\min_{x_2} 1 - 3x_2^2$.

This function is not bounded.

$\rightarrow\ \min = -\infty$!

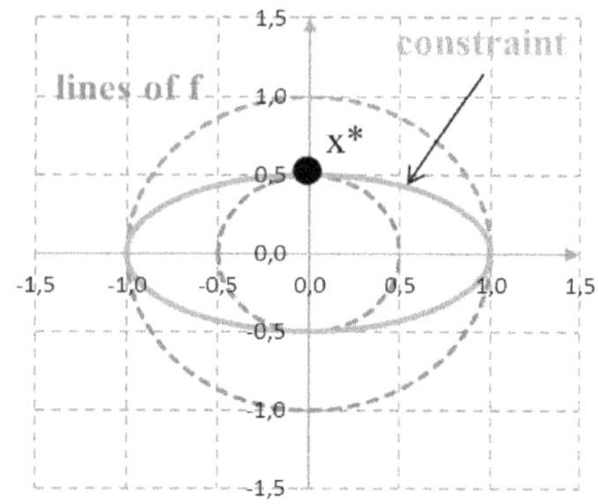

The error comes from the definition domain implicitly linked to the constraint:
$\left\{-1 \le x_1 \le 1\ ;\ -\frac{1}{2} \le x_2 \le \frac{1}{2}\right\}$. This domain is "forgotten" during the reduction.

In practical applications, the constraints are evaluated by simulation processes and they do not have an analytical expression. Direct elimination is then not feasible. Instead, generalized reduction can be used, which consists in locally linearizing the active constraints and applying the linear reduction to them (section 1.3.1). The linearized solution must then be corrected to restore a feasible point. This process involves the three stages detailed below.

Selection of active constraints

The active constraints in the initial point x_0 are equalities and inequalities taking a zero value. Strictly negative inequalities in x_0 do not restrict the moves in the vicinity of x_0 as seen in section 1.1.6.

The problem restricted to active constraints in x_0 is formulated as

$$\min_{x \in \mathbb{R}^n} f(x) \ \text{ s.t. } \ c(x) = 0 \tag{1.81}$$

Local linearization

The active constraints are linearized in the vicinity of x_0. A move $p_1 \in \mathbb{R}^n$ from x_0 is feasible if

$$c(x_0 + p_1) \approx c(x_0) + \nabla c(x_0)^T p_1 = 0 \tag{1.82}$$

The problem with linearized constraints is formulated as

$$\min_{p_1 \in \mathbb{R}^n} f(x_0 + p_1) \text{ s.t. } A_0 p_1 = b_0 \text{ with } \begin{cases} A_0 = \nabla c(x_0)^T \\ b_0 = -c(x_0) \end{cases} \tag{1.83}$$

We recover a formulation similar to (1.58), with the matrix A_0 being the Jacobian matrix of the constraints in x_0. The linear reduction technique with a (Y,Z) basis then gives a move p_1 (1.64). This move minimizes the function assuming linear constraints (1.82), but it is not necessarily feasible for the real problem (1.81).

It must be supplemented by a move p_2 in order to restore the constraints as shown in figure 1-12.

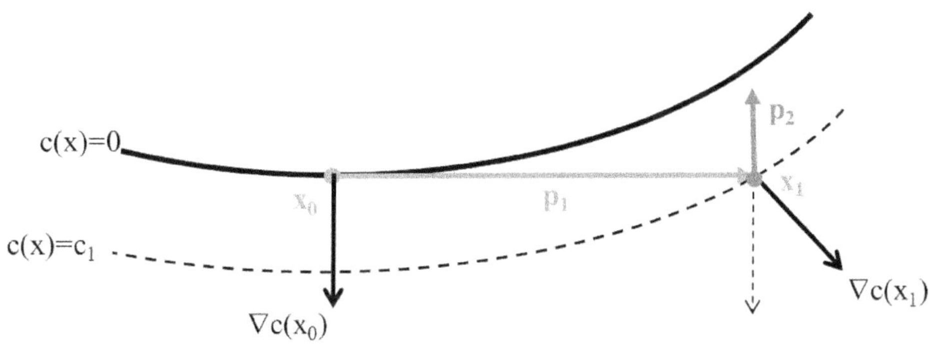

Figure 1-12: Linearized move and restoration.

Restoration of feasibility

The constraints take the value $c(x_1) \neq 0$ in the point $x_1 = x_0 + p_1$ resulting of the reduced problem. The restoration aims to find a new point $x_2 = x_1 + p_2$ that is feasible and "close" to the point x_1. The restoration move p_2 is based on the constraints linearized in x_1.

$$c(x_1 + p_2) \approx c(x_1) + \nabla c(x_1)^T p_2 = 0 \tag{1.84}$$

A first issue comes from the Jacobian constraint matrix that must be calculated in the point x_1. The approximation $\nabla c(x_1) \approx \nabla c(x_0)$ shown in figure 1-12 (dotted arrow in x_1) is justified if the move p_1 is small and it avoids the expensive calculation of this matrix.

A second issue comes from the underdetermination of the system (1.84) with n unknowns and m < n equations. Two approaches are favored with the objective of degrading as little as possible the solution x_1 of the linearized problem.

- Restoration by move of minimum norm

The minimum norm move is obtained by solving the problem

$$\min_{p_2 \in \mathbb{R}^n} \|p_2\| \text{ s.t. } A_1 p_2 = b_1 \quad \text{with } \begin{cases} A_1 = A_0 = \nabla c(x_0)^T \\ b_1 = -c(x_1) \end{cases} \tag{1.85}$$

This projection problem is solved by choosing a basis Z of the kernel and with $Y = A_1^T$ as the complementary basis (theorem 1-5).

The move $p_2 = A_1^T p_Y + Z p_Z$ must satisfy

$$A_1 p_2 = b_1 \quad \Rightarrow \quad A_1 A_1^T p_Y = b_1 \quad \Rightarrow \quad p_Y = \left(A_1 A_1^T\right)^{-1} b_1 \tag{1.86}$$

Since the component p_Y is fixed by (1.86), the minimization (1.85) only concerns p_Z, and the minimum is obtained for $p_Z = 0$. The minimum norm move is

$$p_2 = A_1^T \left(A_1 A_1^T\right)^{-1} b_1 \tag{1.87}$$

- Restoration by out-of-kernel move

The restoration move is sought in the form (1.61) : $p_2 = Y p_Y + Z p_Z$

by fixing the component p_Z to zero. Indeed, this component represents the free variables used for the minimization (1.83) and the objective is to keep it unchanged. The calculation similar to (1.64) gives

$$p_2 = Y\left(A_0 Y\right)^{-1} b_1 \tag{1.88}$$

As the move p_2 (1.87) or (1.88) is defined from the linearized constraints and with the Jacobian matrix not recalculated in x_1 there is no guarantee that the point $x_2 = x_1 + p_2$ is feasible. If we still have $c(x_2) \neq 0$, then either a new move p_3 must be calculated from x_2 using the same approach (1.84), or the move p_1 must be reduced in the hope of coming back to the "linearity domain" of the constraints. These situations are illustrated in figure 1-13.

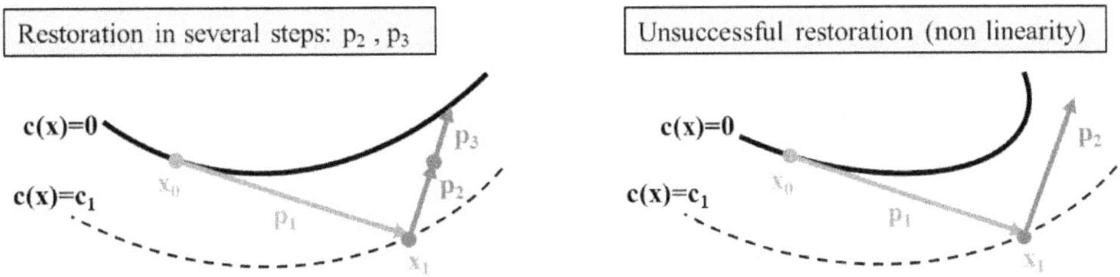

Figure 1-13: Effect of non-linearities on the restoration.

The restoration can result in a feasible point x_2 : $c(x_2)=0$, but it modifies the solution x_1 of the linearized problem. This restoration can degrade the cost function. It is therefore necessary to check that the point x_2 is better than the initial point x_0 : $f(x_2)<f(x_0)$.

If this is not the case, the move p_1 must be reduced in the hope of mitigating the adverse effect of the restoration. This situation is illustrated in figure 1-14 where the restoration leads to a deviation from the minimum of the function f.

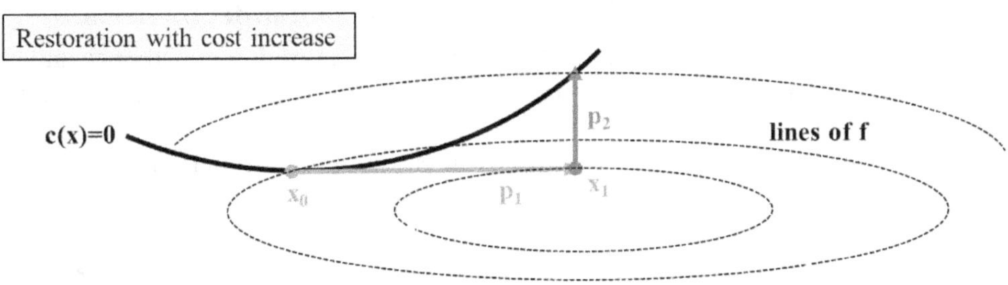

Figure 1-14: Effect of the restoration on the cost function.

The difficulties caused by the non-linearities could be mitigated by recalculating the gradient matrix $\nabla c(x_1)$ in order to improve the direction of restoration (as shown in figure 1-12), but this calculation, which should be carried out in each trial point x_1, would be far too costly.

A more efficient method is to use the direction $p_c = p_1 + p_2$, where p_1 and p_2 are calculated as before by (1.83) and (1.84). The constraints are expanded to the second order in the vicinity of x_0, with their Hessian denoted C_0 .

$$\begin{cases} c(x_0 + p_1) = c(x_0) + A_0 p_1 + \frac{1}{2} p_1^T C_0 p_1 \\ c(x_0 + p_c) = c(x_0) + A_0 p_c + \frac{1}{2} p_c^T C_0 p_c \end{cases} \tag{1.89}$$

Using the relations satisfied by p_1 and p_2

$$\begin{cases} A_0 p_1 = -c(x_0) \\ A_0 p_2 = -c(x_0 + p_1) \end{cases} \tag{1.90}$$

we obtain, after calculation, the values of the constraints in $x_0 + p_1$ and $x_0 + p_c$

$$\begin{cases} c(x_0 + p_1) = \frac{1}{2} p_1^T C_0 p_1 \\ c(x_0 + p_c) = p_1^T C_0 p_2 + \frac{1}{2} p_2^T C_0 p_2 \end{cases} \tag{1.91}$$

The value of p_2 is of order 2 with respect to p_1 : $A_0 p_2 = -c(x_0 + p_1) = -\frac{1}{2} p_1^T C_0 p_1$.

Formula (1.91) shows that the errors on the constraints pass from order 2 in $x_0 + p_1$ to order 3 in $x_0 + p_c$. The direction p_c is called the **second-order direction**.
The direction p_c takes into account the non-linearities of the constraints and it deviates less from the feasible line than the direction p_1 (figure 1-15). By reducing the move along this direction p_c (if necessary), it becomes easier to restore a feasible point while maintaining an improvement in the cost function.

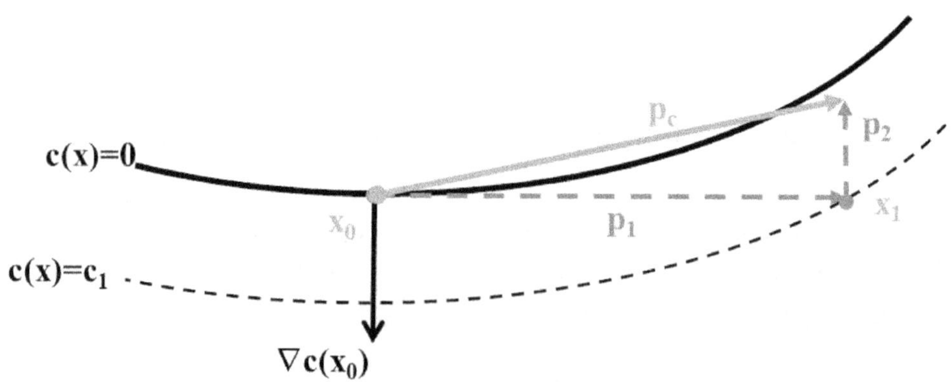

Figure 1-15: Direction of order 2.

1.4 Global optimum

Finding a global optimum is a difficult problem even for a function of one variable (figure 1-6). This section presents conditions that guarantee that a point is a global optimum. These conditions are formulated from the Lagrangian and the dual function.

1.4.1 Dual problem

Consider again the continuous optimization problem in standard form with p equality constraints c_E and q inequality constraints c_I. The domain of definition of the variables x is denoted $x \in \mathbb{R}^n$ (it can be the full set $\mathbb{R}^n$).

$$\min_{x \in X} f(x) \;\text{ s.t. } \begin{cases} c_E(x) = 0 \\ c_I(x) \le 0 \end{cases} \tag{1.92}$$

Dual methods deal with the constraints indirectly by associating with each of them a real number called the **Lagrange multiplier** :

- the vector $\lambda \in \mathbb{R}^p$ represents the multipliers of the equality constraints c_E;
- the vector $\mu \in \mathbb{R}^q$ represents the multipliers of the inequality constraints c_I.

The **Lagrangian** (or Lagrange function) is the function L defined as

$$L(x,\lambda,\mu) = f(x) + \lambda^T c_E(x) + \mu^T c_I(x) \tag{1.93}$$

or by detailing the terms

$$L(x,\lambda,\mu) = f(x) + \sum_{j=1}^{p} \lambda_j c_{Ej}(x) + \sum_{j=1}^{q} \mu_j c_{Ij}(x) \tag{1.94}$$

The Lagrangian is a function of $\mathbb{R}^{n+p+q}$ in $\mathbb{R}$. The variables x are called **primal variables** and the multipliers λ, μ are called **dual variables**.

The dual **function** ψ is defined by the minimum of the Lagrangian function with respect to x.

$$\psi(\lambda,\mu) = \min_{x \in X} L(x,\lambda,\mu) \tag{1.95}$$

Note that this is a global minimum, which can be difficult to find.

This function ψ from $\mathbb{R}^{p+q}$ in $\mathbb{R}$ has the following properties:

- its domain of definition $D=\{\lambda,\mu \ / \ \psi(\lambda,\mu) > -\infty\}$ is convex;
- the function ψ is concave;
- if $\mu \geq 0$, ψ is upper bounded by $f(x^*)$ where x^* is the solution of (1.92).

Demonstration

These properties can be demonstrated from the definitions of the dual function and the Lagrangian. The dual variables are denoted: $v=(\lambda,\mu)$ and α is any real.

- Concavity of ψ

For all x, the Lagrangian L is linear with respect to $v=(\lambda,\mu)$. Indeed:

$$L(x,\alpha v_1+(1-\alpha)v_2)=\alpha L(x,v_1)+(1-\alpha)L(x,v_2)$$

Taking the minimum w.r.t. x of each term, we have

$$\psi(\alpha v_1+(1-\alpha)v_2) \geq \alpha\psi(v_1) + (1-\alpha)\psi(v_2)$$

- Convexity of D

If $v_1, v_2 \in D$, we have: $\psi(v_1)>-\infty$, $\psi(v_2)>-\infty$

Since ψ is concave: $\psi(\alpha v_1+(1-\alpha)v_2) \geq \alpha\psi(v_1)+(1-\alpha)\psi(v_2) > -\infty$

From this, we deduce: $\alpha v_1+(1-\alpha)v_2 \in D$

- Upper bound on ψ

$$\begin{aligned}
\psi(\lambda,\mu) &= \min_{x\in X} L(x,\lambda,\mu) \\
&\leq L(x^*,\lambda,\mu) = f(x^*)+\lambda^T c_E(x^*)+\mu^T c_I(x^*) \\
&= f(x^*)+\mu^T c_I(x^*) \quad \text{because } x^* \text{ feasible} \Rightarrow c_E(x^*)=0 \\
&\leq f(x^*) \quad \text{because } x^* \text{ feasible} \Rightarrow c_I(x^*)\leq 0 \\
&\qquad \text{and } \mu\geq 0 \text{ (by assumption)}
\end{aligned}$$

Since the dual function is bounded by f(x*), it is natural to look for its maximum. Problem (1.92) is called the **primal problem**. The **dual problem** associated with problem (1.92) is to maximize the dual function.

$$\max_{\lambda,\mu} \psi(\lambda,\mu) \quad \text{s.t.} \quad \mu \geq 0 \tag{1.96}$$

Its solution is denoted (λ^*,μ^*) and the maximum value $\psi(\lambda^*,\mu^*)$ is a lower bound of the minimum cost $f(x^*)$. This property constitutes the weak **duality theorem.**

$$\psi(\lambda^*,\mu^*) \le f(x^*) \tag{1.97}$$

The difference $f(x^*)-\psi(\lambda^*,\mu^*)$ between the cost of the primal problem and the cost of the dual problem is called the **duality gap**.

The following example illustrates these notions linked to the duality.

Example 1-9: Dual function and dual problem

Consider the two variable problem: $\min\limits_{x_1,x_2} f(x_1,x_2)=\frac{1}{2}(x_1^2+x_2^2)$ s.t. $x_1=1$

whose solution is $(x_1,x_2)^*=(1,0)$. This problem is the primal problem.

The Lagrangian is a function of 3 variables: $L(x_1,x_2,\lambda)=\frac{1}{2}(x_1^2+x_2^2)+\lambda(x_1-1)$

where λ is the equality constraint multiplier.

The dual function is defined by

$$\psi(\lambda)=\min_{x_1,x_2} L(x_1,x_2,\lambda)=\min_{x_1,x_2}\frac{1}{2}(x_1^2+x_2^2)+\lambda(x_1-1)$$

Deriving L with respect to x_1 and x_2: $\frac{\partial L}{\partial x}=0 \Rightarrow \begin{cases} x_1+\lambda=0 \\ x_2=0 \end{cases} \Rightarrow \begin{cases} x_1=-\lambda \\ x_2=0 \end{cases}$,

we obtain the explicit expression of ψ: $\psi(\lambda)=-\frac{1}{2}\lambda^2-\lambda$ with $\begin{cases} x_1=-\lambda \\ x_2=0 \end{cases}$

The dual problem can then be solved.

$$\max_{\lambda}\psi(\lambda)=-\frac{1}{2}\lambda^2-\lambda \rightarrow \lambda^*=-1 \rightarrow \begin{cases} x_1^*=1 \\ x_2^*=0 \end{cases}$$

We retrieve here the solution to the primal problem. This is not always the case.

1.4.2 Saddle point

A saddle point of the Lagrangian is a triplet (x^*,λ^*,μ^*) that minimizes the Lagrangian with respect to x and maximizes it with respect to $(\lambda,\mu \geq 0)$.

$$(x^*,\lambda^*,\mu^*) \text{ saddle point } \Leftrightarrow \begin{cases} L(x^*,\lambda^*,\mu^*) \leq L(x,\lambda^*,\mu^*) \,,\ \forall x \in X \\ L(x^*,\lambda^*,\mu^*) \geq L(x^*,\lambda,\mu) \quad ,\ \forall \lambda, \forall \mu \geq 0 \end{cases} \tag{1.98}$$

Figure 1-16 shows a saddle point in two dimensions (x and λ are simple reals).

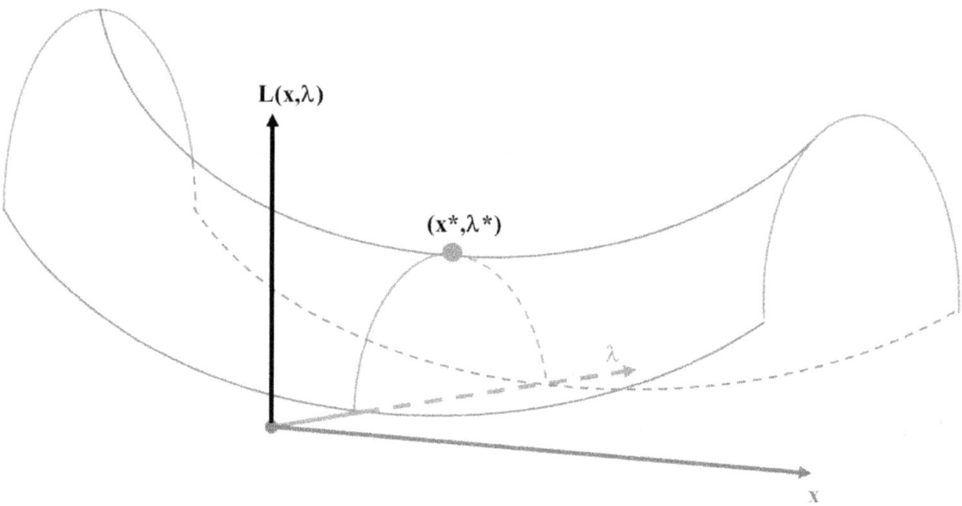

Figure 1-16: Lagrangian saddle point.

A saddle point of the Lagrangian is characterized by the following conditions.

$$(x^*,\lambda^*,\mu^*) \text{ saddle point } \Leftrightarrow \begin{cases} x^* \rightarrow \min\limits_{x \in X} L(x,\lambda^*,\mu^*) \\ c_E(x^*) = 0 \\ c_I(x^*) \leq 0 \\ \mu^* \geq 0 \\ \mu^* c_I(x^*) = 0 \end{cases} \tag{1.99}$$

Demonstration of the implication $\Rightarrow$

By definition of a saddle point: $\begin{cases} L(x^*,\lambda^*,\mu^*) \le L(x,\lambda^*,\mu^*) & , \forall x \in X \\ L(x^*,\lambda^*,\mu^*) \ge L(x^*,\lambda,\mu) & , \forall \lambda, \forall \mu \ge 0 \end{cases}$.

The conditions on x* and μ* are already satisfied by definition of the saddle point. We have to demonstrate the conditions on c_E and c_I. Let us start from the second inequality above.

$$L(x^*,\lambda^*,\mu^*) \ge L(x^*,\lambda,\mu) \quad , \forall \lambda, \forall \mu \ge 0$$

$$\Rightarrow \lambda^{*T} c_E(x^*) + \mu^{*T} c_I(x^*) \ge \lambda^T c_E(x^*) + \mu^T c_I(x^*) \quad , \forall \lambda, \forall \mu \ge 0$$

$$\Rightarrow (\lambda^* - \lambda)^T c_E(x^*) + (\mu^* - \mu)^T c_I(x^*) \ge 0 \quad , \forall \lambda, \forall \mu \ge 0$$

If we take $\mu = \mu^*$, then $(\lambda^* - \lambda)^T c_E(x^*) \ge 0$, $\forall \lambda$,
which can be satisfied only if $c_E(x^*) = 0$

If we take $\lambda = \lambda^*$, then $(\mu^* - \mu)^T c_I(x^*) \ge 0$, $\forall \mu \ge 0$,
which can be satisfied for μ large only if $c_I(x^*) \le 0$.

If we take $\lambda = \lambda^*$ and $\mu = 0$, then $\mu^{*T} c_I(x^*) \ge 0$.
Since we have otherwise $\mu^* \ge 0$ and $c_I(x^*) \le 0$, we deduce that $\mu^{*T} c_I(x^*) = 0$.

Demonstration of the implication $\Leftarrow$

We have to prove the 2nd inequality: $L(x^*,\lambda^*,\mu^*) \ge L(x^*,\lambda,\mu)$, $\forall \lambda, \forall \mu \ge 0$

Let us evaluate each member of this inequality.

$$\begin{cases} L(x^*,\lambda,\mu) = f(x^*) + \lambda^T c_E(x^*) + \mu^T c_I(x^*) = f(x^*) + \mu^T c_I(x^*) \\ L(x^*,\lambda^*,\mu^*) = f(x^*) + \lambda^{*T} c_E(x^*) + \mu^{*T} c_I(x^*) = f(x^*) \end{cases}$$

Using $c_E(x^*) = 0$ and $\mu^{*T} c_I(x^*) = 0$.

Since $\mu \ge 0$ and $c_I(x^*) \le 0$, we obtain the expected inequality.

The interest of a saddle point comes from the following two theorems on strong duality and on global optimality. The proofs of these theorems are detailed in reference [R11].

Theorem 1-6: Strong duality

The existence of a saddle point is equivalent to a zero duality gap.

Demonstration of the implication $\Rightarrow$

Let us assume that a saddle point (x^*,λ^*,μ^*) exists.

$$\begin{cases} L(x^*,\lambda^*,\mu^*) \le L(x,\lambda^*,\mu^*) & , \ \forall x \in X \\ L(x^*,\lambda^*,\mu^*) \ge L(x^*,\lambda,\mu) & , \ \forall \lambda, \ \forall \mu \ge 0 \end{cases}$$

We first calculate $L(x^*,\lambda^*,\mu^*)$ using the saddle point properties (1.99).

$L(x^*,\lambda^*,\mu^*)=f(x^*)+\lambda^{*T}c_E(x^*)+\mu^{*T}c_I(x^*)=f(x^*)$ since $c_E(x^*)=0$ and $\mu^{*T}c_I(x^*)=0$

By definition of the dual function (1.95) : $\psi(\lambda^*,\mu^*)=\min\limits_{x\in X} L(x,\lambda^*,\mu^*)$.

From the first saddle point inequality:

$L(x^*,\lambda^*,\mu^*) \le L(x,\lambda^*,\mu^*) \ , \ \forall x \in X$

Therefore, L reaches its minimum in $x = x^*$ and $\psi(\lambda^*,\mu^*) = L(x^*,\lambda^*,\mu^*) = f(x^*)$.

We have thus shown that f and ψ take the same value in the saddle point.

It remains to show that this value is the optimum of the primal and dual problems.

- For the primal problem

From the first saddle point inequality:

$L(x^*,\lambda^*,\mu^*) \le L(x,\lambda^*,\mu^*) \ , \ \forall x \in X$

$\Rightarrow f(x^*) \le f(x)+\lambda^{*T}c_E(x)+\mu^{*T}c_I(x) \ , \ \forall x \in X$

If x is a feasible point: $\begin{cases} c_E(x)=0 \\ c_I(x) \le 0 \end{cases} \Rightarrow f(x^*) \le f(x)$.

x* is therefore the solution of the primal problem (minimization of f).

- For the dual problem

By definition of the dual function (1.95) : $\psi(\lambda,\mu)=\min\limits_{x\in X} L(x,\lambda,\mu) \le L(x^*,\lambda,\mu)$.

According to the second saddle point inequality:

$$L(x^*,\lambda,\mu) \le L(x^*,\lambda^*,\mu^*) \quad , \quad \forall\lambda, \forall\mu \ge 0$$

$$\Rightarrow \quad \psi(\lambda,\mu) \le L(x^*,\lambda^*,\mu^*) = \psi(\lambda^*,\mu^*)$$

(λ^*,μ^*) is therefore the solution of the dual problem (maximization of ψ).

Demonstration of the implication $\Leftarrow$

Assume that the duality gap is zero and let us note x^* the solution of the primal problem, and (λ^*,μ^*) the solution of the dual problem.

We have to show that (x^*,λ^*,μ^*) satisfies the saddle point conditions (1.99).

By definition of the dual function (1.95)

$$\psi(\lambda^*,\mu^*) = \min_{x\in X} L(x,\lambda^*,\mu^*) \le L(x^*,\lambda^*,\mu^*)$$

and since $\psi(\lambda^*,\mu^*) = f(x^*)$, we get the inequality: $f(x^*) \le L(x^*,\lambda^*,\mu^*)$.

Let us calculate $L(x^*,\lambda^*,\mu^*)$.

$$\begin{aligned} L(x^*,\lambda^*,\mu^*) &= f(x^*) + \lambda^{*T} c_E(x^*) + \mu^{*T} c_I(x^*) \\ &= f(x^*) + \mu^{*T} c_I(x^*) \quad \text{because } c_E(x^*) = 0 \\ &\le f(x^*) \quad \text{because } c_I(x^*) \le 0 \text{ and } \mu^* \ge 0 \end{aligned}$$

We obtain the double inequality: $f(x^*) \le L(x^*,\lambda^*,\mu^*) = f(x^*) + \mu^{*T} c_I(x^*) \le f(x^*)$

from which we deduce: $L(x^*,\lambda^*,\mu^*) = f(x^*)$ and $\mu^{*T} c_I(x^*) = 0$.

Let us take the dual function again: $\psi(\lambda^*,\mu^*) = \min_{x\in X} L(x,\lambda^*,\mu^*)$

with $\psi(\lambda^*,\mu^*) = f(x^*) = L(x^*,\lambda^*,\mu^*) \Rightarrow L(x^*,\lambda^*,\mu^*) = \min_{x\in X} L(x,\lambda^*,\mu^*)$

The point (x^*,λ^*,μ^*) therefore satisfies all saddle point conditions (1.99).

This first theorem helps proving the existence of a saddle point by solving separately the primal and the dual problem, and by comparing their costs.

Once the existence has been proved, the following theorem can be used.

Theorem 1-7: Global optimum

If there exists a saddle point (x^*, λ^*, μ^*) for the Lagrangian function, then x^* is the global minimum of the problem (1.92).

Demonstration

Let us start with the first inequality defining the saddle point.

$L(x^*, \lambda^*, \mu^*) \le L(x, \lambda^*, \mu^*) \ , \ \forall x \in X$

We evaluate each member of the inequality for any feasible point x: $\begin{cases} c_E(x) = 0 \\ c_I(x) \le 0 \end{cases}$.

$$\begin{cases} L(x^*, \lambda^*, \mu^*) = f(x^*) + \lambda^{*T} c_E(x^*) + \mu^{*T} c_I(x^*) = f(x^*) \\ L(x, \lambda^*, \mu^*) \ = f(x) \ + \lambda^{*T} c_E(x) \ + \mu^{*T} c_I(x) \ \le f(x) \end{cases}$$

Using $c_E(x^*) = 0$, $\mu^{*T} c_I(x^*) = 0$ and $\mu^* \ge 0$.

For any feasible point x, we obtain thus the inequality: $f(x^*) \le f(x)$.

These two theorems allow us to prove global optimality by finding a saddle point. They apply in the very important case of linear programming (section 1.4.3).

But for most nonlinear problems, there exists no saddle point, as shown by the following example.

Example 1-10: Problem without saddle point

Consider the problem: $\min_{x \in [0;2]} f(x) = -x^2$ s.t. $x - 1 \le 0$.

The domain of definition is here $X = [0\,;\,2]$ and the solution to the problem is : $x^* = 1 \rightarrow f(x^*) = -1$

The Lagrangian is: $L(x, \mu) = -x^2 + \mu(x-1)$ with the multiplier $\mu \ge 0$.
The dual function ψ is defined by minimizing the Lagrangian on the domain of definition X.

$$\min_{x \in [0;2]} L(x, \mu) = \min_{x \in [0;2]} -x^2 + \mu(x-1)$$

The value of x depends on μ. The dual function is then defined by

$$\begin{cases} x=2 & \text{if } \mu<2 \\ x=0 \text{ or } 2 & \text{if } \mu=2 \\ x=0 & \text{if } \mu>2 \end{cases} \Rightarrow \psi(\mu)=\begin{cases} \mu-4 & \text{if } \mu<2 \\ -2 & \text{if } \mu=2 \\ -\mu & \text{if } \mu>2 \end{cases}$$

The dual function is maximum for $\mu^*=2 \rightarrow \psi(\mu^*)=-2$ and $x=0$ or 2.
The solutions of the dual problem ($x=0$ or 2) and those of the primal problem ($x^*=1$) are different.
There is therefore no saddle point and the duality gap is: $f(x^*)-\psi(\mu^*)=1$.

1.4.3 Linear programming

Linear programming concerns problems with linear cost function and constraints. The standard form of a linear problem is as follows:

$$\min_{x\in\mathbb{R}^n} c^T x \text{ s.t. } \begin{cases} Ax=b \\ x\ge 0 \end{cases}, \quad A\in\mathbb{R}^{m\times n}, b\in\mathbb{R}^m, c\in\mathbb{R}^n \tag{1.100}$$

The Lagrangian is defined with multipliers $\lambda\in\mathbb{R}^m$ for the equality constraints put into the form: $b-Ax=0$, and multipliers $\mu\in\mathbb{R}^n$ for the inequality constraints put into the form: $-x\le 0$.

$$\begin{aligned} L(x,\lambda,\mu) &= c^T x+\lambda^T(b-Ax)+\mu^T(-x) \\ &= (c-A^T\lambda-\mu)^T x+\lambda^T b \end{aligned} \tag{1.101}$$

The dual function $\psi(\lambda,\mu)$ is defined as the minimum of the Lagrangian when this one is bounded. The Lagrangian (1.101) is a linear function in x which is bounded only if the coefficient of x is zero. The domain of definition D of the dual function is therefore

$$D=\left\{(\lambda,\mu)\ /\ A^T\lambda+\mu=c\right\} \tag{1.102}$$

On this definition domain, the Lagrangian reduces to

$$L(x,\lambda,\mu)=\lambda^T b \tag{1.103}$$

The dual function is identical to the Lagrangian, because the latter no longer depends on x.

$$\psi(\lambda,\mu)=\min_x L(x,\lambda,\mu)=\lambda^T b \tag{1.104}$$

The dual problem is the maximization of the dual function.

$$\max_{\lambda,\mu} \psi(\lambda,\mu) \text{ s.t. } \begin{cases}(\lambda,\mu)\in D \\ \mu \ge 0\end{cases} \Leftrightarrow \max_{\lambda,\mu} \lambda^T b \text{ s.t. } \begin{cases}A^T\lambda+\mu = c \\ \mu \ge 0\end{cases} \tag{1.105}$$

This dual problem is reformulated by posing $y=\lambda$ and grouping the constraints.

$$\min_{y\in\mathbb{R}^m} -b^T y \text{ s.t. } A^T y - c \le 0 \tag{1.106}$$

This is a new linear problem with m variables y and n inequality constraints. This linear problem is analyzed in the same way as the original problem (1.100).

Its Lagrangian noted L_d is defined with multipliers $\nu\in\mathbb{R}^n$ for the inequality constraints.

$$\begin{aligned} L_d(y,\nu) &= -b^T y + \nu^T (A^T y - c) \\ &= (A\nu - b)^T y - \nu^T c \end{aligned} \tag{1.107}$$

This Lagrangian being linear in y, the dual function noted $\psi_d(\nu)$ is only defined if the coefficient of y is zero. We obtain thus

$$\psi_d(\nu) = -\nu^T c \quad \text{with} \quad A\nu - b = 0 \tag{1.108}$$

and the dual problem is formulated as

$$\max_{\nu\in\mathbb{R}^n} -\nu^T c \text{ s.t. } \begin{cases}A\nu - b = 0 \\ \nu \ge 0\end{cases} \tag{1.109}$$

This linear problem is identical to the original problem (1.100) by posing $x=\nu$.

Therefore, the dual of the dual of a linear problem is identical to the primal.

The inequality (1.97) applied to the primal problem and to the dual problem shows that their costs are equal. The duality gap is zero, which guarantees the existence of a saddle point achieving the global optimum. It is therefore equivalent to solve the primal problem (1.100) or the dual problem (1.106). The simpler of the two can be chosen, depending on the number of variables and constraints.

The above calculations apply to a linear problem in standard form (which can always be reduced to). In the case of a general linear problem with equality constraints, inequality constraints $\le$ or $\ge$ and positive, negative or free variables, the primal-dual transformation is performed as follows:

Primal problem

$$\min_{\substack{x_1\in\mathbb{R}^{n_1}\\ x_2\in\mathbb{R}^{n_2}\\ x_3\in\mathbb{R}^{n_3}}} c_1^T x_1 + c_2^T x_2 + c_3^T x_3 \quad \text{s.t.} \begin{cases} A_1x_1 + B_1x_2 + C_1x_3 = b_1\,, & b_1 \in \mathbb{R}^{m_1} \\ A_2x_1 + B_2x_2 + C_2x_3 \le b_2\,, & b_2 \in \mathbb{R}^{m_2} \\ A_3x_1 + B_3x_2 + C_3x_3 \ge b_3\,, & b_3 \in \mathbb{R}^{m_3} \\ x_1 \ge 0 \\ x_2 \le 0 \\ x_3 \in \mathbb{R}^{n_3} \end{cases} \tag{1.110}$$

Dual problem

$$\max_{\substack{y_1\in\mathbb{R}^{m_1}\\ y_2\in\mathbb{R}^{m_2}\\ y_3\in\mathbb{R}^{m_3}}} b_1^T y_1 + b_2^T y_2 + b_3^T y_3 \quad \text{s.t.} \begin{cases} A_1^T y_1 + A_2^T y_2 + A_3^T y_3 \le c_1\,, & c_1 \in \mathbb{R}^{n_1} \\ B_1^T y_1 + B_2^T y_2 + B_3^T y_3 \ge c_2\,, & c_2 \in \mathbb{R}^{n_2} \\ C_1^T y_1 + C_2^T y_2 + C_3^T y_3 = c_3\,, & c_3 \in \mathbb{R}^{n_3} \\ y_1 \in \mathbb{R}^{m_1} \\ y_2 \le 0 \\ y_3 \ge 0 \end{cases} \tag{1.111}$$

Demonstration

The Lagrangian of the primal problem (1.110) has the expression

$$\begin{aligned} & L(x_1,x_2,x_3,\lambda_1,\lambda_2,\lambda_3,\mu_1,\mu_2) \\ &= c_1^T x_1 + c_2^T x_2 + c_3^T x_3 \\ &\quad + \lambda_1^T (b_1 - A_1x_1 - B_1x_2 - C_1x_3) && \rightarrow m_1 \text{ multipliers } \lambda_1 \\ &\quad + \lambda_2^T (A_2x_1 + B_2x_2 + C_2x_3 - b_2) && \rightarrow m_2 \text{ multipliers } \lambda_2 \ge 0 \\ &\quad + \lambda_3^T (b_3 - A_3x_1 - B_3x_2 - C_3x_3) && \rightarrow m_3 \text{ multipliers } \lambda_3 \ge 0 \\ &\quad - \mu_1^T x_1 + \mu_2^T x_2 && \rightarrow n_1 \text{ multipliers } \mu_1 \ge 0 \\ & && \rightarrow n_2 \text{ multipliers } \mu_2 \ge 0 \end{aligned}$$

Let us group the terms with x_1, x_2 and x_3.

$$\begin{aligned} L(x_1,x_2,x_3,\lambda_1,\lambda_2,\lambda_3,\mu_1,\mu_2) = &\; b_1^T\lambda_1 - b_2^T\lambda_2 + b_3^T\lambda_3 \\ & + (c_1 - A_1^T\lambda_1 + A_2^T\lambda_2 - A_3^T\lambda_3 - \mu_1)^T x_1 \\ & + (c_2 - B_1^T\lambda_1 + B_2^T\lambda_2 - B_3^T\lambda_3 + \mu_2)^T x_2 \\ & + (c_3 - C_1^T\lambda_1 + C_2^T\lambda_2 - C_3^T\lambda_3)^T x_3 \end{aligned}$$

The Lagrangian is linear in x_1, x_2, x_3. It is bounded only if the coefficients of x_1, x_2, x_3 are zero. These conditions give the constraints of the dual problem.

By posing: $\begin{cases} y_1 = \lambda_1 \\ y_2 = -\lambda_2 \le 0 \\ y_3 = \lambda_3 \ge 0 \end{cases}$, the dual function is $\psi(y_1, y_2, y_3,) = b_1^T y_1 + b_2^T y_2 + b_3^T y_3$

and we find the formulation of the dual problem (1.111).

The transformation rules are summarized in the following table.

Number	Primal	Dual
m_1	Constraints equality = value b_1	Free variables y_1 of cost b_1
m_2	Constraints inequality $\le$ value b_2	Negative variables y_2 of cost b_2
m_3	Constraints inequality $\ge$ value b_3	Positive variables y_3 of cost b_3
n_1	Positive variables x_1 of cost c_1	Constraints inequality $\le$ value c_1
n_2	Negative variables x_2 of cost c_2	Constraints inequality $\ge$ value c_2
n_3	Free variables x_3 cost c_3	Constraints equality = value c_3

Table 1-4: Transition from primal to dual for a linear problem.

Example 1-11: Transition to the dual problem

Consider the following linear problem (primal problem):

$$\min_{x_1,x_2,x_3} x_1 + 2x_2 + 3x_3 \quad \text{s.t.} \begin{cases} -x_1 + 3x_2 = 5 \\ 2x_1 - x_2 + 3x_3 \ge 6 \\ x_3 \le 4 \\ x_1 \ge 0,\ x_2 \le 0,\ x_3 \in \mathbb{R} \end{cases}$$

Its dual has the following formulation:

$$\min_{y_1,y_2,y_3} -5y_1 - 6y_2 - 4y_3 \quad \text{s.t.} \begin{cases} y_1 - 2y_2 \ge -1 \\ -3y_1 + y_2 \le -2 \\ -3y_2 - y_3 = -3 \\ y_1 \in \mathbb{R},\ y_2 \ge 0,\ y_3 \le 0 \end{cases}$$

1.5 Local optimum

In most cases, the global optimality of a solution cannot be proved. One has to be satisfied with checking local optimality conditions. This section presents the conditions of Karush, Kuhn and Tucker and their interpretation.

1.5.1 Feasible directions

A point x* is a local optimum if there are no better feasible points in its neighborhood. To establish the local optimality, one must examine the set of "small" feasible moves from point x*.

Consider any sequence of feasible points (x_k) with limit x*. These points satisfy the equality and inequality constraints of the optimization problem.

$$\begin{cases} c_E(x_k) = 0 \\ c_I(x_k) \le 0 \end{cases} \tag{1.112}$$

From these points, let us form the series of directions (d_k) defined by

$$d_k = \frac{x_k - x^*}{\|x_k - x^*\|} \tag{1.113}$$

If a subset of the directions (d_k) has a limit d as shown in figure 1-17, then this direction d is called a **direction feasible to the limit**. Indeed, an infinitesimal move along such a direction remains feasible.

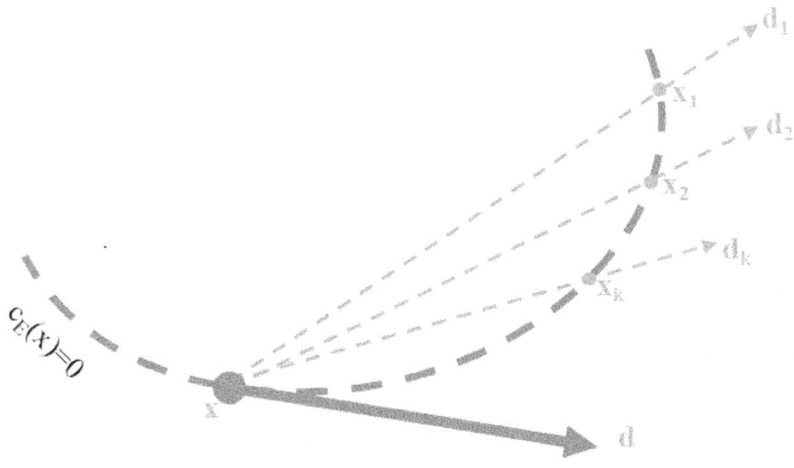

Figure 1-17: Feasible direction to the limit.

The set of feasible directions to the limit in the point x* is denoted D(x*). To establish the local optimality of x*, all directions of D(x*) must be examined.

Let us consider a direction $d \in D(x^*)$ and a series of feasible points (x_k) used to define it. From (1.113), these points are given by

$$x_k = x^* + s_k d_k \quad \text{with} \quad s_k = \|x_k - x^*\| \text{ and } \lim_{k\to\infty} s_k = 0 \tag{1.114}$$

Let us express the constraints in x_k by an expansion of order 1 for s_k small.

$$\begin{cases} c_E(x_k) = c_E(x^* + s_k d_k) = c_E(x^*) + s_k \nabla c_E(x^*)^T d_k + o(s_k) \\ c_I(x_k) = c_I(x^* + s_k d_k) = c_I(x^*) + s_k \nabla c_I(x^*)^T d_k + o(s_k) \end{cases} \tag{1.115}$$

If an inequality constraint is inactive in x*, it does not forbid any move in the vicinity of x* (section 1.1.6) and it does not reduce the set D(x*). Therefore, we consider only the <u>active constraints</u> in x* : $c_E(x^*) = 0$, $c_I(x^*) = 0$.

Since the points x_k are feasible: $c_E(x_k) = 0$, $c_I(x_k) \le 0$, we deduce from (1.115)

$$\begin{cases} \nabla c_E(x^*)^T d_k = \dfrac{c_E(x_k) - c_E(x^*) + o(s_k)}{s_k} = \dfrac{o(s_k)}{s_k} \\ \nabla c_I(x^*)^T d_k = \dfrac{c_I(x_k) - c_I(x^*) + o(s_k)}{s_k} = \dfrac{c_I(x_k) + o(s_k)}{s_k} \le 0 \end{cases} \tag{1.116}$$

Then by passing to the limit with $\lim_{k\to\infty} d_k = d$ and $\lim_{k\to\infty} s_k = 0$, we obtain conditions satisfied by any direction d belonging to D(x*).

$$\begin{cases} \nabla c_E(x^*)^T d = 0 \\ \nabla c_I(x^*)^T d \le 0 \quad \text{if} \quad c_I(x^*) = 0 \end{cases} \tag{1.117}$$

The set of directions satisfying the conditions (1.117) forms the **tangent cone** in the point x*. This set noted T(x*) is indeed a cone, because if $d \in T(x^*)$, then $\alpha d \in T(x^*)$ for any real $\alpha \ge 0$.

The conditions (1.117) indicate that the direction d is tangent to the equality constraints (normal to their gradient), and tangent or interior to the inequality constraints active in x* (forming an obtuse angle with the gradient).

Any feasible direction $d \in D(x^*)$ belongs to the tangent cone $T(x^*)$, but the converse is false. This is the case especially if the gradient of a constraint is zero in x*, as the following example shows.

Example 1-12: Non-feasible direction belonging to the tangent cone

Consider the equality constraint: $c(x_1, x_2) = (x_1^2 + x_2^2 - 1)^2 = 0$.

The feasible points form the circle of center O and of radius 1.

Let us place in the point A (1; 0).

The feasible directions to the limit are obtained by considering a series of points of the circle converging to A.
The tangent to the circle in A is obtained.

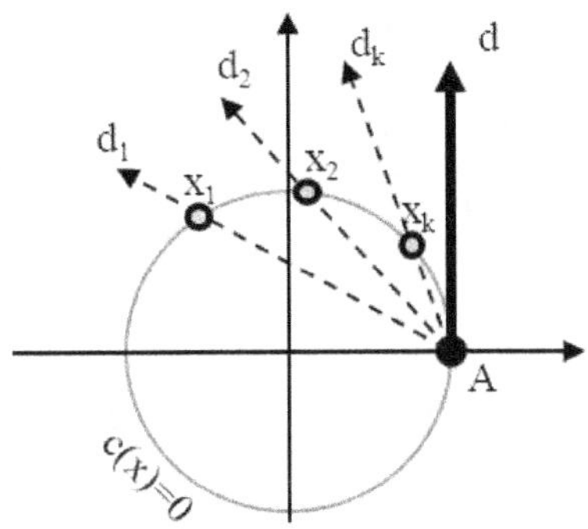

Now let us determine the tangent cone in A : $T(A) = \left\{ d \in \mathbb{R}^2 \ / \ \nabla c(A)^T d = 0 \right\}$.

The gradient of c: $\nabla c(x_1, x_2) = 2(x_1^2 + x_2^2 - 1)\begin{pmatrix} x_1 \\ x_2 \end{pmatrix}$ is zero in the point A (1; 0).

Therefore, any direction d belongs to the cone T(A). In that case, the tangent cone is not equivalent to the set of feasible directions to the limit.

The tangent cone $T(x^*)$ is defined by simple conditions (1.117), but it does not characterize the set of feasible directions $D(x^*)$. To establish the local optimality conditions (section 1.5.2), the following assumption is made.

Qualification of the constraints

The constraints satisfy the **condition of qualification** in the feasible point x* if any direction of the tangent cone is a feasible direction to the limit.

$$D(x^*) = T(x^*) \tag{1.118}$$

The qualification condition can be proved if we assume that the Jacobian matrix of the active constraints is of full rank in the point x*.

The constraints are then said to be **linearly independent** in the point x*, which means that their gradients are linearly independent in that point.

Demonstration (see [R3])

Let us show that if the constraints are linearly independent in x*, then any direction d of the tangent cone T(x*) is a feasible direction to the limit.

Let A be the Jacobian matrix of the active constraints in x*.

$$A = \nabla c(x^*)^T \quad \Leftrightarrow \quad \begin{pmatrix} A_E \\ A_I \end{pmatrix} = \begin{pmatrix} \nabla c_E(x^*)^T \\ \nabla c_I(x^*)^T \end{pmatrix}$$

Since the matrix $A \in \mathbb{R}^{m\times n}$ has full rank, we can form a basis of $\mathbb{R}^n$ of the form $(A^T\ Z)$, where $Z \in \mathbb{R}^{n\times(n-m)}$ is a basis of the kernel of A (theorem 1-5).

Let $d \in T(x^*)$ be any tangent direction. d satisfies (1.117) : $\begin{cases} A_E d = 0 \\ A_I d \le 0 \end{cases}$.

Consider a real $s \ge 0$ and the system (S) of n equations and unknowns $x \in \mathbb{R}^n$.

$$(S) \quad \begin{cases} c(x) = sAd \\ Z^T(x - x^* - sd) = 0 \end{cases}$$

The solution x(s) of this system parameterized by s has the following properties:

- the Jacobian matrix in $s = 0$ of this system is $(A^T\ Z)$. Since this matrix has full rank (basis of $\mathbb{R}^n$), the solution is unique in $s = 0$, and also in the neighborhood of $s = 0$ (implicit function theorem);
- the solution x(s) is feasible: $\begin{cases} c_E[x(s)] = sA_E d = 0 \\ c_I[x(s)] = sA_I d \le 0 \end{cases}$ since $d \in T(x^*)$ and $s \ge 0$;
- The solution in $s = 0$ is $x(0) = x^*$.

Let $s_k \ge 0$ be a series of limit 0. The series $x_k = x(s_k)$ is feasible with limit x*.

Let us then show that the series of directions $d_k = \dfrac{x_k - x^*}{s_k}$ has the limit d.

For that purpose, let us expand the constraints to order 1 in $x_k = x^* + s_k d_k$:
$c(x_k) = c(x^*) + s_k A d_k + o(s_k) = s_k A d_k + o(s_k)$ since $c(x^*) = 0$ (active constraints) and replace them in the system (S) :

$$\begin{cases} s_k A\,(d_k - d) = o(s_k) \\ s_k Z^T(d_k - d) = 0 \end{cases} \Rightarrow \begin{pmatrix} A \\ Z^T \end{pmatrix}(d_k - d) = \begin{pmatrix} o(s_k)/s_k \\ 0 \end{pmatrix}$$

Since the matrix $(A^T\ Z)$ has full rank, this linear system has a unique solution. When $s_k \to 0$, the 2nd member tends to 0. The limit direction satisfies $d_k - d \to 0$. We get a series of directions of the type (1.113) with limit d, thus $d \in D(x^*)$.

The linear independence condition is widely used, as it is simple to check. However, it is a sufficient condition and not a necessary one. We can have the qualification of the constraints without this condition as in the following example.

Example 1-13: Linear independence of constraints

Consider the two constraints

$$\begin{cases} c_1(x) = x_2 - x_1^2 = 0 \\ c_2(x) = x_1^2 + (x_2 - 1)^2 - 1 \le 0 \end{cases}$$

Their 0-level lines are drawn opposite.

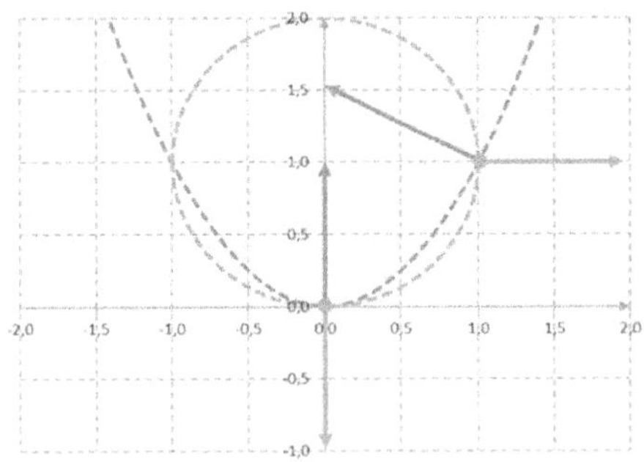

- Let us place in $x = (1\,;1)$.

The constraint gradients are linearly independent:

$$\nabla c_1 = \begin{pmatrix} -2 \\ 1 \end{pmatrix} \;,\; \nabla c_2 = \begin{pmatrix} 2 \\ 0 \end{pmatrix}$$

The qualification condition is therefore met in this point: $D(x) = T(x)$.

The tangent cone defined by $\begin{cases} -2d_1 + d_2 = 0 \\ 2d_1 \le 0 \end{cases}$ reduces to the directions of the form

$$\alpha \begin{pmatrix} -1 \\ -2 \end{pmatrix} , \alpha \ge 0 .$$

- Let us place in $x = (0\,;0)$.

The constraint gradients are not linearly independent:

$$\nabla c_1 = \begin{pmatrix} 0 \\ 1 \end{pmatrix} \;,\; \nabla c_2 = \begin{pmatrix} 0 \\ -2 \end{pmatrix}$$

Nothing can be said about the qualification of the constraints in this point.

In this case, the feasible directions are of the form $\alpha \begin{pmatrix} 1 \\ 0 \end{pmatrix}$ and correspond to the tangent cone of the constraint c_1.

1.5.2 Conditions of Karush, Kuhn and Tucker

Let us take again the optimization problem in standard form with n real variables, p equality constraints c_E and q inequality constraints c_I.

$$\min_{x \in \mathbb{R}^n} f(x) \quad \text{s.t.} \quad \begin{cases} c_E(x) = 0 \\ c_I(x) \le 0 \end{cases} \tag{1.119}$$

As in section 1.4.1, the **Lagrangian** (or Lagrange function) is defined by

$$\begin{aligned} L(x,\lambda,\mu) &= f(x) + \lambda^T c_E(x) + \mu^T c_I(x) \\ &= f(x) + \sum_{j=1}^{p} \lambda_j c_{Ej}(x) + \sum_{j=1}^{q} \mu_j c_{Ij}(x) \end{aligned} \tag{1.120}$$

The vector $\lambda \in \mathbb{R}^p$ represents the multipliers of the p equality constraints c_E.

The vector $\mu \in \mathbb{R}^q$ represents the multipliers of the q inequality constraints c_I.

The following two theorems due to Karush (1939) and Kuhn and Tucker (1951) give respectively necessary and sufficient conditions for local optimality.

Theorem 1-5: Necessary conditions for a constrained local minimum

Assume that x^* is a local minimum of problem (1.119) and that the active constraints at x^* are linearly independent at this point.

Let us note $c_a(x)$ the active constraints and $T_a(x^*)$ their tangent cone at x^*.

$$T_a(x^*) = \left\{ d \in \mathbb{R}^n \,/\, \nabla c_a(x^*)^T d = 0 \right\} \quad \text{with} \quad c_a(x^*) = 0$$

Then there exists a unique vector $\lambda^* \in \mathbb{R}^p$ and a unique vector $\mu^* \in \mathbb{R}^q$ satisfying the following conditions of order 1 and order 2.

- Conditions of order 1 : $$\begin{cases} \nabla_x L(x^*,\lambda^*,\mu^*) = 0 \\ \nabla_\lambda L(x^*,\lambda^*,\mu^*) = 0 \\ \nabla_\mu L(x^*,\lambda^*,\mu^*) \le 0 \\ \mu^* \ge 0 \\ \mu_k^* c_{Ik}(x^*) = 0 \;,\; k=1 \text{ to } q \end{cases}$$
- Conditions of order 2 : $$d^T \nabla^2_{xx} L(x^*,\lambda^*,\mu^*) d \ge 0 \;,\; \forall d \in T_a(x^*)$$

The condition $\mu_k * c_{Ik}(x^*) = 0$ is called **the complementarity condition.**
It requires that either the inequality constraint is active or its multiplier is zero.

The condition of order 2 indicates that the Hessian $\nabla^2_{xx}L$ **reduced to the tangent directions** is positive semidefinite.

The KKT conditions are interpreted geometrically in section 1.5.3.

Elements of the demonstration

Different demonstrations are proposed in references [R3, R10, R11].
Here, we take up the main ideas of reference [R10].

Assume that x* is a local minimum of problem (1.119) and define the series of problems by: $\min_{x \in V} f_k(x) = f(x) + \frac{1}{2}k\|c_E(x)\|^2 + \frac{1}{2}k\|c_I^+(x)\|^2 + \|x - x^*\|$, $k \in \mathbb{N}$.

The function: $c_I^+(x) = \max(0, c_I(x))$ measures the inequality constraint violation.

The domain of minimization noted V is a closed neighborhood of x*.

<u>Minimum of</u> f_k

The function f_k is bounded on V (because it is continuous and V is compact).
It therefore admits a minimum $x_k \in V$ (by Weierstrass theorem).
This minimum satisfies: $f_k(x_k) \le f_k(x^*) = f(x^*)$ from which we deduce the following inequality noted I_1.

$$\|c_E(x_k)\|^2 + \|c_I^+(x_k)\|^2 \le \frac{2}{k}\left[f(x^*) - f(x_k) - \|x_k - x^*\|^2\right] \qquad \textbf{(inequality } \mathbf{I_1}\textbf{)}$$

<u>Limit of</u> (x_k)

The domain V being compact, we can extract from the series (x_k) a converging sub-series whose limit is noted $\bar{x}$. This limit satisfies the inequality I_1.
For $k \to \infty$, the second member has for limit 0 (because the continuous function f on compact V is bounded).

The inequality I_1 leads to: $c_E(\bar{x}) = c_I^+(\bar{x}) = 0$, so that the point $\bar{x} \in V$ is feasible for the problem (1.119). It therefore satisfies the following inequality I_2.

$f(x^*) \le f(\bar{x})$ **(inequality I_2)**

Moreover, the inequality I_1 gives: $f(x^*) - f(x_k) - \|x_k - x^*\|^2 \geq 0$, and by passing to the limit: $f(\bar{x}) + \|\bar{x} - x^*\|^2 \leq f(x^*)$. With the inequality I_2, we deduce that: $\bar{x} = x^*$.

Since this result holds for any sub-series of (x_k), we have: $\lim_{k \to \infty} x_k = x^*$.

Condition of order 1

For k large enough, the minimum x_k lies within V and it satisfies the condition of order 1: $\nabla f_k(x_k) = 0$. By deriving the function f_k, this implies

$$\nabla f(x_k) + k\nabla c_E(x_k)c_E(x_k) + k\nabla c_I^+(x_k)c_I^+(x_k) + 2(x_k - x^*) = 0 \quad \textbf{(equality E}_1\textbf{)}$$

Dividing by: $C_k = \sqrt{1 + k^2 \|c_E(x_k)\|^2 + k^2 \|c_I^+(x_k)\|^2}$,

and posing: $\lambda_k = \dfrac{kc_E(x_k)}{C_k} \in \mathbb{R}^p$, $\mu_k = \dfrac{kc_I^+(x_k)}{C_k} \in \mathbb{R}^q$, $\nu_k = \dfrac{1}{C_k} \in \mathbb{R}$,

the equality E_1 becomes: $\nu_k \nabla f(x_k) + \nabla c_E(x_k)\lambda_k + \nabla c_I^+(x_k)\mu_k + \dfrac{2}{C_k}(x_k - x^*) = 0$

The vector $(\lambda_k, \mu_k, \nu_k)$ is of norm 1. As the series (λ_k), (μ_k), (ν_k) are bounded, we can extract sub-series having the respective limits λ, μ, ν. The passage to the limit in the above equation then yields (the gradients being continuous)

$$\nu \nabla f(x^*) + \nabla c_E(x^*)\lambda + \nabla c_I^+(x^*)\mu = 0 \quad \textbf{(equality E}_2\textbf{)}$$

Existence and uniqueness of multipliers

Let us assume by contradiction that $\nu = 0$. Then (E_2) implies $\lambda = \mu = 0$, because the constraints are assumed to be linearly independent in x*. This contradicts the norm of (λ, μ, ν) being equal to 1. The multiplier ν is therefore non-zero and we can set $\lambda^* = \dfrac{\lambda}{\nu}$, $\mu^* = \dfrac{\mu}{\nu}$ to obtain the first order KKT: $\nabla_x L(x^*, \lambda^*, \mu^*) = 0$.

The existence of the multipliers λ^*, μ^* has thus been shown. The gradient of f is a linear combination (equality E_2) of the gradients of the constraints, which are assumed to be independent in x*. This demonstrates the uniqueness of λ^*, μ^*.

Condition of complementarity

The reals μ_k are positive, because $c_I^+(x_k) = \max(0, c_I(x_k)) \geq 0$.
If an inequality c_{Ij} is inactive in x*, then from some rank $(k > k_0)$:

$$c_{Ij}(x_k) < 0 \Rightarrow c_{Ij}^+(x_k) = 0 \Rightarrow (\mu_k)_j = 0 \xrightarrow{k\to\infty} \mu_j{}^* = 0$$

This demonstrates the condition of complementarity.

Condition of order 2

Any direction d of the tangent cone is the limit of a series of directions of the form $d_k = \dfrac{x_k - x^*}{s_k}$, where the feasible series (x_k) converges to x* and $\lim\limits_{k\to\infty} s_k = 0$, $s_k > 0$.

Let us expand the Lagrangian to order 2.

$$\begin{aligned} L(x_k, \lambda^*, \mu^*) &= L(x^* + s_k d_k, \lambda^*, \mu^*) \\ &= L(x^*, \lambda^*, \mu^*) + s_k \nabla_x L(x^*, \lambda^*, \mu^*)^T d_k + \frac{1}{2} s_k^2 d_k^T \nabla_{xx}^2 L(x^*, \lambda^*, \mu^*) d_k + o(s_k^2) \end{aligned}$$

In this expression:

$L(x_k, \lambda^*, \mu^*) = f(x_k) + \mu^{*T} c_I(x_k)$ (because x_k is feasible: $c_E(x_k) = 0$)

$L(x^*, \lambda^*, \mu^*) = f(x^*) + \mu^{*T} c_I(x^*) = f(x^*)$ (complementary condition)

$\nabla_x L(x^*, \lambda^*, \mu^*) = 0$ (condition of order 1)

There remains: $f(x_k) + \mu^{*T} c_I(x_k) = f(x^*) + \dfrac{1}{2} s_k^2 d_k^T \nabla_{xx}^2 L(x^*, \lambda^*, \mu^*) d_k + o(s_k^2)$

$$\Rightarrow d_k^T \nabla_{xx}^2 L(x^*, \lambda^*, \mu^*) d_k = 2\frac{f(x_k) - f(x^*) + \mu^{*T} c_I(x_k)}{s_k^2} + 2\frac{o(s_k^2)}{s_k^2} \quad \textbf{(equality E}_3\textbf{)}$$

If a constraint c_{Ij} is inactive in x*, its multiplier $\mu_j{}^*$ is zero.

If a constraint c_{Ij} is active in x*, the tangent direction d_k is defined from points x_k such that $c_{Ij}(x_k) = 0$. In both cases, we have: $\mu_j{}^* c_{Ij}(x_k) = 0$.
Furthermore, x_k being feasible and x* being the minimum: $f(x_k) \geq f(x^*)$.

By replacing into the equality E_3 and passing to the limit, we obtain the condition of order 2: $d^T \nabla_{xx}^2 L(x^*, \lambda^*, \mu^*) d \geq 0$.

The following example illustrates the importance of the linear independence assumption.

Example 1-14: Importance of the linear independence assumption

Let us consider the following optimization problem with a single constraint.

$\min_{x \in \mathbb{R}^n} f(x)$ s.t. $c(x)=0$.

An equivalent formulation of this problem is: $\min_{x \in \mathbb{R}^n} f(x)$ s.t. $C(x)=0$

where the constraint C is defined by: $C(x)=[c(x)]^2$.

It is obvious that both problems have the same solutions.

Let us try to apply the KKT conditions to the second formulation.

The Lagrangian has the expression: $L(x,\lambda)=f(x)+\lambda C(x)$.

The KKT conditions of order 1 are

$$\begin{cases} \nabla_x L = \nabla f(x) + \lambda \nabla C(x) = \nabla f(x) + 2\lambda c(x)\nabla c(x) = 0 \\ \nabla_\lambda L = C(x) = [c(x)]^2 = 0 \end{cases}$$

This leads to the system: $\begin{cases} \nabla f(x) = 0 \\ c(x) \quad = 0 \end{cases}$, which does no longer depend on λ.

This system of n+1 equations with n unknowns has no solution in the general case. The issue comes from the constraint Jacobian matrix $\nabla C(x) = 2c(x)\nabla c(x)$, which vanishes in any feasible point $c(x)=0$. The linear independence assumption (used to prove the KKT conditions) is not satisfied, so that there is no guarantee that the KKT conditions apply.

This example shows that reformulations of the constraints such as: $C(x)=[c(x)]^2$ should be avoided.

The following example illustrates the application of the necessary conditions to an optics problem.

Example 1-15: Descartes' law of refraction

Fermat's principle states that the path of a light ray minimizes the travel time.

Let us consider a light ray going from a point A situated in the homogeneous medium 1 to a point B situated in the homogeneous medium 2. The speed of propagation in a homogeneous medium is constant and Fermat's principle leads to a straight line propagation. The path from A to B therefore consists of two segments connecting in an interface point I between the two media.

Let us note v_1 and v_2 the respective speeds in each medium, d_1 and d_2 the distances travelled, t_1 and t_2 the travel times, α_1 and α_2 the incidence angles at the interface and l_0 the distance AB parallel to the interface (figure 1-18).

We have the relations: $d_1 = \dfrac{l_1}{\cos\alpha_1}$, $d_2 = \dfrac{l_2}{\cos\alpha_2}$, $t_1 = \dfrac{d_1}{v_1}$, $t_2 = \dfrac{d_2}{v_2}$.

By choosing α_1 and α_2 as unknowns, Fermat's principle leads to the following optimization problem, where T is the total travel time and L is the distance travelled parallel to the interface.

$$\min_{\alpha_1,\alpha_2} T = \frac{l_1}{v_1 \cos\alpha_1} + \frac{l_2}{v_2 \cos\alpha_2} \quad \text{s.t.} \quad L = l_1 \tan\alpha_1 + l_2 \tan\alpha_2 = l_0$$

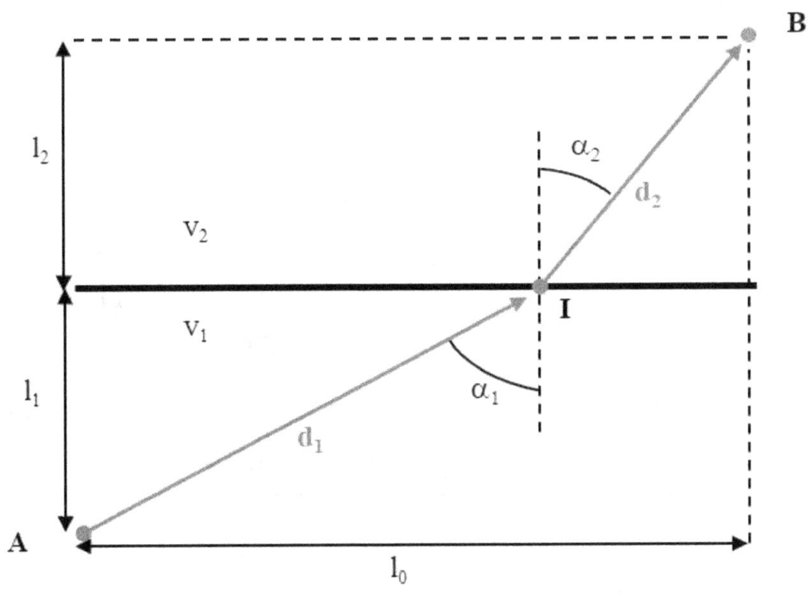

Figure 1-18: Law of refraction.

The Lagrangian has the expression

$$L(\alpha_1,\alpha_2,\lambda)=\frac{l_1}{v_1\cos\alpha_1}+\frac{l_2}{v_2\cos\alpha_2}+\lambda\left(l_1\tan\alpha_1+l_2\tan\alpha_2-l_0\right)$$

The KKT conditions of order 1 are

$$\begin{cases}\nabla_{\alpha_1}L=\dfrac{l_1\sin\alpha_1}{v_1\cos^2\alpha_1}+\lambda l_1\dfrac{1}{\cos^2\alpha_1}=0\\ \nabla_{\alpha_2}L=\dfrac{l_2\sin\alpha_2}{v_2\cos^2\alpha_2}+\lambda l_2\dfrac{1}{\cos^2\alpha_2}=0\\ \nabla_\lambda L=l_1\tan\alpha_1+l_2\tan\alpha_2-l_0=0\end{cases}\Rightarrow\begin{cases}\sin\alpha_1+\lambda v_1=0\\ \sin\alpha_2+\lambda v_2=0\\ l_1\tan\alpha_1+l_2\tan\alpha_2=l_0\end{cases}$$

The angles of incidence at the interface satisfy thus Descartes' law of refraction:

$$\frac{\sin\alpha_1}{\sin\alpha_2}=\frac{v_1}{v_2}$$

The sufficient conditions for local optimality are very similar.

Theorem 1-6: Sufficient conditions for a constrained local minimum

Assume that there exists $x^*\in\mathbb{R}^n$, $\lambda^*\in\mathbb{R}^p$, $\mu^*\in\mathbb{R}^q$ satisfying the following conditions of order 1 and order 2.

- Conditions of order 1 : $$\begin{cases}\nabla_x L(x^*,\lambda^*,\mu^*)=0\\ \nabla_\lambda L(x^*,\lambda^*,\mu^*)=0\\ \nabla_\mu L(x^*,\lambda^*,\mu^*)\le 0\\ \mu_k^*>0 \text{ if } c_{Ik}(x^*)=0\ ,\ \text{k=1 to q}\\ \mu_k^*=0 \text{ if } c_{Ik}(x^*)<0\ ,\ \text{k=1 to q}\end{cases}$$
- Conditions of order 2 : $d^T\nabla^2_{xx}L(x^*,\lambda^*,\mu^*)d>0\ ,\ \forall d\in T_a(x^*)\ ,\ d\neq 0$

Then x* is a strict local minimum of problem (1.119).

<u>Note</u>

The sufficient conditions do not make any assumptions about the qualification of the constraints.

The differences with the necessary conditions are:

- the condition of strict complementarity: either $\mu_k^* = 0$ or $c_{Ik}(x^*) = 0$;
- the condition on the strictly positive reduced Hessian.

Elements of the demonstration (see [R3], [R10], [R11])

Let (x^*, λ^*, μ^*) satisfy the above sufficient conditions. Assume by contradiction that x* is not a strict local minimum of problem (1.119).

We can then construct a series of feasible points (x_k) having the limit x* and being better than x* : $f(x_k) \le f(x^*)$.

These points define a series of directions: $d_k = \dfrac{x_k - x^*}{s_k}$, where the series of positive reals (s_k) has for limit 0 : $\lim_{k\to\infty} s_k = 0$, $s_k > 0$.

Directions of the tangent cone

Let us show that the directions (d_k) belong to the active constraints tangent cone. The active inequality constraints are denoted by c_{Ia} . We expand to order 1.

$$\begin{cases} f(x_k) = f(x^* + s_k d_k) = f(x^*) + s_k \nabla f(x^*)^T d_k + o(s_k) \\ c_E(x_k) = c_E(x^* + s_k d_k) = c_E(x^*) + s_k \nabla c_E(x^*)^T d_k + o(s_k) \\ c_{Ia}(x_k) = c_{Ia}(x^* + s_k d_k) = c_{Ia}(x^*) + s_k \nabla c_{Ia}(x^*)^T d_k + o(s_k) \end{cases}$$

Using the assumptions on x_k (feasible point, better than x*), we have

$$f(x_k) \le f(x^*) \quad \Rightarrow \quad \nabla f(x^*)^T d_k \le 0$$

$$c_E(x_k) = 0 \text{ and } c_E(x^*) = 0 \quad \Rightarrow \quad \nabla c_E(x^*)^T d_k = 0 \qquad \textbf{(inequalities } \mathbf{I_1}\textbf{)}$$

$$c_{Ia}(x_k) \le 0 \text{ and } c_{Ia}(x^*) = 0 \quad \Rightarrow \quad \nabla c_{Ia}(x^*)^T d_k \le 0$$

Let us use these inequalities in the first order condition by taking the scalar product with d_k.

$$\begin{aligned} & \nabla f(x^*) + \nabla c_E(x^*)\lambda^* + \nabla c_I(x^*)\mu^* = 0 \\ \Rightarrow\ & \nabla f(x^*)^T d_k + \lambda^* \nabla c_E(x^*)^T d_k + \mu^* \nabla c_I(x^*)^T d_k = 0 \end{aligned}$$

The first member of the equation has only negative or zero terms. Each term must therefore be zero. By the strict complementarity assumption made in the sufficient conditions, the multipliers of the inactive inequalities are zero and those of the active inequalities are strictly positive. For the corresponding term to be zero, we must have: $\nabla c_{Ia}(x^*)^T d_k = 0$.

The directions d_k satisfy: $\begin{cases} \nabla c_E(x^*)^T d_k = 0 \\ \nabla c_{Ia}(x^*)^T d_k = 0 \end{cases}$, and they belong therefore to $T_a(x^*)$.

Expansion to order 2

Let us now use the second order assumption by expanding the Lagrangian.

$$\begin{aligned} L(x_k,\lambda^*,\mu^*) &= L(x^*+s_k d_k,\lambda^*,\mu^*) \\ &= L(x^*,\lambda^*,\mu^*) + s_k \nabla_x L(x^*,\lambda^*,\mu^*)^T d_k + \frac{1}{2} s_k^2 d_k^T \nabla_{xx}^2 L(x^*,\lambda^*,\mu^*) d_k + o(s_k^2) \end{aligned}$$

In this expression:

$L(x_k,\lambda^*,\mu^*) = f(x_k) + \mu^{*T} c_I(x_k)$ (because x_k is feasible: $c_E(x_k)=0$)

$L(x^*,\lambda^*,\mu^*) = f(x^*) + \mu^{*T} c_I(x^*) = f(x^*)$ (complementary condition)

$\nabla_x L(x^*,\lambda^*) = 0$ (condition of order 1)

This leads to

$$f(x_k) = f(x^*) + \frac{1}{2} s_k^2 d_k^T \nabla_{xx}^2 L(x^*,\lambda^*,\mu^*) d_k - \mu^{*T} c_I(x_k) \qquad \text{for } s_k \text{ small}$$

$$\Rightarrow f(x_k) \geq f(x^*) + \frac{1}{2} s_k^2 d_k^T \nabla_{xx}^2 L(x^*,\lambda^*,\mu^*) d_k \qquad \text{(because } x_k \text{ feasible} \to c_I(x_k) \leq 0\text{)}$$

$\Rightarrow f(x_k) > f(x^*)$ using the second order assumption: $d_k^T \nabla_{xx}^2 L(x^*,\lambda^*,\mu^*) d_k > 0$

because it was shown earlier that $d_k \in T_a(x^*)$

This result contradicts the initial assumption about the series (x_k) : $f(x_k) \leq f(x^*)$.

The following example illustrates the application of the sufficient conditions and it also shows the importance of the strict complementarity condition.

Example 1-16: Application of the sufficient conditions

Consider the problem

$$\min_{x_1,x_2} \frac{1}{2}(x_2^2 - x_1^2) \quad \text{s.t.} \quad x_1 \le 1$$

The level lines are hyperbolas, which are shown on the right.

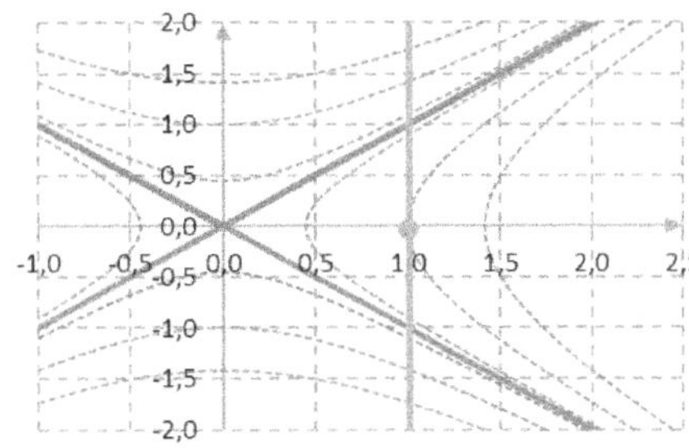

The function is minimized for $x_2 = 0$ and it decreases when $x_1 \to \pm\infty$.

The Lagrangian has the expression: $L(x_1,x_2,\mu) = \frac{1}{2}(x_2^2 - x_1^2) + \mu(x_1 - 1)$.

The necessary KKT conditions of order 1: $\begin{cases} \nabla_{x_1} L = -x_1 + \mu = 0 \\ \nabla_{x_2} L = x_2 = 0 \\ \nabla_{\mu} L = x_1 - 1 \le 0 \\ \mu(x_1 - 1) = 0 \ , \ \mu \ge 0 \end{cases}$

are satisfied in the point $\begin{cases} x_1{}^* = 1 \\ x_2{}^* = 0 \\ \mu^* = 1 \end{cases}$ and in the point $\begin{cases} x_1{}^* = 0 \\ x_2{}^* = 0 \\ \mu^* = 0 \end{cases}$.

<u>Examination of the first point</u>: $\begin{cases} x_1{}^* = 1 \\ x_2{}^* = 0 \\ \mu^* = 1 \end{cases}$

This point satisfies the sufficient conditions of order 1 (<u>strict</u> complementarity):
$\begin{cases} c(x^*) = 0 \\ \mu^* = 1 \ne 0 \end{cases}$

The constraint is active in x*. The second order KKT condition concerns the tangent directions which satisfy: $\nabla c(x^*)^T d = 0 \Rightarrow \begin{pmatrix} 1 \\ 0 \end{pmatrix}^T \begin{pmatrix} d_1 \\ d_2 \end{pmatrix} = d_1 = 0$.

For any non-zero tangent direction, we have

$$d^T \nabla_{xx}^2 L(x^*,\mu^*) d = \begin{pmatrix} d_1 \\ d_2 \end{pmatrix}^T \begin{pmatrix} -1 & 0 \\ 0 & 1 \end{pmatrix} \begin{pmatrix} d_1 \\ d_2 \end{pmatrix} = d_2^2 > 0$$

The point $x^*(1\,;\,0)$ satisfies the sufficient conditions of order 1 and order 2.

This point is therefore a local minimum.

Examination of the second point: $\begin{cases} x_1^* = 0 \\ x_2^* = 0 \\ \mu^* = 0 \end{cases}$

This point satisfies the sufficient conditions of order 1 (strict complementarity):
$\begin{cases} c(x^*) = -1 \neq 0 \\ \mu^* = 0 \end{cases}$

No constraint is active in x*. All directions are feasible.

For any direction: $d^T \nabla^2_{xx} L(x^*, \mu^*) d = \begin{pmatrix} d_1 \\ d_2 \end{pmatrix}^T \begin{pmatrix} -1 & 0 \\ 0 & 1 \end{pmatrix} \begin{pmatrix} d_1 \\ d_2 \end{pmatrix} = -d_1^2 + d_2^2$.

The condition of order 2 is not met if $d_1 \neq 0$. Indeed, a move along x_1 from the point $(0\,;0)$ causes the function to decrease.
This point actually corresponds to a maximum in the x_1 direction.

Shift of the inequality constraint

Suppose we shift the inequality constraint to $x_1 \leq 0$.

The optimization problem becomes: $\min_{x_1, x_2} \frac{1}{2}(x_2^2 - x_1^2)$ s.t. $x_1 \leq 0$.

The KKT conditions of order 1: $\begin{cases} \nabla_{x_1} L = -x_1 + \mu = 0 \\ \nabla_{x_2} L = x_2 = 0 \\ \nabla_{\mu} L = x_1 \leq 0 \\ \mu \geq 0 \\ \mu x_1 = 0 \end{cases}$ are satisfied in $\begin{cases} x_1^* = 0 \\ x_2^* = 0 \\ \mu^* = 0 \end{cases}$.

This point satisfies the sufficient conditions of order 1 and order 2, except the strict complementarity: the inequality constraint $x_1 \leq 0$ being active, the multiplier μ should not be zero. This point is indeed not a local minimum, because the function decreases along the direction $x_1 \to \infty$.

This example illustrates the importance of the strict complementarity condition.

Inequality constraints may be either active or inactive. For each inequality, the complementarity condition $\mu c_1(x) = 0$ leads to two possibilities to be examined separately. The resolution of the KKT conditions thus has a combinatorial aspect due to the inequality constraints as illustrated by the following example.

Example 1-17: Combinatorial aspect due to inequality constraints

Consider the problem: $\min_{x_1,x_2} f(x)=x_1+x_2$ s.t. $\begin{cases} c_1(x)=x_1^2+(x_2-1)^2-1\le 0 \\ c_2(x)=1-x_2\le 0 \end{cases}$.

The level lines of the constraints are shown opposite (a circle and a straight line).

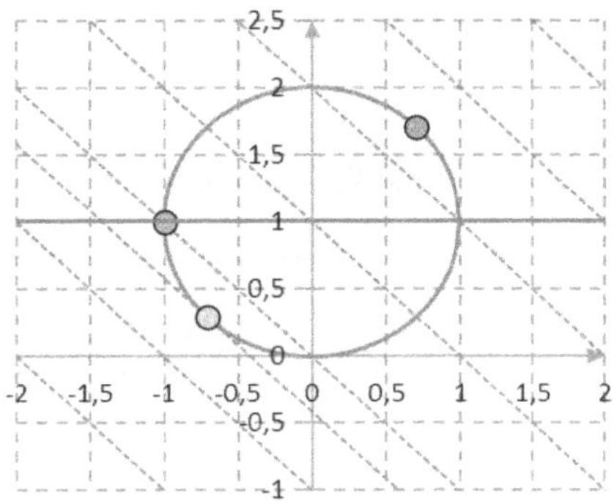

The function decreases when $\begin{cases} x_1\to-\infty \\ x_2\to-\infty \end{cases}$.

Resolution of KKT conditions

The Lagrangian has the expression

$L(x,\mu)=x_1+x_2+\mu_1\left(x_1^2+(x_2-1)^2-1\right)+\mu_2(1-x_2)$

The KKT conditions of order 1: $\begin{cases} 1+2\mu_1x_1=0 \\ 1+2\mu_1(x_2-1)-\mu_2=0 \\ x_1^2+(x_2-1)^2-1\le 0 \\ 1-x_2\le 0 \\ \mu_1c_1(x)=0\ ,\ \mu_1\ge 0 \\ \mu_2c_2(x)=0\ ,\ \mu_2\ge 0 \end{cases}$

give 4 possibilities: $\begin{cases} \mu_1=0 \text{ or } c_1(x)=0 \\ \mu_2=0 \text{ or } c_2(x)=0 \end{cases}$

- First constraint

If $\mu_1=0$ $\rightarrow$ incompatible with the first equation: $1+2\mu_1x_1=0$

We deduce that the constraint c_1 is active: $x_1^2+(x_2-1)^2-1=0$.

- Second constraint

If $\mu_2=0 \Rightarrow 1+2\mu_1(x_2-1)=0$ $\rightarrow$ incompatible inequalities: $\begin{cases} 1-x_2\le 0 \\ \mu_1\ge 0 \end{cases}$.

We deduce that the constraint c_2 is active: $1-x_2=0$.

- Selected combination

$$\begin{cases} c_1(x)=0 \\ c_2(x)=0 \end{cases} \Rightarrow \begin{cases} x_1=\pm 1 \\ x_2=1 \end{cases} \Rightarrow \begin{cases} \mu_1=\mp 0,5 \text{ and } \mu_1\geq 0 \to x_1=-1 \\ \mu_2=1 \end{cases}$$

The solution to the first order conditions is $\begin{cases} x_1=-1 \ , \ \mu_1=0,5 \\ x_2=\ 1 \ , \ \mu_2=1 \end{cases}$.

The two constraints being active and linearly independent, the tangent cone is empty. The condition of order 2 is then satisfied. This point is a local minimum.

<u>Change of sign of the second inequality constraint</u>

Let us change the sign of the second inequality. The new problem is

$$\min_{x_1,x_2} f(x)=x_1+x_2 \ \text{ s.t. } \begin{cases} c_1(x)=x_1^2+(x_2-1)^2-1\leq 0 \\ c_2(x)=1-x_2\geq 0 \end{cases}$$

The KKT conditions of order 1 become: $\begin{cases} 1+2\mu_1 x_1=0 \\ 1+2\mu_1(x_2-1)-\mu_2=0 \\ x_1^2+(x_2-1)^2-1\leq 0 \\ 1-x_2\geq 0 \\ \mu_1 c_1(x)=0 \ , \ \mu_1\geq 0 \\ \mu_2 c_2(x)=0 \ , \ \mu_2\geq 0 \end{cases}$

- First constraint

If $\mu_1=0 \to$ incompatible with the first equation: $1+2\mu_1 x_1=0$

We deduce that the constraint c_1 is active: $x_1^2+(x_2-1)^2-1=0$.

- Second constraint

If $c_2(x)=0 \Rightarrow \begin{cases} x_1=\pm 1 \\ x_2=1 \end{cases} \Rightarrow \begin{cases} \mu_1=\mp 0,5 \\ \mu_2=-1 \end{cases} \to$ incompatible with $\mu_2\geq 0$.

We deduce that the constraint c_2 is inactive: $\mu_2=0$.

- Selected combination

$$\begin{cases} 1+2\mu_1 x_1=0 \\ 1+2\mu_1(x_2-1)=0 \\ x_1^2+(x_2-1)^2-1=0 \end{cases} \Rightarrow \begin{cases} x_1=-1/(2\mu_1) \\ x_2=1-1/(2\mu_1) \\ \mu_1=1/\sqrt{2}>0 \end{cases} \Rightarrow \begin{cases} x_1=-1/\sqrt{2} \\ x_2=1-1/\sqrt{2} \\ \mu_1=1/\sqrt{2} \end{cases}$$

The solution to the first order conditions is $\begin{cases} x_1=-1/\sqrt{2} \ , \ \mu_1=0,5 \\ x_2=1-1/\sqrt{2} \ , \ \mu_2=0 \end{cases}$.

The Hessian of the Lagrangian: $\nabla^2_{xx}L(x,\mu)=\begin{pmatrix} 2\mu_1 & 0 \\ 0 & 2\mu_1 \end{pmatrix}$ is positive.

The condition of order 2 on the tangent directions is satisfied. This point is a local minimum.

Change the first inequality constraint into an equality

Let us modify the problem by turning the first constraint to an equality.

$$\min_{x_1,x_2} f(x)= x_1 + x_2 \quad \text{s.t.} \quad \begin{cases} c_1(x) = x_1^2 + (x_2-1)^2 - 1 = 0 \\ c_2(x) = 1 - x_2 \le 0 \end{cases}$$

The KKT conditions of order 1 become $\begin{cases} 1+2\lambda_1 x_1 = 0 \\ 1+2\lambda_1(x_2-1)-\mu_2 = 0 \\ x_1^2+(x_2-1)^2-1=0 \\ 1-x_2 \le 0 \\ \mu_2 c_2(x) = 0 \;,\; \mu_2 \ge 0 \end{cases}$

- First possibility: $\mu_2 = 0$

$$\Rightarrow \begin{cases} 1+2\lambda_1 x_1 = 0 \\ 1+2\lambda_1(x_2-1) = 0 \\ x_1^2+(x_2-1)^2-1=0 \end{cases} \Rightarrow \begin{cases} x_1 = -1/(2\lambda_1) \\ x_2 = 1-1/(2\lambda_1) \\ \lambda_1 = \pm 1/\sqrt{2} \end{cases} \Rightarrow \begin{cases} x_1 = 1/\sqrt{2} \\ x_2 = 1+1/\sqrt{2} \;\text{ since }\; 1-x_2 \le 0 \\ \lambda_1 = -1/\sqrt{2} \end{cases}$$

The Hessian of the Lagrangian: $\nabla^2_{xx} L(x,\lambda,\mu) = \begin{pmatrix} 2\lambda_1 & 0 \\ 0 & 2\lambda_1 \end{pmatrix}$ is negative.

The condition of order 2 on the tangent directions is not satisfied. This point is in fact a local maximum (negative Hessian).

- Second possibility: $c_2(x) = 0 \Rightarrow \begin{cases} x_1 = \pm 1 \\ x_2 = 1 \end{cases} \Rightarrow \begin{cases} \lambda_1 = \mp 0,5 \\ \mu_2 = 1 \end{cases}$

The Hessian of the Lagrangian is: $\nabla^2_{xx} L(x,\lambda,\mu) = \begin{pmatrix} 2\lambda_1 & 0 \\ 0 & 2\lambda_1 \end{pmatrix}$.

The two constraints being active and linearly independent, the tangent cone is empty. The condition of order 2 is then satisfied. This point is a local minimum.

Two local minima are obtained in $\begin{cases} x_1 = -1 \;,\; \lambda_1 = 0,5 \\ x_2 = 1 \;,\; \mu_2 = 1 \end{cases} \rightarrow f(x) = 0$

and in $\begin{cases} x_1 = 1 \;,\; \lambda_1 = -0,5 \\ x_2 = 1 \;,\; \mu_2 = 1 \end{cases} \rightarrow f(x) = 2$

These two minima are diametrically opposed on the circle. The best one is the one located in $(-1\,;1)$.

After these examples illustrating the sufficient conditions of theorem 1-9, we turn to conditions based on a reduction approach.

Reduced optimality conditions

The KKT conditions on the gradient and the Hessian of the Lagrangian are

$$\begin{cases} \nabla_x L(x^*,\lambda^*,\mu^*) = 0 \\ d^T \nabla^2_{xx} L(x^*,\lambda^*,\mu^*) d \geq 0 \ , \ \forall d \in T_a(x^*) \end{cases} \tag{1.121}$$

where the tangent cone to the active constraints $T_a(x^*)$ is defined by

$$T_a(x^*) = \left\{ d \in \mathbb{R}^n \ / \ \nabla c_a(x^*)^T d = 0 \right\} \quad \text{with} \quad c_a(x^*) = 0 \tag{1.122}$$

Consider the Jacobian matrix of the active constraints in x*: $A = \nabla c_a(x^*)^T$, and a basis Z of its null space (section 1.3.1) : $AZ = 0$. The directions of the tangent cone are then of the form: $d = Zd_Z$, where the vector $d_Z \in \mathbb{R}^{n-m}$ can be chosen freely.

Let us put together in λ_a^* the multipliers of the active constraints in x*.

$$\nabla_x L(x^*,\lambda_a^*) = \nabla f(x^*) + \nabla c_a(x^*)\lambda_a = \nabla f(x^*) + A^T \lambda_a \tag{1.123}$$

Then take the scalar product with a direction $d = Zd_z$ belonging to $T_a(x^*)$.

$$\nabla_x L(x^*,\lambda_a^*)^T d = \nabla f(x^*)^T Z d_Z \quad \text{since} \quad AZ = 0 \tag{1.124}$$

According to the first KKT condition (1.121), this scalar product is zero for any vector $d_Z \in \mathbb{R}^{n-m}$. This implies that the reduced **gradient** of f is zero.

$$g_Z(x^*) = Z^T \nabla f(x^*) = 0 \tag{1.125}$$

Let us then replace the direction $d = Zd_Z$ into the second KKT condition (1.121).

$$d_Z^T Z^T \nabla^2_{xx} L(x^*,\lambda_a^*) Z d_Z \geq 0 \ , \ \forall d_Z \in \mathbb{R}^{n-m} \tag{1.126}$$

This implies that the reduced **Hessian of** the Lagrangian is positive semidefinite.

$$H_Z = Z^T \nabla^2_{xx} L(x^*,\lambda_a^*) Z \geq 0 \tag{1.127}$$

Note that these reduced conditions concern the reduced gradient g_Z of the cost function (1.125) on the one hand and the reduced Hessian H_Z of the Lagrangian (1.127) on the other.

Example 1-18: Box problem (reduced conditions)

Let us go back to the box problem (example 1-7) formulated as

$$\min_{h,r} f(h,r) = r^2 + rh \quad \text{s.t. } c(h,r) = r^2h - 2v_0 = 0$$

The Lagrangian is $L(h,r,\lambda) = r^2 + rh + \lambda(r^2h - 2v_0)$. The KKT conditions are

$$\begin{cases} r + \lambda r^2 = 0 \\ 2r + h + 2\lambda rh = 0 \\ r^2h = 2v_0 \end{cases} \Rightarrow \begin{cases} \lambda r = -1 \\ h = 2r \\ r^3 = v_0 \end{cases} \Rightarrow r = v_0^{\frac{1}{3}}, \ h = 2v_0^{\frac{1}{3}}, \ \lambda = -v_0^{-\frac{1}{3}}$$

Let us check the reduced conditions. The useful matrices are:

- the gradient of the cost function: $\nabla f(h,r) = \begin{pmatrix} r \\ 2r+h \end{pmatrix}$;
- the Hessian of the Lagrangian: $\nabla^2_{xx} L(h,r) = \begin{pmatrix} 0 & 1+2\lambda r \\ 1+2\lambda r & 2+2\lambda h \end{pmatrix}$;
- the Jacobian of the constraint: $A = \nabla c^T = \begin{pmatrix} r^2 & 2rh \end{pmatrix}$.

Let us choose as basis (1.67) the first column of the Jacobian.

$$A = \begin{pmatrix} \overset{h}{B} & \overset{r}{N} \end{pmatrix} \text{ with } \begin{cases} B = (r^2) \\ N = (2rh) \end{cases}$$

A basis Z of the null space is given by (1.69) : $Z = \begin{pmatrix} -B^{-1}N \\ I \end{pmatrix} = \begin{pmatrix} -2h/r \\ 1 \end{pmatrix}$.

We can then calculate on this basis Z:

- the reduced gradient of the cost function

$$g_Z(h,r) = Z^T \nabla f(h,r) = \begin{pmatrix} -2h/r \\ 1 \end{pmatrix}^T \begin{pmatrix} r \\ 2r+h \end{pmatrix} = 2r - h;$$

- the reduced Hessian of the Lagrangian

$$H_Z(h,r) = Z^T H(h,r) Z = \begin{pmatrix} -2h/r \\ 1 \end{pmatrix}^T \begin{pmatrix} 0 & 1+2\lambda r \\ 1+2\lambda r & 2+2\lambda h \end{pmatrix} \begin{pmatrix} -2h/r \\ 1 \end{pmatrix} = -2 - 2\lambda h.$$

At the solution point $\begin{cases} h = 2r \\ \lambda r = -1 \end{cases}$, we obtain $\begin{cases} g_Z(h,r) = 2r - h = 0 \\ H_Z(h,r) = -2 - 2\lambda h = 2 > 0 \end{cases}$.

The reduced optimality conditions are thus satisfied.

The KKT conditions can be extended by introducing a multiplier $\nu \geq 0$ associated with the cost function (abnormal multiplier). The Lagrangian (1.120) becomes

$$L(x,\lambda,\mu) = \nu f(x) + \lambda^T c_E(x) + \mu^T c_I(x) \tag{1.128}$$

The KKT conditions are unchanged, with the additional unknown ν.

The problem is said to be **normal** if there exists a solution (x^*,λ^*,μ^*) for any strictly positive value of ν. We can then set arbitrarily $\nu = 1$, which gives the usual KKT conditions.

The problem is said to be **abnormal** if the KKT solution imposes $\nu = 0$.

This situation corresponds to a feasible domain reduced to isolated points. The solution satisfies the constraints, but it is not possible to minimize the cost.
In this case, the constraint qualification assumption is not satisfied.

Example 1-19: Abnormal problem

Consider the problem: $\min_{x_1,x_2} x_1$ s.t. $x_1^2 + x_2^2 = 0$.

The Lagrangian has the expression: $L(x_1,x_2,\nu,\lambda) = \nu x_1 + \lambda(x_1^2 + x_2^2)$
where the multiplier ν has been introduced for the cost function.

The KKT conditions $\begin{cases} \nabla_{x_1} L = \nu + 2\lambda x_1 = 0 \\ \nabla_{x_2} L = \quad 2\lambda x_2 = 0 \\ \nabla_\lambda L = x_1^2 + x_2^2 = 0 \end{cases}$ have no solution if $\nu > 0$.

The solution is $x_1 = x_2 = 0$ for $\nu = 0$ and it is the only feasible point.
We note that this point does not satisfy the qualification hypothesis ($\nabla c = 0$).

1.5.3 Geometric interpretation

The KKT conditions of order 1 indicate that the gradient of the cost function is a linear combination of the gradients of the constraints. This section presents some geometric illustrations in dimension 2.

<u>Problem with one equality constraint</u>

$$\min_{x_1,x_2} f(x_1,x_2) \ \text{ s.t. } \ c_E(x_1,x_2)=0 \tag{1.129}$$

Figure 1-19 shows the feasible level line $c_E = 0$ as a solid line, the gradient ∇c_E pointing inwards, the level lines of the cost function as dashed lines and the opposite of the gradient $-\nabla f$ directed to the left (direction of steepest descent). The cost function plotted here is: $f(x_1,x_2)=x_1$.

Let us look at the directions of descent and the feasible directions:

- the directions of descent (satisfying $\nabla f^T d \le 0$) are pointing to the left;
- on the curve $c_E = 0$, the feasible directions are tangent: $\nabla c_E^T d = 0$.

At the point x* satisfying the KKT condition: $\nabla f(x^*) + \lambda \nabla c_E(x^*) = 0$, there exists no feasible direction of descent.

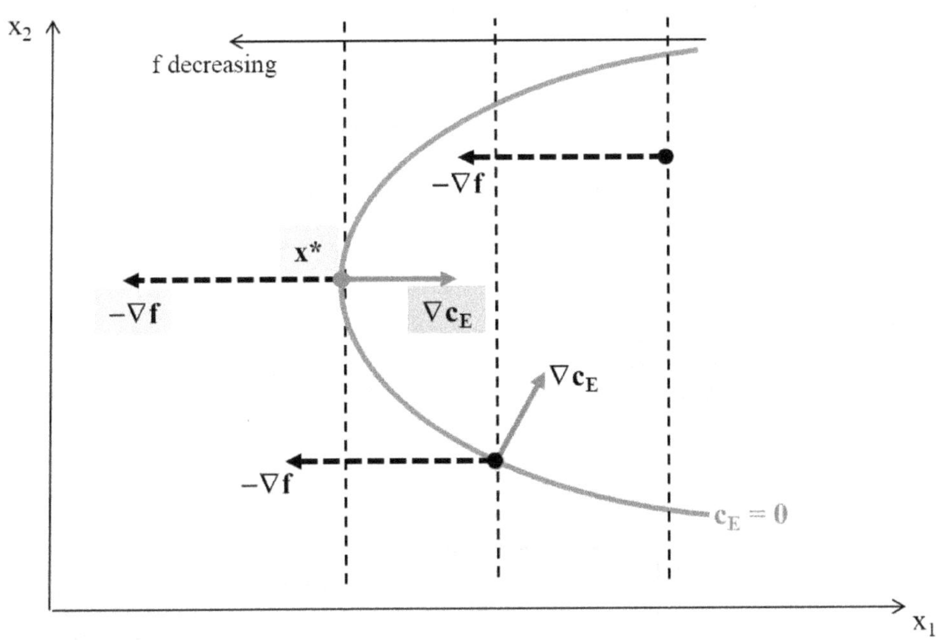

Figure 1-19: Linear cost – One equality constraint.

Problem with one inequality constraint

$$\min_{x_1,x_2} f(x_1,x_2) \quad \text{s.t.} \quad c_I(x_1,x_2) \le 0 \tag{1.130}$$

Figure 1-20 shows the feasible domain $c_I \le 0$ bounded by the level line $c_I = 0$ and the gradient ∇c_I directed outwards from the curve (c_I increasing direction).

The cost function is the same as previously: $f(x_1,x_2) = x_1$.

Let us look at the directions of descent and the feasible directions:

- the directions of descent (satisfying $\nabla f^T d \le 0$) are pointing to the left;
- within the feasible domain $c_I < 0$, all directions are feasible;
- at the boundary of the feasible domain $c_I = 0$, the feasible directions are tangent or inwards directed: $\nabla c_I^T d \le 0$.

At the point x* satisfying the KKT condition: $\nabla f(x^*) + \mu \nabla c_I(x^*) = 0$ with $\mu \ge 0$, there exists no feasible direction of descent.

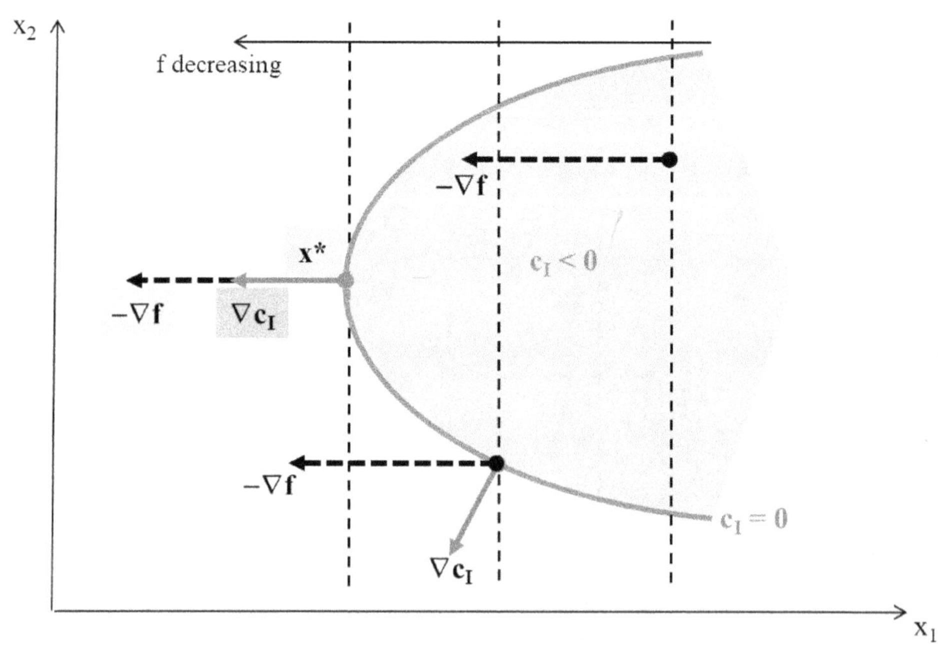

Figure 1-20: Linear cost – One inequality constraint.

Problem with two inequality constraints

$$\min_{x_1,x_2} f(x_1,x_2) \ \text{ s.t. } \begin{cases} c_{I1}(x_1,x_2) \le 0 \\ c_{I2}(x_1,x_2) \le 0 \end{cases} \tag{1.131}$$

The feasible domain is the intersection of the domains $c_{I1} \le 0$ and $c_{I2} \le 0$ as shown in figure 1-21. It is bounded by the curves $c_{I1} = 0$ and $c_{I2} = 0$.

The cost function is the same as previously: $f(x_1,x_2) = x_1$.

Let us look at the directions of descent and the feasible directions:

- the directions of descent (satisfying $\nabla f^T d \le 0$) are pointing to the left;
- within the feasible domain, all directions are feasible;
- at the boundary $c_{I1} = 0$, the feasible directions satisfy: $\nabla c_{I1}^T d \le 0$;
- at the boundary $c_{I2} = 0$, the feasible directions satisfy: $\nabla c_{I2}^T d \le 0$.

At the point x* satisfying the KKT condition: $\nabla f(x^*) + \mu_1 \nabla c_{I1}(x^*) + \mu_2 \nabla c_{I2}(x^*) = 0$ with $\mu_1 \ge 0$, $\mu_2 \ge 0$, there exists no feasible direction of descent.

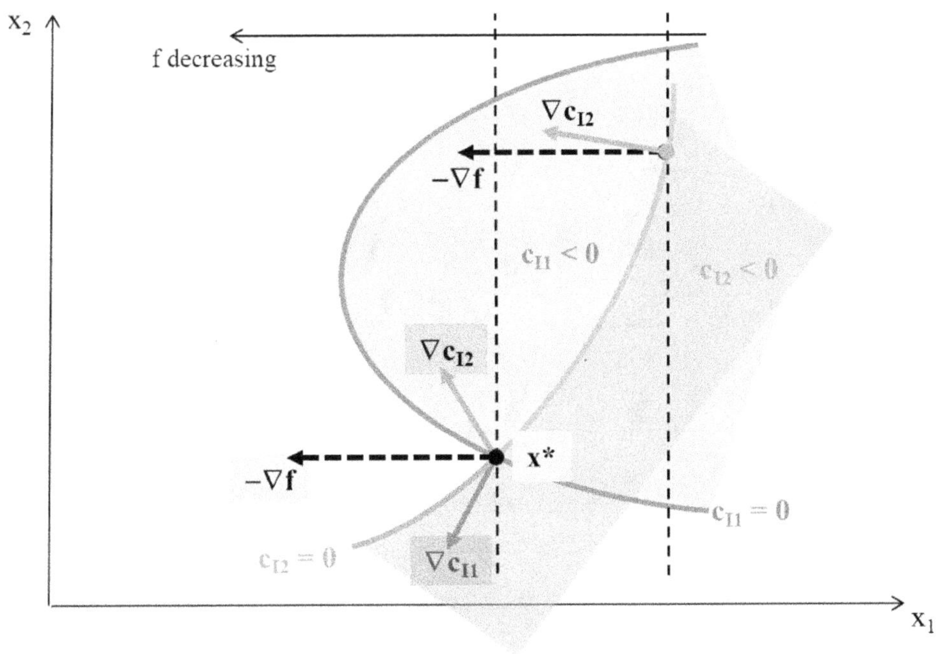

Figure 1-21: Linear cost – Two inequality constraints.

The previous illustrations assume a linear cost function whose level lines are parallel straight lines. In the general case, the cost function has a minimum in x_0 surrounded by concentric level lines. These can be approximated by a quadratic model formed from the Taylor expansion to order 2 in the vicinity of x_0.

Figure 1-22 illustrates the case of an inequality constraint $c_I \le 0$.

At the point x* satisfying the KKT conditions, the gradient ∇f is collinear and opposite to ∇c_I. In this point, the boundary of the feasible domain $c_I = 0$ is tangent to a level line of the cost function. This line is the one of lowest level that has at least one point in the feasible domain. All directions of descent (satisfying $\nabla f^T d \le 0$) lead out of the feasible domain.

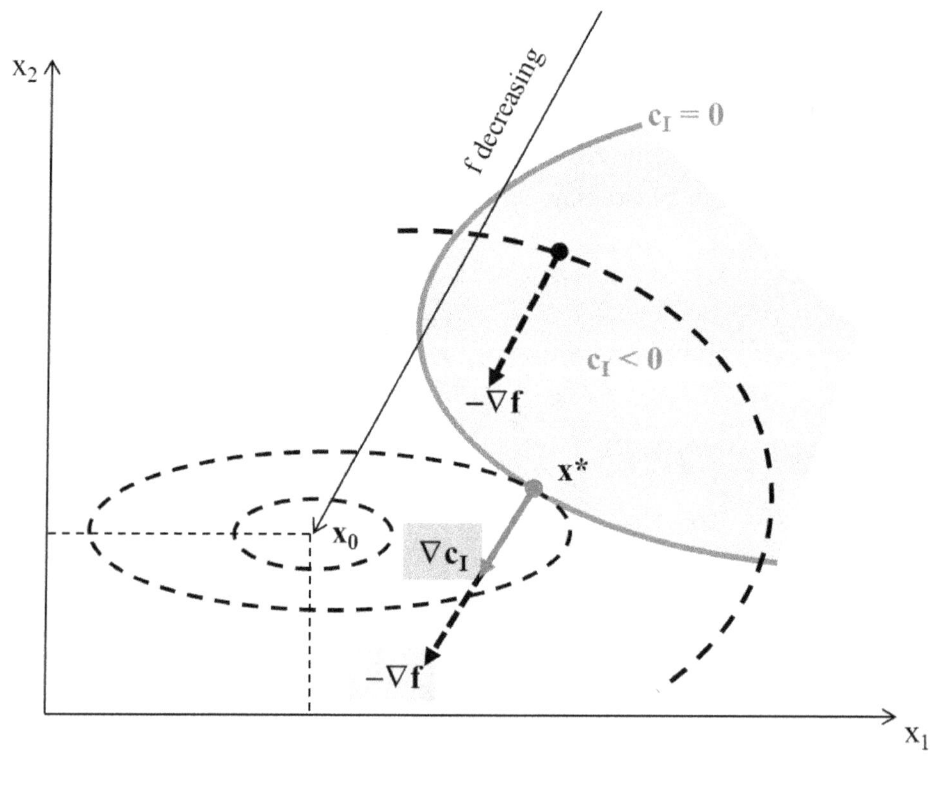

Figure 1-22: Quadratic cost – One inequality constraint.

Figure 1-23 illustrates the case of an equality constraint $c_E = 0$ and an inequality constraint $c_I \le 0$. The feasible domain is the portion of the curve $c_E = 0$ included in the domain $c_I \le 0$.

At the point x* satisfying the KKT conditions, the direction $-\nabla f$ is a linear combination of ∇c_E and ∇c_I with a positive coefficient μ on ∇c_I.

The point x* is on the lowest level line having at least one point on the feasible curve section.

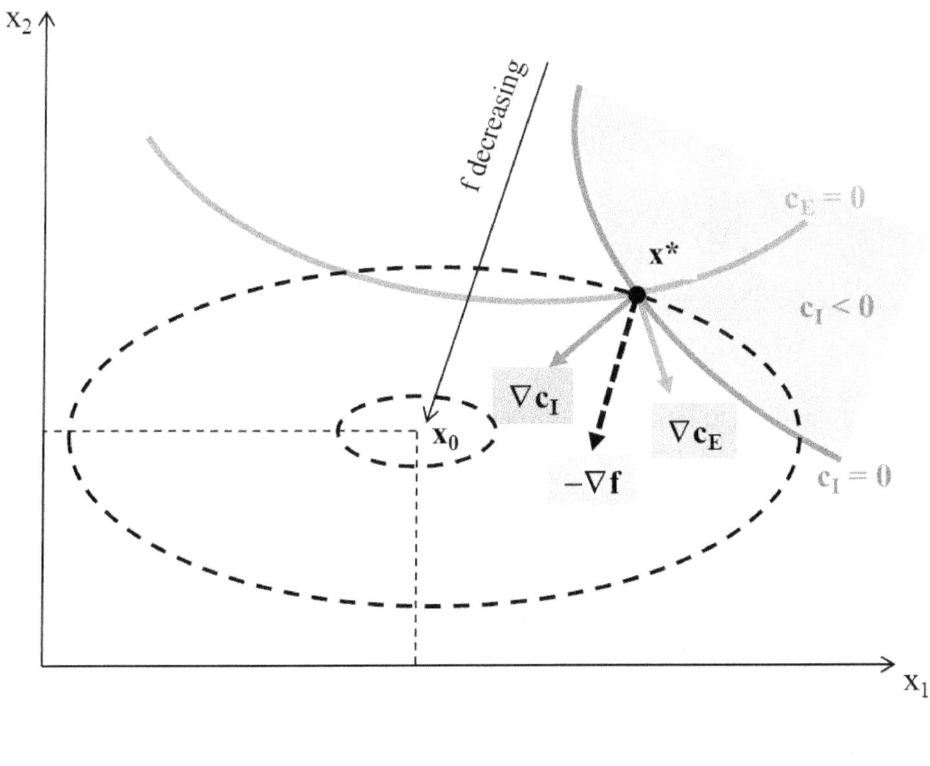

Figure 1-23: Quadratic cost – One equality and one inequality constraint.

1.5.4 Quadratic-linear problem

A standard quadratic-linear problem is of the form

$$\min_{x\in\mathbb{R}^n} \frac{1}{2}x^TQx + c^Tx \quad \text{s.t. } Ax = b \tag{1.132}$$

with $A \in \mathbb{R}^{m\times n}$, $b \in \mathbb{R}^m$, $c \in \mathbb{R}^n$, $Q \in \mathbb{R}^{n\times n}$ and Q being symmetric invertible.

The Lagrangian is defined with the multiplier $\lambda \in \mathbb{R}^m$ of the equality constraints.

$$L(x,\lambda) = \frac{1}{2}x^TQx + c^Tx + \lambda^T(b - Ax) \tag{1.133}$$

The KKT conditions of order 1 give

$$\begin{cases} Qx - A^T\lambda = -c \\ Ax = b \end{cases} \Rightarrow \begin{cases} x = Q^{-1}(A^T\lambda - c) \\ Ax = b \end{cases} \tag{1.134}$$

To solve this system, we pre-multiply the first equation by A, then eliminate Ax.

$$\begin{cases} \lambda = (AQ^{-1}A^T)^{-1}(AQ^{-1}c + b) \\ x = Q^{-1}A^T(AQ^{-1}A^T)^{-1}(AQ^{-1}c + b) - Q^{-1}c \end{cases} \tag{1.135}$$

The KKT conditions of order 2 are expressed with a basis Z of the kernel of A.

$$\left(\forall d \,,\, Ad = 0 \Rightarrow d^TQd \ge 0\right) \Leftrightarrow Z^TQZ \ge 0 \tag{1.136}$$

This result applies to finding the projection of a point on a hyperplane.

Example 1-20: Projection on a hyperplane

The projection of $x_0 \in \mathbb{R}^n$ on the hyperplane of equation $Ax = b$ is the hyperplane point x_P closest to x_0.
This point is the solution to the problem:

$$\min_{x\in\mathbb{R}^n} \frac{1}{2}\|x - x_0\|^2 \quad \text{s.t. } Ax = b$$

This is a quadratic problem of the form (1.132) with $Q = I$, $c = -x_0$.

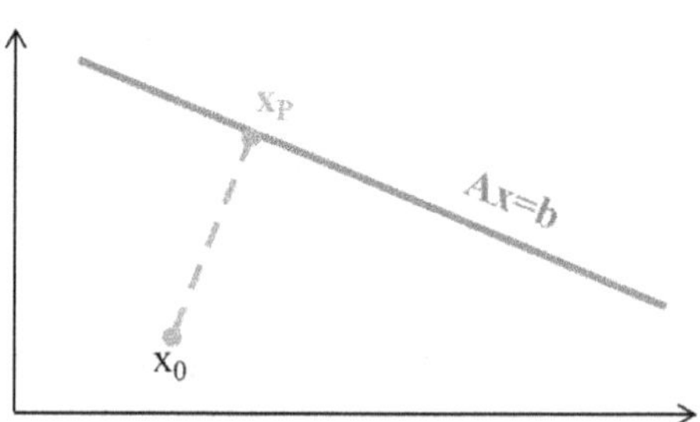

The solution is: $x_P = \left(I - A^T(AA^T)^{-1}A\right)x_0 + A^T(AA^T)^{-1}b$.

We retrieve the formula (1.64) with $Y = A^T$. The matrix $P = I - A^T(AA^T)^{-1}A$ is called the **projection matrix** on the hyperplane of equation $Ax = b$.

1.5.5 Sensitivity analysis

The Lagrange multipliers provide information on the sensitivity of the cost when the constraint level or the model parameters change.

Sensitivity to constraint level

Consider a minimization problem under equality constraints.

$$\min_{x \in \mathbb{R}^n} f(x) \text{ s.t. } c(x) = 0 \tag{1.137}$$

Suppose that (x^*, λ^*) satisfies the KKT conditions of order 1 and gives a cost f*.

$$\nabla_x L(x^*, \lambda^*) = 0 \ , \ c(x^*) = 0 \ , \ f(x^*) = f^* \tag{1.138}$$

We consider the perturbed problem where the constraints level is δc instead of 0.

$$\min_{x \in \mathbb{R}^n} f(x) \text{ s.t. } c(x) = \delta c \tag{1.139}$$

The solution to this problem is denoted $x^* + \delta x$ and its optimal cost is $f^* + \delta f$.

We are trying to express the cost variation δf as a function of the constraint variation δc. By expanding the cost and the constraints to order 1

$$\begin{cases} f(x^* + \delta x) = f(x^*) + \nabla f(x^*)^T \delta x = f(x^*) + \delta f \\ c(x^* + \delta x) = c(x^*) + \nabla c(x^*)^T \delta x = c(x^*) + \delta c \end{cases} \Rightarrow \begin{cases} \delta f = \nabla f(x^*)^T \delta x \\ \delta c = \nabla c(x^*)^T \delta x \end{cases} \tag{1.140}$$

then combining the two relations with the multiplier λ^*, we obtain

$$\delta f + \lambda^{*T} \delta c = \left(\nabla f(x^*)^T + \lambda^{*T} \nabla c(x^*)^T \right) \delta x = \nabla_x L(x^*, \lambda^*)^T \delta x = 0 \tag{1.141}$$

because the gradient with respect to x of the Lagrangian is zero in (x^*, λ^*).

δf is thus expressed as a function of δc.

$$\delta f = -\lambda^{*T} \delta c = -\sum_{j=1}^{m} \lambda_j^* \delta c_j \tag{1.142}$$

If the problem is subject to m equality constraints, each multiplier λ_j^* represents the sensitivity of the optimal cost to a level variation of the constraint number j.

<u>*Note 1*</u>

The multiplier gives the opposite of the sensitivity. This explains why the opposite sign convention is sometimes adopted for the Lagrangian: $L(x,\lambda) = f(x) - \lambda^T c(x)$.

<u>*Note 2*</u>

Inactive inequality constraints have a zero multiplier. The optimal cost is indeed indifferent to a variation of the level of the constraint, as long as this variation δc *is small and satisfies:* $c(x^*) + \delta c < 0$ *(the constraint remains inactive).*

The Lagrange multipliers thus give the sensitivity of the optimum cost to the constraint levels. If the variation is small, it is not necessary to solve the optimization problem again to find the cost. Figure 1-24 illustrates the effect on the solution of a variation of the constraint level.

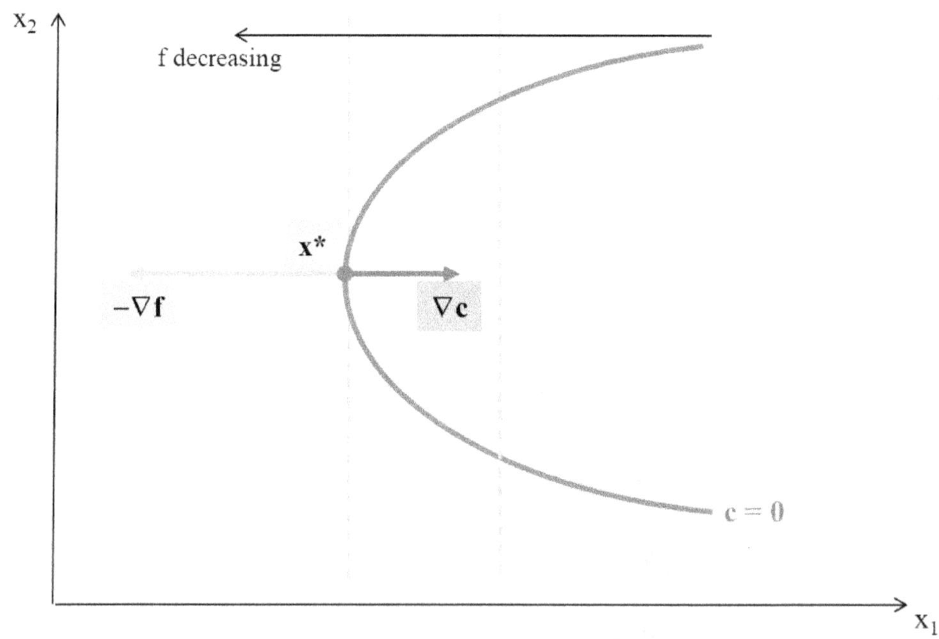

Figure 1-24: Sensitivity to the constraint level.

Example 1-21: Box problem (sensitivity)

Let us apply the sensitivity calculation (1.142) to the box problem (example 1-7) formulated as

$$\min_{h,r} f(h,r) = r^2 + rh \quad \text{s.t.} \quad c(h,r) = r^2 h - 2v_0 = 0 .$$

The solution for a volume v_0 is: $r = v_0^{\frac{1}{3}}$, $h = 2v_0^{\frac{1}{3}}$, $\lambda = -v_0^{-\frac{1}{3}}$

and it yields the minimum cost: $f = 3v_0^{\frac{2}{3}}$.

Let us assume that the required volume changes by δv_0 and calculate the cost change, first directly and then by using the constraint multiplier.

Direct calculation

The cost formula is directly derived: $f = 3v_0^{\frac{2}{3}}$,

which yields the cost change: $\delta f = \dfrac{df}{dv_0}\delta v_0 = 2v_0^{-\frac{1}{3}}\delta v_0$.

Calculation with the multiplier

This calculation is based on the formula (1.142) recalled here: $\delta f = -\lambda \delta c$.

The change of volume δv_0 produces a change δc of the constraint level.

The new constraint is formulated as

$$r^2 h - 2(v_0 + \delta v_0) = 0 \;\Rightarrow\; r^2 h - 2v_0 = 2\delta v_0 \;\Rightarrow\; c(r,h) = \delta c = 2\delta v_0$$

Formula (1.142) then gives the cost change: $\delta f = -\lambda \delta c = v_0^{-\frac{1}{3}}(2\delta v_0)$.

We retrieve the variation obtained by direct calculation.

Sensitivity to a model parameter

Consider now a problem with model parameters noted α.

$$\min_{x\in\mathbb{R}^n} f(x,\alpha) \quad \text{s.t.} \quad c(x,\alpha)=0 \tag{1.143}$$

Suppose that (x^*,λ^*) satisfies the KKT conditions of order 1 and gives a cost f*.

$$\nabla_x L(x^*,\lambda^*,\alpha)=0 \quad , \quad c(x^*,\alpha)=0 \quad , \quad f(x^*,\alpha)=f^* \tag{1.144}$$

Consider the perturbed problem where the model parameters becomes $\alpha+\delta\alpha$.

$$\min_{x\in\mathbb{R}^n} f(x,\alpha+\delta\alpha) \quad \text{s.t.} \quad c(x,\alpha+\delta\alpha)=0 \tag{1.145}$$

The solution to this problem is denoted $x^*+\delta x$ and its optimal cost is $f^*+\delta f$.

We are trying to express the cost variation δf as a function of the parameter variation $\delta\alpha$. By expanding the cost and constraints to the first order in x and α

$$\begin{cases} f(x^*+\delta x,\alpha+\delta\alpha)=f(x^*,\alpha)+\nabla_x f(x^*,\alpha)^T\delta x+\nabla_\alpha f(x^*,\alpha)^T\delta\alpha \\ \qquad\qquad\qquad\quad =f(x^*,\alpha)+\delta f \\ c(x^*+\delta x,\alpha+\delta\alpha)=c(x^*,\alpha)+\nabla_x c(x^*,\alpha)^T\delta x+\nabla_\alpha c(x^*,\alpha)^T\delta\alpha \\ \qquad\qquad\qquad\quad =0 \end{cases} \tag{1.146}$$

and using $c(x^*,\alpha)=0$ in the second equation, we get

$$\begin{cases} \nabla_x f(x^*,\alpha)^T\delta x+\nabla_\alpha f(x^*,\alpha)^T\delta\alpha=\delta f \\ \nabla_x c(x^*,\alpha)^T\delta x+\nabla_\alpha c(x^*,\alpha)^T\delta\alpha=0 \end{cases} \tag{1.147}$$

These two relations are combined with the multiplier λ^* to give the Lagrangian.

$$\begin{aligned} &\left(\nabla_x f(x^*,\alpha)^T+\lambda^{*T}\nabla_x c(x^*,\alpha)^T\right)\delta x \\ &+\left(\nabla_\alpha f(x^*,\alpha)^T+\lambda^{*T}\nabla_\alpha c(x^*,\alpha)^T\right)\delta\alpha=\delta f \\ \Leftrightarrow\ &\nabla_x L(x^*,\lambda^*,\alpha)^T\delta x+\nabla_\alpha L(x^*,\lambda^*,\alpha)^T\delta\alpha=\delta f \end{aligned} \tag{1.148}$$

The gradient with respect to x of the Lagrangian being zero in (x^*,λ^*), we get

$$\delta f=\nabla_\alpha L(x^*,\lambda^*,\alpha)^T\delta\alpha \tag{1.149}$$

This gives the variation δf as a function of the parameter variation $\delta\alpha$.

Example 1-22: Ballistic range (sensitivity)

The ballistic range with a flat earth and a constant gravity is given by

$$R = \frac{v^2 \sin 2\gamma}{g}$$

where g is the gravity, v is the initial velocity and γ is the initial slope.

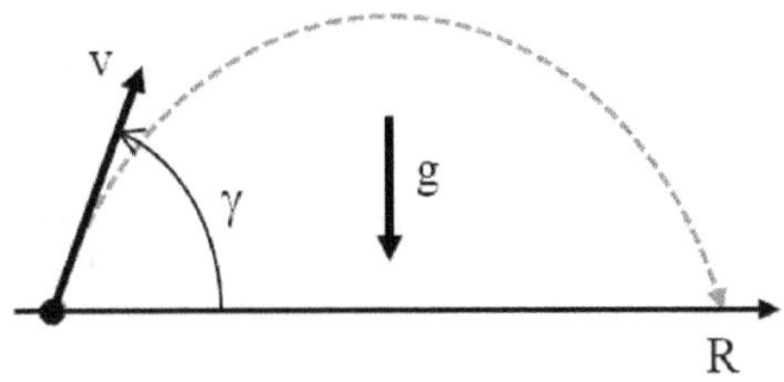

Let us look for the minimum velocity to reach a given range R_c.

$$\min_{v,\gamma} f(v,\gamma) = v \quad \text{s.t.} \quad c(v,\gamma) = \frac{v^2 \sin 2\gamma}{g} - R_c = 0$$

The Lagrangian is: $L(v,\gamma,\lambda) = v + \lambda\left(\frac{v^2 \sin 2\gamma}{g} - R_c\right)$.

The KKT conditions $\begin{cases} g + 2\lambda v \sin 2\gamma = 0 \\ 2\lambda v^2 \cos 2\gamma = 0 \\ v^2 \sin 2\gamma - gR_c = 0 \end{cases}$

give the solution $\begin{cases} \gamma = 45 \text{ deg} \\ v = \sqrt{gR_c} \end{cases}$ and $\lambda = -\frac{1}{2}\sqrt{\frac{g}{R_c}}$.

The result is a 45 deg shot that maximizes the range for a given velocity.

Let us calculate the sensitivity with respect to the range using formula (1.142).

$$\frac{\partial f}{\partial c} = -\lambda \quad \Rightarrow \quad \frac{\partial v}{\partial R_c} = \frac{1}{2}\sqrt{\frac{g}{R_c}}$$

Let us calculate the sensitivity to gravity using formula (1.149).

$$\frac{\partial f}{\partial \alpha} = \nabla_\alpha L \quad \Rightarrow \quad \frac{\partial v}{\partial g} = \frac{\partial}{\partial g}\left[v + \lambda\left(\frac{v^2 \sin 2\gamma}{g} - R_c\right)\right] = -\lambda \frac{v^2 \sin 2\gamma}{g^2} = \frac{1}{2}\sqrt{\frac{R_c}{g}}$$

In both cases, we retrieve the sensitivities that would be obtained by directly deriving the solution: $v = \sqrt{gR_c}$.

1.6 Conclusion

1.6.1 The key points

- One can seldom guarantee that the global optimum of a nonlinear optimization problem has been found;
- the KKT (Karush, Kuhn and Tucker) conditions allow to determine a local optimum by solving a system of equations and inequations;
- the resolution is easier if we know in advance the active constraints (that can be changed into equalities) and inactive constraints (which can be ignored);
- scaling variables and functions (multiplying them by a factor to bring them closer to unity) reduces numerical errors;
- a relative increment equal to the root of the machine accuracy (10^{-8} in double precision) is recommended for calculating finite difference derivatives.

1.6.2 To go further

- *Programmation mathématique (M. Minoux, Lavoisier 2008, 2e édition)*

  Sections 4.1, 4.2 and 6.2 present the KKT conditions and the theory of duality (saddle point) with proofs of most results.

- *Introduction à l'optimisation différentiable (M. Bierlaire, Presses polytechniques et universitaires normandes 2006)*

  Chapter 3 presents the concepts related to constraints. Chapter 6 presents the KKT conditions and their demonstration. Numerous examples are discussed in detail throughout the text.

- *Les mathématiques du mieux-faire (J.B. Hiriart-Urruty, Ellipses 2008)*

  The first volume explains step by step the establishment of the KKT conditions with all demonstrations.

- *Optimisation continue (F.J. Bonnans, Dunod 2006)*

  The book is devoted to the theory of continuous optimization. Chapter 3 presents the optimality conditions with all the proofs.

- *Numerical optimization (J. Nocedal, S.J. Wright, Springer 2006)*

  Chapter 8 presents the numerical derivation methods and the principles of automatic derivation. Chapter 12 presents the KKT conditions with detailed demonstrations.

- *Practical optimization (P.E. Gill, W. Murray, M.H. Wright, Elsevier 2004)*

  Section 2.1 and chapters 7 and 8 discuss in detail the difficulties associated with numerical errors. A lot of practical advice is given on this subject for the implementation of the algorithms.

- *Practical methods of optimization (R. Fletcher, Wiley 1987, 2nd edition)*

  Chapter 9 presents the KKT conditions and their demonstration.

2. Gradient-free optimization

This chapter is devoted to gradient-free optimization methods. These methods are especially applicable to difficult problems with discrete variables or with many local minima.

Section 1 presents some reformulation techniques and introduces methods for difficult optimization. These methods can be local or global, deterministic or random.

Section 2 deals with one-dimensional optimization based on dichotomy methods. The optimal strategy is the golden ratio method which minimizes the number of function evaluations.

Section 3 presents the DIRECT algorithm which is a deterministic multidimensional optimization method based on interval splitting. This method is used to locate the global minimum of a continuous function.

Section 4 presents the Nelder-Mead algorithm, which is a deterministic method of local exploration. This method consists in moving a set of points following decreasing values of the function. Although it has no theoretical justification, it is very efficient for many applications.

Sections 5 to 11 are devoted to methods with a random part. These methods are generally called metaheuristics, as they are based on empirical search principles ("heuristics") applicable to a wide variety of problems ("meta"). These metaheuristics can be local or global in nature, and they can apply to discrete or continuous problems. Among the main metaheuristics are simulated annealing, particle swarms, ant colonies and evolutionary algorithms.

The presentation of these methods in sections 5 to 11 is limited to the essentials and it is inspired by *Métaheuristiques pour l'optimization difficile (J. Dréo, A. Pétrowski, P. Siarry, E. Taillard, Eyrolles 2003)*. The reader who wishes to go deeper into the subject of metaheuristics will find in this reference book a very clear presentation of the principles and possibilities of application.

2.1 Difficult optimization

An optimization problem is considered difficult if it has discrete variables and/or an unknown number of local minima. This section presents some reformulation techniques and introduces gradient-free methods suitable for difficult optimization.

2.1.1 Discrete variables

Consider the minimization of a function of one variable which belongs to a finite definition domain such as $D = \{d_1 ; d_2 ; \cdots ; d_p\}$.

$$\min_{x \in D} f(x) \tag{2.1}$$

This problem can be tackled by minimizing the penalized function f_ρ.

$$\min_{x \in \mathbb{R}} f_\rho(x) = f(x) + \rho C(x) \tag{2.2}$$

The penalty function C vanishes when $x \in D$ and it is positive when $x \notin D$. In order to have a derivable function, one can use a polynomial penalty of the form

$$C(x) = -(x - d_j)(x - d_{j+1}) \quad \text{if} \quad d_j \le x < d_{j+1} \tag{2.3}$$

or a trigonometric penalty of the form

$$C(x) = \sin\left(\pi \frac{x - d_j}{d_{j+1} - d_j}\right) \quad \text{if} \quad d_j \le x < d_{j+1} \tag{2.4}$$

The penalty coefficient $\rho > 0$ weights the cost function f and the distance of x to the finite set D. The function f_ρ has local minima around each value d_j. These minima are sharper when the penalty ρ increases.

Example 2-1: Penalization of a discrete variable

We seek the minimum of the function: $f(x) = x^4 - \frac{8}{3}x^3 - 2x^2 + 8x$.

Suppose that the definition domain is the discrete set: $D=\{-2;-1;0;1;2;3\}$.

Figure 2-1 shows in solid line the function f to be minimized and in dotted line the function f_ρ with a polynomial penalization $(\rho=30)$. Figure 2-2 shows the function f_ρ with a trigonometric penalization $(\rho=10)$.

We check that the penalized functions coincide with the true function in all feasible values of x. The penalization creates lobes between these values. These lobes are all the more pronounced as the penalization is high.

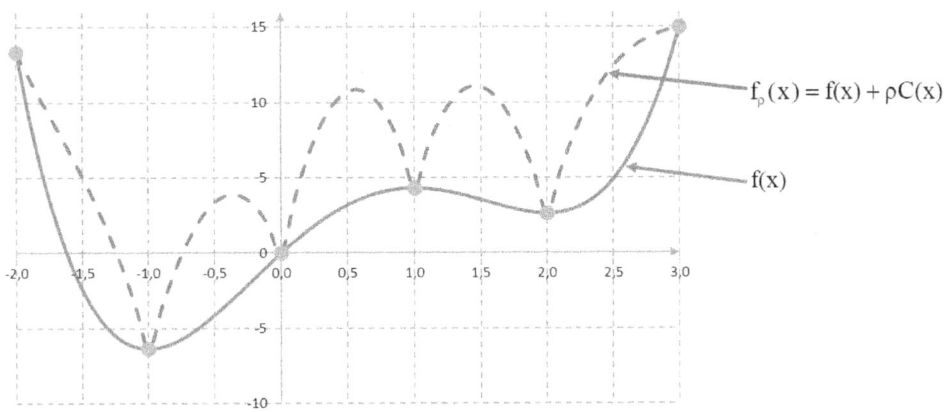

Figure 2-1: Polynomial penalization with $\rho = 30$.

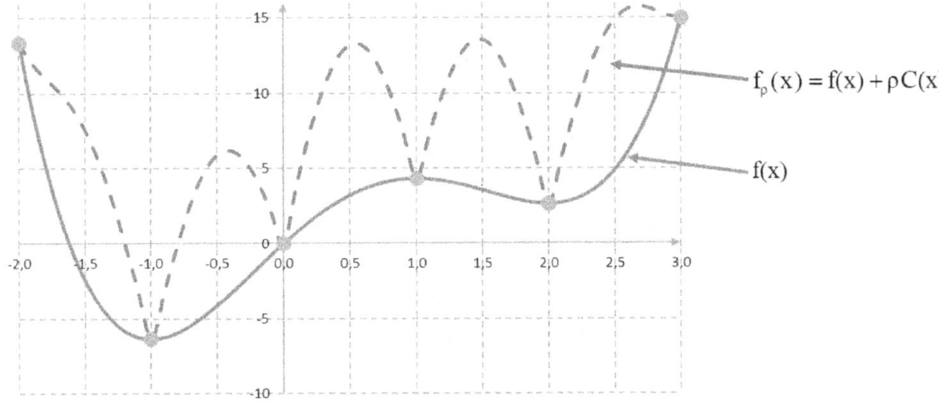

Figure 2-2: Trigonometric penalization with $\rho = 10$.

This technique can be applied to any problem mixing discrete and real variables. The global penalty function is then the sum of functions such as (2.3) or (2.4) associated with each discrete variable.

Since the penalized function f_p has a large number of local minima, a global method should preferably be used to minimize it.

2.1.2 Local minima

Consider the minimization of a function of n real variables.

$$\min_{x \in \mathbb{R}^n} f(x) \tag{2.5}$$

When the function to be minimized has an unknown number of local minima, it is possible to progressively eliminate these minima by a dilatation process.

Assume that a local minimum has already been located in x_m giving the value $f_m = f(x_m)$ for the cost function. We then define the dilated function F by

$$F(x) = \begin{cases} f(x) + c_1 \|x - x_m\| + \dfrac{c_2}{\|f(x) - f_m\|} & \text{if } f(x) > f_m \\ f(x) & \text{if } f(x) \le f_m \end{cases} \tag{2.6}$$

The dilated function F retains all local minima of f below f_m:

- the term in c_1 is intended to penalize local minima far from x_m and greater than f_m. The neighborhood of x_m is nearly unchanged;
- the term in c_2 aims to penalize the function in x_m (infinite value) and in its neighborhood in order to discard it and favor more distant domains.

By setting appropriately the values of c_1 and c_2, it is possible to favor the exploration of domains where $f(x) < f_m$.

Minimizing the dilated function allows one to find a new local minimum of f lower than f_m. By repeating the procedure, one can locate a set of decreasing local minima and finally isolate the global minimum.

Example 2-2: Elimination of local minima by dilatation

We are looking for the minimum of the Griewank function of one variable:

$$f(x) = \frac{x^2}{4000} - \cos x + 1$$

This function is plotted in figure 2-3.

It admits local minima in all points x such that: $\sin x + \frac{x}{2000} = 0$.

For $x \ll 2000$, the local minima are close to $k.2\pi$. The global minimum is located in $x = 0$. The values of the local minima are given in table 2-1.

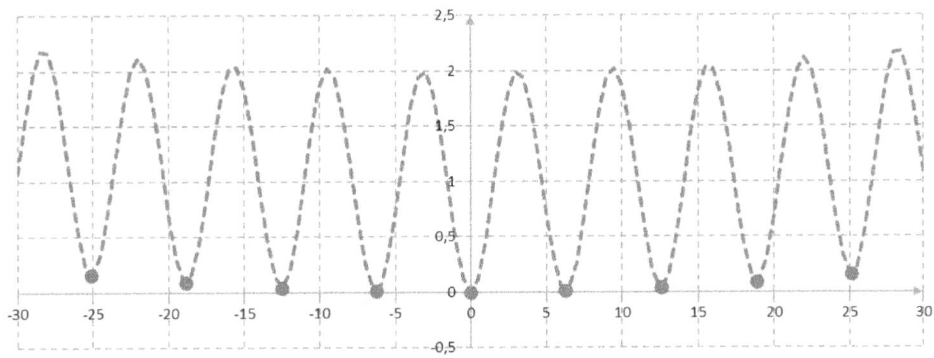

Figure 2-3: Griewank function.

x_m	−25.12	−18.84	−12.56	−6.28	0.00	6.28	12.56	18.84	25.12
f_m	1.578	0.887	0.394	0.098	0.0000	0.0986	0.3946	0.8878	1.578

Table 2-1: Values of local minima.

Let us assume that we have found the local minimum located in $x_m = 6.28$ and of cost value $f_m = 0.0986$. We look for a better local minimum (if it exists).

By applying the first dilatation term with a coefficient $c_1 = 2$, all local minima above $f_m = 0.0986$ are penalized. The known local minimum in $x_m = 6.28$ and the (unknown) global minimum in $x_m = 0$ are unchanged by this first dilatation.

The dilated function looks like the one shown in figure 2-4.

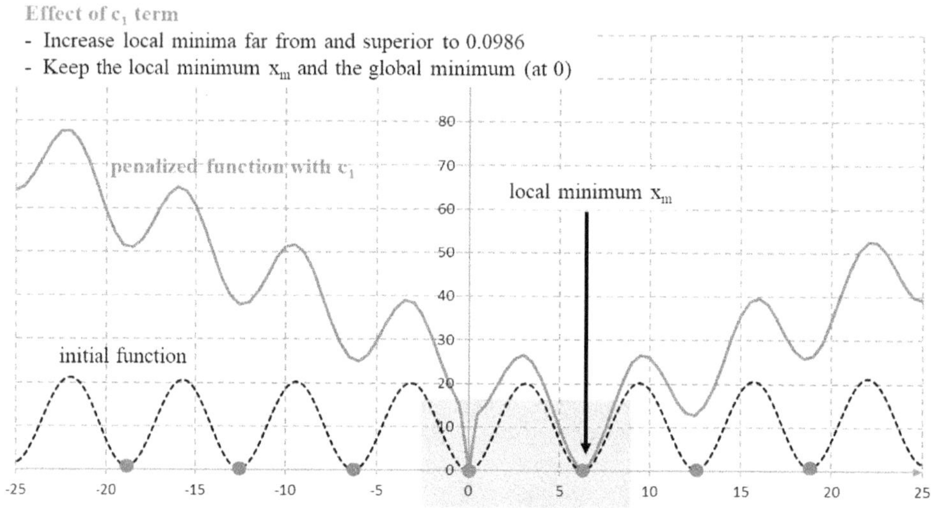

Figure 2-4: Effect of the first dilatation.

By applying the second dilatation term with a coefficient $c_2 = 5$, the known local minimum in $x_m = 6.28$ is strongly penalized. The global (unknown) minimum in $x_m = 0$ is unchanged.

The new dilated function looks like the one shown in figure 2-5.

Figure 2-6 shows a zoom of the dilated function over the interval $[-5\,;10]$.

It can be seen that the two dilatations have created a sharp minimum in $x_m = 0$. A minimization of this dilated function starting from the known point $x_m = 6.28$ allows us to find the global minimum.

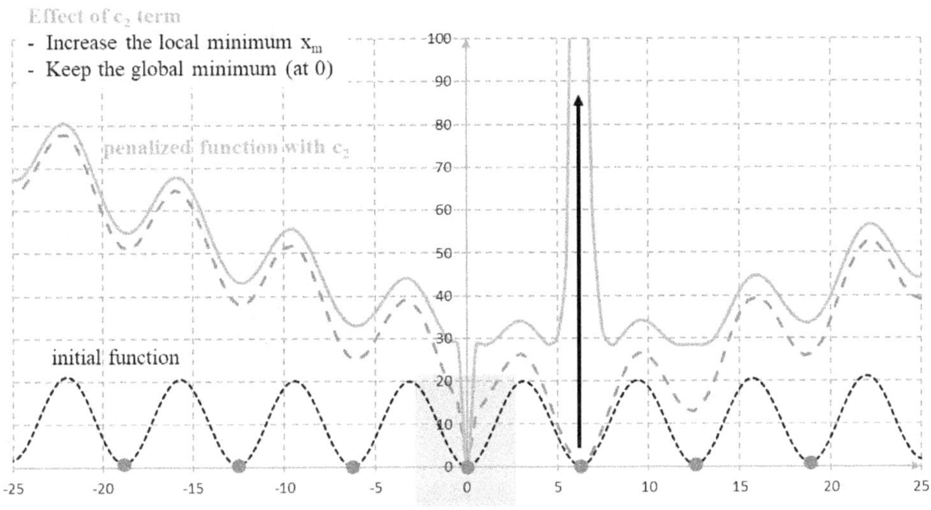

Figure 2-5: Effect of the second dilatation.

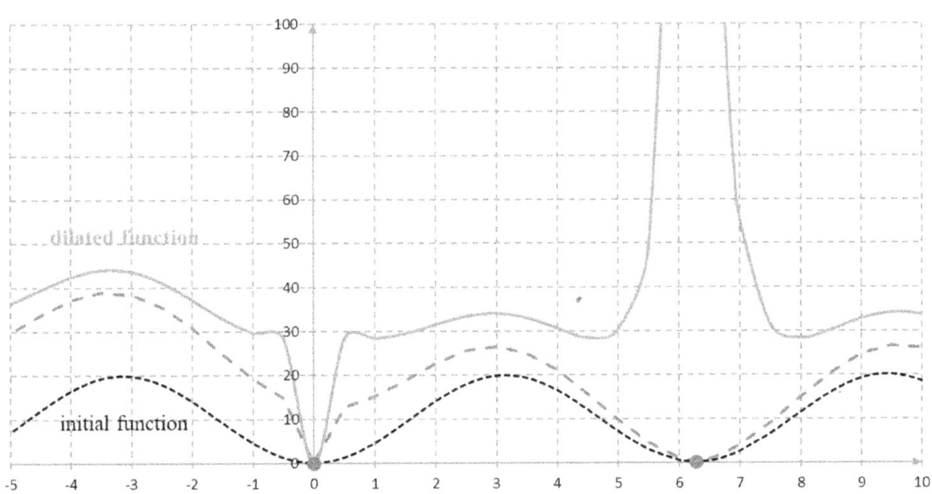

Figure 2-6: Dilated function and global minimum.

The dilated function (2.6) is written as

$$F(x) = f(x) + \frac{1}{2}\left(1 + \text{sgn}\left[f(x) - f_m\right]\right)\left(c_1 \|x - x_m\| + \frac{c_2}{\|f(x) - f_m\|}\right) \quad (2.7)$$

where the sign function is defined by: $\text{sgn}(y) = \begin{cases} +1 & \text{if } y > 0 \\ -1 & \text{if } y \le 0 \end{cases}$.

The function F is discontinuous in all points where $f(x) = f_m$. For some algorithms, it is preferable for the function to be continuous. This can be achieved by approximating the sign function by a sigmoid function of the form

$$\text{sgn}(y) \approx \frac{2}{1+e^{-\alpha y}} - 1 \quad \text{or} \quad \text{sgn}(y) \approx \tanh(\beta y) = \frac{e^{\beta y} - e^{-\beta y}}{e^{\beta y} + e^{-\beta y}} \quad (2.8)$$

The sigmoid function shown in figure 2-7 can be made as steep as desired (and as close as desired to the sign function) by increasing the coefficient α or β.

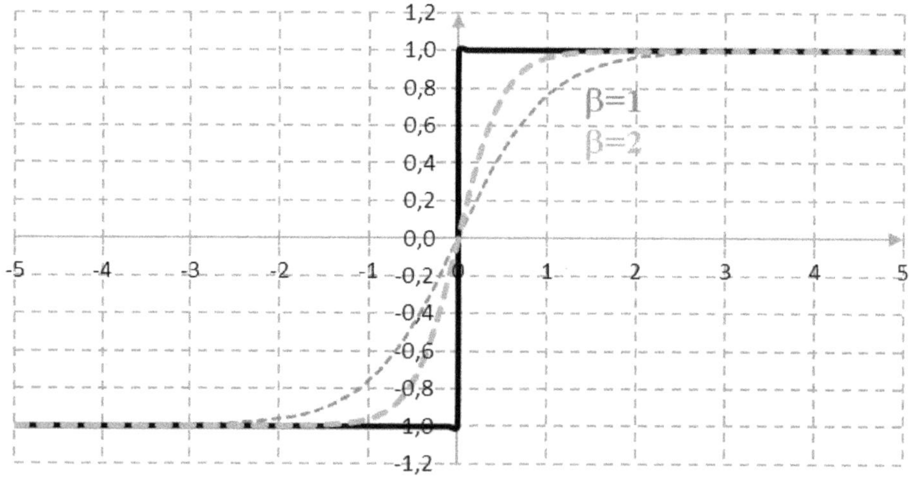

Figure 2-7: Sigmoid function.

Local minima may arise from the cost function itself, but also from constraints that partition the space into feasible domains with distinct minima. The use of a penalization method (section 4.2) is also likely to create local minima in the vicinity of feasible points.

2.1.3 Local and global methods

Gradient methods (presented in chapters 3 and 4) allow the determination of a precise minimum in the vicinity of an initial point, provided that the cost function is differentiable. A theoretically differentiable problem is often not numerically differentiable. When a simulation software contains conditional branchings with thresholds, interpolations or numerical noise (due to many successive operations), the calculated outputs are not differentiable or even not continuous.

Gradient methods are usable for problems that are well known in terms of computational processes (continuity, derivability) and existence of local minima.

Gradient-free methods aim to minimize non-differentiable or non-continuous functions, to deal with continuous or discrete variables and to explore the search domain widely to locate the global minimum.

These methods are local or global, deterministic or random, and apply to continuous or non-continuous problems. An optimization problem is said to be continuous if the cost function and the variables are continuous.

Local methods

Local methods start from an initial point and explore its neighborhood. Their result is a unique solution that depends on the initial point and the search domain. These methods are called "local" because the exploration is performed around the initial point. However, they may include mechanisms that allow them to escape local minima and eventually find a global minimum.

The main local methods are the following:

- the Nelder-Mead method is deterministic and it applies to continuous problems;
- the Affine Shaker method is similar to the Nelder-Mead method with a random component;
- the CMAES (*Covariance Matrix Adaptation Evolution Strategy*) method is based on a statistical exploration of the neighborhood and it applies to continuous problems;
- the Simulated Annealing method is based on a random exploration of the neighborhood and it is applicable to continuous and discrete problems;
- the Tabu Search method is based on a random exploration of the neighborhood and it applies to discrete problems.

Global methods

Global methods make a set of points move in a large search domain. The result is a set of solutions corresponding to local minima. These methods are called "global" because they explore simultaneously several areas of the search domain.

Most of these methods are metaheuristics. A heuristic is an empirical search technique specific to a given optimization problem. A "meta" heuristic is a general empirical search technique that can be applied to any optimization problem.

The metaheuristics make a set of points called **individuals** evolve, forming a **population**. The evolution of individuals follows rules inspired by a natural phenomenon. The interactions between individuals include random disturbances favoring on the one hand diversification (to spread the individuals over the whole search domain), and on the other hand intensification (to concentrate individuals in interesting areas).

The main global methods are the following:

- the DIRECT (*Dividing Rectangles*) method is deterministic and it applies to continuous problems;
- the Particle Swarm method is a metaheuristic inspired by the movements of groups of animals (bees, fish, birds);
- the Ant Colony method is a metaheuristic inspired by the exploration behavior of ants;
- the Genetic Algorithm method is a metaheuristic inspired by Darwin' law of natural evolution.

Figure 2-8 shows the general principle of metaheuristics.

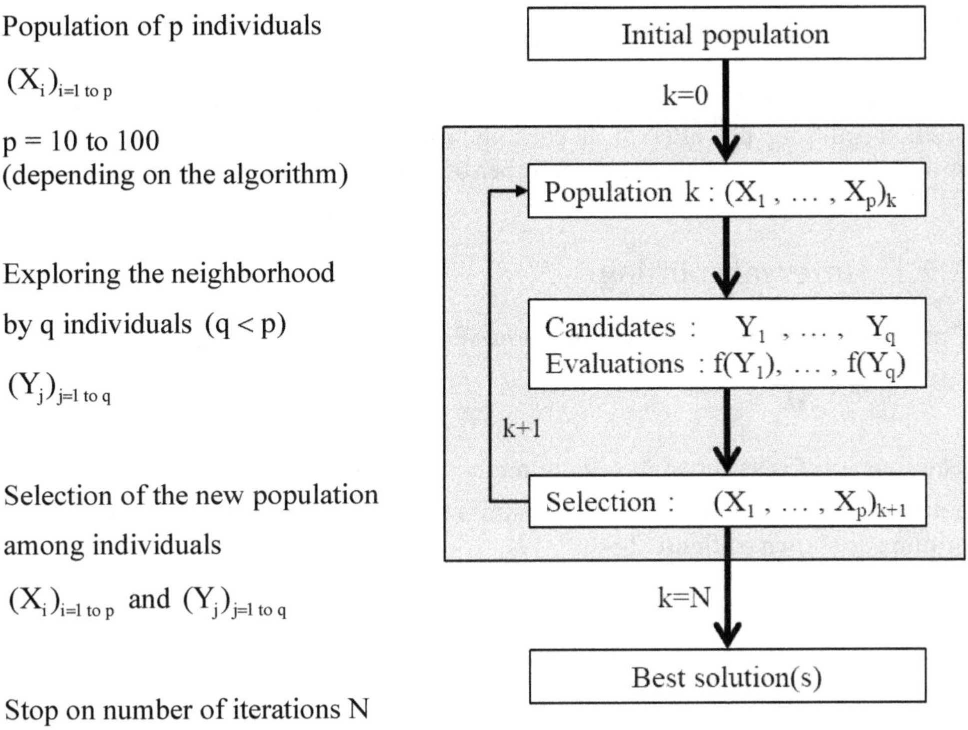

Figure 2-8: Principle of metaheuristics.

The following sections present the different gradient-free optimization methods, beginning with continuous problems of a single variable.

2.2 One-dimensional optimization

A one-dimensional optimization (single variable) can be done by a dichotomy method splitting the interval at each iteration. The objective is to define the splitting in order to minimize the number of evaluations of the function.

2.2.1 Interval splitting

Consider the minimization of a function of one variable on an interval.

$$\min_{x\in[a,d]} f(x) \tag{2.9}$$

The function f is assumed to be continuous on the interval $[a;d]$. It therefore has a minimum on the interval (by Weierstrass theorem). There may be several local minima as shown in figure 2-9.

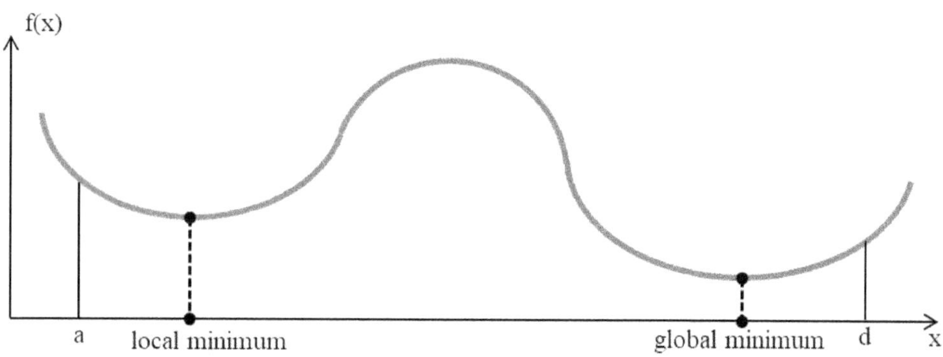

Figure 2-9: Minimizing a function of one variable.

The search for the minimum is carried out by dichotomy. The interval $[a;d]$ is divided into three sub-intervals by choosing two intermediate points b and c as shown in figure 2-10. The function is evaluated in points a, b, c and d:

- if $f(b) < f(c)$, we can assume that the minimum is around b. We then reduce the search interval to $[a;c]$;
- if $f(b) > f(c)$, we can assume that the minimum is around c. We then reduce the search interval to $[b;d]$.

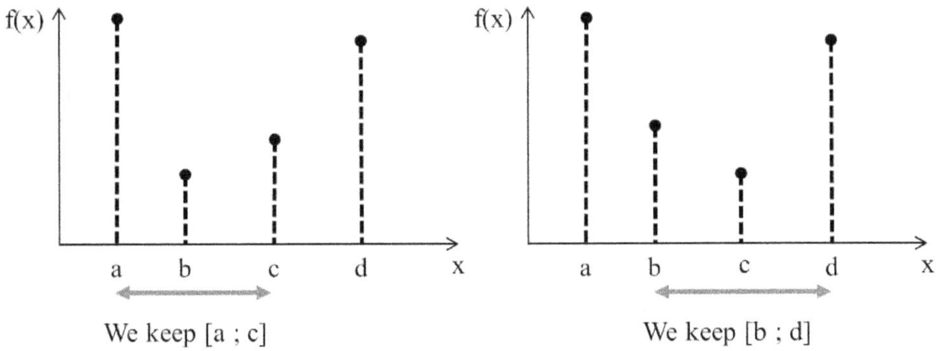

Figure 2-10: Reduction of the search interval.

2.2.2 Split points positioning

The dichotomy efficiency depends on the position of the points b and c. Their placement aims to reduce as much as possible the interval at each iteration and also to calculate the function f as little as possible.

The initial interval $[a;d]$ has the length: $\Delta_1 = d - a$.

The selected interval is either $[a;c]$ or $[b;d]$. In order to control the reduction of the interval, it is imposed that both intervals have the same length Δ_2.

$$\Delta_2 = c - a = d - b \tag{2.10}$$

Points b and c are placed symmetrically with respect to the middle of $[a;d]$, as in figure 2-11. Furthermore, we wish to evaluate only one new point e for the next iteration. Let us assume that the selected interval is $[a;c]$. This point e must then satisfy, as shown in figure 2-11,

$$\Delta_3 = b - a = c - e \tag{2.11}$$

The successive intervals $\Delta_1 = d - a$, $\Delta_2 = d - b$ and $\Delta_3 = b - a$ are linked by

$$\Delta_1 = \Delta_2 + \Delta_3 \tag{2.12}$$

We thus define a sequence of decreasing intervals whose lengths satisfy

$$\Delta_k = \Delta_{k+1} + \Delta_{k+2} \tag{2.13}$$

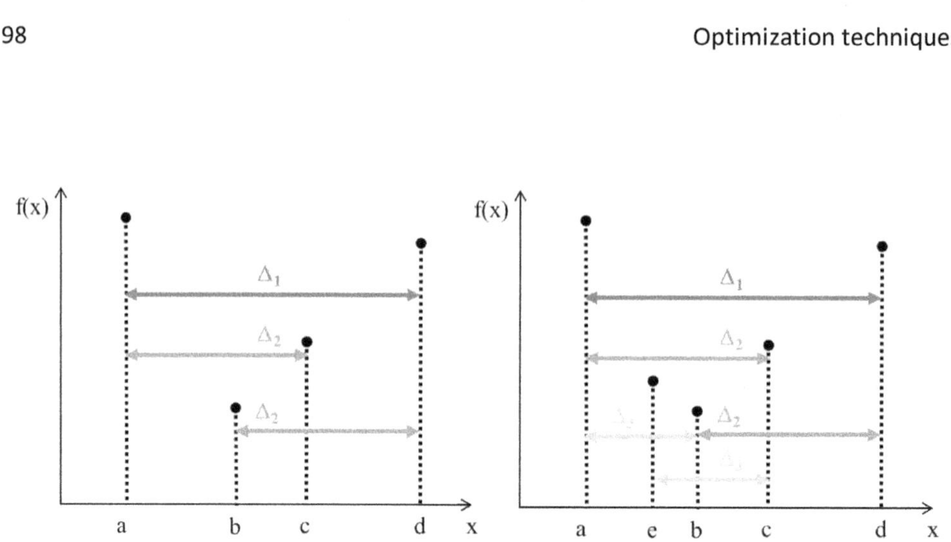

Figure 2-11: Successive intervals.

Let us assume that we perform N iterations starting from an interval of length Δ_0 to result in an interval of length Δ_N.

Let us define the real series $(F_k)_{k=0 \text{ to } N}$ from the length ratios.

$$F_{N-k} = \frac{\Delta_k}{\Delta_N} \quad , \quad k = 0 \text{ to } N \tag{2.14}$$

Dividing the relation (2.13) by Δ_N, and then posing $n = N - k$, we obtain

$$F_{N-k} = F_{N-k-1} + F_{N-k-2} \quad \Rightarrow \quad F_n = F_{n-1} + F_{n-2} \quad , \quad n = 2 \text{ to } N \tag{2.15}$$

The series (F_N) is completely defined by its first two terms:

- the first term is (2.14) : $F_0 = 1$;
- the second term F_1 can be chosen freely and determines the series.

The dichotomy objective is to minimize the number of iterations N to achieve a given accuracy ε starting from an interval of length Δ_0.

The number N of iterations to be performed is determined by

$$\Delta_N = \frac{\Delta_0}{F_N} \le \varepsilon \quad \Rightarrow \quad F_N \ge \frac{\Delta_0}{\varepsilon} \tag{2.16}$$

Therefore F_1 should be chosen such that the (F_n) series increases as quickly as possible.

The maximum value of $F_1 = \frac{\Delta_{N-1}}{\Delta_N}$ is 2 (taking b = c in the middle of the interval).

The numbers (F_n) then form the Fibonacci series (1, 2, 3, 5, 8, 13, 21, ...) and the **Fibonacci method** consists in placing the points according to the ratio

$$\frac{\Delta_k}{\Delta_{k+1}} = \frac{F_{N-k}}{F_{N-k-1}} \tag{2.17}$$

The positions of the points and the interval reduction ratio depend on the total number of iterations to be performed and they change at each iteration.

We observe that the ratio (2.17) tends towards a limit φ when $k \to \infty$. Indeed, dividing (2.13) by Δ_{k+1}, we have

$$\Delta_k = \Delta_{k+1} + \Delta_{k+2} \quad \Rightarrow \quad \frac{\Delta_k}{\Delta_{k+1}} = 1 + \frac{\Delta_{k+2}}{\Delta_{k+1}} \quad \xrightarrow[k\to\infty]{} \quad \varphi = 1 + \frac{1}{\varphi} \tag{2.18}$$

The limiting reduction ratio φ is equal to the **golden ratio** (so called because it gives a geometric proportion close to 16:9 that is pleasing to the human eye).

$$\varphi = \frac{1+\sqrt{5}}{2} \approx 1,618034 \tag{2.19}$$

2.2.3 Golden ratio method

The golden ratio method consists of simplifying the point placement by imposing the same reduction ratio at each iteration.

$$\frac{\Delta_k}{\Delta_{k+1}} = \varphi \tag{2.20}$$

This method is not optimal in terms of iteration number, but it is simpler to implement than the Fibonacci method and still very efficient.

The reduction ratio of the interval after N iterations is:

- $\frac{\Delta_0}{\Delta_N} = F_N$ for the Fibonacci method;
- $\frac{\Delta_0}{\Delta_N} = \varphi^N$ for the golden ratio method.

Table 2-2 compares the number of assessments needed to achieve different interval reduction ratios.

Reduction ratio of the interval	Number of iterations by Fibonacci	Number of iterations by golden ratio
10^{-2}	11	13
10^{-3}	15	18
10^{-4}	20	22
10^{-5}	25	27
10^{-6}	30	31

Table 2-2: Number of iterations by Fibonacci and golden ratio.

The algorithm of the golden ratio method is as follows.

Golden ratio algorithm

- **Initialization**

Choose the search interval $[a;d]$ of length: $\Delta_1 = d - a$.

Place the points b and c: $\begin{cases} b = a + r^2\Delta_1 \\ c = a + \ r\Delta_1 \end{cases}$ with $r = \dfrac{1}{\varphi} = \dfrac{\sqrt{5}-1}{2} \approx 0.618034$.

Evaluate $f(a), f(b), f(c), f(d)$.

- **Iterations**

If f(b) < f(c), replace d, c, Δ_1 and b with $\begin{cases} d \leftarrow c \\ c \leftarrow b \\ \Delta_1 \leftarrow \Delta_2 = r\Delta_1 \\ b = a + r^2\Delta_1 \end{cases}$ and calculate f(b).

If f(c) < f(b), replace a, b, Δ_1 and c with $\begin{cases} a \leftarrow b \\ b \leftarrow c \\ \Delta_1 \leftarrow \Delta_2 = r\Delta_1 \\ c = a + r\Delta_1 \end{cases}$ and calculate f(c).

- **Stop**

On a precision ε_x on the variable: $\Delta_1 = d - a \le \varepsilon_x$.

On a precision ε_f on the function: $|f(c) - f(b)| \le \varepsilon_f$.

Example 2-3: Golden ratio method

The minimum of the function $f(x) = -x\cos x$ is sought in the interval $\left[0\,;\,\dfrac{\pi}{2}\right]$.

The first five iterations of the golden ratio method are shown below. The values of a, b, c, d are marked below the line and the values of f(a), f(b), f(c), f(d) are marked above. The selected point (b or c) is boxed and an arrow indicates its position for the next iteration.

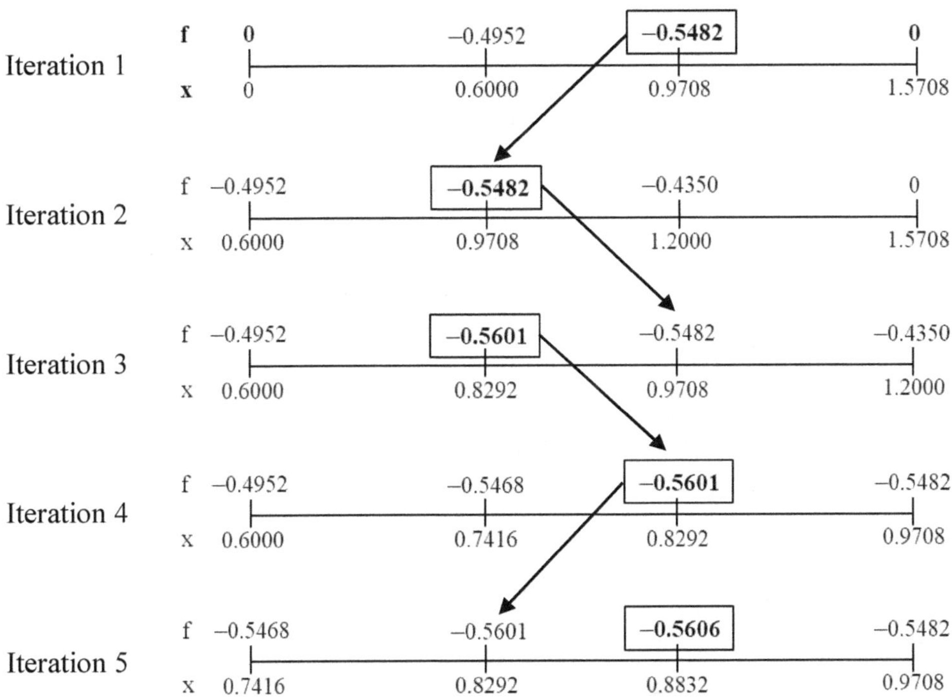

The approximate minimum after 5 iterations is: $x \approx 0.8832 \;\rightarrow\; f(x) \approx -0.5606$

to be compared with the exact solution: $x \approx 0.8603 \;\rightarrow\; f(x) \approx -0.5611$

The golden ratio method gains one significant digit on the solution every five iterations $(\varphi^5 \approx 10)$.
It is possible to speed up the convergence using **quadratic interpolation**.

2.2.4 Quadratic interpolation

Let us assume that the function is known in three points x_1, x_2, x_3 and takes the respective values y_1, y_2, y_3 . These points are shown in figure 2-12.

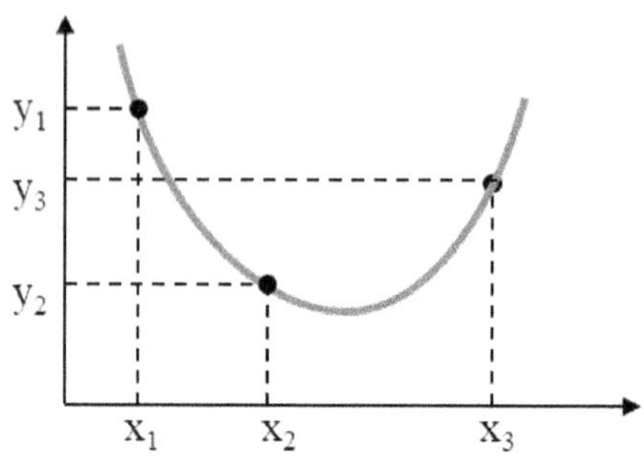

Notations

$$\begin{cases} y_1 = f(x_1) \\ y_2 = f(x_2) \\ y_3 = f(x_3) \end{cases} \text{ and } \begin{cases} s_{ij} = x_i - x_j \\ r_{ij} = x_i^2 - x_j^2 \end{cases}$$

Figure 2-12: Quadratic interpolation.

The polynomial q(x) of degree 2 passing through the three points $(x_1;y_1)$, $(x_2;y_2)$ and $(x_3;y_3)$ has the expression

$$q(x) = y_1 \frac{(x-x_2)(x-x_3)}{(x_1-x_2)(x_1-x_3)} + y_2 \frac{(x-x_3)(x-x_1)}{(x_2-x_3)(x_2-x_1)} + y_3 \frac{(x-x_1)(x-x_2)}{(x_3-x_1)(x_3-x_2)} \quad (2.21)$$

Its derivative given by

$$q'(x) = y_1 \frac{(2x-x_2-x_3)}{(x_1-x_2)(x_1-x_3)} + y_2 \frac{(2x-x_3-x_1)}{(x_2-x_3)(x_2-x_1)} + y_3 \frac{(2x-x_1-x_2)}{(x_3-x_1)(x_3-x_2)} \quad (2.22)$$

is zero at

$$x_m = \frac{1}{2} \frac{y_1 r_{23} + y_2 r_{31} + y_3 r_{12}}{y_1 s_{23} + y_2 s_{31} + y_3 s_{12}} \quad \text{with} \quad \begin{cases} s_{ij} = x_i - x_j \\ r_{ij} = x_i^2 - x_j^2 \end{cases} \quad (2.23)$$

By calculating the value of the function f in this point: $y_m = f(x_m)$ and comparing it with the previous values y_1, y_2, y_3, the new search interval can be determined.

Table 2-3 shows the possibles cases that can occur, with in the first row the four possibilities of location of x_m calculated by (2.23) and in the first column the four possibilities of location of the minimum among the values y_1, y_2, y_3, y_m.

The cases of divergence correspond to an extremum x_m located outside the interval $[x_1;x_3]$ and a negative curvature of the function q(x). The zero of the derivative (2.22) may indeed correspond to a maximum. This interpolation method is therefore not as robust as the golden ratio method, especially if the shape of the function is not well known and the search interval is large. It should preferably be used at the end of convergence to improve the accuracy of the solution.

Minimum observed	$x_m < x_1$	$x_1 < x_m < x_2$	$x_2 < x_m < x_3$	$x_3 < x_m$
at y_m	$(x_m\,;x_1\,;x_2)$	$(x_1\,;x_m\,;x_2)$	$(x_2\,;x_m\,;x_3)$	$(x_2\,;x_3\,;x_m)$
at y_1	$(x_m\,;x_1\,;x_2)$	$(x_1\,;x_m\,;x_2)$	$(x_1\,;x_2\,;x_m)$	divergence
at y_2	divergence	$(x_m\,;x_2\,;x_3)$	$(x_1\,;x_2\,;x_m)$	divergence
at y_3	divergence	$(x_m\,;x_2\,;x_3)$	$(x_2\,;x_m\,;x_3)$	$(x_2\,;x_3\,;x_m)$

Table 2-3: Interval reduction by quadratic interpolation.

Example 2-4: Quadratic interpolation method

We seek the minimum of the function: $f(x)=(x-1)(x+1)^2$ in the interval $[0\ ;\ 2]$
The three initial points are in $x_1=0$, $x_2=1$, $x_3=2$.

The first five iterations of the quadratic interpolation method are shown below. The values of x are marked below the line, the values of f(x) are marked above. The interpolated point x_m (2.23) is in bold, the minimum is boxed and an arrow indicates its position for the next iteration.

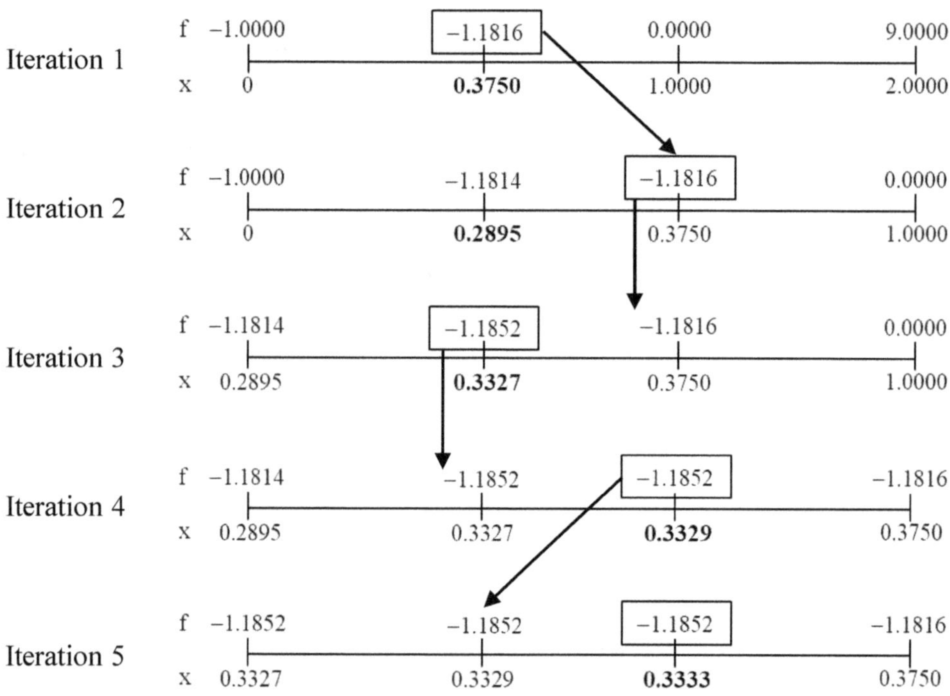

The approximate minimum after 5 iterations is: $x \approx 0.3333 \ \rightarrow\ f(x) \approx -1.1852$

It is very close to the exact solution: $x=\frac{1}{3} \ \rightarrow\ f(x)=-\frac{32}{27}$

2.3 DIRECT method

The DIRECT (DIviding RECTangles, 1993) performs a multidimensional optimization by interval splitting. This deterministic method identifies several local minima and possibly finds the global minimum.

2.3.1 Lipschitzian function

Consider the minimization of a function on a bounded domain $D \subset \mathbb{R}^n$.

$$\min_{x \in D} f(x) \tag{2.24}$$

The function f is assumed to be Lipschitzian on D, meaning that it satisfies

$$\exists K > 0 , \forall x, y \in D , \|f(y) - f(x)\| \le K \|y - x\| \tag{2.25}$$

The (unknown) Lipschitz constant of f on the domain D is the smallest real K satisfying (2.25). This assumption stronger than continuity is satisfied by most functions encountered in applications. Figure 2-13 shows the two half-lines of respective slopes ±K which lower bound the function from point c.

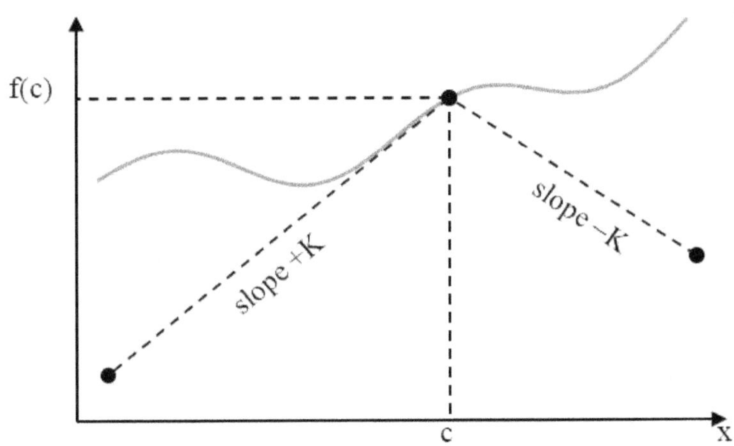

Figure 2-13: Lipschitzian function.

The domain D is an hyperrectangle of $\mathbb{R}^n$ whose largest dimension is denoted d.

From any point $c \in \mathbb{R}^n$ where the function takes the value f(c), the property (2.25) allows bounding f(x) over the entire domain D.

$$\forall x \in D \ , \quad \|f(x)-f(c)\| \le K\|x-c\| \le Kd \\ \Rightarrow f(c)-Kd \ \le \ f(x) \ \le \ f(c)+Kd \tag{2.26}$$

In particular $f(c)-Kd$ is a minorant of the function on D.

Example 2-5: Minorant of a Lipschitzian function

Considerer the function $f(x)=\sin\left[4\pi\left(x-\frac{1}{2}\right)\right]+6x^2+2$ on $D=[0\ ;1]$.

Its derivative $f'(x)=4\pi\cos\left[4\pi\left(x-\frac{1}{2}\right)\right]+12x$ is bounded by $|f'(x)|\le 4\pi+12$ on the interval $D=[0\ ;1]$. This function is therefore Lipschitzian and it admits for example $K=25$ as a Lipschitz constant on $D=[0\ ;1]$.
Place ourselves in the point $c=0{,}5 \rightarrow f(c)=3{,}5$, with the interval half-length $d=0{,}5$ as shown below. We have then as a minorant of f

$$f(x)\ge f(c)-Kd=3{,}5-25\times 0{,}5=-9$$

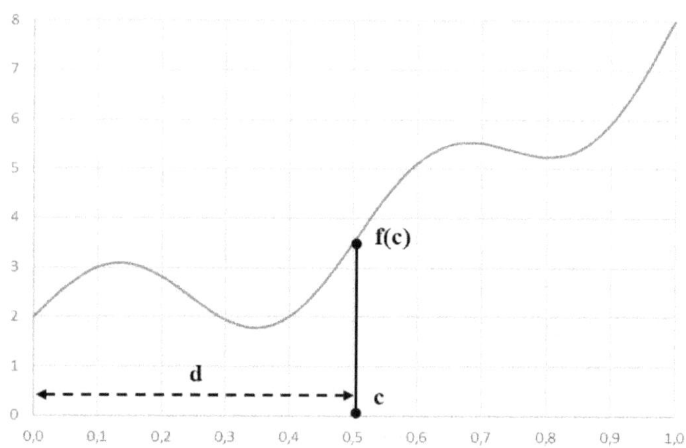

The DIRECT method uses the property (2.26) to partition the domain by detecting the most interesting areas. The objective is to find the global minimum of the function on a hypercube of $\mathbb{R}^n$.
The algorithm is presented in dimension 1 and then generalized to dimension n.

2.3.2 Algorithm in dimension 1

Consider a function f of variable x defined on an interval D. The function is assumed to be Lipschitzian with a constant K (unknown) on D. The interval D is divided into three equal intervals D_1, D_2, D_3 as in figure 2-14.

Each interval D_j is associated with a center c_j and a half-length d_j:

- the value of f in the center is: $f_j = f(c_j)$;
- a minorant of f on D_j is: $f_{bj} = f_j - Kd_j$.

The value of K will be determined below from the known points.

The "best" sub-interval is the one having the lowest minorant. It can be assumed that this interval is the one most likely to contain the minimum of f. In figure 2-14, this would be the interval D_1 .

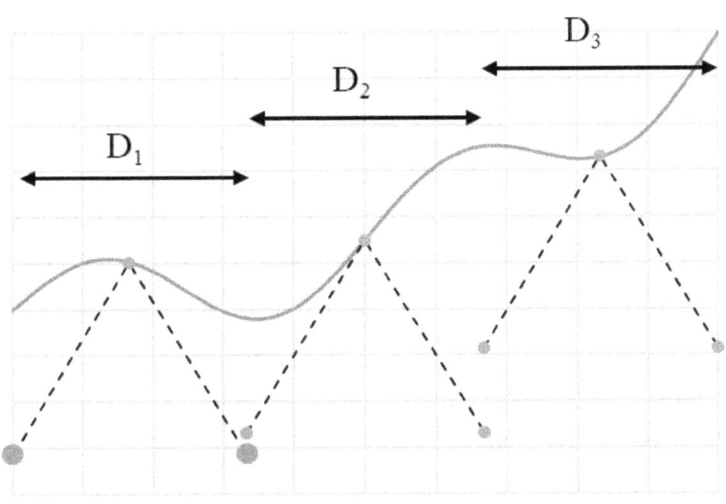

Figure 2-14: Estimating the minorant per interval.

The best interval D_1 is selected to be divided in turn into three equal intervals. At the end of this splitting, five intervals are available whose minorants can be estimated with the constant K. The best one will be selected for the next iteration. Figure 2-15 shows the first three iterations of this interval splitting process.

Iteration 1: 3 intervals

Best interval:

number 1

Iteration 2: 5 intervals

Best interval:

number 4

Iteration 3: 7 intervals

Best interval:

number 4

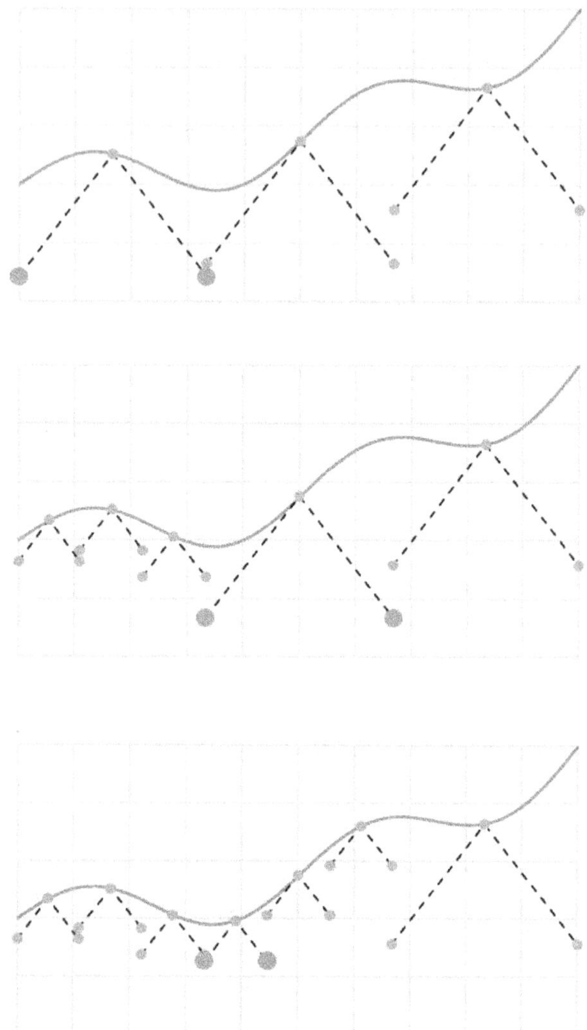

Figure 2-15: Successive interval divisions.

To ease the determination of the best interval, we draw for each of them the line of equation: $y = K(x - d_j) + f_j = Kx + f_{bj}$ as in figure 2-16. This straight line of slope K passes through the point of coordinates $(d_j ; f_j)$, f_j being the value of f at the center of the interval D_j. The y axis-intercept makes it possible to compare the intervals and identify the one with the lowest minorant.

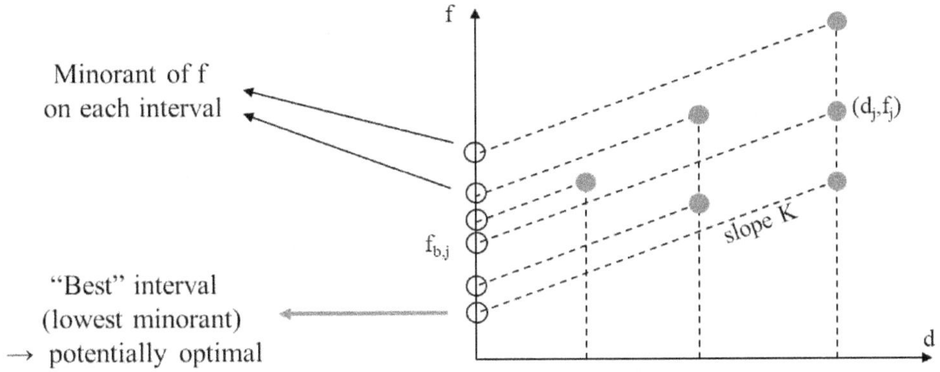

Figure 2-16: Comparison of minorants.

The selected interval depends on the value of K, which is currently arbitrary. For example, consider the six intervals shown in figure 2-17 for two different values of K (K_1 on the left, K_2 on the right).

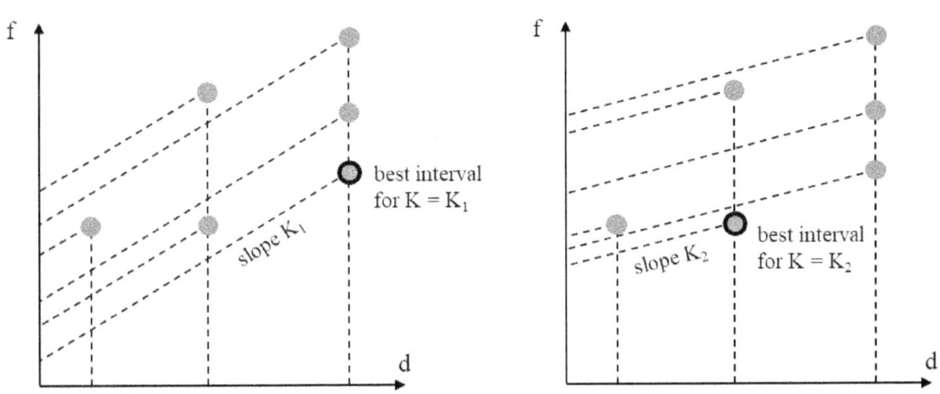

Figure 2-17: Best interval depending on the constant K.

An interval is said to be "potentially optimal" if there exists a positive K value such that this interval gives the lowest minorant.

Formally, the interval D_j associated with $(d_j ; f_j)$ is **potentially optimal** if

$$\begin{aligned} & \exists K \geq 0 \,,\, \forall i \,,\, f_{b,j} \leq f_{b,i} \\ \Leftrightarrow \quad & \exists K \geq 0 \,,\, \forall i \,,\, f_j - K d_j \leq f_i - K d_i \\ \Leftrightarrow \quad & \exists K \geq 0 \,,\, \forall i \,,\, f_j - f_i \leq K(d_j - d_i) \end{aligned} \tag{2.27}$$

This inequality can be interpreted differently depending on the sign of $d_j - d_i$:

- for all intervals i such that $d_i = d_j$, the condition to be satisfied is

$$f_j \leq f_i \tag{2.28}$$

- for all intervals i such that $d_i < d_j$, K must satisfy the condition

$$K \geq K_1 = \max_{i/d_i<d_j} \frac{f_j - f_i}{d_j - d_i} \tag{2.29}$$

- for all intervals i such that $d_i > d_j$, K must satisfy the condition

$$K \leq K_2 = \min_{i/d_i>d_j} \frac{f_j - f_i}{d_j - d_i} \tag{2.30}$$

These conditions express that the point $(d_j ; f_j)$ belongs to the lower convex envelope of the set of points $(d_i ; f_i)$ as illustrated in figure 2-18.

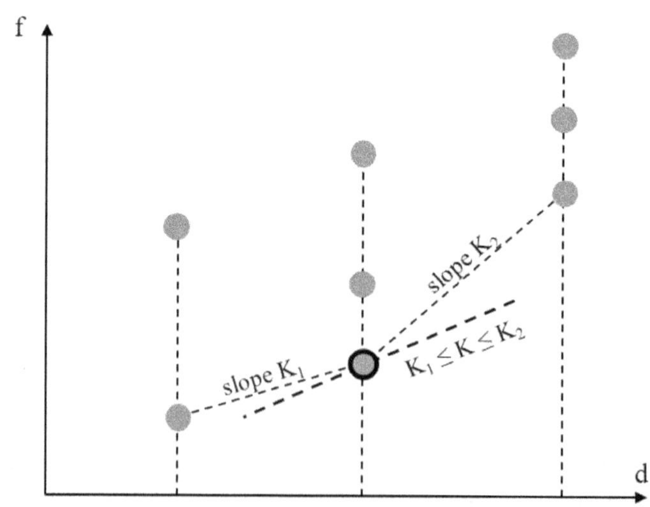

Figure 2-18: Potentially optimal interval.

The conditions for the interval D_j associated with $(d_j ; f_j)$ to be potentially optimal are as follows.

$$\begin{cases} f_j \le f_i \quad , \quad \forall i / d_i = d_j \\ \max\limits_{i/d_i<d_j} K_i = K_1 \le K_2 = \min\limits_{i/d_i>d_j} K_i \quad \text{with} \quad K_i = \dfrac{f_j - f_i}{d_j - d_i} \end{cases} \tag{2.31}$$

The second condition ensures the existence of a value of K satisfying (2.27).

The number of potentially optimal intervals increases with each division. To limit their number, the smallest intervals are eliminated, either directly (if $d_j > d_{min}$), or by a sufficient decrease condition such as

$$f_j - K_2 d_j \le f_{min} - \varepsilon |f_{min}| \quad \text{with} \quad f_{min} = \min_i f_i \quad , \quad \varepsilon = 10^{-3} \tag{2.32}$$

This condition prevents the search from becoming too local. The minimum size of a potentially optimal interval is then given by

$$d_j \ge \frac{f_j - f_{min} + \varepsilon |f_{min}|}{K_2} \tag{2.33}$$

The DIRECT algorithm in dimension 1 is depicted in figure 2-19.

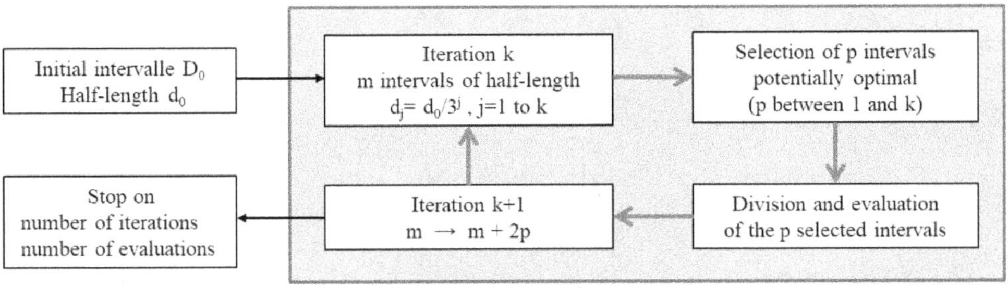

Figure 2-19: DIRECT algorithm in dimension 1.

Each iteration consists of two stages:

- selecting p intervals among the potentially optimal ones;
- dividing them into 3 equal intervals.

The number p of selected intervals is free. In general, the p largest ones are chosen so that the search is as global as possible. The number of intervals increases by 2p at each iteration. The initial interval has a half-length d_0.

At iteration k, the partition comprises intervals of half-length $d_j = d_0 / 3^j$ with $1 \le j \le k$. There are at most k potentially optimal intervals (one per dimension d_j), of which p will be divided at the next iteration.

Example 2-6: Minimization by the DIRECT algorithm in dimension 1

Let us try to minimize the function: $f(x) = \sin\left[4\pi\left(x - \frac{1}{2}\right)\right] + 6x^2 + 2$

in the interval $D = [-1 ; 1]$.

This function plotted in figure 2-20 has 4 local minima given in table 2-4.

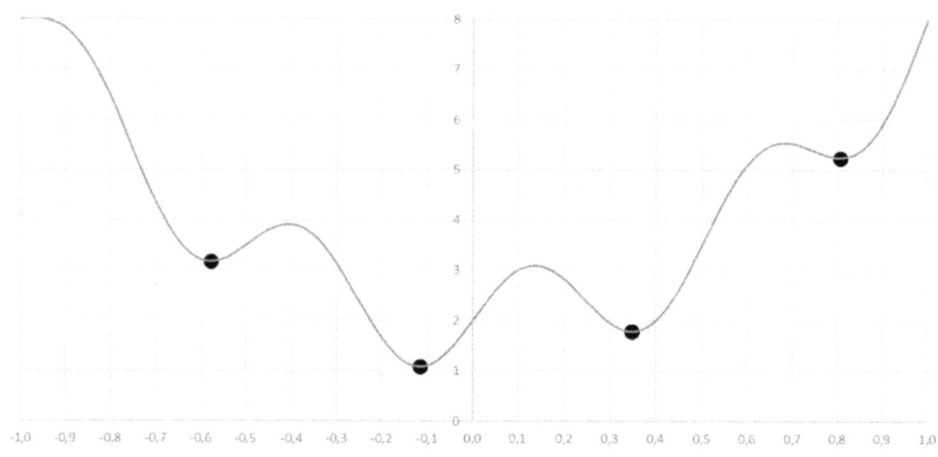

Figure 2-20: Function of one variable with several local minima.

Minimum	local	global	local	local
x	−0.57843	−0.11615	0.34804	0.80520
f(x)	3.17389	1.08712	1.78363	5.25072

Table 2-4: Coordinates of the minima.

The following tables and graphs detail the first 7 iterations of the algorithm. For each iteration, the tables present the set of intervals already evaluated with their bounds (x_{inf} ; x_{sup}), their center (x_{mid} ; f_{mid}), their length d, the evaluation of the minorant f_b and the associated value of K (this value is indicated on the right-hand side of the table for each potentially optimal interval).

The intervals are ranked by decreasing length, with a separation line at each length change. The best point for each length (column f_{mid}) is highlighted.

The best minorants (column f_b) are also highlighted. Among the potentially optimal intervals, the 2 largest ones are selected and divided for the next iteration. Each of them give 3 new highlighted intervals (columns x_{inf} ; x_{sup}). The graphs on the right show the points (d_j ; f_j) already evaluated.

x inf	x sup	x mid	Length	fmid	fb	K
-1.000	-0.333	-0.667	0.667	3.801	3.801	0
-0.333	0.333	0.000	0.667	2.000	2.000	
0.333	1.000	0.667	0.667	5.533	5.533	

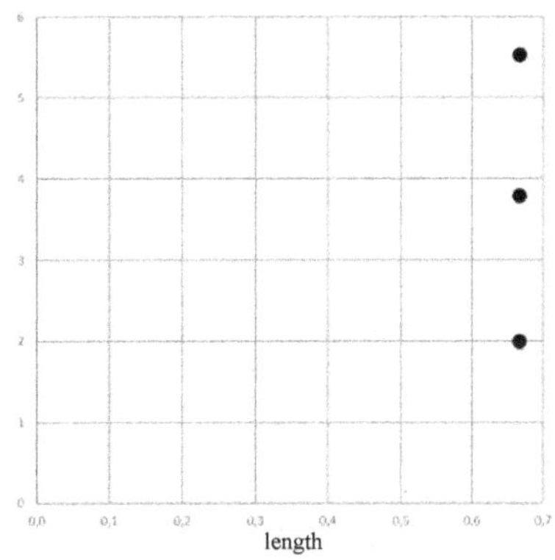

Iteration 1: 3 intervals

The interval $D=[-1 ; 1]$ is divided into 3.

The function is evaluated in the center x_{mid}.

The minimum at this iteration is $f_{mid} = 2.000$.

The best interval is the second one ($f_b = 2.000$).

This interval is selected to be divided for the next iteration.

x inf	x sup	x mid	Length	fmid	fb	K
-1.000	-0.333	-0.667	0.667	3.801	1.031	4.155
0.333	1.000	0.667	0.667	5.533	2.763	
-0.333	-0.111	-0.222	0.222	1.954	1.031	
-0.111	0.111	0.000	0.222	2.000	1.077	
0.111	0.333	0.222	0.222	2.638	1.715	

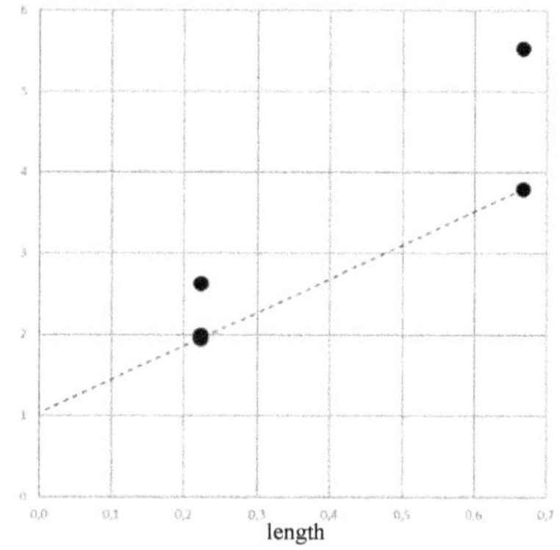

Iteration 2: 5 intervals

The 3 intervals resulting from the division are highlighted in the columns giving x_{inf} and x_{sup}. The minimum at this iteration is $f_{mid} = 1.954$.

Two intervals are potentially optimal (the 1st and the 3rd) with the same minorant $f_b = 1.0311$. They are both selected to be divided for the next iteration.

x inf	x sup	x mid	Length	fmid	fb	K
0.333	1.000	0.667	0.667	5.533	0.234	7.949
-0.111	0.111	0.000	0.222	2.000	0.234	5.578
0.111	0.333	0.222	0.222	2.638	1.399	
-1.000	-0.778	-0.889	0.222	7.726	6.486	
-0.778	-0.556	-0.667	0.222	3.801	2.561	
-0.556	-0.333	-0.444	0.222	3.828	2.589	
-0.333	-0.259	-0.296	0.074	3.076	2.663	
-0.259	-0.185	-0.222	0.074	1.954	1.541	
-0.185	-0.111	-0.148	0.074	1.174	0.761	

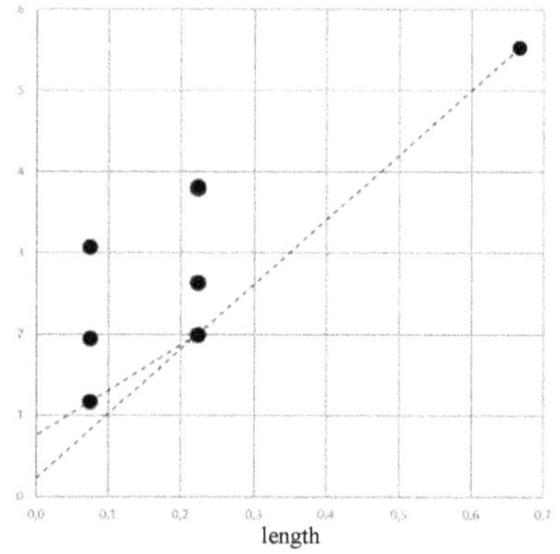

Iteration 3: 9 intervals

The minimum at this iteration is $f_{mid} = 1.174$.

x inf	x sup	x mid	Length	fmid	fb	K
0.111	0.333	0.222	0.222	2.638	0.585	
-1.000	-0.778	-0.889	0.222	7.726	5.672	
-0.778	-0.556	-0.667	0.222	3.801	1.748	
-0.556	-0.333	-0.444	0.222	3.828	1.775	
0.333	0.556	0.444	0.222	2.542	0.489	9.239
0.556	0.778	0.667	0.222	5.533	3.480	
0.778	1.000	0.889	0.222	5.756	3.703	
-0.333	-0.259	-0.296	0.074	3.076	2.392	
-0.259	-0.185	-0.222	0.074	1.954	1.270	
-0.185	-0.111	-0.148	0.074	1.174	0.489	
-0.111	-0.037	-0.074	0.074	1.231	0.546	
-0.037	0.037	0.000	0.074	2.000	1.316	
0.037	0.111	0.074	0.074	2.835	2.151	

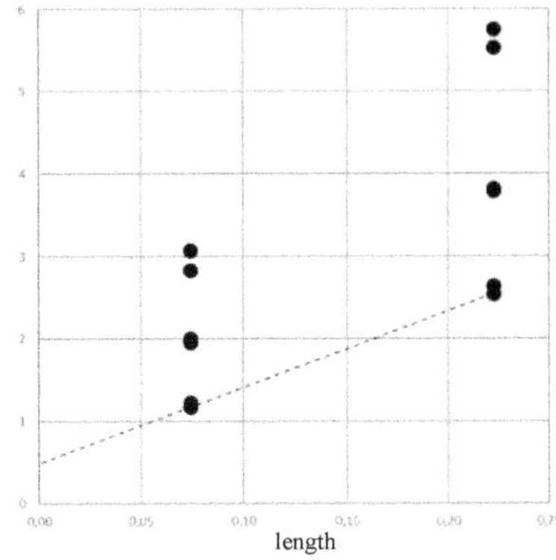

Iteration 4 : 13 intervals

Minimum at this iteration:

$f_{mid} = 1.174$

x inf	x sup	x mid	Length	fmid	fb	K
0.111	0.333	0.222	0.222	2.638	0.527	9.501
-1.000	-0.778	-0.889	0.222	7.726	5.614	
-0.778	-0.556	-0.667	0.222	3.801	1.689	
-0.556	-0.333	-0.444	0.222	3.828	1.717	
0.556	0.778	0.667	0.222	5.533	3.421	
0.778	1.000	0.889	0.222	5.756	3.645	
-0.333	-0.259	-0.296	0.074	3.076	2.372	
-0.259	-0.185	-0.222	0.074	1.954	1.251	
-0.111	-0.037	-0.074	0.074	1.231	0.527	2.818
-0.037	0.037	0.000	0.074	2.000	1.791	
0.037	0.111	0.074	0.074	2.835	2.626	
0.333	0.407	0.370	0.074	1.825	1.616	
0.407	0.481	0.444	0.074	2.542	2.334	
0.481	0.556	0.519	0.074	3.844	3.635	
-0.185	-0.160	-0.173	0.025	1.355	1.285	
-0.160	-0.136	-0.148	0.025	1.174	1.104	
-0.136	-0.111	-0.123	0.025	1.092	1.022	

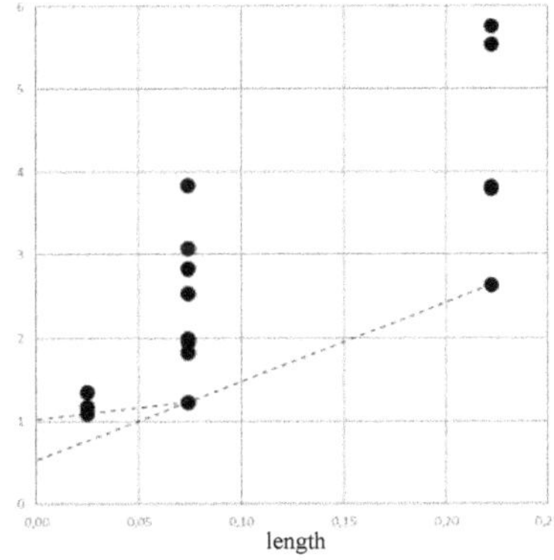

Iteration 5 : 17 intervals

Minimum at this iteration:

$f_{mid} = 1.092$

x inf	x sup	x mid	Length	fmid	fb	K
-1.000	-0.778	-0.889	0.222	7.726	4.762	
-0.778	-0.556	-0.667	0.222	3.801	0.837	13.338
-0.556	-0.333	-0.444	0.222	3.828	0.864	
0.556	0.778	0.667	0.222	5.533	2.569	
0.778	1.000	0.889	0.222	5.756	2.792	
-0.333	-0.259	-0.296	0.074	3.076	2.088	
-0.259	-0.185	-0.222	0.074	1.954	0.966	
-0.037	0.037	0.000	0.074	2.000	1.012	
0.037	0.111	0.074	0.074	2.835	1.847	
0.333	0.407	0.370	0.074	1.825	0.725	14.846
0.407	0.481	0.444	0.074	2.542	1.443	
0.481	0.556	0.519	0.074	3.844	2.744	
0.111	0.185	0.148	0.074	3.090	1.990	
0.185	0.259	0.222	0.074	2.638	1.539	
0.259	0.333	0.296	0.074	1.977	0.878	
-0.185	-0.160	-0.173	0.025	1.355	0.988	
-0.160	-0.136	-0.148	0.025	1.174	0.807	
-0.136	-0.111	-0.123	0.025	1.092	0.725	
-0.111	-0.086	-0.099	0.025	1.112	0.746	
-0.086	-0.062	-0.074	0.025	1.231	0.864	
-0.062	-0.037	-0.049	0.025	1.433	1.067	

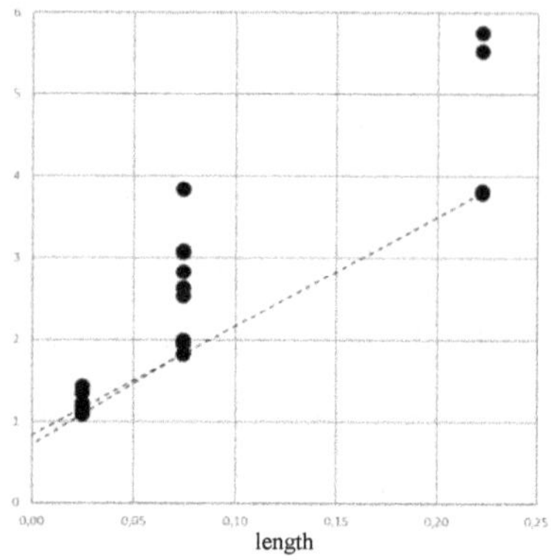

<u>Iteration 6 : 21 intervals</u>

Minimum at this iteration:

$f_{mid} = 1.092$

x inf	x sup	x mid	Length	fmid	fb	K
-1.000	-0.778	-0.889	0.222	7.726	4.956	
-0.778	-0.556	-0.667	0.222	3.801	1.031	12.463
-0.556	-0.333	-0.444	0.222	3.828	1.058	
0.556	0.778	0.667	0.222	5.533	2.763	
0.778	1.000	0.889	0.222	5.756	2.986	
-0.333	-0.259	-0.296	0.074	3.076	2.153	
-0.259	-0.185	-0.222	0.074	1.954	0.691	17.049
-0.037	0.037	0.000	0.074	2.000	0.737	
0.037	0.111	0.074	0.074	2.835	1.572	
0.407	0.481	0.444	0.074	2.542	1.280	
0.481	0.556	0.519	0.074	3.844	2.581	
0.111	0.185	0.148	0.074	3.090	1.827	
0.185	0.259	0.222	0.074	2.638	1.375	
0.259	0.333	0.296	0.074	1.977	0.714	
-0.185	-0.160	-0.173	0.025	1.355	0.934	
-0.160	-0.136	-0.148	0.025	1.174	0.753	
-0.111	-0.086	-0.099	0.025	1.112	0.691	1.530
-0.086	-0.062	-0.074	0.025	1.231	1.193	
-0.062	-0.037	-0.049	0.025	1.433	1.395	
0.333	0.358	0.346	0.025	1.784	1.746	
0.358	0.383	0.370	0.025	1.825	1.787	
0.383	0.407	0.395	0.025	1.968	1.930	
-0.136	-0.128	-0.132	0.008	1.108	1.095	
-0.128	-0.119	-0.123	0.008	1.092	1.079	
-0.119	-0.111	-0.115	0.008	1.087	1.075	

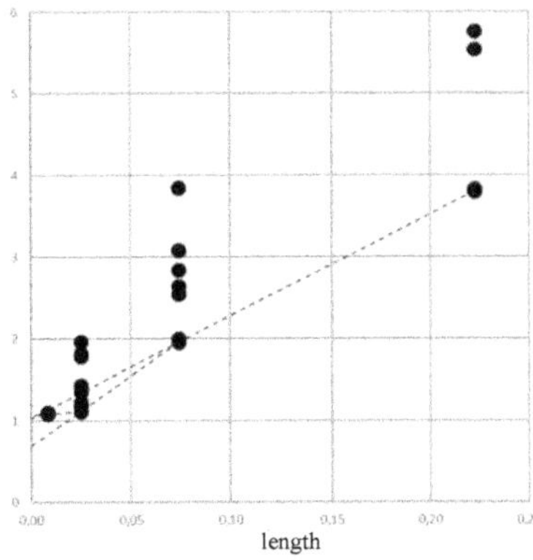

Iteration 7 : 25 intervals

Minimum at this iteration:

$f_{mid} = 1.087$

After 7 iterations, the best point is: $x_{mid} = -0.115 \rightarrow f_{mid} = 1.087$.

This approximate solution obtained with 25 function evaluations is very close to the global minimum: $x^* = -0.116$ given in table 2-4.

Figure 2-21 shows the progression of the best point over the seven iterations.

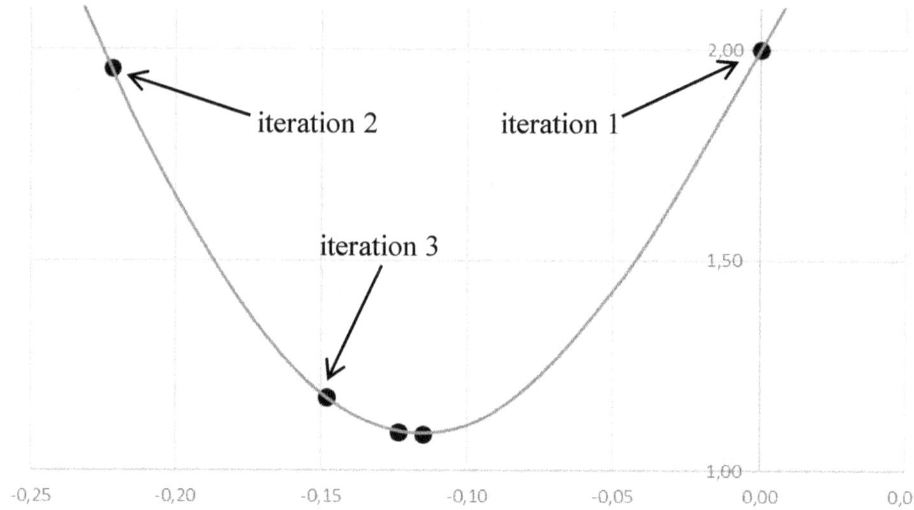

Figure 2-21: Progression towards the global minimum.

2.3.3 Algorithm in dimension n

The previous algorithm can be generalized to the minimization of a function of n variables. The intervals are replaced by hyperrectangles in $\mathbb{R}^n$.

We start by first presenting the method for dividing a hyperrectangle and then defining the criterion for selecting a potentially optimal hyperrectangle.

Method of division

Figure 2-22 shows a hypercube of center c_0 and side length δ. The representation is in $\mathbb{R}^3$, but it generalizes without difficulty to $\mathbb{R}^n$.

The unit vectors along each coordinate $(x_i)_{i=1 \text{ to } n}$ are noted $(e_i)_{i=1 \text{ to } n}$.

Two points are positioned on each axis e_i. These points noted c_i^+ and c_i^- are placed on either side of the center c_0 at a distance $\delta/3$. The function is evaluated in these 2n points.

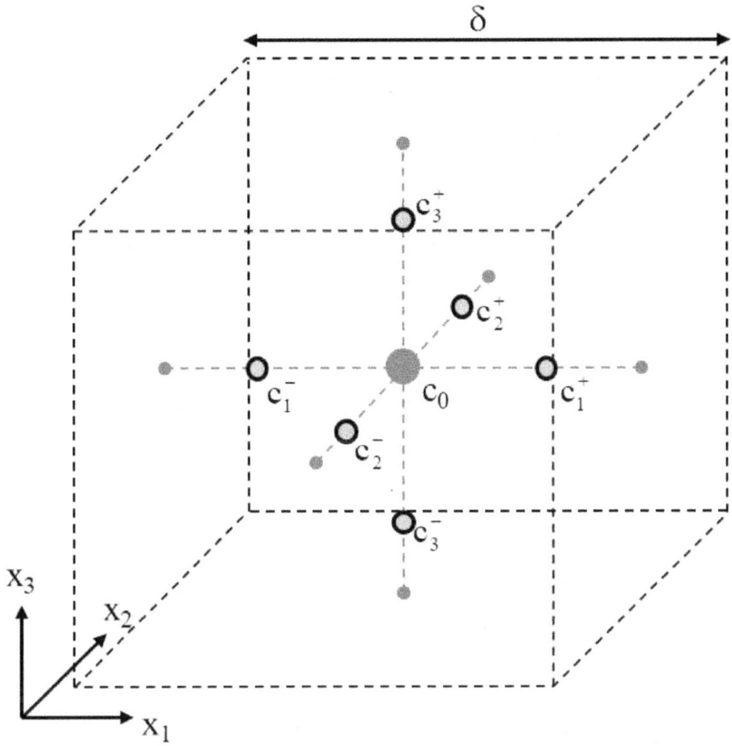

Figure 2-22: Evaluation points along each axis of the hypercube.

The values of the function in the 2n points $(c_i^{\pm})_{i=1 \text{ to } n}$ are noted $(\varphi_i^{\pm})_{i=1 \text{ to } n}$.

$$\begin{cases} c_i^- = c_0 - \dfrac{1}{3}\delta e_i & \rightarrow \varphi_i^- = f(c_i^-) \\ c_i^+ = c_0 + \dfrac{1}{3}\delta e_i & \rightarrow \varphi_i^+ = f(c_i^+) \end{cases} \tag{2.34}$$

On each axis, the best evaluation is noted φ_i.

$$\varphi_i = \min(\varphi_i^-, \varphi_i^+), \quad i = 1 \text{ to } n \tag{2.35}$$

The n values $(\varphi_i)_{i=1 \text{ to } n}$ are ranked by increasing values.

$$\varphi_{i1} \leq \varphi_{i2} \leq \cdots \leq \varphi_{in} \tag{2.36}$$

This order determines the best direction e_{i1} and the worst one e_{in}.

The hypercube is then divided direction after direction in this order:

- first, the hypercube is divided into 3 parts along the best direction e_{i1}.

  This results in 3 hyperrectangles of respective centers c_{i1}^-, c_0, c_{i1}^+ with one side of length δ/3 and n−1 sides of length δ (figure 2-23a);

- we select the central hyperrectangle (of center c_0) and divide it into 3 along the 2nd best direction e_{i2}.

  This results in 3 hyperrectangles of respective centers c_{i2}^-, c_0, c_{i2}^+ with 2 sides of length δ /3 and n−2 sides of length δ (figure 2-23b);

- we continue in the same way by dividing the central hyperrectangle along the successive directions: $e_{i3},\cdots,e_{in}$. The last division gives a hypercube of center c_0 and side length δ /3 (figure 2-23c).

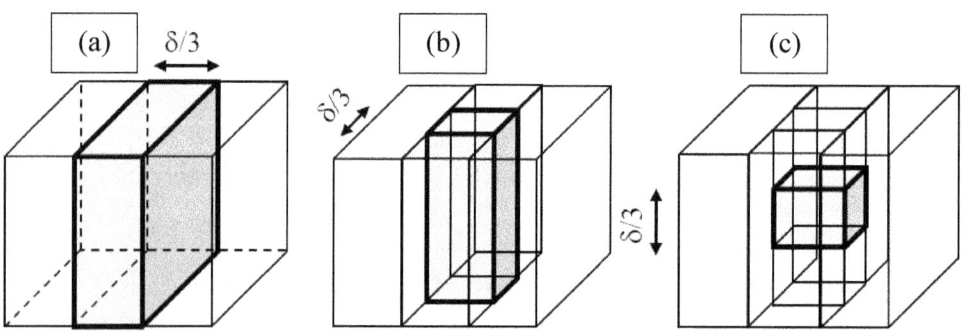

Figure 2-23: Division of the hypercube along each direction.

Each division generates 2 additional hyperrectangles whose side lengths are either δ /3 or δ. The result of the divisions along the n directions is finally:

- 2n hyperrectangles of centers $(c_i^{\pm})_{i=1\text{ to }n}$;
- 1 central hypercube of center c_0 and of side length δ/3.

The best point c_{i1}^+ or c_{i1}^- is at the center of one of the two largest hyperrectangles resulting from the first division.

Subdividing a hyperrectangle follows exactly the same procedure as for a hypercube, but the division is restricted to its longest sides. If the hyperrectangle has p sides of length δ and n−p sides of length $\delta/3$, this will result in 2p additional hyperrectangles having sides of length $\delta/3$ or δ.

The following example illustrates this division process in $\mathbb{R}^2$.

Example 2-7: Division process in $\mathbb{R}^2$

The following figures show the positions of the points and the value of the function in these points.

The value of the function in the center of the initial hypercube is 9. The best point is on the e_2 axis, and the value of the function in this point c_2^- is 2. The division is done first along e_2, which gives 3 horizontal hyperrectangles, then along e_1 for the central rectangle, which gives 3 hypercubes.

Figure 2-24c shows the 5 hyperrectangles obtained after the first iteration.

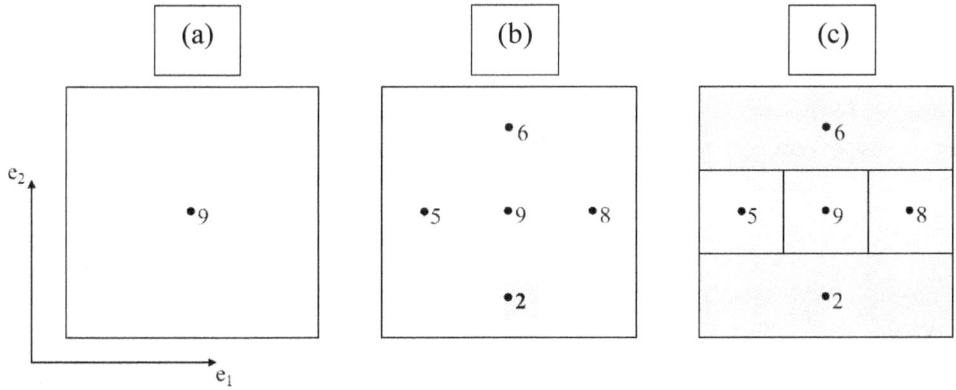

Figure 2-24: Divisions during the first iteration.

The method for selecting a potentially optimal hyperrectangle will be detailed below, after this example. This method is not made explicit at this time.

In figure 2-25, let us assume that the selected hyperrectangle is the one at the bottom (shaded in figure 2-25a).

This hyperrectangle is divided only along its longest sides starting with the best direction. In this example in dimension 2, only one division is made along e_1, which leads to the middle diagram with the new values 3 and 6.

Figure 2-25c shows the 7 hyperrectangles obtained after the second iteration, the 3 new hyperrectangles being those colored at the bottom.

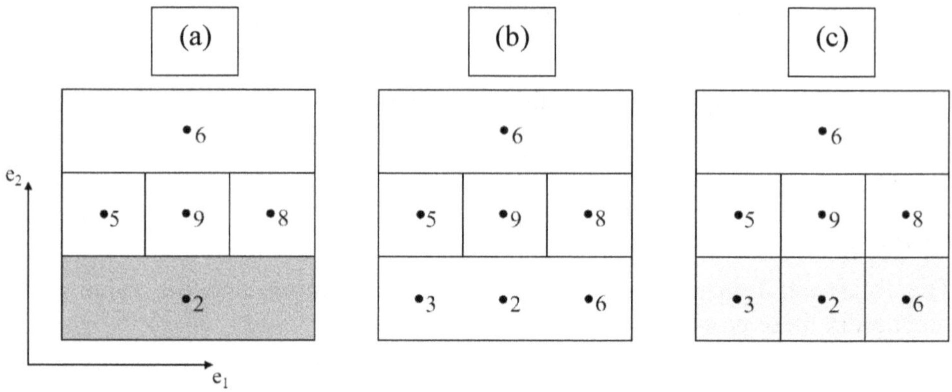

Figure 2-25: Divisions during the second iteration.

Now suppose that two hyperrectangles are selected (shaded in figure 2-26a) : one hyperrectangle at the top (with value f = 6) and one hypercube at the bottom (with value f = 2).

The hyperrectangle is divided only along e_1 and gives 3 hypercubes (with the new values 7 and 9). The hypercube is divided along e_1 and e_2. Since the best value is along e_1 (value f = 1), the division is done first along e_1, then along e_2. This results in 5 hyperrectangles. Figure 2-26b shows the values resulting from these divisions. Figure 2-26c shows the 13 hyperrectangles obtained after the third iteration.

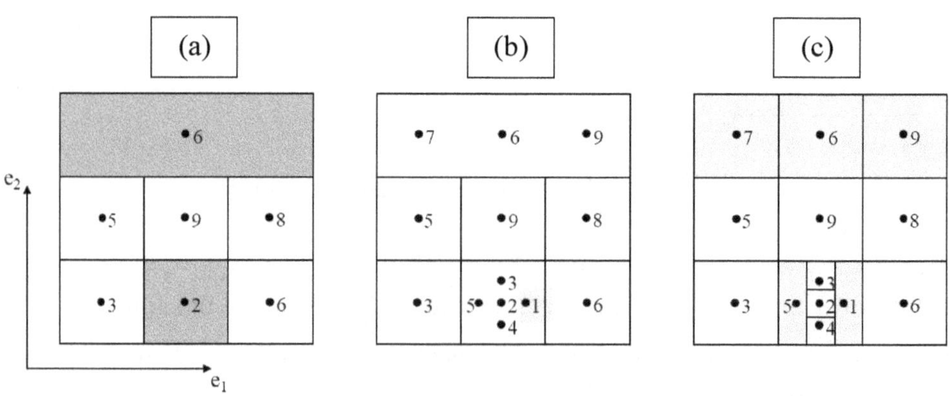

Figure 2-26: Divisions during the third iteration.

Selection method

The choice of a potentially optimal hyperrectangle relies on the estimation of a minorant of the function on the hyperrectangle. This estimation is very similar to the one developed in dimension 1. The difference lies in the way the size of a hyperrectangle is defined.

The initial hypercube is assumed to have a side length δ_0. It is always possible to start from this situation by an affine variable changes of the form

$$a_i \le x_i \le b_i \quad \Leftrightarrow \quad 0 \le x_i' \le \delta_0 \quad \text{with} \quad x_i' = \delta_0 \frac{x_i - a_i}{b_i - a_i} \tag{2.37}$$

The division method described above leads to hyperrectangles with short sides of length $\delta_0/3^{k+1}$ and long sides of length $\delta_0/3^k$.
The dimension of a hyperrectangle is measured by its L_2 norm or its L_∞ norm.
If the hyperrectangle D_j has p short sides and n−p long sides, its L_2 norm is

$$\|D_j\|_2 = \sqrt{\sum_{i=1}^{n} (\delta_i)^2} = \sqrt{\sum_{i=1}^{p} (\delta_0/3^{k+1})^2 + \sum_{i=p+1}^{n} (\delta_0/3^k)^2} = \delta_0/3^k \sqrt{n - \frac{8}{9}p} \tag{2.38}$$

and its L_∞ norm is the length of the longest side.

$$\|D_j\|_\infty = \max_{i=1 \text{ to } n} \delta_i = \delta_0/3^k \tag{2.39}$$

The minorant of the function f on the hyperrectangle D_j is estimated by

$$f_{b,j} = f_j - K d_j \quad \text{with} \quad \begin{cases} f_j = f(c_j) \\ d_j = \frac{1}{2}\|D_j\|_2 \quad \text{or} \quad d_j = \frac{1}{2}\|D_j\|_\infty \end{cases} \tag{2.40}$$

where f_j is the value of f in the center of the hyperrectangle, d_j is the half-size of the hyperrectangle and K is the (unknown) Lipschitz constant of the function f.

We can now define a potentially optimal hyperrectangle as in dimension 1.
The hyperrectangle D_j associated to $(d_j;f_j)$ is said **potentially optimal** if

$$\exists K \ge 0 \,,\ \forall i \,,\ f_{b,j} \le f_{b,i} \tag{2.41}$$

By the same development (2.27-2.33) as in dimension 1, the conditions for the hyperrectangle D_j $(d_j;f_j)$ to be potentially optimal are

$$\begin{cases} f_j \le f_i \quad , \quad \forall i / d_i = d_j \\ \max\limits_{i/d_i<d_j} K_i = K_1 \le K_2 = \min\limits_{i/d_i>d_j} K_i \quad \text{with} \quad K_i = \dfrac{f_j - f_i}{d_j - d_i} \end{cases} \tag{2.42}$$

to which we add the following condition of sufficient decrease.

$$f_j - K_2 d_j \le f_{min} - \varepsilon |f_{min}| \quad \text{with } f_{min} = \min_i f_i \ , \ \varepsilon = 10^{-3}$$
$$\Rightarrow d_j \ge \frac{f_j - f_{min} + \varepsilon |f_{min}|}{K_2} \tag{2.43}$$

The DIRECT algorithm in dimension n is depicted in figure 2-27.

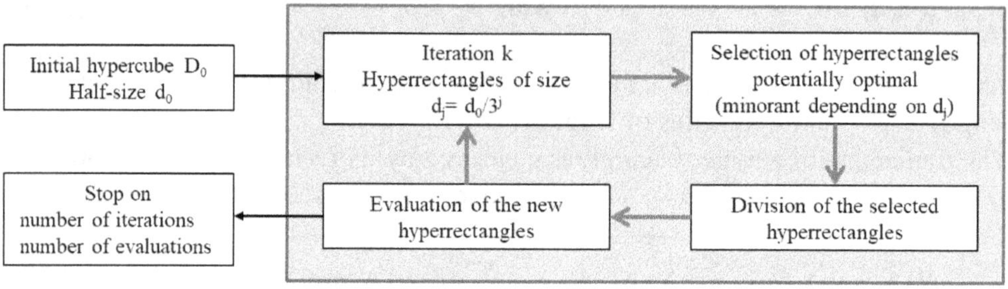

Figure 2-27: DIRECT algorithm in dimension n.

The following example in dimension 2 illustrates the iterations of the algorithm.

Example 2-8: Minimization by the DIRECT algorithm in dimension 2

Let the following function be minimized in the domain $\{-1 \le x \le 1 \,;\, -1 \le y \le 1\}$:

$$f(x,y) = (x-0,2)^2 + 2(y-0,1)^2 - 0,3\cos[3\pi(x-0,2)] - 0,4\cos[4\pi(y-0,1)] + 0,7$$

The level lines for this function are drawn in figure 2-28.

There are 9 local minima, 4 local maxima and 12 saddle points whose coordinates and values are given in figure 2-28a.

The global minimum is in $\begin{cases} x = 0.2 \\ y = 0.1 \end{cases} \rightarrow f = 0$.

(a)

0.570	0.883	1.190	0.470	1.190	0.883
0.367	1.346	1.654	0.933	1.654	1.346
0.100	0.413	0.720	**0.000**	0.720	0.413
-0.167	1.346	1.654	0.933	1.654	1.346
-0.370	0.883	1.190	0.470	1.190	0.883
y / x	-0.419	-0.161	0.200	0.561	0.819

(b)

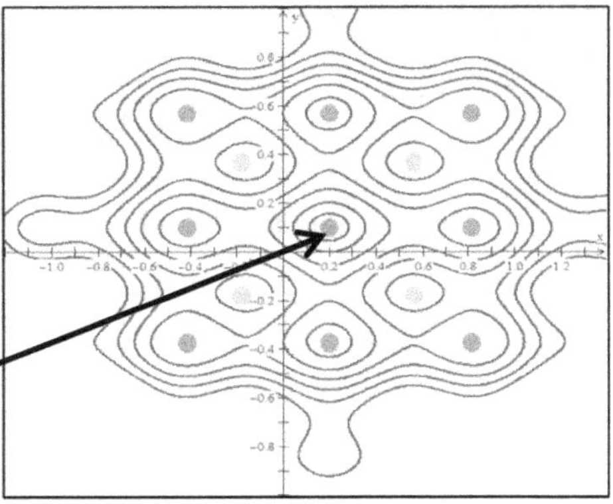

global minimum

Figure 2-28: Function of two variables with several local minima.

The tables and graphs on the following pages detail the first 6 iterations of the algorithm. The tables show all the hyperrectangles previously evaluated at each iteration.

For each hyperrectangle, the coordinates of the center (x_{mid} ; y_{mid}), its evaluation (f_{mid}), its size (dx ; dy ; dmax), the evaluation of the minorant f_b and the associated value of K are given. The graphs on the right depict the previously evaluated hyperrectangles.

The hyperrectangles are ranked by decreasing size, with a separation line at each change of dimension. For each dimension the best point (column f_{mid}) and the best minorant (column f_b) are highlighted. The potentially optimal hyperrectangles are divided at the next iteration. They yield new highlighted hyperrectangles (columns x_{mid} and y_{mid}).

x	y	f
-0.667	0.000	1.440
0.667	0.000	0.907
0.000	-0.667	2.400
0.000	0.667	1.207
0.000	0.000	0.729

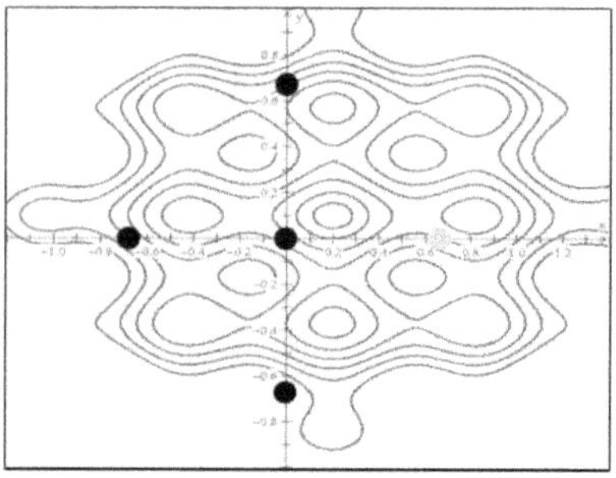

Initial hypercube

The center is in $x = 0$, $y = 0$.

The function is evaluated in $\left(\pm\frac{2}{3}; 0\right)$ and $\left(0; \pm\frac{2}{3}\right)$.

The best value ($f_{min} = 0.907$) is on the x axis.

The hypercube is divided first along x, then along y → 5 hyperrectangles

x mid	y mid	dx	dy	d max	fmid	fb	K
-0.667	0.000	0.333	1.000	1.000	1.440	1.174	
0.667	0.000	0.333	1.000	1.000	0.907	0.640	0.267
0.000	-0.667	0.333	0.333	0.333	2.400	2.311	
0.000	0.667	0.333	0.333	0.333	1.207	1.118	
0.000	0.000	0.333	0.333	0.333	0.729	0.640	

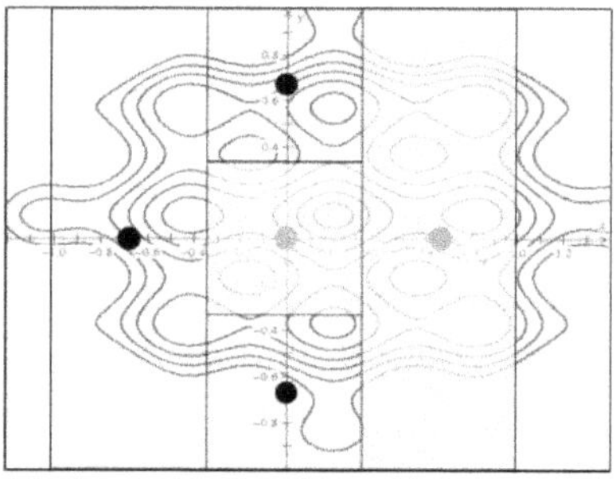

Iteration 1: 5 hyperrectangles

Two hyperrectangles are selected ($f_b = 0.640$).

The former is divided along its longest size (y):

→ +2 hyperrectangles along y

The second is a hypercube divided along its two sides:

→ +2 hyperrectangles along x and +2 hyperrectangles along y

This iteration generates a total of 6 additional hyperrectangles to be evaluated.

x mid	y mid	dx	dy	d max	fmid	fb	K
-0.667	0.000	0.333	1.000	1.000	1.440	-0.265	1.705
0.000	-0.667	0.333	0.333	0.333	2.400	1.831	
0.000	0.667	0.333	0.333	0.333	1.207	0.639	
0.667	-0.667	0.333	0.333	0.333	2.577	2.009	
0.667	0.667	0.333	0.333	0.333	1.385	0.817	
0.667	0.000	0.333	0.333	0.333	0.907	0.338	
-0.222	0.000	0.111	0.333	0.333	0.975	0.407	
0.222	0.000	0.111	0.333	0.333	0.303	-0.265	-1.915
0.000	-0.222	0.111	0.111	0.111	1.287	1.499	
0.000	0.222	0.111	0.111	0.111	0.849	1.061	
0.000	0.000	0.111	0.111	0.111	0.729	0.942	

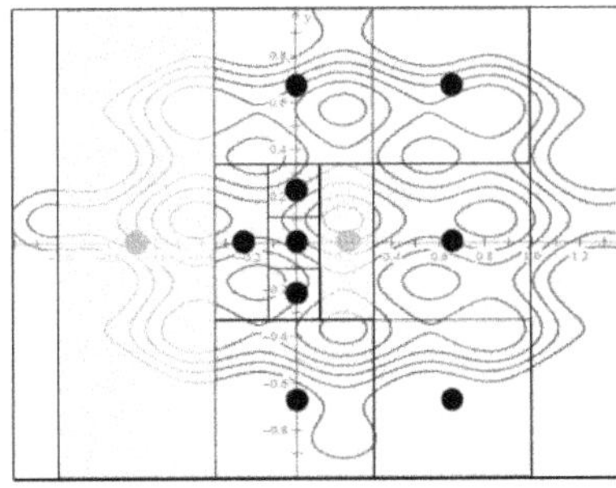

Iteration 2 : 11 hyperrectangles

Minimum : $f_{mid} = 0.303$

x mid	y mid	dx	dy	d max	fmid	fb	K
0.000	-0.667	0.333	0.333	0.333	2.400	1.494	
0.000	0.667	0.333	0.333	0.333	1.207	0.302	
0.667	-0.667	0.333	0.333	0.333	2.577	1.672	
0.667	0.667	0.333	0.333	0.333	1.385	0.480	
0.667	0.000	0.333	0.333	0.333	0.907	0.002	2.715
-0.222	0.000	0.111	0.333	0.333	0.975	0.070	
-0.667	-0.667	0.333	0.333	0.333	3.111	2.205	
-0.667	0.667	0.333	0.333	0.333	1.918	1.013	
-0.667	0.000	0.333	0.333	0.333	1.440	0.535	
0.000	-0.222	0.111	0.111	0.111	1.287	0.985	
0.000	0.222	0.111	0.111	0.111	0.849	0.547	
0.222	-0.222	0.111	0.111	0.111	0.861	0.559	
0.222	0.222	0.111	0.111	0.111	0.423	0.121	
0.222	0.000	0.111	0.111	0.111	0.303	0.002	
0.000	0.000	0.111	0.111	0.111	0.729	0.942	

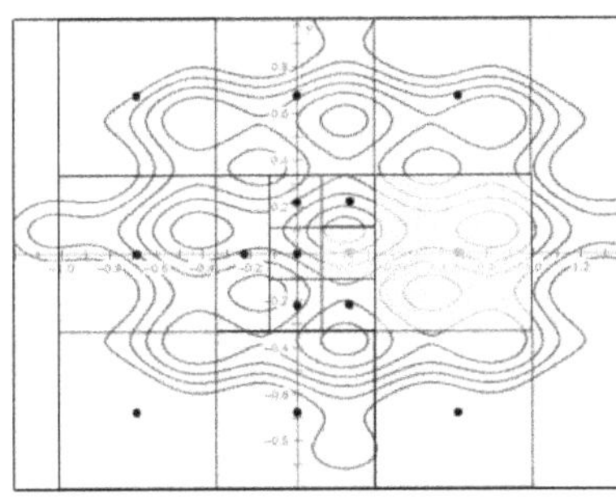

Iteration 3 : 15 hyperrectangles

Minimum : $f_{mid} = 0.303$

x mid	y mid	dx	dy	d max	fmid	fb	K
0.000	-0.667	0.333	0.333	0.333	2.400	1.277	
0.000	0.667	0.333	0.333	0.333	1.207	0.085	
0.667	-0.667	0.333	0.333	0.333	2.577	1.455	
0.667	0.667	0.333	0.333	0.333	1.385	0.263	
-0.222	0.000	0.111	0.333	0.333	0.975	-0.147	
-0.667	-0.667	0.333	0.333	0.333	3.111	1.989	
-0.667	0.667	0.333	0.333	0.333	1.918	0.796	
-0.667	0.000	0.333	0.333	0.333	1.440	0.318	
0.444	0.000	0.111	0.333	0.333	0.857	-0.265	
0.889	0.000	0.111	0.333	0.333	0.778	-0.345	3.366
0.667	-0.222	0.111	0.111	0.111	1.464	1.090	
0.667	0.222	0.111	0.111	0.111	1.026	0.652	
0.667	0.000	0.111	0.111	0.111	0.907	0.533	
0.000	-0.222	0.111	0.111	0.111	1.287	0.913	
0.000	0.222	0.111	0.111	0.111	0.849	0.475	
0.222	-0.222	0.111	0.111	0.111	0.861	0.487	
0.222	0.222	0.111	0.111	0.111	0.423	0.049	
0.000	0.000	0.111	0.111	0.111	0.729	0.355	
0.222	-0.074	0.111	0.037	0.111	0.699	0.325	
0.222	0.074	0.111	0.037	0.111	0.029	-0.345	-3.699
0.148	0.000	0.037	0.037	0.037	0.334	0.471	
0.296	0.000	0.037	0.037	0.037	0.421	0.558	
0.222	0.000	0.037	0.037	0.037	0.303	0.440	

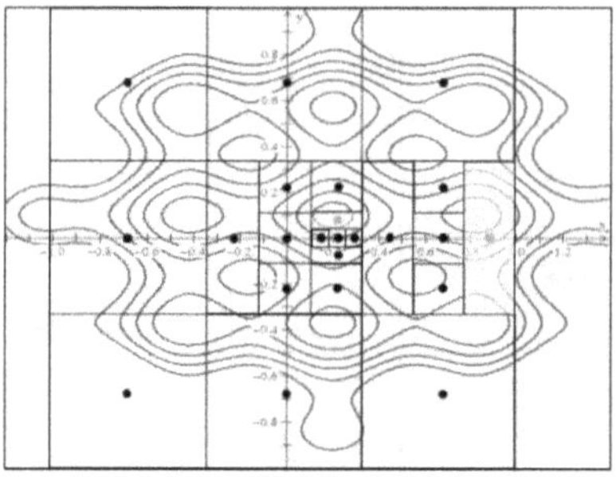

Iteration 4 : 23 hyperrectangles

Minimum : $f_{mid} = 0.029$

x mid	y mid	dx	dy	d max	fmid	fb	K
0.000	-0.667	0.333	0.333	0.333	2.400	1.749	
0.000	0.667	0.333	0.333	0.333	1.207	0.556	
0.667	-0.667	0.333	0.333	0.333	2.577	1.926	
0.667	0.667	0.333	0.333	0.333	1.385	0.734	
-0.222	0.000	0.111	0.333	0.333	0.975	0.325	
-0.667	-0.667	0.333	0.333	0.333	3.111	2.460	
-0.667	0.667	0.333	0.333	0.333	1.918	1.268	
-0.667	0.000	0.333	0.333	0.333	1.440	0.789	
0.444	0.000	0.111	0.333	0.333	0.857	0.206	1.953
0.667	-0.222	0.111	0.111	0.111	1.464	1.247	
0.667	0.222	0.111	0.111	0.111	1.026	0.809	
0.667	0.000	0.111	0.111	0.111	0.907	0.690	
0.000	-0.222	0.111	0.111	0.111	1.287	1.070	
0.000	0.222	0.111	0.111	0.111	0.849	0.632	
0.222	-0.222	0.111	0.111	0.111	0.861	0.644	
0.222	0.222	0.111	0.111	0.111	0.423	-0.167	5.313
0.000	0.000	0.111	0.111	0.111	0.729	0.512	
0.222	-0.074	0.111	0.037	0.111	0.699	0.109	
0.889	-0.222	0.111	0.111	0.111	1.335	0.745	
0.889	0.222	0.111	0.111	0.111	0.897	0.307	
0.889	0.000	0.111	0.111	0.111	0.778	0.187	
0.148	0.000	0.037	0.037	0.037	0.334	0.471	
0.296	0.000	0.037	0.037	0.037	0.421	0.558	
0.148	0.074	0.037	0.037	0.037	0.060	-0.137	
0.296	0.074	0.037	0.037	0.037	0.147	-0.050	
0.222	0.074	0.037	0.037	0.037	0.029	-0.167	
0.222	0.000	0.037	0.037	0.037	0.303	0.107	

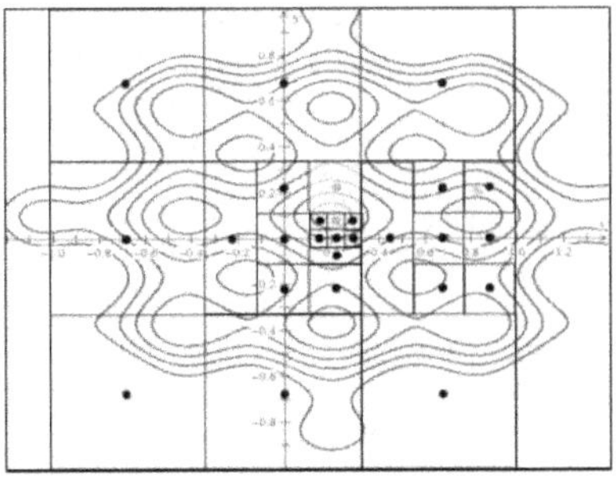

Iteration 5 : 27 hyperrectangles

Minimum : $f_{mid} = 0.029$

x mid	y mid	dx	dy	d max	fmid	fb	
0.000	-0.667	0.333	0.333	0.333	2.400	1.238	
0.000	0.667	0.333	0.333	0.333	1.207	0.046	
0.667	-0.667	0.333	0.333	0.333	2.577	1.416	
0.667	0.667	0.333	0.333	0.333	1.385	0.224	
-0.222	0.000	0.111	0.333	0.333	0.975	-0.186	
-0.667	-0.667	0.333	0.333	0.333	3.111	1.949	
-0.667	0.667	0.333	0.333	0.333	1.918	0.757	
-0.667	0.000	0.333	0.333	0.333	1.440	0.279	
0.444	0.000	0.111	0.333	0.333	0.857	-0.304	3.484
0.667	-0.222	0.111	0.111	0.111	1.464	1.077	
0.667	0.222	0.111	0.111	0.111	1.026	0.639	
0.667	0.000	0.111	0.111	0.111	0.907	0.520	
0.000	-0.222	0.111	0.111	0.111	1.287	0.900	
0.000	0.222	0.111	0.111	0.111	0.849	0.462	
0.222	-0.222	0.111	0.111	0.111	0.861	0.474	
0.000	0.000	0.111	0.111	0.111	0.729	0.342	
0.222	-0.074	0.111	0.037	0.111	0.699	0.312	
0.889	-0.222	0.111	0.111	0.111	1.335	0.948	
0.889	0.222	0.111	0.111	0.111	0.897	0.510	
0.889	0.000	0.111	0.111	0.111	0.778	0.390	
0.222	0.148	0.111	0.037	0.111	0.083	-0.304	1.021
0.222	0.296	0.111	0.037	0.111	0.796	0.409	
0.148	0.222	0.037	0.037	0.037	0.454	0.416	
0.296	0.222	0.037	0.037	0.037	0.540	0.503	
0.222	0.222	0.037	0.037	0.037	0.423	0.385	
0.148	0.000	0.037	0.037	0.037	0.334	0.137	
0.296	0.000	0.037	0.037	0.037	0.421	0.224	
0.148	0.074	0.037	0.037	0.037	0.060	0.197	
0.296	0.074	0.037	0.037	0.037	0.147	0.284	
0.222	0.000	0.037	0.037	0.037	0.303	0.107	
0.222	0.049	0.037	0.012	0.037	0.090	0.053	
0.222	0.099	0.037	0.012	0.037	0.007	-0.031	-0.623
0.198	0.074	0.012	0.012	0.012	0.022	0.030	
0.247	0.074	0.012	0.012	0.012	0.053	0.061	
0.222	0.074	0.012	0.012	0.012	0.029	0.037	

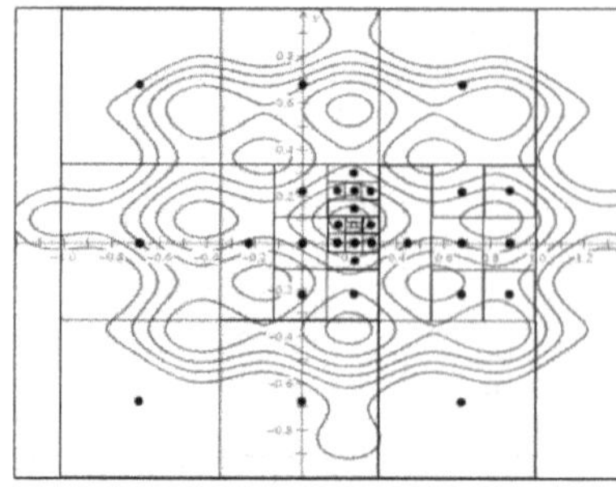

Iteration 6 : 35 hyperrectangles

Minimum : $f_{mid} = 0.007$

Table 2-5 shows the progression of the best point through the iterations.

Iteration	x	y	f
1	0.000	0.000	0.729
2	0.222	0.000	0.303
3	0.222	0.000	0.303
4	0.222	0.074	0.029
5	0.222	0.074	0.029
6	0.222	0.099	0.007

Table 2-5: Progression of the best point.

The solution $\begin{cases} x_{min} = 0.222 \\ y_{min} = 0.099 \end{cases}$ is obtained with 35 evaluations of the function.

It is very close to the global minimum: $\begin{cases} x^* = 0.2 \\ y^* = 0.1 \end{cases}$.

The DIRECT algorithm is efficient in locating the global minimum of a function of n variables. Its main limitation is the memory space required to store all the evaluated hyperrectangles. When the number of variables exceeds 10, it is necessary to "sacrifice" hyperrectangles, for example on a size criterion. The algorithm can be used to locate areas of interest and initialize a local descent algorithm such as Nelder-Mead.

2.4 Nelder-Mead method

The Nelder-Mead method (1965) looks for the minimum of a function of n real variables by moving a set of n + 1 points.

2.4.1 Polytope

Consider an unconstrained minimization problem.

$$\min_{x \in \mathbb{R}^n} f(x) \tag{2.44}$$

The Nelder-Mead method uses a set P of n+1 points (or vertices) called a **polytope** (or simplex) of $\mathbb{R}^n$.

$$P=\{x_0,x_1,\cdots,x_{n-1},x_n\} \ , \ x_k \in \mathbb{R}^n \tag{2.45}$$

These points are ranked from best to worst.

$$f(x_0)\le f(x_1)\le\cdots\le f(x_{n-1})\le f(x_n) \tag{2.46}$$

Figure 2-29 shows a polytope in $\mathbb{R}^2$ with its 3 points $\{x_0,x_1,x_2\}$ ranked from best to worst (according to their position on the level lines of the function).

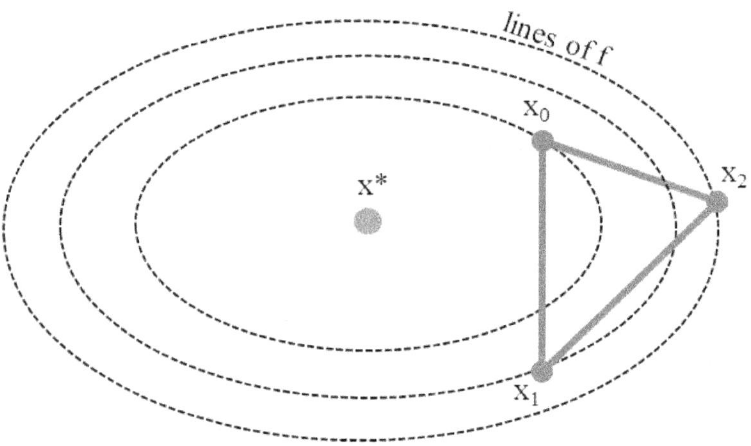

Figure 2-29: Polytope in dimension 2.

At each iteration, the worst point x_n will be replaced by a new better point x_{new}. The new polytope P' will then reranked from best to worst point.

$$\begin{aligned} P=\{x_0,x_1,\cdots,x_{n-1},x_n\} &\rightarrow P'=\{x_0,x_1,\cdots,x_{n-1},x_{new}\} \\ &\rightarrow P'=\{x'_0,x'_1,\cdots,x'_{n-1},x'_n\} \end{aligned} \tag{2.47}$$

The move of the worst point x_n is directed towards the barycenter x_c of the n best points $\{x_0,x_1,\cdots,x_{n-1}\}$.

$$x_c=\frac{1}{n}\sum_{i=0}^{n-1}x_i \tag{2.48}$$

The direction from x_n to x_c is assumed to be a descent direction, as it starts from the worst known point. Four candidate points are tested on this half-line D. They correspond to a reflection (x_r), an expansion (x_e), an external contraction (x_{ce}) and an internal contraction (x_{ci}). They are expressed in terms of x_n and x_c.

$$\begin{cases} x_r = x_c + (x_c - x_n) \\ x_e = x_c + 2(x_c - x_n) \\ x_{ce} = x_c + 0,5(x_c - x_n) \\ x_{ci} = x_c - 0,5(x_c - x_n) \end{cases} \quad (2.49)$$

These points are shown in figure 2-30 in $\mathbb{R}^2$. The two best points $\{x_0, x_1\}$ have their midpoint x_c as barycenter. The half-line D runs from x_2 to x_c.

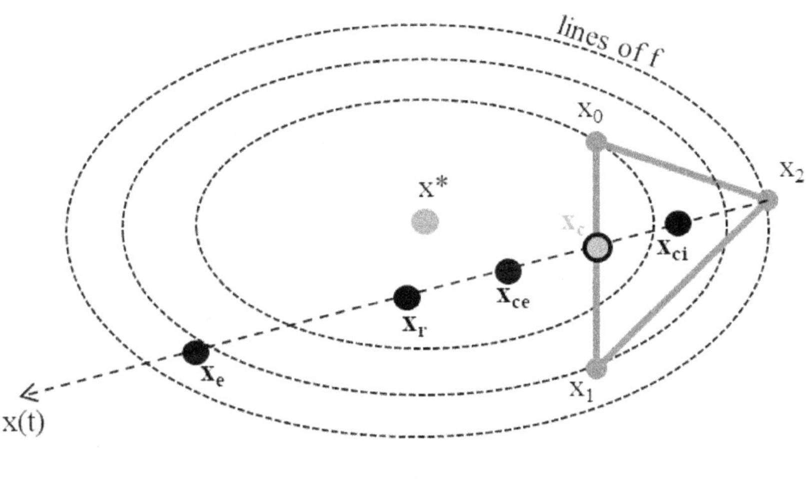

Figure 2-30: Candidate points.

The first point tested is the reflected point x_r, symmetrical of x_n relative to the barycenter x_c. Depending on the value of $f(x_r)$, four situations can occur:

- very good improvement: $f(x_r) < f(x_0)$
  The point x_r is better than all the points of the polytope.
  Since the direction D is favorable, we try the point x_e located further along D.
  The point x_n is replaced by the better of the two points x_r or x_e;
- good improvement: $f(x_0) < f(x_r) < f(x_{n-1})$
  The point x_r is better than the penultimate point x_{n-1}.
  The point x_n is replaced by x_r;

- low improvement: $f(x_{n-1}) < f(x_r) < f(x_n)$
  The point x_r is better than the point x_n, but worse than all the others.
  We try the point x_{ce} situated less far on the half-line D.
  The point x_n is replaced by the better of the two points x_r or x_{ce};
- no improvement: $f(x_n) < f(x_r)$
  We try the point x_{ci} located in the middle of the segment $[x_n ; x_c]$.
  If the point x_{ci} is better than x_n, it replaces x_n.
  Otherwise, all points of the polytope are reduced towards x_0.
  This reduction is a homothety of center x_0 and ratio 1/2.

In order to achieve some progress in the first few iterations, it is preferable that the initial polytope is large enough and that its vertices are in different directions. A good choice is to position the points $\{x_1, \cdots, x_n\}$ around x_0 along the canonical directions of $\mathbb{R}^n$.

When a local minimum x* is found, the polytope shrinks around x* and its size decreases. The stopping criteria are a sufficient size of the polytope: $\max_{i,j} |x_i - x_j|$ (distance between the vertices), a sufficient gap $|f(x_0) - f(x_n)|$ between the best and worst point (case of a "flat" function) and a maximum number of iterations or evaluations of the function.

It may happen that the polytope degenerates, which means that its vertices no longer form a basis of $\mathbb{R}^n$. In $\mathbb{R}^2$, this results in three aligned points. The moves are then restricted to the subspace generated by the vertices of the polytope and the other directions can no longer be explored. In such cases, the polytope can be reset around its best point x_0 by redefining the points $\{x_1, \cdots, x_n\}$ along the axes of $\mathbb{R}^n$.

2.4.2 Calculation stages

The calculation stages at iteration k are as follows.

Polytope at iteration k

$$P^k = \{x_0^k, x_1^k, \cdots, x_{n-1}^k, x_n^k\} \quad \text{with} \quad f(x_0^k) \le f(x_1^k) \le \cdots \le f(x_{n-1}^k) \le f(x_n^k)$$

1. Barycenter

$$x_c = \frac{1}{n}\sum_{i=0}^{n-1} x_i^k \rightarrow d = x_c - x_n^k$$

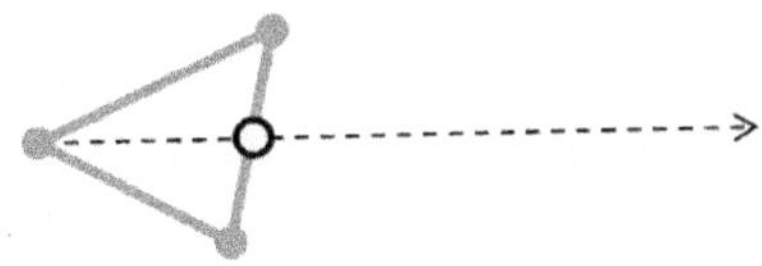

2. Reflection

$$x_r = x_c + d \qquad \rightarrow f(x_r)$$

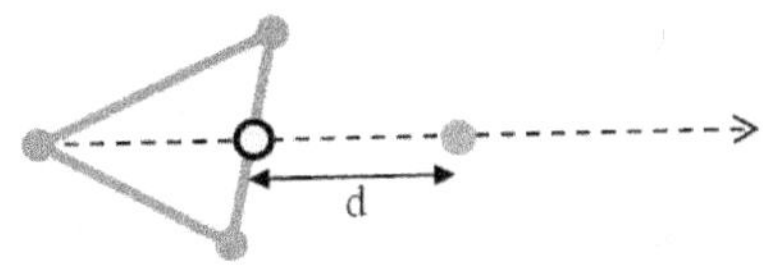

3. If $f(x_r) < f(x_0)$

Expansion $x_e = x_c + 2d \qquad \rightarrow f(x_e)$

If $f(x_e) < f(x_r) \qquad \rightarrow x' = x_e$

Otherwise $\qquad \rightarrow x' = x_r$

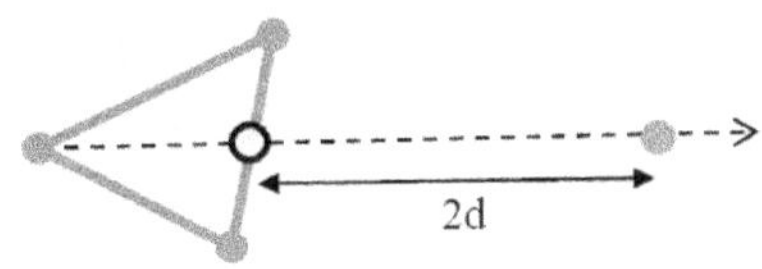

4. If $f(x_0) < f(x_r) < f(x_{n-1}) \qquad \rightarrow x' = x_r$

5. If $f(x_{n-1}) < f(x_r) < f(x_n)$

External contraction

$x_{ce} = x_c + d/2 \qquad \rightarrow f(x_{ce})$

If $f(x_{ce}) < f(x_r) \qquad \rightarrow x' = x_{ce}$

Otherwise $\qquad \rightarrow x' = x_r$

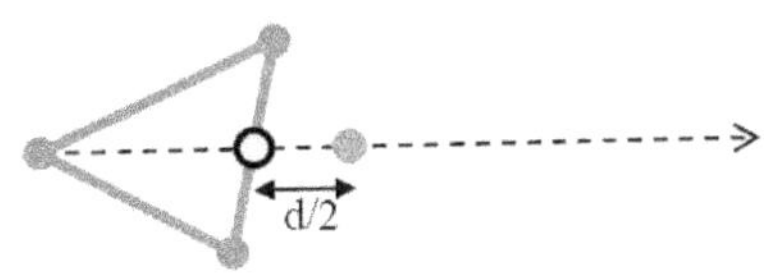

6. If $f(x_n) < f(x_r)$

Internal contraction

$x_{ci} = x_c - d/2 \qquad \rightarrow f(x_{ci})$

If $f(x_{ci}) < f(x_n) \qquad \rightarrow x' = x_{ci}$

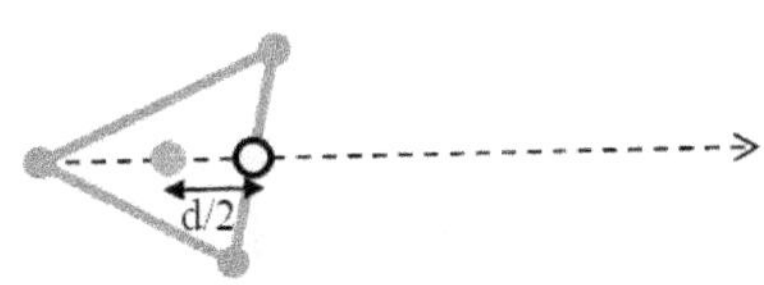

Otherwise reduction of P towards x_0

$$\rightarrow x_i^{k+1} = \frac{1}{2}\left(x_0 + x_i^k\right)$$

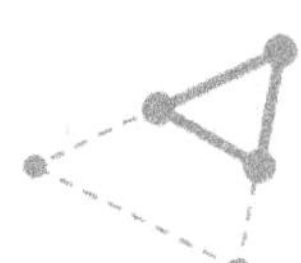

Reranking of the polytope with the new point x':

$$P^{k+1} = \left\{x_0^{k+1}, x_1^{k+1}, \cdots, x_{n-1}^{k+1}, x_n^{k+1}\right\} \quad \text{with} \quad f(x_0^{k+1}) \le f(x_1^{k+1}) \le \cdots \le f(x_{n-1}^{k+1}) \le f(x_n^{k+1})$$

2.4.3 Improvements

Several improvements have been proposed to the original Nelder-Mead method. The most effective ones concern the adaptive adjustment of the coefficients and the estimation of a pseudo-gradient.

Adaptive adjustment

The formulas (2.49) defining the test points are of the form

$$\begin{cases} x_r = x_c + \alpha(x_c - x_n) \\ x_e = x_c + \beta(x_c - x_n) \\ x_{ce} = x_c + \gamma(x_c - x_n) \\ x_{ci} = x_c - \gamma(x_c - x_n) \end{cases} \tag{2.50}$$

The reduction coefficient of the polytope in case of failure is noted.

The standard Nelder-Mead method uses the coefficients

$$\alpha = 1 \;,\; \beta = 2 \;,\; \gamma = \frac{1}{2} \;,\; \delta = \frac{1}{2} \tag{2.51}$$

When the dimension n of the problem is large, it is observed that the points tested (reflection, expansion, contraction) with the standard settings (2.51) deviate strongly from the polytope. This has the effect of deforming it rapidly and often leads to degeneration.

The **adaptive coefficients** are depending on the dimension n of the problem.

$$\alpha = 1 \;,\; \beta = 1 + \frac{2}{n} \;,\; \gamma = \frac{3}{4} - \frac{1}{2n} \;,\; \delta = 1 - \frac{1}{n} \tag{2.52}$$

In dimension 2, we retrieve the coefficients of the standard Nelder-Mead method. In dimension greater than 2, the adaptive coefficients are smaller than the standard coefficients and attenuate the moves of the points tested. The polytope thus moves progressively while maintaining a "well-conditioned" shape.

Pseudo-gradient

The **pseudo-gradient** is evaluated from the n+1 values of the function in the vertices of the polytope.

$$P = \{x_0, x_1, \cdots, x_{n-1}, x_n\} \quad \text{with} \quad f(x_0) \le f(x_1) \le \cdots \le f(x_{n-1}) \le f(x_n) \tag{2.53}$$

By expanding the function f to order 1 in the vicinity of x_0, we define a system of n linear equations linking the variations of x to the variations of f.

$$f(x_i) \approx f(x_0) + \nabla f(x_0)^T (x_i - x_0) \quad \text{for } i = 1 \text{ to } n$$
$$\Rightarrow \delta f_i \approx (\delta x_i)^T g \quad \text{with} \quad \begin{cases} \delta f_i = f(x_i) - f(x_0) \\ \delta x_i = x_i - x_0 \\ g = \nabla f(x_0) \end{cases} \tag{2.54}$$

This system can be put in matrix form:

$$\delta F \approx (\delta X)^T g$$
$$\text{with } \delta F = \begin{pmatrix} \delta f_1 \\ \vdots \\ \vdots \\ \vdots \\ \delta f_n \end{pmatrix} , \quad \delta X = \begin{pmatrix} (\delta x_1)_1 & \cdots & (\delta x_i)_1 & \cdots & (\delta x_n)_1 \\ \vdots & & \vdots & & \vdots \\ \vdots & & \vdots & & \vdots \\ \vdots & & \vdots & & \vdots \\ (\delta x_1)_n & \cdots & (\delta x_i)_n & \cdots & (\delta x_n)_n \end{pmatrix} \tag{2.55}$$

- the matrix $\delta X \in \mathbb{R}^{n \times n}$ gives the components of the vectors $\delta x \in \mathbb{R}^n$;
- the vector $\delta F \in \mathbb{R}^n$ gives the variations of f between the vertices;
- the vector $g \in \mathbb{R}^n$ represents an approximation of the gradient of f.

If the polytope P is not degenerate, the n vectors $(\delta x_i)_{i=1 \text{ to } n}$ originating from x_0 form a basis of $\mathbb{R}^n$. The matrix δX is then invertible, which makes it possible to determine an approximation of the gradient g of f at x_0 : $g \approx (\delta X)^{-T} \delta F$. Since the function is not necessarily derivable, the vector g is called "pseudo-gradient".

The search direction is then corrected by a weighting factor ρ between the standard direction d (from x_n to x_c) and the pseudo-gradient g.

$$d_g = (1-\rho)d - \rho g \quad \text{with} \quad \begin{cases} d = x_c - x_n \\ 0 \le \rho \le 1 \end{cases} \tag{2.56}$$

The weighting factor ρ can be set at 0.5.

The corrected direction (2.56) is generally better than the standard direction. The points tested (2.50) along this corrected direction allow a faster progression towards the minimum.

The directions d, g and d_g are shown in dimension 2 in figure 2-31.

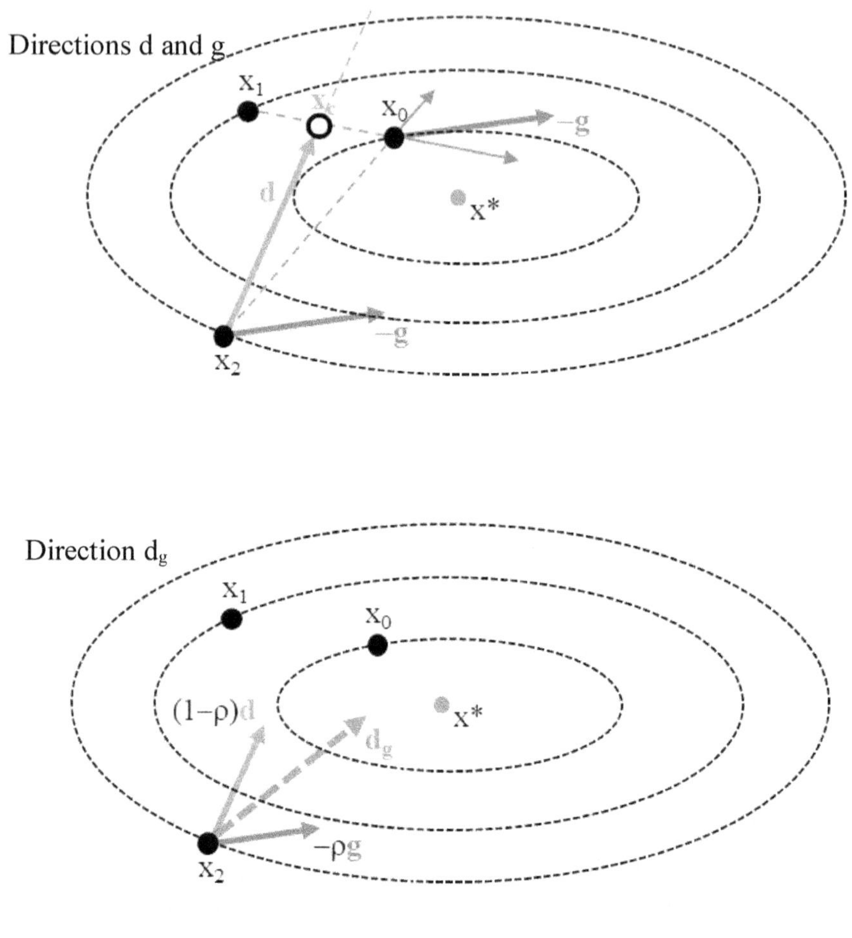

Figure 2-31: Direction corrected by the pseudo-gradient.

When the Nelder-Mead method works, the polytope shrinks on a local minimum of f. The following example illustrates the displacement of the polytope for two problems in $\mathbb{R}^2$. Although the Nelder-Mead method is limited to finding a local minimum, and has also no firm theoretical basis, it performs remarkably well in practice on any type of function and it does not suffer from the memory limitation problem inherent in the DIRECT method.

Example 2-9: Nelder-Mead method

Let us apply the Nelder-Mead method to a quadratic function and to the Rosenbrock function.

Quadratic function: $f(x_1, x_2) = \frac{1}{2}x_1^2 + \frac{9}{2}x_2^2$

The level lines are ellipses centered on the minimum in (0; 0).

Figure 2-32 shows the displacement of the polytope during the first 9 iterations with the position of the best vertex.

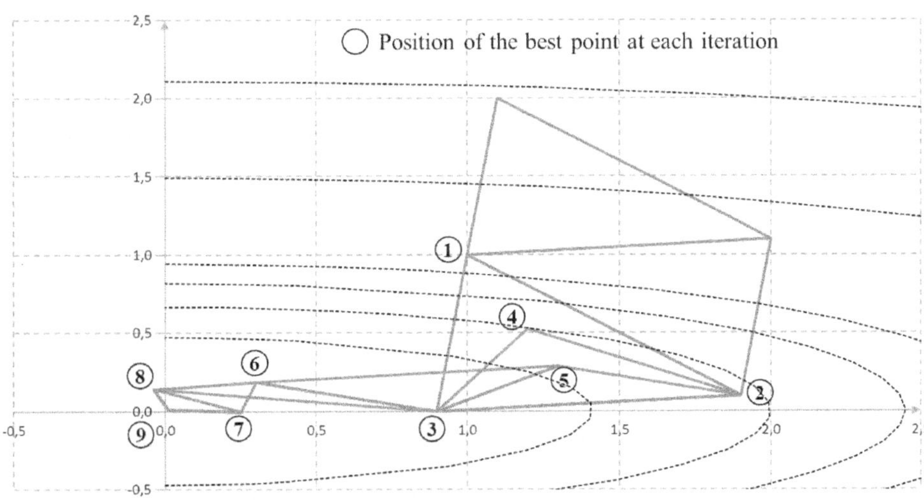

Figure 2-32: Minimization of a quadratic function by Nelder-Mead.

Rosenbrock function: $f(x_1, x_2) = 100\left(x_2 - x_1^2\right)^2 + \left(1 - x_1\right)^2$

Figure 2-33 shows the displacement of the best vertex during the iterations. There is a convergence towards the center of the valley and then a following of the valley until reaching the minimum in (1; 1).

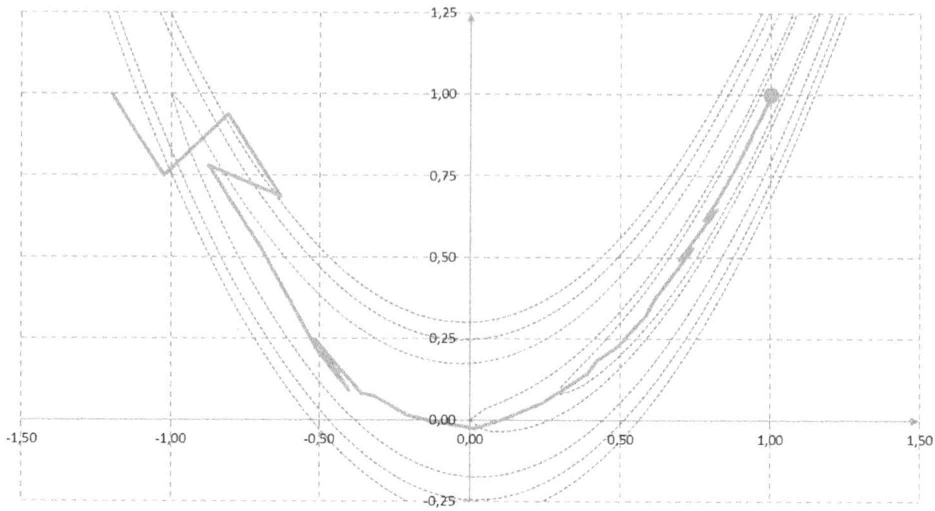

Figure 2-33: Minimization of the Rosenbrock function by Nelder-Mead.

2.5 Affine shaker

The "affine shaker" algorithm (Battiti and Tecchiolli, 1994) is similar to the Nelder-Mead algorithm, but it introduces a random move component.

2.5.1 Principle

We seek to minimize a function of n real variables.

$$\min_{x \in \mathbb{R}^n} f(x) \tag{2.57}$$

The search domain D is centered on the current point x_0 of value $f(x_0)$.

A random move $d \in \mathbb{R}^n$ is generated in the search domain and a new point is built as follows:

- first, the point $x_d^+ = x_0 + d$ is evaluated.

If this point is better than the point x_0: $f(x_d^+) < f(x_0)$, the direction $+d$ is favorable. The point x_d^+ is accepted and becomes the new current point: $x_1 = x_d^+$. The search domain D is expanded along the direction $+d$;

- otherwise, the point $x_d^- = x_0 - d$ is evaluated.

If this point is better than the point x_0: $f(x_d^-) < f(x_0)$, the direction $-d$ is favorable. The point x_d^- is accepted and becomes the new current point: $x_1 = x_d^-$. The search domain D is expanded along the direction $-d$;

- otherwise, the initial point is retained: $x_1 = x_0$.

As the directions $\pm d$ are unfavorable, the search domain D is contracted along the direction d.

The name "affine shaker" expresses that the move is random (shaker) and that the search domain is modified by an affine transformation detailed in the next section.

Figure 2-34 shows the evolution of the search domain from an initial point A or an initial point A'. The level lines of the cost function show the existence of three local minima. The iterations (B , C) or (B' , C') progress to different local minima depending on the starting point.

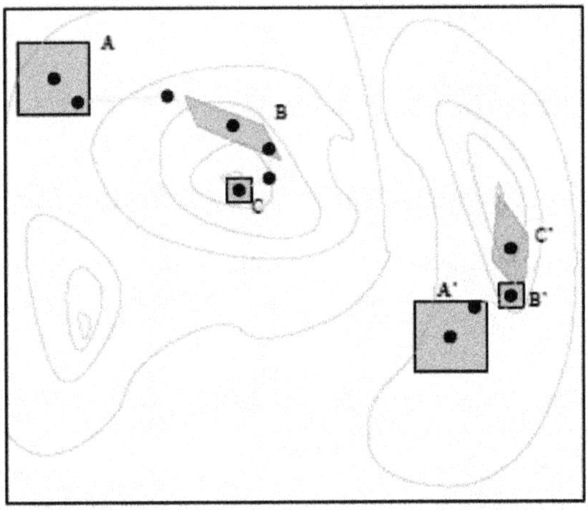

Figure 2-34: Evolution of the search domain.

2.5.2 Affine transformation

The search domain D centered on the current point x_0 is defined by n independent vectors of $\mathbb{R}^n$ denoted $(b_1,\cdots,b_n)$.

The move $d \in \mathbb{R}^n$ is a linear combination of vectors b_i of the form

$$d=\sum_{i=1}^{n}\beta_i b_i \quad \text{with} \quad -1 \le \beta_i \le 1 \tag{2.58}$$

The coefficients $(\beta_1,\cdots,\beta_n)$ are drawn uniformly from the interval $\left[-1\,;1\right]$.

Depending on the values of $f(x_0+d)$ and $f(x_0-d)$, the domain D is either expanded or contracted with a coefficient ρ along the direction d.

The transformation consists in adding to each basis vector b a component proportional to the scalar product $d^T b$, as shown in figure 2-35.

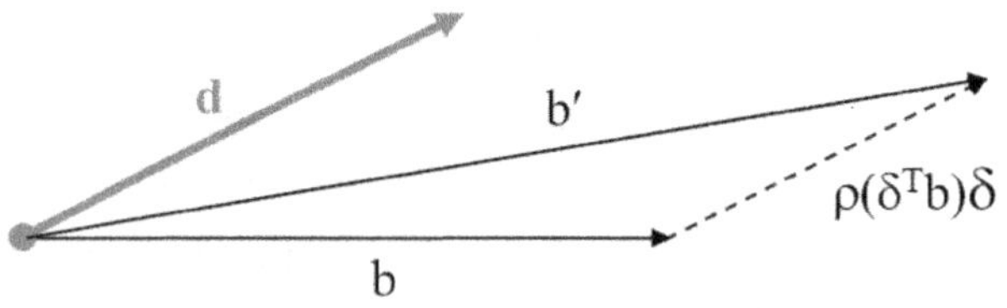

Figure 2-35: Affine transformation of basis vectors.

The vector b yields the vector b' defined by

$$b'=b+\rho\frac{d^T b}{\|d\|^2}d=b+\rho(\delta^T b)\delta \quad \text{with} \quad \delta=\frac{d}{\|d\|} \tag{2.59}$$

This transformation is applied to each of the basis vectors $(b_1,\cdots,b_n)$.

Let B be the matrix of basis vectors and δ the unit vector along d.

$$B=\begin{pmatrix} b_1 & b_2 & \cdots & \cdots & b_n \end{pmatrix}=\begin{array}{c} \begin{array}{ccccc} b_1 & b_2 & & & b_n \end{array} \\ \begin{pmatrix} b_{11} & b_{21} & \cdots & \cdots & b_{n1} \\ b_{12} & b_{22} & \cdots & \cdots & b_{n2} \\ \vdots & \vdots & & & \vdots \\ \vdots & \vdots & & & \vdots \\ b_{1n} & b_{2n} & \cdots & \cdots & b_{nn} \end{pmatrix} \end{array} \quad \text{and} \quad \delta=\begin{pmatrix} \delta_1 \\ \delta_2 \\ \vdots \\ \vdots \\ \delta_n \end{pmatrix} \tag{2.60}$$

The transformation (2.59) is expressed in matrix form

$$\begin{cases} b'_1 = b_1 + \rho\delta(\delta^T b_1) \\ b'_2 = b_2 + \rho\delta(\delta^T b_2) \\ \vdots \\ \vdots \\ b'_n = b_n + \rho\delta(\delta^T b_n) \end{cases} \Leftrightarrow \quad B' = B + \rho\delta(\delta^T B) = (I + \rho\delta\delta^T)B \tag{2.61}$$

which can be written with the transformation matrix noted P

$$B' = PB \quad \text{with} \quad P = I + \rho\delta\delta^T = I + \rho\frac{dd^T}{\|d\|^2} \tag{2.62}$$

Let us illustrate this transformation in dimension 2 for a square domain. The domain D centered on the point x is defined by the orthogonal vectors (b_1, b_2).

The random move $d = \beta_1 b_1 + \beta_2 b_2$ yields the new point x'. The vectors (b_1, b_2) become the vectors (b'_1, b'_2) defined by

$$\begin{cases} b'_1 = b_1 + \rho(\delta^T b_1)\delta \\ b'_2 = b_2 + \rho(\delta^T b_2)\delta \end{cases} \tag{2.63}$$

These new vectors are no longer orthogonal. The new domain D' shown in figure 2-36 is a parallelogram centered in x' and expanded along the direction d.

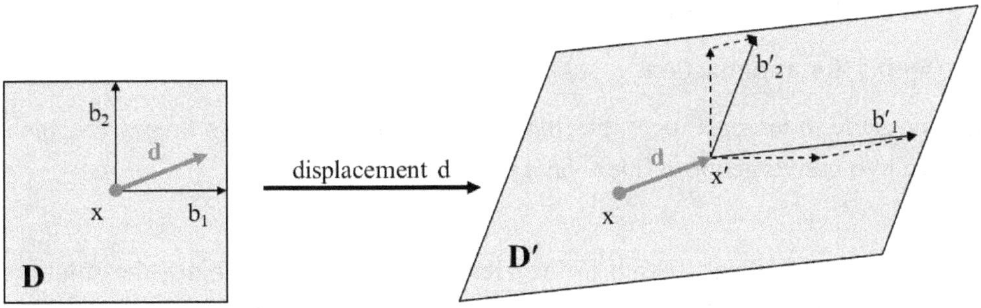

Figure 2-36: Dilatation of the search domain.

2.5.3 Algorithm

Figure 2-37 depicts the stages of the affine shaker algorithm.

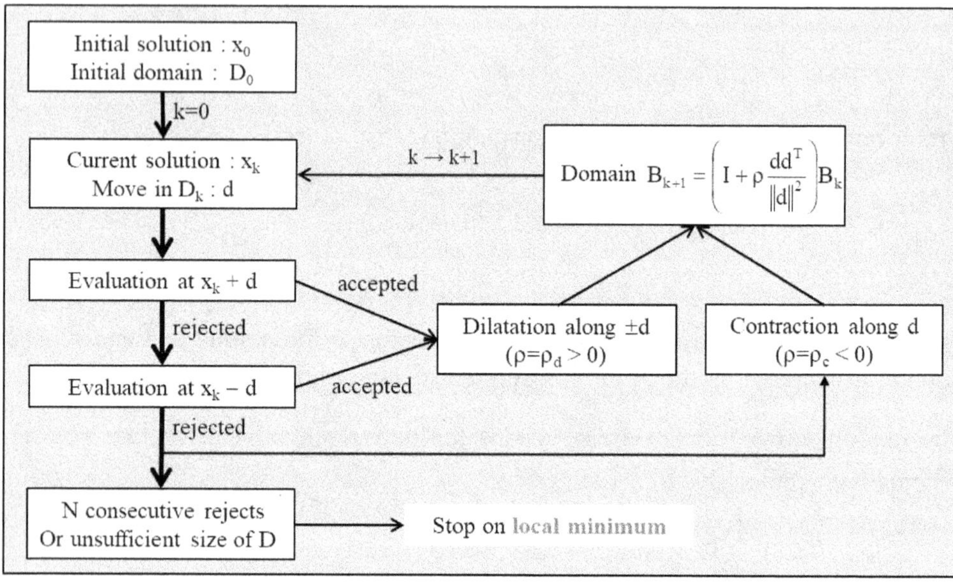

Figure 2-37: Affine shaker algorithm.

The initial domain D_0 can be defined by the canonical basis of $\mathbb{R}^n$.

The coefficient of expansion or contraction can be set at 0.5 for example:

- $\rho = +0.5$ for an expansion;
- $\rho = -0.5$ for a contraction.

It is possible to increase ρ in the first few iterations to speed up the moves, then reduce it to converge more finely on a minimum.

As the method is local (search in the vicinity of the current point), the solution obtained depends on the initial point x_0 and of the size of the initial domain D_0.

The following example illustrates the progression of the algorithm on the Rosenbrock function.

Example 2-10: Minimization of the Rosenbrock function by affine shaker

Let us apply the affine shaker algorithm to the function

$$f(x_1,x_2)=100\left(x_2-x_1^2\right)^2+\left(1-x_1\right)^2$$

The initial point is in (– 5; 5), the initial domain is a square defined by the canonical basis of $\mathbb{R}^2$.

Table 2-6 shows the evolution of the solution and the number of evaluations of the function. The progression is quite fast at the beginning: the point (0.99; 0.98) very close to the solution (1; 1) is reached after 1200 evaluations of the function. However, it is difficult to obtain a more accurate solution in a reasonable number of evaluations

Figure 2-38 shows the progression of the algorithm. The valley floor is reached in 24 evaluations, then the algorithm oscillates in a zone close to the minimum.

N_f	f	x_1	x_2
1	40036.0	-5.0000	5.0000
3	4842.0	-2.5123	-0.6379
6	1517.8	-1.0533	5.0000
9	971.7	2.5591	3.4356
15	932.0	1.6944	-0.1809
24	51.5	1.4993	1.5322
39	1.2779	1.3401	1.9036
48	1.0386	0.6511	0.3282
402	0.1641	0.5952	0.3527
1206	0.000097	0.9903	0.9806
28944	0.000023	1.0040	1.0082
95676	0.000021	0.9978	0.9960

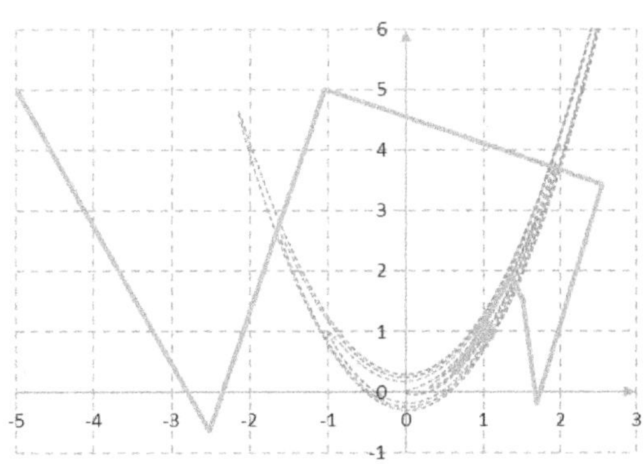

Figure 2-38: Affine shaker progression.

Table 2-6: Affine shaker iterations.

2.6 CMAES

The CMAES algorithm (*Covariance Matrix Adaptation Evolution Strategy*, Hansen and Ostermeyer, 2001) is a descent method including an evolutionary aspect based on a population of individuals.

2.6.1 Principle

We seek to minimize a function of n real variables.

$$\min_{x \in \mathbb{R}^n} f(x) \tag{2.64}$$

The pieces of information used at each iteration are the position of the current point and a covariance matrix.

An iteration consists of three stages illustrated in figure 2-39:

- drawing a sample of p candidates according to the normal distribution;

- selecting the q best candidates with respect to the minimization of f;

- updating the normal distribution parameters with respect to the selection.

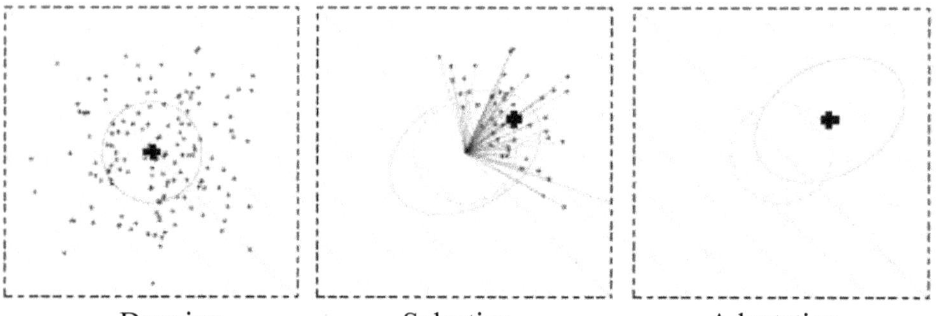

Drawing Selection Adaptation

Figure 2-39: Stages of a CMAES iteration.

The random generation of candidates promotes exploration (diversification). The selection of the best ones orientates the overall move towards a minimum (intensification).

2.6.2 Covariance adaptation

The distribution parameters at the beginning of the iteration are the mean position noted m, the covariance matrix noted C and a step length noted σ.

The sample of p candidates $(x_1, x_2, \cdots, x_p)$ is of the form

$$x_i = m + \sigma y_i \tag{2.65}$$

where the moves $(y_1, y_2, \cdots, y_p)$ follow the centered normal law $N(0\,;C)$.

The p candidates are ranked by increasing values of f.

$$f(x_1) \le f(x_2) \le \cdots \le f(x_q) \le \cdots \le f(x_p) \tag{2.66}$$

The top q candidates $(x_1, x_2, \cdots, x_q)$ are selected to update the mean position m, the covariance matrix C and a step length σ.

Updating the average

A direction of descent d is defined from the moves $(y_1, y_2, \cdots, y_q)$ associated with the q best candidates.

$$d = \sum_{j=1}^{q} w_j y_j \tag{2.67}$$

The weights $(w_1, w_2, \cdots, w_q)$ are either equal or decreasing in order to favor the best candidates. The average position m is updated by

$$m' = m + \sigma d \tag{2.68}$$

This update is shown in figure 2-40a.

Update of the step length

The step length σ is updated by

$$\sigma' = \sigma e^{\frac{1}{3}\cdot\frac{\tau-\tau_s}{1-\tau_s}} \tag{2.69}$$

The real τ represents the sample success rate, defined as the part of candidates satisfying $f(x_i) < f(m)$. This rate is compared to the threshold: $\tau_s = \frac{1}{5}$.

- if $\tau > \tau_s$, the success rate in the direction d is good and the step σ is increased;
- if $\tau < \tau_s$, the success rate in the direction d is low and the step σ is decreased.

<u>Updating the covariance</u>

The covariance update uses two matrices noted C_q and C_d.

The matrix C_q is the weighted covariance of the q best moves.

$$C_q = \sum_{j=1}^{q} w_j \left(y_j y_j^T \right) \tag{2.70}$$

The matrix C_d is a rank 1 matrix defined from the direction d.

$$C_d = d\,d^T \tag{2.71}$$

The covariance matrix C is updated by combining the matrices C_q and C_d with weights α_q and α_d comprised between 0 and 1.

$$C' = (1 - \alpha_q - \alpha_d)C + \alpha_q C_q + \alpha_d C_d \tag{2.72}$$

This update is depicted in figure 2-40b.

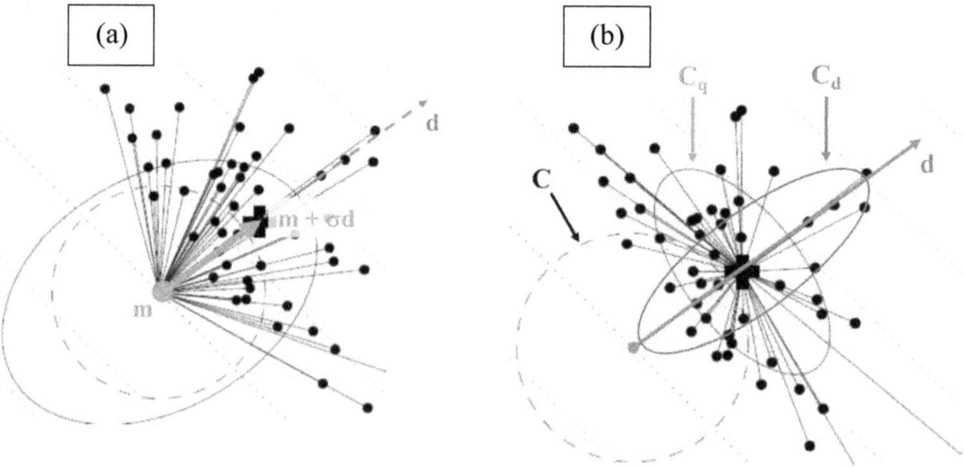

Figure 2-40: Updating the mean and covariance.

Figure 2-41 shows the mean and the updated covariance.

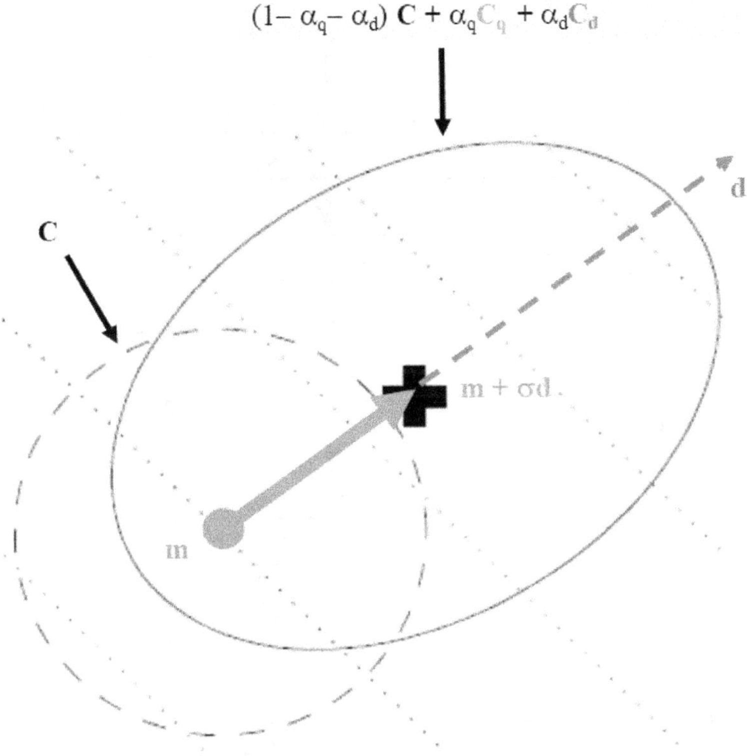

Figure 2-41: Updated mean and covariance.

Updating based on a random draw of the sample may lead to oscillatory moves in successive iterations.

In order to stabilize the algorithm, the current descent direction d can be combined with the direction d_p obtained in the previous iteration: $d' = (1-\alpha_p)d + \alpha_p d_p$.

This allows the direction of move to be averaged over several iterations.

2.6.3 Algorithm

Figure 2-42 depicts the stages of the CMAES algorithm.

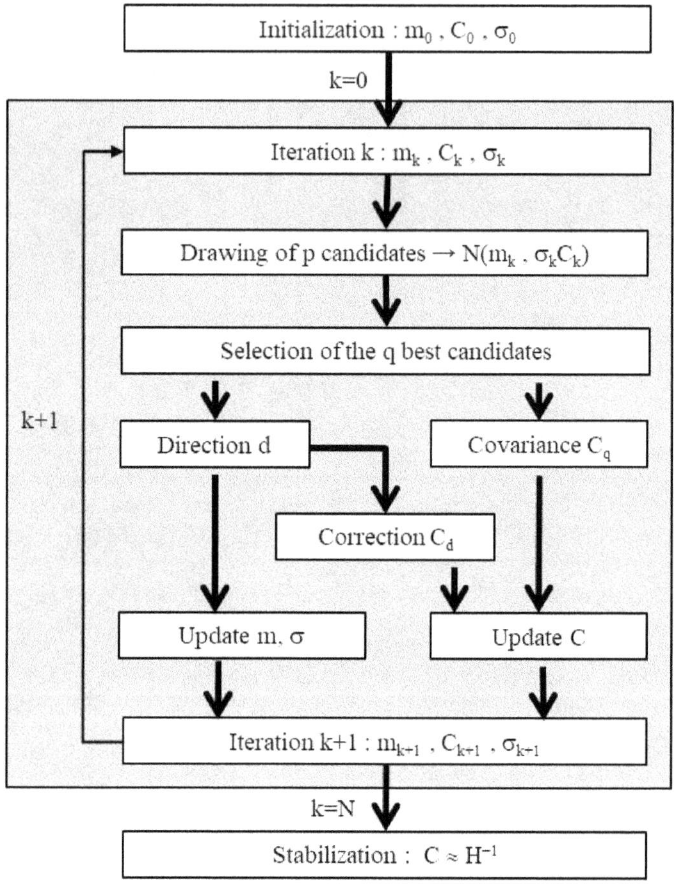

Figure 2-42: CMAES algorithm.

The setting parameters are the sample size p, the number of selected candidates q, the weights $(w_1, \cdots, w_q)$ defining the descent direction, the weights α_q and α_d of the covariance matrices, the success threshold τ_s and the maximum number of iterations or evaluations of the function. These settings are to be adapted according to the problem and to the number of variables.

The initial covariance matrix can be the identity. As the search is performed in the neighborhood of the current point, the method is local and the solution will depend on the initial point x_0.

The covariance matrix in the end point is related to the Hessian of the function. Indeed, if the final average position m is a minimum: $\nabla f(m)=0$, the expansion to order 2 of the function f in the neighborhood of m is

$$f(x) \approx f(m)+\frac{1}{2}(x-m)^T H(x-m) \quad \text{with} \quad H=\nabla^2 f(m) \tag{2.73}$$

and its level lines have the equation

$$(x-m)^T H(x-m) = C^{te} \tag{2.74}$$

Furthermore, the density of the normal distribution $N(0\,;C)$ is

$$p(x) = \frac{1}{(2\pi)^{\frac{n}{2}}\sqrt{\det C}} e^{-\frac{1}{2}(x-m)^T C^{-1}(x-m)} \tag{2.75}$$

and its iso-density lines have the equation

$$(x-m)^T C^{-1}(x-m) = C^{te} \tag{2.76}$$

If the covariance C is close to H^{-1} (within a multiplicative constant), the sample around the position m (coinciding with the minimum) is distributed on the level lines of f and it does not exhibit a clear direction of descent. The move is then zero $(d \approx 0)$ and the mean and covariance stabilize. In practice, we observe a convergence to a matrix of the form: $C \approx \alpha H^{-1}$, as illustrated in figure 2-43.

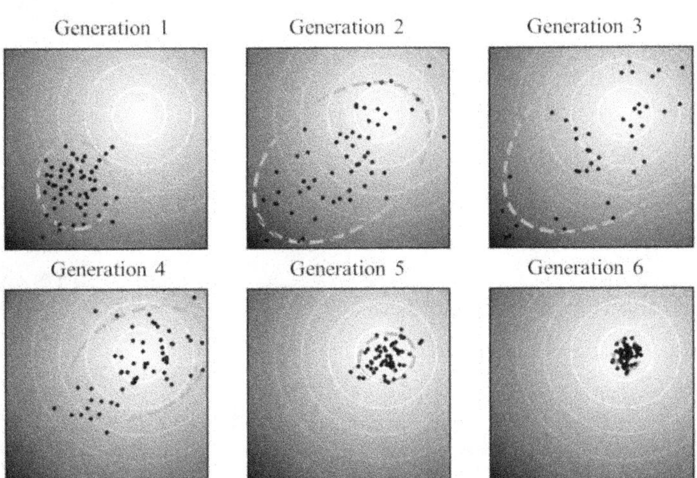

Figure 2-43: Covariance and Hessian.

The following example illustrates the progression of the algorithm on the Rosenbrock function.

Example 2-11: Minimization of the Rosenbrock function by CMAES

Let us apply the CMAES algorithm to the function

$$f(x_1,x_2)=100\left(x_2-x_1^2\right)^2+\left(1-x_1\right)^2$$

The sample size is p = 5, the number of selected candidates is q = 2.

The initial point is placed in (12 ; −10).

Figure 2-44 shows the progression of the algorithm. The first iterations (in figure 2-44a) make large moves and quickly reach the valley floor, then the algorithm progresses following the valley to the minimum (in figure 2-44b).

The solution in (1; 1) is reached very precisely (to 10^{-7}) with 140 iterations and 800 evaluations of the function.

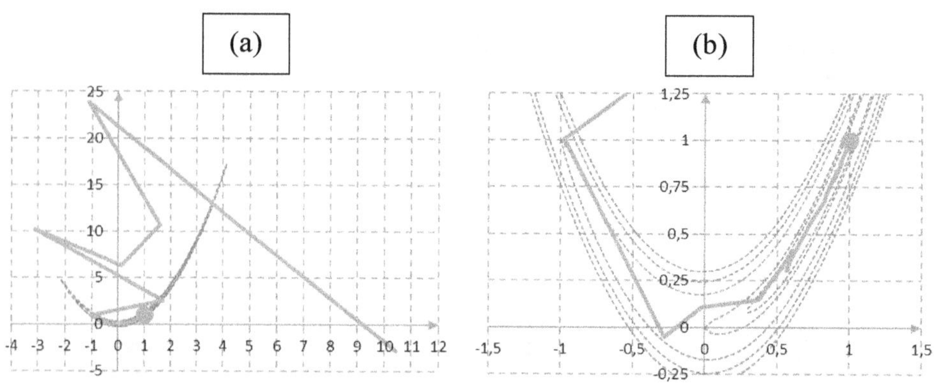

Figure 2-44: CMAES progression.

The inverse of the final covariance is: $C^{-1}=\begin{pmatrix} 240.5 & -119.4 \\ -119.4 & 59.5 \end{pmatrix}\cdot 10^9$

to be compared with the Hessian in (1;1): $H=\begin{pmatrix} 802 & -400 \\ -400 & 200 \end{pmatrix}$

These two matrices are almost proportional: $C^{-1}\approx 3\cdot 10^8 H$.

2.7 Simulated annealing

The simulated annealing method (Kirkpatrick et al, 1983) is inspired by the thermodynamics of a particle system.

2.7.1 Principle

Annealing is a process used in metallurgy to obtain a defect-free alloy. At very high temperatures, the metal is in a liquid state and the atoms move freely. It is then cooled to a solid state:

- if the cooling is rapid (quenching), the atoms freeze in a disordered state. The resulting alloy has a high energy. Its structure is irregular and has defects (in figure 2-45a);
- if the cooling is slow (annealing), the atoms organize themselves in a regular way. The resulting alloy has a low energy. Its structure is crystalline without defects (in figure 2-45c).

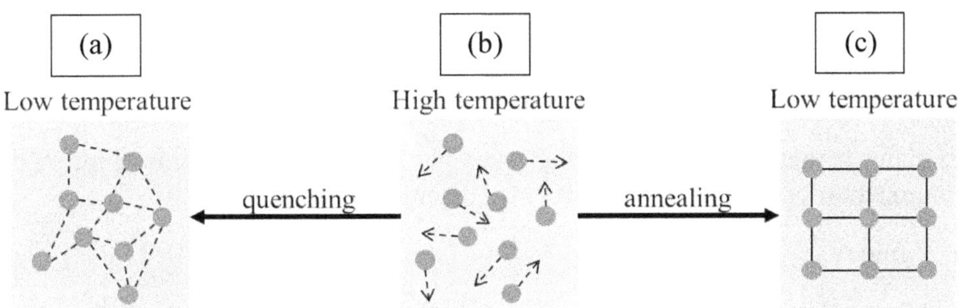

Figure 2-45: Metallurgical quenching and annealing.

The energy level E of the system depends on the atom arrangement. This energy level is a continuous variable between 0 and $+\infty$.

The probability that the energy of the system is equal to E at the temperature T is expressed by Gibbs law. The probability density of this law is

$$P_T(E) = c e^{-\frac{E}{kT}} \tag{2.77}$$

k is the Boltzmann constant and $c = \int_E e^{-\frac{E}{kT}} dE$ is the normalization coefficient.

Figure 2-46 shows the accessible energy states depending on the temperature.

At high temperatures, all energy states have almost the same probability. At low temperatures, the high energy states have a very low probability.

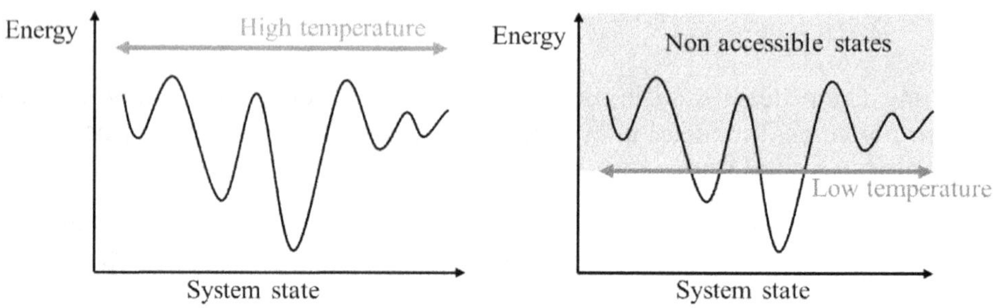

Figure 2-46: Accessible states depending on the temperature.

2.7.2 Probability of transition

Simulated annealing transposes the principle of metallurgical annealing to the minimization of a function f of n real (or discrete) variables.

$$\min_{x \in \mathbb{R}^n} f(x) \tag{2.78}$$

The variables x correspond to the atom arrangement, the function f represents the system energy: $E(x) = f(x)$.

We start from an initial state of the system noted x_0 of energy $E_0 = f(x_0)$.

The state of the system is randomly perturbed: $x_0 \to x = x_0 + \delta x$, which leads to a change in energy: $E_0 \to E = E_0 + \delta E$.

- if the energy decreases, the new state x is accepted;
- if the energy increases, the new state x is accepted with the probability:

$$P(x) = e^{-\frac{\delta E}{T}} = e^{-\frac{f(x)-f(x_0)}{T}} \tag{2.79}$$

The temperature parameter T regulates the acceptance of energy increase:

- at high temperatures, a higher energy state can be accepted with a high probability. This allows the system to explore all possible states;
- at low temperatures, a higher energy state is almost always rejected. The system will stabilize in a low energy state corresponding to thermodynamic equilibrium.

The acceptance of worse solutions at high temperatures allows the solution space to be widely explored. The system can escape from a local minimum provided that a sufficient number of trials are allowed (figure 2-47).

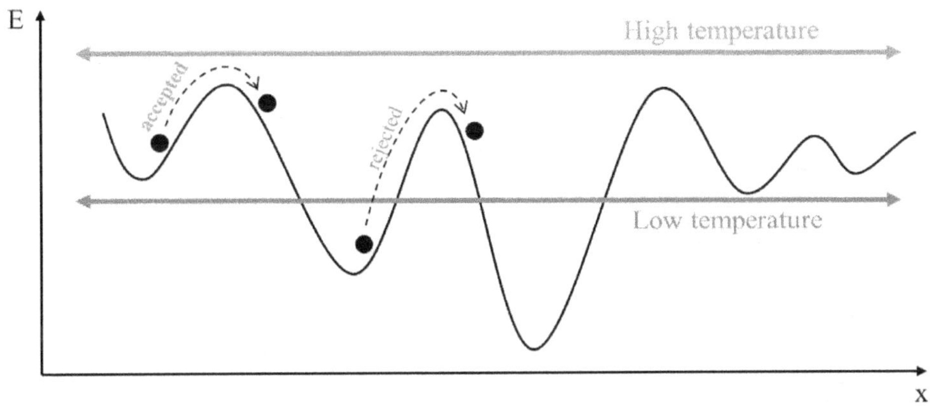

Figure 2-47: Escape from a local minimum.

2.7.3 Algorithm

Figure 2-48 depicts the stages of the simulated annealing algorithm (also called the Metropolis algorithm). The algorithm can be applied to discrete or continuous problems. The random perturbations are to be defined according to the problem features (nature of the variables x and the function f). These perturbations must generate a solution that is "locally close" to the current solution so as to explore its immediate neighborhood.

The convergence to a global minimum requires a high initial temperature to allow access to all states and a slow temperature decrease to escape local minima. For example, the initial temperature T_0 can be tuned so that 90% of the random moves generated from x_0 are accepted.

This automatic tuning method can be easily implemented and it allows to start with a temperature that is neither too high (which would waste computation time) nor too low (which would lead to premature convergence).

The temperature decrease is of the form $T_{k+1} = \alpha T_k$, with a coefficient very close to 1 (*e.g.* $\alpha = 0.999$). The decrease is carried out by successive stages after a number M of unsuccessful trials at each stage. The algorithm stops on a maximum number of trials N or on a number of stages without improvement. The quality of the solution and the calculation time depend on the values of T_0, α, M, N.

These settings are to be adjusted experimentally on a case-by-case basis.

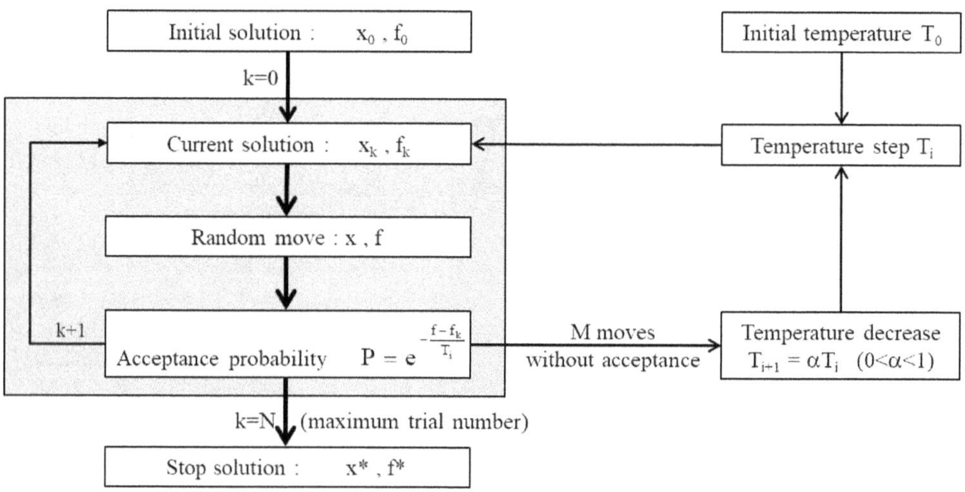

Figure 2-48: Simulated annealing algorithm.

The convergence of the algorithm can be observed by measuring the acceptance rate, the expectation and variance of the function at each temperature step. We show below how to use this information.

At a fixed temperature T, the expectation of the function f with the acceptance probability defined by (2.79) is given by

$$E[f] = \frac{\int_X f(x)P(x)dx}{\int_X P(x)dx} \quad \text{with} \quad P(x) = e^{-\frac{f(x)-f(x_0)}{T}} \tag{2.80}$$

Let us calculate the derivative of this expectation with respect to the temperature. The variable x is omitted to simplify the formulae.

$$\frac{d}{dT}E[f]=\frac{\left(\int f\frac{dP}{dT}dx\right)\left(\int Pdx\right)-\left(\int fPdx\right)\left(\int\frac{dP}{dT}dx\right)}{\left(\int Pdx\right)^2} \tag{2.81}$$

By expressing the derivative of the probability (2.79)

$$\frac{d}{dT}P(x)=\frac{f(x)-f(x_0)}{T^2}P(x) \tag{2.82}$$

and replacing in (2.81), we obtain after eliminating the terms in $f(x_0)$

$$\frac{d}{dT}E[f]=\frac{1}{T^2}\left[\frac{\int f^2Pdx}{\int Pdx}-\left(\frac{\int fPdx}{\int Pdx}\right)^2\right]=\frac{E[f^2]-E[f]^2}{T^2} \tag{2.83}$$

This formula shows that the derivative of the expectation of f with respect to the temperature is equal to its variance divided by the square of the temperature.

$$\frac{d}{dT}E[f]=\frac{V[f]}{T^2} \tag{2.84}$$

At each temperature step, a number of trials are performed producing solutions that are accepted with probability (2.79). By measuring the expectation and variance of f and plotting the evolution of $E[f]$ and $V[f]/T^2$ as a function of temperature, one can appreciate the good behavior of the algorithm.

The expectance $E[f]$ generally follows a sigmoidal (S-shaped) decrease, with the decrease first accelerating and then slowing down to stabilize.

The variance $V[f]/T^2$ is maximum for a so-called stabilization temperature, at which the progression is the fastest (largest derivative of $E[f]$).

The shape of these curves indicates whether the settings (T_0, α, M, N) are correct or can be improved by changing the initial temperature (lower or higher), the cooling rate (faster or slower) and the number of trials per stage.

The example 2-12 illustrates the behavior of the simulated annealing algorithm on the travelling salesman problem (TSP).

Example 2-12: Travelling salesman problem by simulated annealing

The Travelling Salesman Problem (TSP) consists in finding the path of minimum length passing once and only once through n given cities. The distances of the cities from each other are fixed.

The current state is the ordered list of the n cities to visit. The objective is to find the optimal permutation among the n! possible solutions.

Three basic perturbations are defined for this problem, as shown in figure 2-49.

Insertion consists in randomly drawing an element from the list and randomly inserting it at a new position.

Swapping consists in randomly drawing two elements from the list and exchanging their positions.

Inversion consists in randomly drawing two elements from the list and reversing the order of the string between these two elements.

Insertion

2 is inserted after 4.

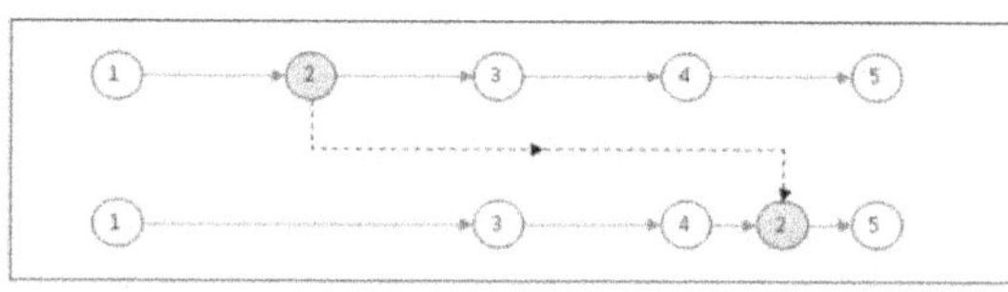

Swapping

2 and 3 are swapped.

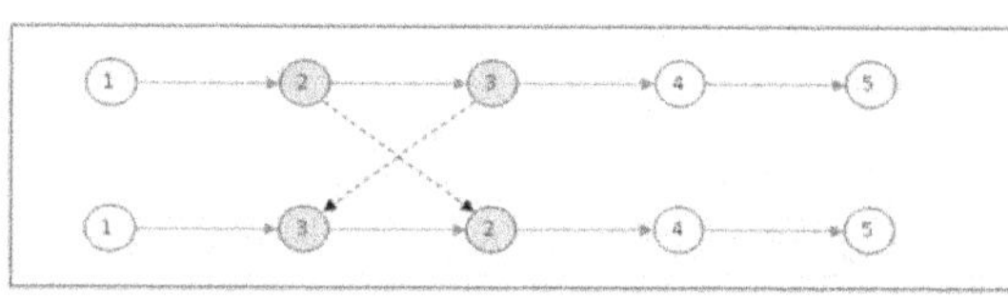

Inversion

The path from 2 to 5 is reversed.

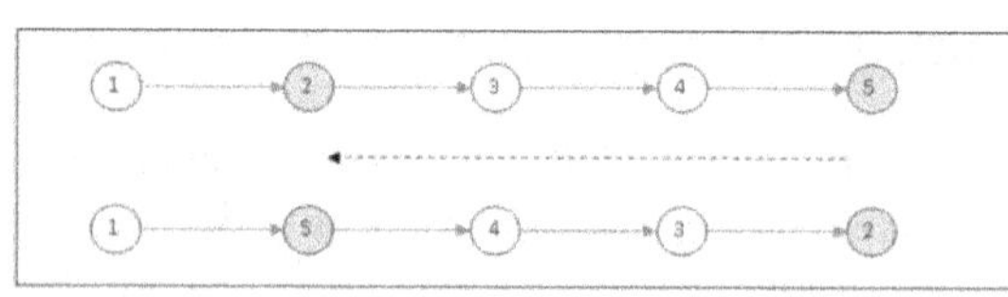

Figure 2-49: Elementary perturbations of a path.

Three different instances of the TSP are presented below:

- the Challenge250 problem is a fictitious problem with 250 cities;
- the Pcb442 problem is a 442-component printed circuit board;
- the Att532 problem is the problem of the 532 largest cities in the US.

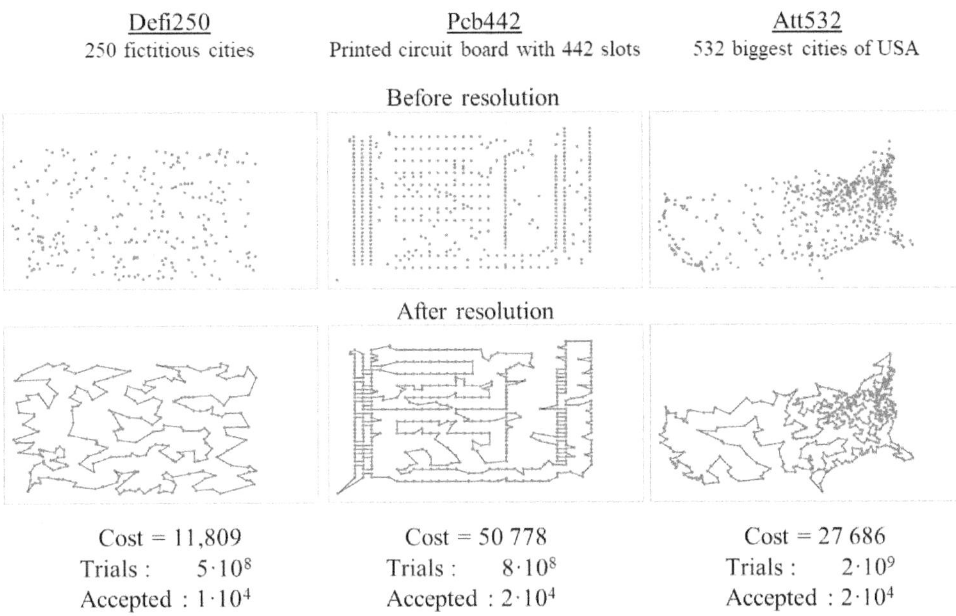

Figure 2-50: Instances of the travelling salesman problem.

Figure 2-50 shows the positions of the cities (top row before resolution) and the optimal path found (bottom row after resolution). The cost is the total path length. The number of trials is in the order of one billion, with about 10 000 accepted trials, which represents an acceptance rate of about 0.001%.

The evolution of the acceptance rate depending on the temperature is plotted in figure 2-51. The temperature is in logarithmic scale. The diagram is read from right to left (decreasing temperature). The acceptance rate is almost 100% at high temperature, then decreases and becomes zero at low temperature.

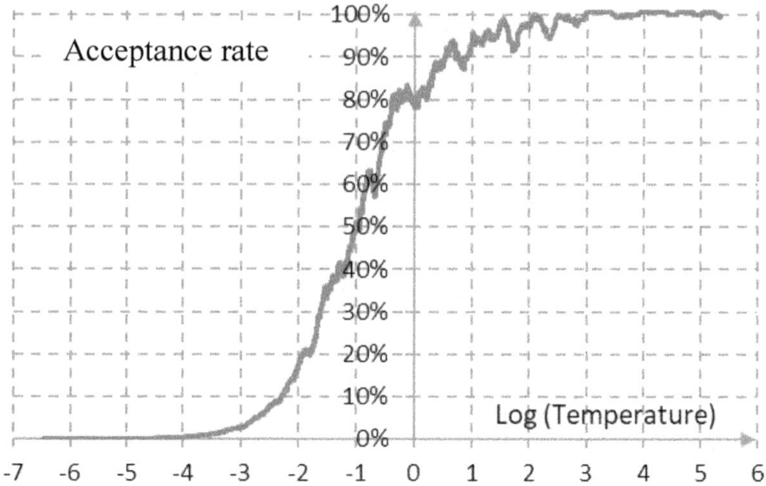

Figure 2-51: Acceptance rate depending on the temperature.

Figure 2-52 shows the evolution of the expectation and the variance/T² ratio of the drawn solutions as a function of temperature.

The expectation decreases according to a sigmoid curve and stabilizes for a temperature of $T_s \approx 10^{-4}$.

The variance/T² ratio is maximum at this temperature where most of the moves leading to the optimal solution take place.

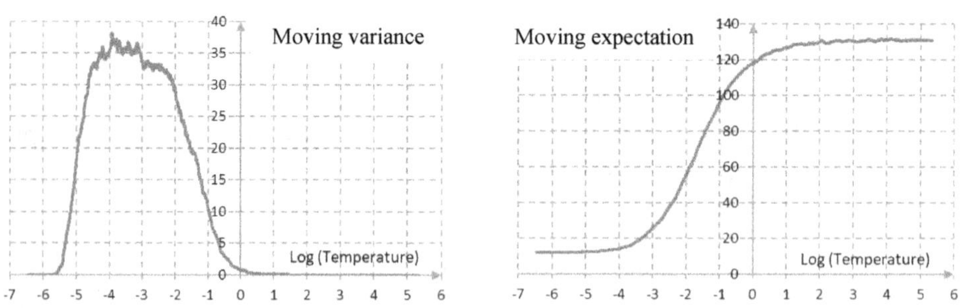

Figure 2-52: Expectation and variance of accepted solutions depending on the temperature.

2.8 Research with tabu

The Tabu Search method (Glover, 1986) is a local method that keeps a list of tried solutions in memory to avoid returning to them. This method is mainly applied to discrete problems.

2.8.1 Principle

The Tabu Search algorithm randomly explores solutions in the neighborhood of the current point x_k. The list of the last p iterations is stored in memory and constitutes the "taboo" list: $L = \{x_{k-p}, x_{k-p+1}, \cdots, x_{k-1}\}$.

Any solution in the neighborhood of x_k can be accepted if it is not part of the taboo list. The objective is to prohibit in the short term a return to these solutions, in order to escape a local minimum.

Figure 2-53 illustrates this mechanism. The first iterations (figure 2-53a) freely descend towards a local minimum. After passing the minimum, the taboo list prohibits a return to the minimum and forces uphill moves (figure 2-53b). If the taboo list is large enough, the algorithm moves away and switches to the global minimum (figure 2-53c).

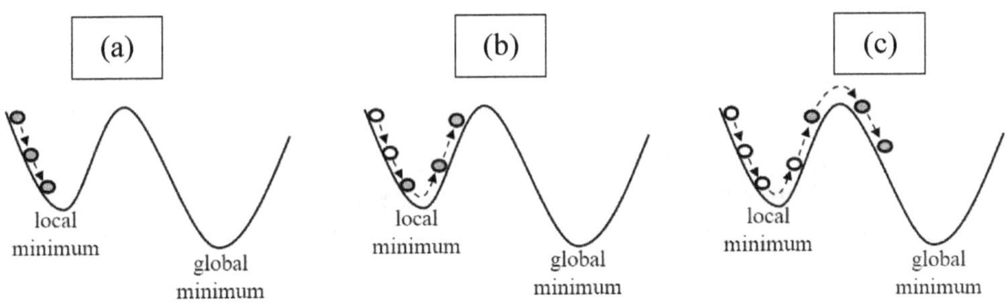

Figure 2-53: Escaping a local minimum by taboo list.

2.8.2 Taboo list and neighborhood

The purpose of the taboo list is to temporarily prohibit a return to the solutions already visited. The size of the list must be adapted to the problem:

- a too short list may not allow to escape a local minimum;
- a too long list may block access to some solutions.

Figure 2-54 shows blocking situations due to a too long list. Each point represents a solution. It is assumed that only adjacent point can be explored. The arrows show the taboo lists associated with four different paths. It can be seen that the last point of each path becomes disconnected from the minimum, as the taboo list "locks in" this point and prevents a return to the interesting area.

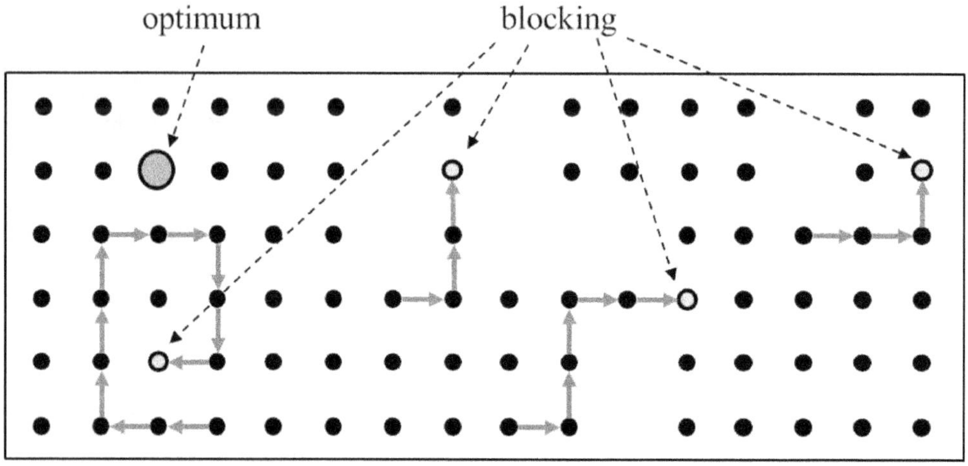

Figure 2-54: Blocking due to a too long taboo list.

The method efficiency also depends on the way of exploring the neighborhood.

The neighborhood is defined as the set of solutions reachable by elementary moves. Many discrete problems involve finding the optimal permutation of a finite set, such as the traveling salesman problem (example 2-12). For this type of problem, two elementary moves are usually considered.

Swapping consists in randomly drawing two elements from the list and exchanging their positions. For an ordered list of n elements, this move defines a neighborhood containing $n(n-1)/2$ possibilities.

Figure 2-55 shows an ordered list of 4 items and the 6 different possibilities after swapping (greyed out solutions are redundant).

1 2 3 4	→	2 1 3 4	2 1 3 4	3 2 1 4	4 2 3 1	→	Neighborhood of size
		3 2 1 4	1 3 2 4	1 3 2 4	1 4 3 2		$\frac{n(n-1)}{2}$
		4 2 3 1	1 4 3 2	1 2 4 3	1 2 4 3		

Figure 2-55: Swapping move.

Insertion consists in randomly drawing an element from the list and randomly inserting it in another place. For an ordered list of n elements, this move defines a neighborhood containing $n(n-2)+1$ possibilities.

Figure 2-56 shows an ordered list of 4 elements and the 9 different possibilities after insertion (greyed out solutions are redundant).

1 2 3 4 ⟶	2 1 3 4	2 1 3 4	3 1 2 4	4 1 2 3	⟶	Neighborhood of size $n(n-2)+1$
	2 3 1 4	1 3 2 4	1 3 2 4	1 4 2 3		
	2 3 4 1	1 3 4 2	1 2 4 3	1 2 4 3		

Figure 2-56: Insertion move.

The stopping criterion can be a maximum number of moves, a number of moves without improvement or a threshold value of the cost.

2.8.3 Quadratic assignment

The Tabu Search is applied to solve discrete problems such as the Quadratic Assignment Problem noted QAP. This problem consists in assigning n objects to n sites, knowing that:

- the objects i and j exchange a fixed flow f_{ij};
- the sites u and v are at a fixed distance d_{uv};
- the cost is the product of the flow by the distance $f_{ij}d_{uv}$.

The objects and sites are numbered from 1 to n. The number of the site to which object number i is assigned is denoted by p(i). Assigning the object i to the site $u = p(i)$ and the object j to the site $v = p(j)$ results in a cost: $f_{ij}d_{p(i)p(j)}$.

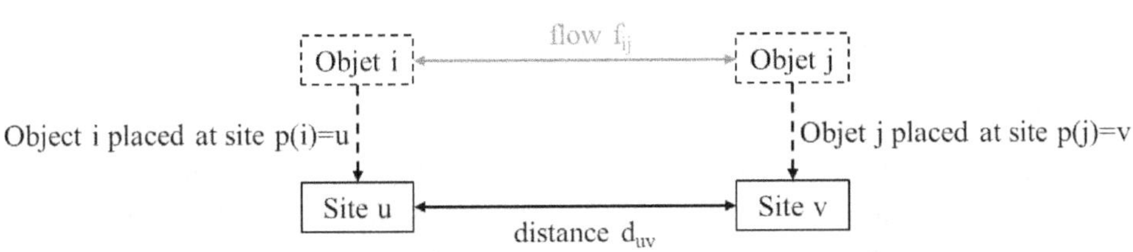

Figure 2-57: Quadratic assignment problem.

The assignment of the n objects to the n sites represents a permutation P of the set $(1,2,\cdots,n)$.

$$P=(p(1)\,,\,p(2)\,,\,\cdots,\,p(n)) \tag{2.85}$$

The cost of this permutation P is

$$C(P)=\sum_{i\neq j} f_{ij} d_{p(i)p(j)} \tag{2.86}$$

The solution of the problem is the permutation P of minimum cost.

The quadratic assignment problem has many practical applications:

- distribution of buildings or services.
  The flow is the number of people moving between buildings or services;
- establishment of production plants, delivery centers.
  The flow is the quantity of product exchanged between plants;
- assignment of gates at an airport.
  The flow is the number of people going from one gate to another;
- placement of electronic components.
  The flow is the number of connections between modules;
- storage of files in a database.
  The flow is the probability of consecutive accesses between 2 files;
- computer or typewriter keyboard.
  The flow is the frequency between 2 successive letters in a given language.

The systematic examination of the $n!$ possible permutations is not feasible when n is large. To apply the Tabu Search algorithm, we define the neighborhood of a solution as the swapping of 2 objects in the list.

The two lists P and Q below are neighboring solutions.

$$\begin{aligned} P &= (p(1),\cdots,p(u)\cdots,p(v),\cdots,p(n)) \\ Q &= (p(1),\cdots,p(v)\cdots,p(u),\cdots,p(n)) \end{aligned} \tag{2.87}$$

The size of this neighborhood is of the order of n^2.

The taboo list contains the previous solutions P_k and their banning duration t_k. The duration t_k is the number of iterations during which the solution P_k is taboo. This duration can be fixed or randomly drawn at each iteration.

The evaluation of a solution is the most time-consuming part of the calculation. To avoid completely recalculating the cost of a solution, we only evaluate the variation in cost between two neighboring solutions.

Let F be the fixed matrix of flows between the n objects and D(P) the matrix of distances between objects when they undergo the permutation P.
The cost (2.86) is noted in a compact way: $C(P) = F \times D(P)$.

Consider the solution P next to P' obtained by swapping the objects u and v.
The cost of this solution is: $C(P') = F \times D(P')$.

The matrices D(P) and D(P') differ only in the rows and columns of the objects u and v, as shown in figure 2-58c.

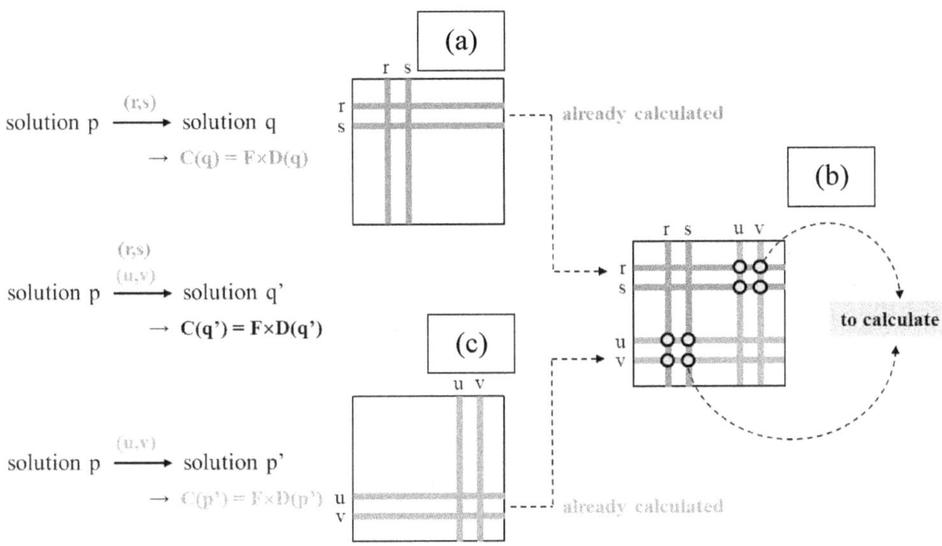

Figure 2-58: Cost variation between neighboring solutions.

The cost variation between P and P' concerns only the rows and columns of objects u and v and is calculated by

$$
\begin{aligned}
\Delta C = & \sum_{j=1\,\text{to}\,n, j\neq u} \left(f_{uj} d_{p(v)p(j)} - f_{uj} d_{p(u)p(j)} \right) && \rightarrow \text{edges starting from } p(v) \text{ instead of } p(u) \\
+ & \sum_{j=1\,\text{to}\,n, j\neq v} \left(f_{vj} d_{p(u)p(j)} - f_{vj} d_{p(v)p(j)} \right) && \rightarrow \text{edges starting from } p(u) \text{ instead of } p(v) \\
+ & \sum_{i=1\,\text{to}\,n, i\neq u} \left(f_{iu} d_{p(i)p(v)} - f_{iu} d_{p(i)p(u)} \right) && \rightarrow \text{edges arriving at } p(v) \text{ instead of } p(u) \\
+ & \sum_{i=1\,\text{to}\,n, i\neq v} \left(f_{iv} d_{p(i)p(u)} - f_{iv} d_{p(i)p(v)} \right) && \rightarrow \text{edges arriving at } p(v) \text{ instead of } p(u)
\end{aligned}
\tag{2.88}
$$

Suppose that we have already evaluated the cost of the solution Q obtained from P by swapping the objects r and s (figure 2-58a) and the cost of the solution P' obtained from P by swapping the objects u and v (figure 2-58c). The evaluation of the new solution Q' obtained from P by swapping the objects r and s and then the objects u and v only requires the recalculation of 8 elements of the matrix D (figure 2-58b).

By storing the solutions already examined with their costs, a very significant computational saving can be achieved and the algorithm can explore more solutions within a given computation time.

The example 2-13 illustrates the solution of a quadratic assignment problem by Tabu Search. This example is presented in reference [R7].

Example 2-13: Quadratic assignment by Tabu Search (from [R7])

Consider the quadratic assignment problem called NUG5 in test problem bases. This problem is of dimension 5 and it is defined by its flow matrix F and its distance matrix D.

$$F = \begin{array}{c} \\ 1 \\ 2 \\ 3 \\ 4 \\ 5 \end{array} \!\!\begin{array}{c} \begin{array}{ccccc} 1 & 2 & 3 & 4 & 5 \end{array} \\ \begin{pmatrix} 0 & 5 & 2 & 4 & 1 \\ 5 & 0 & 3 & 0 & 2 \\ 2 & 3 & 0 & 0 & 0 \\ 4 & 0 & 0 & 0 & 5 \\ 1 & 2 & 0 & 5 & 0 \end{pmatrix} \end{array} \qquad D = \begin{array}{c} \\ 1 \\ 2 \\ 3 \\ 4 \\ 5 \end{array} \!\!\begin{array}{c} \begin{array}{ccccc} 1 & 2 & 3 & 4 & 5 \end{array} \\ \begin{pmatrix} 0 & 1 & 1 & 2 & 3 \\ 1 & 0 & 2 & 1 & 2 \\ 1 & 2 & 0 & 1 & 2 \\ 2 & 1 & 1 & 0 & 1 \\ 3 & 2 & 2 & 1 & 0 \end{pmatrix} \end{array}$$

We are looking for the permutation p of the objects $(1,2,3,4,5)$ minimizing the total cost $C(p) = F \times D(p)$. The total number of permutations is: $5! = 120$.

By examining all possible permutations, we find 2 optimal solutions:

- the permutation $P = (2,4,5,1,3)$ has cost $C = 50$;

- the permutation $Q = (3,4,5,1,2)$ has a cost $C = 50$ as well.

The neighborhood of a solution is the set of solutions accessible by swapping the positions of 2 elements. This neighborhood of size 10 is completely explored at each iteration and the best solution is retained, if it is not in the taboo list.

The taboo list contains the last accepted solutions. These solutions are stored for 3 iterations and then deleted. The list therefore remains of length 3 from iteration 3 onwards.

Initialization

The Tabu Search is initialized with the solution $p_0 = (5,4,3,2,1)$ of cost $C_0 = 64$. The tabu list is initially empty.

The exploration of the neighborhood of p_0 gives the following results.

Move

Solution		1-2	1-3	1-4	1-5	2-3	2-4	2-5	3-4	3-5	4-5
(5,4,3,2,1)		(4,5,3,2,1)	(3,4,5,2,1)	(2,4,3,5,1)	(1,4,3,2,5)	(5,3,4,2,1)	(5,2,3,4,1)	(5,1,3,2,4)	(5,4,2,3,1)	(5,4,1,2,3)	(5,4,3,1,2)
64	Cost	60	60	80	68	66	78	80	64	78	66
	Variation	-4	-4	16	4	2	14	16	0	14	2

The best solution in the neighborhood is: $p_1 = (4,5,3,2,1)$ of cost $C_1 = 60$.
The solution p_1 better than p_0 is accepted.
The solution p_0 enters the taboo list for the next 3 iterations.

Iteration 1

The taboo list contains one solution.

Taboo solutions	Banning duration
$p_0 = (5,4,3,2,1)$	$t_0 = 3$

The exploration of the neighborhood of p_1 gives the following results.

Move

Solution		1-2	1-3	1-4	1-5	2-3	2-4	2-5	3-4	3-5	4-5
(4,5,3,2,1)		(5,4,3,2,1)	(3,5,4,2,1)	(2,5,3,4,1)	(1,5,3,2,4)	(4,3,5,2,1)	(4,2,3,5,1)	(4,1,3,2,5)	(4,5,2,3,1)	(4,5,1,2,3)	(4,5,3,1,2)
60	Cost	64	70	82	72	52	72	72	60	74	62
	Variation	4	10	22	12	-8	12	12	0	14	2

The exchange (1-2) is not allowed, as it would return p_0 which is in the taboo list.
The best solution in the neighborhood is: $p_2 = (4,3,5,2,1)$ of cost $C_2 = 52$.
The solution p_2 better than p_0 is accepted.
The solution p_1 enters the taboo list for the next 3 iterations.

Iteration 2

The taboo list contains two solutions.

Taboo solutions	Banning duration
$p_0 = (5, 4, 3, 2, 1)$	$t_0 = 2$
$p_1 = (4, 5, 3, 2, 1)$	$t_1 = 3$

The exploration of the neighborhood of p_2 gives the following results.

Solution		Move 1-2	1-3	1-4	1-5	2-3	2-4	2-5	3-4	3-5	4-5
(4,3,5,2,1)		(3,4,5,2,1)	(5,3,4,2,1)	(2,3,5,4,1)	(1,3,5,2,4)	(4,5,3,2,1)	(4,2,5,3,1)	(4,1,5,2,3)	(4,3,2,5,1)	(4,3,1,2,5)	(4,3,5,1,2)
52	Cost	60	66	74	60	60	52	76	72	62	62
	Variation	8	14	22	8	8	0	24	20	10	10

The exchange (2-3) is not allowed, as it would return p_1 which is in the taboo list.
The best solution in the neighborhood is: $p_3 = (4,2,5,3,1)$ of cost $C_3 = 52$.
The solution p_3 with the same cost as p_2 is accepted.
The solution p_2 enters the taboo list for the next 3 iterations.

Iteration 3

The taboo list contains three solutions.

Taboo solutions	Banning duration
$p_0 = (5, 4, 3, 2, 1)$	$t_0 = 1$
$p_1 = (4, 5, 3, 2, 1)$	$t_1 = 2$
$p_2 = (4, 3, 5, 2, 1)$	$t_2 = 3$

The exploration of the neighborhood of p_3 gives the following results.

Solution		Move 1-2	1-3	1-4	1-5	2-3	2-4	2-5	3-4	3-5	4-5
(4,2,5,3,1)		(2,4,5,3,1)	(5,2,4,3,1)	(3,2,5,4,1)	(1,2,5,3,4)	(4,5,2,3,1)	(4,3,5,2,1)	(4,1,5,3,2)	(4,2,3,5,1)	(4,2,1,3,5)	(4,2,5,1,3)
52	Cost	60	66	74	60	60	52	76	72	62	62
	Variation	8	14	22	8	8	0	24	20	10	10

The exchange (2-4) is not allowed, as it would return p_2 which is in the taboo list.
The best solution in the neighborhood is: $p_4 = (2,4,5,3,1)$ of cost $C_4 = 60$.
The solution p_4, although not as good as p_3, is accepted.
The solution p_3 enters the taboo list for the next 3 iterations.
The p_0 solution is off the taboo list, as its banning period is over.

Iteration 4

The taboo list contains three solutions.

Taboo solutions	Banning duration
$p_1 = (4, 5, 3, 2, 1)$	$t_1 = 1$
$p_2 = (4, 3, 5, 2, 1)$	$t_2 = 2$
$p_3 = (4, 2, 5, 3, 1)$	$t_3 = 3$

The exploration of the neighborhood of p_4 gives the following results.

		Move									
Solution		1-2	1-3	1-4	1-5	2-3	2-4	2-5	3-4	3-5	4-5
(2,4,5,3,1)		(4,2,5,3,1)	(5,4,2,3,1)	(3,4,5,2,1)	(1,4,5,3,2)	(2,5,4,3,1)	(2,3,5,4,1)	(2,1,5,3,4)	(2,4,3,5,1)	(2,4,1,3,5)	(2,4,5,1,3)
60	Cost	52	64	60	72	70	74	72	80	70	50
	Variation	-8	4	0	12	10	14	12	20	10	-10

The exchange (1-2) is not allowed, as it would return p_3 which is in the taboo list.
The best solution in the neighborhood is: $p_5 = (2,4,5,1,3)$ of cost $C_5 = 50$.
The solution p_5 better than p_4 is accepted.
The solution p_4 enters the taboo list for the next 3 iterations.
The p_1 solution is off the taboo list, as its banning period is over.

Iteration 5

The taboo list contains three solutions.

Taboo solutions	Banning duration
$p_2 = (4, 3, 5, 2, 1)$	$t_2 = 1$
$p_3 = (4, 2, 5, 3, 1)$	$t_3 = 2$
$p_4 = (2, 4, 5, 3, 1)$	$t_4 = 3$

The exploration of the neighborhood of p_5 gives the following results.

		Move									
Solution		1-2	1-3	1-4	1-5	2-3	2-4	2-5	3-4	3-5	4-5
(2,4,5,1,3)		(4,2,5,1,3)	(5,4,2,1,3)	(1,4,5,2,3)	(3,4,5,1,2)	(2,5,4,1,3)	(2,1,5,4,3)	(2,3,5,1,4)	(2,4,1,5,3)	(2,4,3,1,5)	(2,4,5,3,1)
50	Cost	62	66	72	50	60	62	74	70	70	60
	Variation	12	16	22	0	10	12	24	20	20	10

The exchange (4-5) is not allowed, as it would return p_4 which is in the taboo list.
The best solution in the neighborhood is: $p_6 = (3,4,5,1,2)$ of cost $C_5 = 50$.
The solution p_6 with the same cost as p_5 is accepted.
The solution p_5 enters the taboo list for the next 3 iterations.
The p_2 solution is off the taboo list, as its banning period is over.

Solution

The following iterations do not produce better solutions.

The 2 optimal permutations were found in 5 iterations (among the 120 possible permutations).

$$\begin{cases} p_5 = P = (2,4,5,1,3) \\ p_6 = Q = (3,4,5,1,2) \end{cases} \rightarrow C = 50$$

The taboo list forced a degradation at iteration 3 ($\Delta C = +8$).

This resulted in a local minimum $p_3 = (4,2,5,3,1)$ of cost $C_3 = 52$, and then in the next iteration the optimal solution of cost $C = 50$.

Table 2-7 summarizes the iterations of the taboo search on this NUG5 problem.

Iteration	Solution		Move 1-2	1-3	1-4	1-5	2-3	2-4	2-5	3-4	3-5	4-5
	(5,4,3,2,1)		(4,5,3,2,1)	(3,4,5,2,1)	(2,4,3,5,1)	(1,4,3,2,5)	(5,3,4,2,1)	(5,2,3,4,1)	(5,1,3,2,4)	(5,4,2,3,1)	(5,4,1,2,3)	(5,4,3,1,2)
0	64	Cost	60	60	80	68	66	78	80	64	78	66
		Variation	-4	-4	16	4	2	14	16	0	14	2
	(4,5,3,2,1)		(5,4,3,2,1)	(3,5,4,2,1)	(2,5,3,4,1)	(1,5,3,2,4)	(4,3,5,2,1)	(4,2,3,5,1)	(4,1,3,2,5)	(4,5,2,3,1)	(4,5,1,2,3)	(4,5,3,1,2)
1	60	Cost	64	70	82	72	52	72	72	60	74	62
		Variation	4	10	22	12	-8	12	12	0	14	2
	(4,3,5,2,1)		(3,4,5,2,1)	(5,3,4,2,1)	(2,3,5,4,1)	(1,3,5,2,4)	(4,5,3,2,1)	(4,2,5,3,1)	(4,1,5,2,3)	(4,3,2,5,1)	(4,3,1,2,5)	(4,3,5,1,2)
2	52	Cost	60	66	74	60	60	52	76	72	62	62
		Variation	8	14	22	8	8	0	24	20	10	10
	(4,2,5,3,1)		(2,4,5,3,1)	(5,2,4,3,1)	(3,2,5,4,1)	(1,2,5,3,4)	(4,5,2,3,1)	(4,3,5,2,1)	(4,1,5,3,2)	(4,2,3,5,1)	(4,2,1,3,5)	(4,2,5,1,3)
3	52	Cost	60	66	74	60	60	52	76	72	62	62
		Variation	8	14	22	8	8	0	24	20	10	10
	(2,4,5,3,1)		(4,2,5,3,1)	(5,4,2,3,1)	(3,4,5,2,1)	(1,4,5,3,2)	(2,5,4,3,1)	(2,3,5,4,1)	(2,1,5,3,4)	(2,4,3,5,1)	(2,4,1,3,5)	(2,4,5,1,3)
4	60	Cost	52	64	60	72	70	74	72	80	70	50
		Variation	-8	4	0	12	10	14	12	20	10	-10
	(2,4,5,1,3)		(4,2,5,1,3)	(5,4,2,1,3)	(1,4,5,2,3)	(3,4,5,1,2)	(2,5,4,1,3)	(2,1,5,4,3)	(2,3,5,1,4)	(2,4,1,5,3)	(2,4,3,1,5)	(2,4,5,3,1)
5	50	Cost	62	66	72	50	60	62	74	70	70	60
		Variation	12	16	22	0	10	12	24	20	20	10

Table 2-7: Taboo search iterations.

2.9 Particle swarms

The particle swarm method (Eberhart and Kennedy, 1995) is inspired by the social behavior of animals (swarms of birds or bees, schools of fishes).

2.9.1 Principle

A swarm is a group of individuals (or particles) in motion, as illustrated in figure 2-59. The movement of each particle is influenced by:

- its current speed, which encourages it to explore the neighborhood ("adventurous" trend);
- its personal experience which leads him to return to his best past position ("conservative" trend);
- its social experience which leads him to follow a group of neighbors ("panurgic" trend).
  The notion of neighbor is not necessarily geographical.

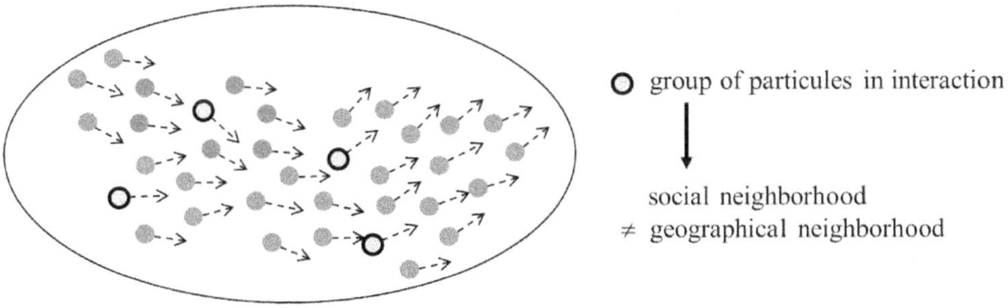

Figure 2-59: Movement of a particle swarm.

2.9.2 Particle movement

The principles of a swarm movement are transposed to the minimization of a function f of n real (or discrete) variables.

$$\min_{x \in \mathbb{R}^n} f(x) \tag{2.89}$$

The x variables are interpreted as the position coordinates of a particle. The value $f(x)$ measures the quality of the x position.

The swarm consists of N moving particles. Each particle has its own position x and moves with its own velocity v.

Let us consider at iteration k, a particle located in position x_k with velocity v_k.

The speed of the particle in the next iteration is influenced by 4 factors:

- inertia component:
  The inertia component reflects the trend to maintain the current velocity v_k;
- personal component:
  The best position encountered by the particle up to iteration k is kept in memory. This "personal" best position is noted p_k.
  The particle is attracted to its best position and its velocity will be deflected by a component v_p oriented towards p_k : $v_p = p_k - x_k$;
- group component:
  The particle is in communication with a group of neighbors V. The best position encountered by the particles in this group is denoted by g_k.
  The particle is attracted to the best position in its group and its velocity will be deflected by a component v_g oriented towards g_k : $v_g = g_k - x_k$;
- random component:
  The random component v_a is added to the three previous components.

The velocity of the particle at iteration k+1 is the sum of the 4 components.

$$v_{k+1} = \omega v_k + c_1 r_p v_p + c_2 r_g v_g + c_3 v_a \quad \text{with} \quad \begin{cases} v_p = p_k - x_k \\ v_g = g_k - x_k \end{cases} \tag{2.90}$$

The coefficients ω, c_1, c_2, c_3 are fixed weights between 0 and 1.
The coefficient ω on the previous velocity v_k is called the coefficient of inertia.
The coefficients r_p, r_g are random weights drawn between 0 and 1.

The position of the particle at iteration k+1 is the sum of its previous position and its updated velocity.

$$x_{k+1} = x_k + v_{k+1} \tag{2.91}$$

Figure 2-60 shows the move of a particle over 7 iterations. The best position encountered until iteration k is x_3 : $p_k = x_3$. Figure 2-61 shows the velocity components and the update of the velocity and position.

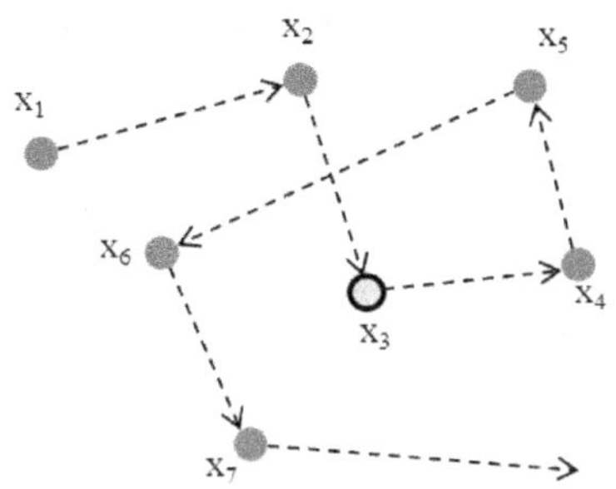

Time	t_1	t_2	t_3	t_4	...	t_k
Position	x_1	x_2	x_3	x_4	...	x_k

$p_k = x_3 =$ best known position at t_k

Figure 2-60: Best personal position.

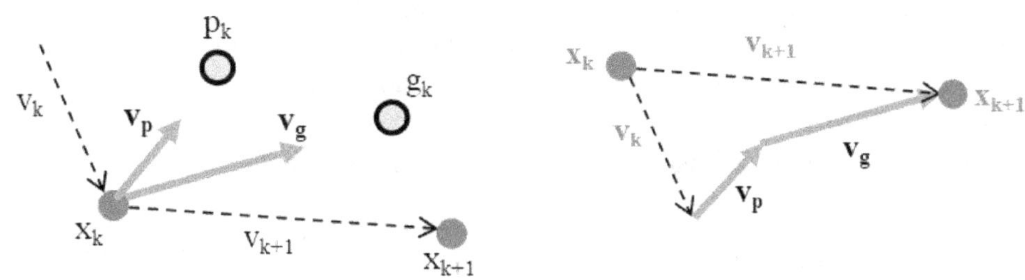

Figure 2-61: Personal and group component of velocity.

2.9.3 Neighborhood

The neighborhood of a particle can be defined in different ways as illustrated in figure 2-62.

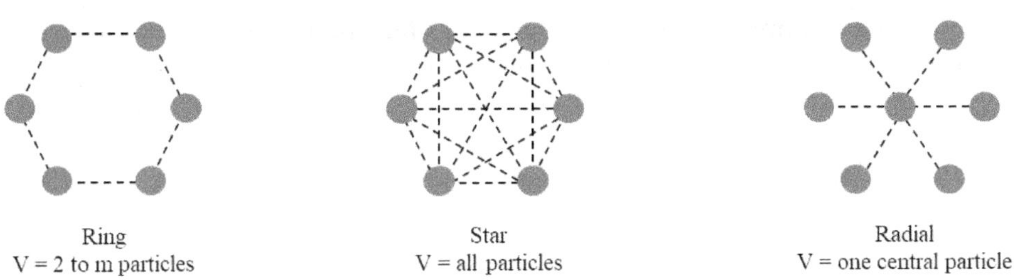

Figure 2-62: Neighborhood topologies.

The **ring** neighborhood consists in ordering the particles in a list: each particle communicates with a fixed number of preceding and following particles. The **star** neighborhood consists in making all the particles communicate with each other. The **radial** neighborhood consists in making each particle communicate with a single central particle.

These different neighborhood topologies each have their own interest and should be considered according to the optimization problem to be addressed.

2.9.4 Algorithm

The algorithm consists of simulating the movement of several particles according to the rules (2.90) and (2.91). The settings of the algorithm are the number of particles, the total number of iterations (or the maximum number of iterations without improvement) and the neighborhood topology. The standard setting is to use a few tens of particles $(N = 10 \text{ to } 100)$ and to consider a ring neighborhood $(m = 2 \text{ to } 10)$. These settings need to be adjusted experimentally.

To favor local exploration, the velocity of the particles can be bounded and a line search can be carried out for each particle along its velocity direction. This local search significantly accelerates the detection of good solutions.

The algorithm applies to both continuous and discrete problems and it has the advantage of being parallelizable, as the movements of the N particles are computed independently at each iteration by the formulas (2.90) and (2.91). The result is a set of N local minima which correspond to the best personal positions of the N particles. These local minima may be distinct or not depending on the neighborhood topology used. If the number of particles and the number of iterations are sufficient, it is possible to locate the global minimum, even for difficult problems.

The following example illustrates the behavior of the particle swarm algorithm on the Griewank function and on the Rosenbrock function.

Example 2-14: Particle swarm minimization

Let us take the function of example 2-2: $f(x) = \frac{x^2}{4000} - \cos x + 1$.

This function is plotted in figure 2-3. The global minimum is in $x = 0$.

The algorithm is run with 100 particles, 1000 iterations and 30 line search trials per particle. Figure 2-63 shows the progress of the algorithm with a zoom on the last few iterations.

The solution $x^* \approx 2 \cdot 10^{-7}$ is obtained after about 650 iterations.

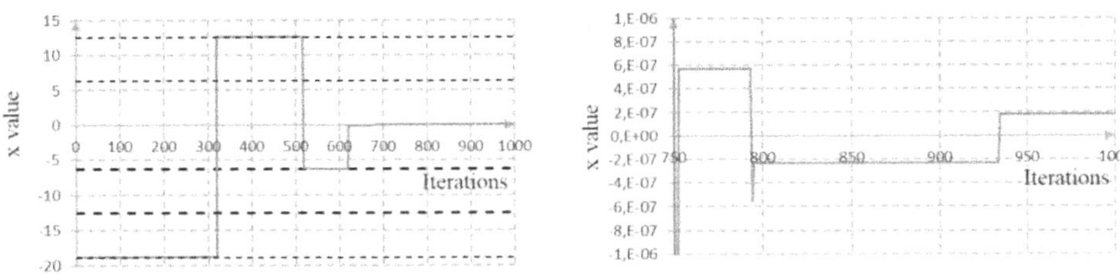

Figure 2-63: Progress of the algorithm on the Griewank function.

The algorithm is applied with the same settings to the Rosenbrock function: $f(x_1, x_2) = 100\left(x_2 - x_1^2\right)^2 + \left(1 - x_1\right)^2$ whose minimum is in (1; 1).

Figure 2-64 shows the progress of the algorithm.

The solution $\begin{cases} x_1 \approx 1{,}0000025 \\ x_2 \approx 1{,}0000048 \end{cases}$ is obtained after about 50 iterations.

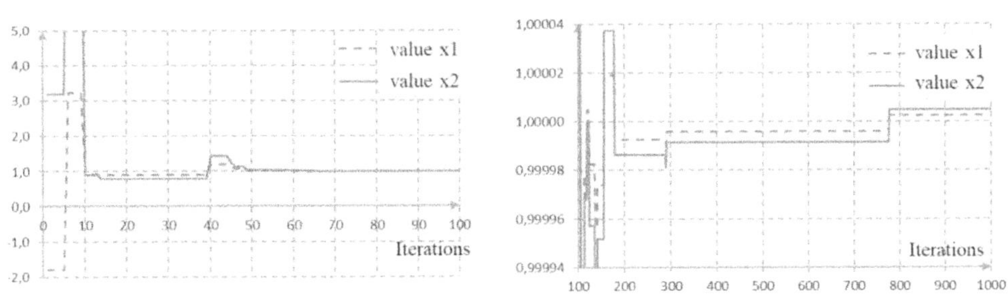

Figure 2-64: Progress of the algorithm on the Rosenbrock function.

2.10 Ant colonies

The ant colony method (Dorigo et al, 1990) is inspired by the collective foraging behavior of ants.

2.10.1 Principle

Ants deposit volatile substances called pheromones on their path. The more pheromones a path is concentrated in, the more attractive it is. This mode of indirect communication by modifying the environment is called stigmergy.

As the shortest path between the anthill and the food is travelled more quickly, it receives more pheromones over time and becomes the most used path as shown in figure 2-65.

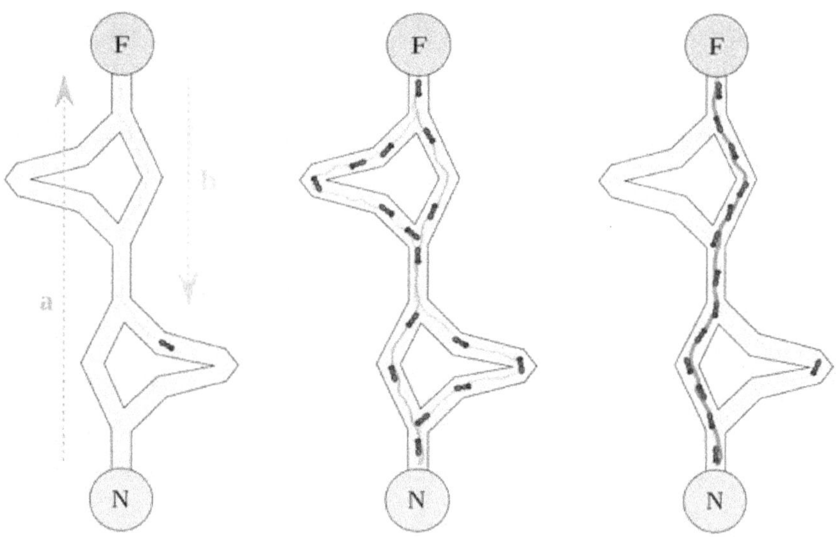

Figure 2-65: Pheromone concentration on the shortest path.

2.10.2 Ant movement

Each ant moves randomly and independently of the other ants. The path of an ant is broken down into a succession of segments. At each branch the ant can choose between several segments. The choice of direction depends on the local pheromone concentration.

The intensity τ_i of segment number i represents its pheromone concentration. The probability p_i of choosing segment i is an increasing function of its intensity.

$$p_i = p(\tau_i)$$
$$\tau_1 < \tau_2 < \tau_3 \;\Rightarrow\; p_1 < p_2 < p_3$$

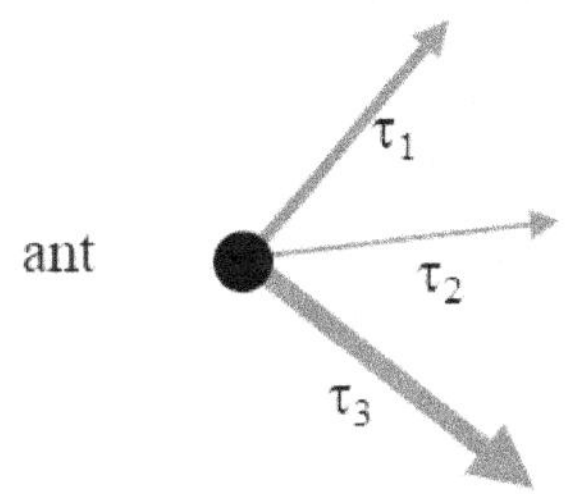

If only one path is marked, all the ants will follow it and the other solutions will not be explored. The evaporation of pheromones over time counteracts the deposition by the ants. This mechanism weakens the initial path and increases the probability of exploring other solutions.

2.10.3 Problem of the travelling salesman

The ant colony method is naturally applied to the Travelling Salesman Problem. This problem consists in finding the path of minimum length passing once and only once through n cities. The set of cities to be visited forms a graph, whose edges are the paths between the cities. The distance between cities i and j is denoted by d_{ij}. If two cities are not connected, an arbitrarily large value is assigned to their distance.

Individual path of an ant

Each ant travels through the graph passing through the n cities to be visited.

When an ant is at city i, it has a set J of unvisited cities left. The probability p_{ij} of choosing city $j \in J$ depends on the distances d_{ij} and the concentrations τ_{ij}.

$$p_{ij} = (\tau_{ij})^{\alpha} (\eta_{ij})^{\beta} \tag{2.92}$$

The **visibility** η_{ij} between cities i and j is a decreasing function of the distance (e.g. $\eta_{ij} = 1/d_{ij}$) so as to favor nearby cities.

The **intensity** τ_{ij} of edge i-j represents the pheromone concentration at the time when the ant has to choose the next edge.

The parameters α and β modulate the influence of visibility and intensity. The probability (2.92) is normalized to 1 over the set J of the remaining cities. At the end of its tour of the n cities, the ant has covered a total distance D. It deposits on the selected i-j edges the quantity of pheromones $\Delta\tau_{ij}$ defined by

$$\Delta\tau_{ij} = \frac{D_0}{D} \tag{2.93}$$

where D_0 is a fixed reference distance. The amount of pheromone deposited is greater the shorter the path.

Iteration of the algorithm

At each iteration, all m ants travel through the n cities. The ant number k increases the intensity τ_{ij} of each segment i-j on its path.

$$\tau_{ij} \rightarrow \tau_{ij} + \Delta\tau_{ij}(k) \tag{2.94}$$

When the m ants have passed, the intensities of all segments are reduced by evaporation in order to weaken the attractiveness of the "bad" solutions.

The evaporation parameter ρ is between 0 and 1.

$$\tau_{ij} \rightarrow (1-\rho)\tau_{ij} \tag{2.95}$$

Figure 2-66 depicts the stages of an iteration of the algorithm.

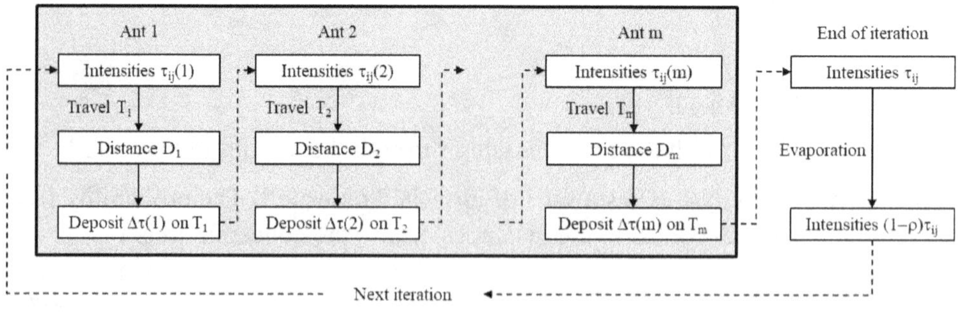

Figure 2-66: Iteration of the ant colony algorithm.

The progress towards the solution consists of 4 phases as shown in figure 2-67. A "scout" ant creates an initial path with little pheromone. Subsequent ants randomly explore neighboring paths. The shortest path becomes progressively stronger in pheromones and more attractive, while evaporation causes the other paths to be forgotten. If the number of iterations and the number of ants are sufficient, the exploration will reveal the optimal path.

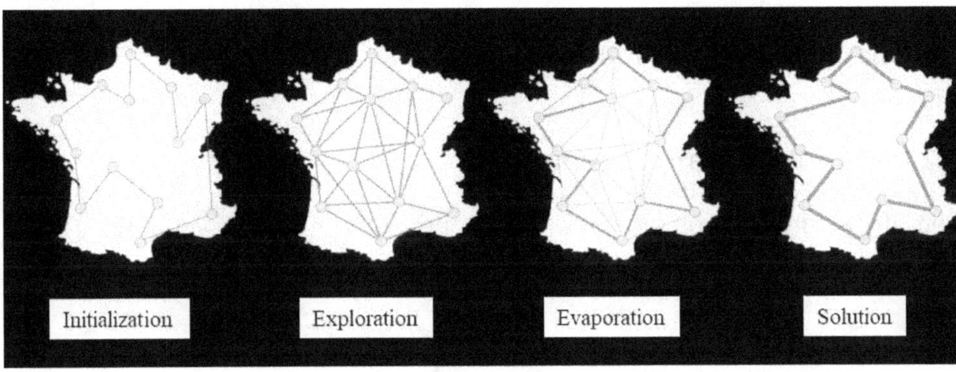

Figure 2-67: Phases of the ant colony algorithm.

The ant colony algorithm can be applied to discrete or continuous problems. It offers parallelization possibilities depending on the strategy chosen for the individual exploration of the ants. By choosing settings (number of ants, intensification and evaporation parameters) adapted to the problem to be solved, one can obtain very good results on difficult problems.

2.11 Evolutionary algorithms

Evolutionary algorithms are inspired by the principles of evolution and Darwin's law of natural selection (1859). Genetic algorithms (Holland, 1975) are a particular form of evolutionary algorithms.

2.11.1 Principle

A population of individuals evolves from generation to generation. Each individual is more or less "successful" depending on its characteristics. Natural evolution favors the characteristics that make an individual perform well.

The renewal of a generation (parents) takes place in 4 stages:
- selection of part of the parents for breeding;
- crossbreeding of selected parents to produce children;
- possible mutation (variation) of some of the children;
- selection between parents and children to form the next generation.

Evolution brings out better and better performing individuals. The speed and efficiency of the process depends on the size of the population, the representation of the individuals (coding in binary, integer or real variables) and the selection and variation operators applied to the population.

2.11.2 Evolutionary mechanisms

Consider a minimization problem with n continuous or discrete variables.

$$\min_{x} f(x) \tag{2.96}$$

The x variables represent the phenotype of an individual, the value f(x) represents its performance also called "fitness".

The population consists of p individuals, of the order of 100 to 1000 depending on the nature of the problem. The mechanisms of evolution include a selection or **intensification** process and a variation or **diversification** process.

Selection operators aim to retain the best solutions. They are based on the performance of individuals. Selection can be deterministic, proportional, by tournaments, etc. It can concern parents (selection for reproduction) and children (selection for replacement).

The purpose of variation operators is to explore new solutions. They include a random component to modify part of the variables. The variation can be done by crossover, mutation, ... The variation method depends on the coding of the individuals. In the case of genetic algorithms, the x variables (phenotype) are represented in binary (genotype) and the variation is performed on the genotype.

2.11.3 Algorithm

The elements of an evolutionary algorithm are the method of selection and variation processes, the form of coding of individuals, the size p of the population, the number q of children generated and the number N of generations.

There are many variants, depending on the selection and variation processes chosen. Application to a particular problem usually requires experimental work to find satisfactory settings.

Figure 2-68 depicts the stages of an evolutionary algorithm.

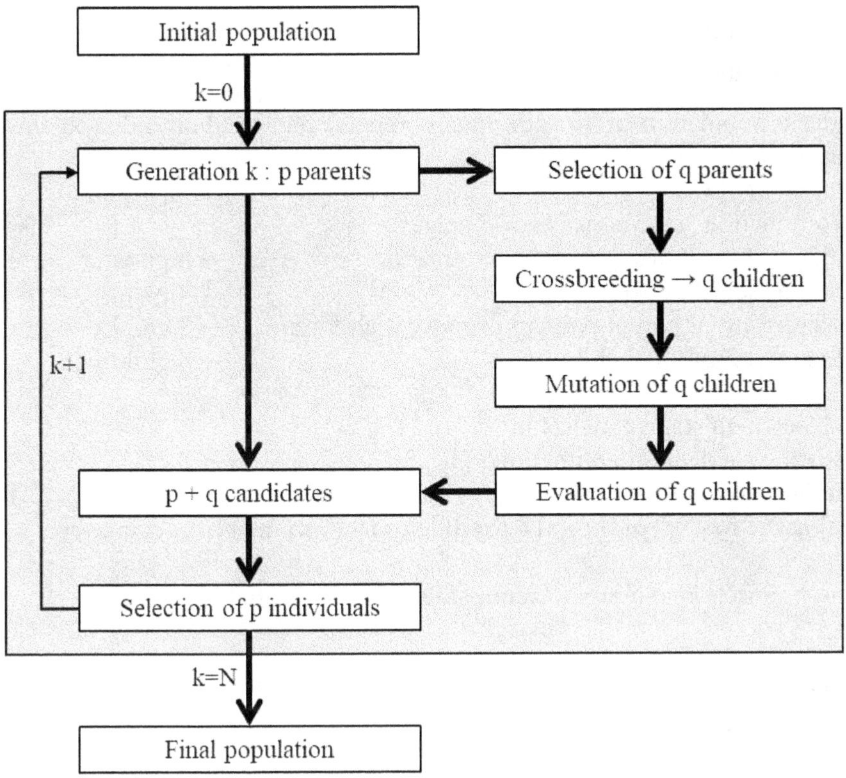

Figure 2-68: Iterations of an evolutionary algorithm.

The implementation of an evolutionary algorithm is illustrated on the very simple example 2-15 (function of a single variable). This example is presented in reference [R7].

Example 2-15: Minimization by evolutionary algorithm (from [R7])

Let us try to minimize the function: $f(x)=x^2$ where the variable x is an integer between −16 and +16.

Consider a population of 10 individuals, represented by their value x_i and their evaluation $f(x_i)$.

The reproduction stage consists of 4 stages:
- selection of the 8 best parents from the 10 → 8 parents;
- pairing of selected parents → 4 pairs;
- crossbreeding of parent pairs to produce 2 children → 8 children;
- mutation of 2 of the 8 children → 8 children.

The replacement stage consists in:
- evaluating the 8 children obtained $(y_j)_{j=1\text{ to }8}$ → $f(y_j)$;
- grouping them with the 10 parents $(x_i)_{i=1\text{ to }10}$ → $f(x_i)$;
- keeping the best 10 of these 18 candidates to form the next generation.

These reproduction and replacement stages are illustrated below.

Initial population: random generation of 10 individuals

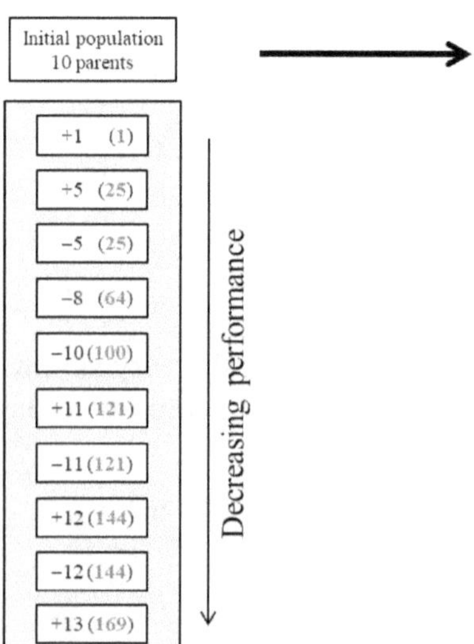

Initial population (generation k).

The initial population comprises 10 individuals (parents).

Each individual is represented by:
- its value $x_{i,\ i=1\text{ to }10}$ comprised between −16 and +16;
- its evaluation or performance $f(x_i)$.

The individuals are ranked by decreasing performance (increasing f value since f is to be minimized).

Parent selection: selection of the 8 best parents and random pairing

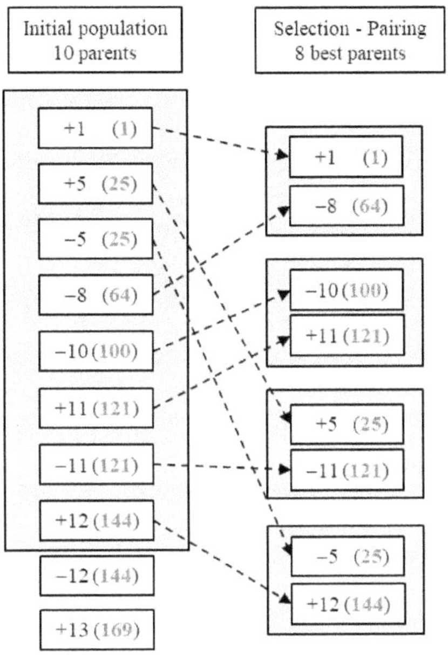

Selection for the reproduction

The 8 best parents are selected,
then randomly paired.
→ deterministic selection

Possible selection methods:
- proportional:
  A parent may be selected several times proportionally to its performance;
- tournaments:
  Several parents are drawn for the tournament. The parent winning the tournament is selected.

Crossing of parents: random generation of 8 children

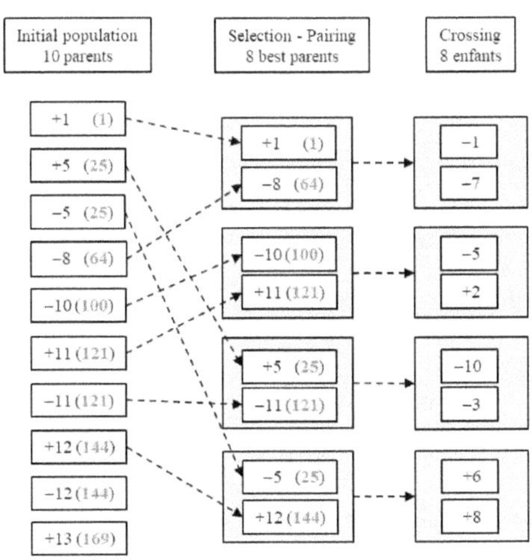

Crossing

2 parents produces 2 children per crossing.
For each child, the value of x is drawn randomly between those of its parents.

The crossing operator must be suited to the problem:
- integer or real variables;
- continuous function or not, local minima or not;
- proximity parents / children.

Mutation of children: random mutations

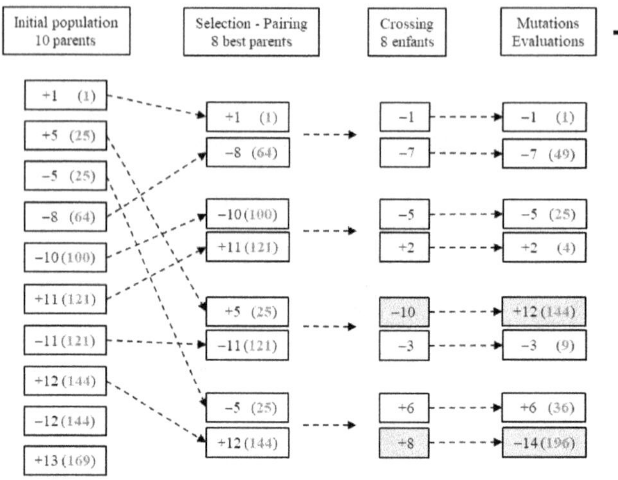

Mutation

2 children are randomly modified.
The 8 children are then evaluated.

The mutation operateur may be uniform or gaussian.

For a gaussian mutation,
the standard deviation σ evolves
wrt the rate τ of favorable mutations.

Rule of 1/5
- $\tau > 1/5$ → increase σ
- $\tau < 1/5$ → decrease σ

Candidates for the next generation: 10 parents + 8 children

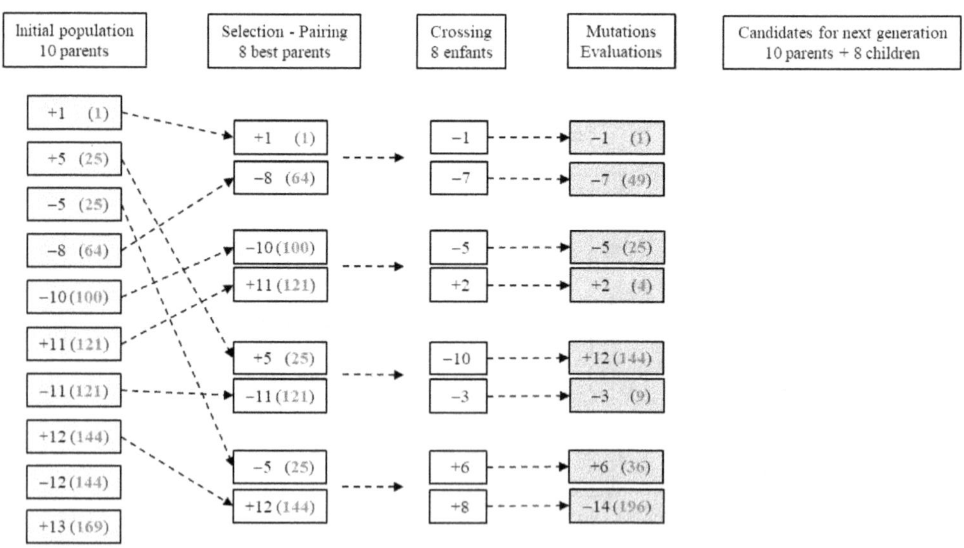

Selection of the 10 best candidates

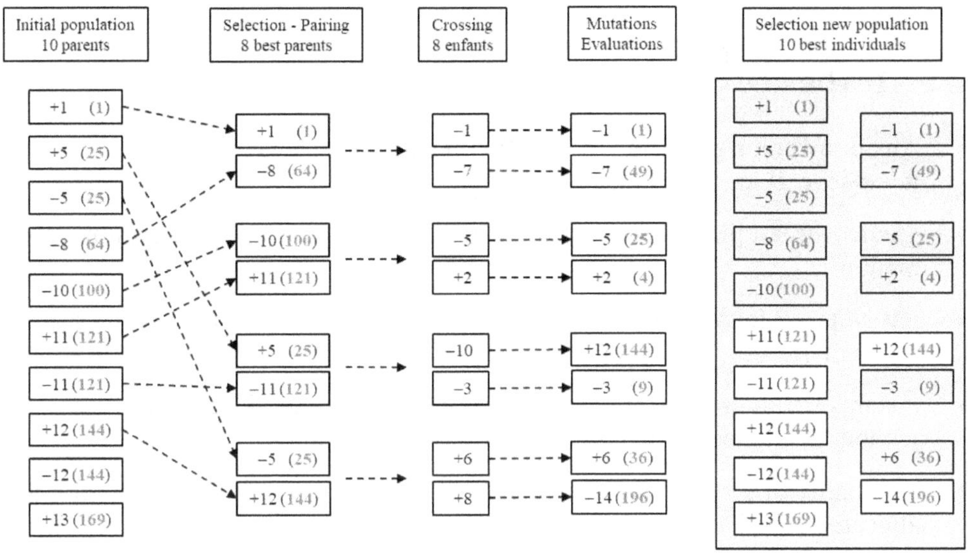

Summary of the iteration

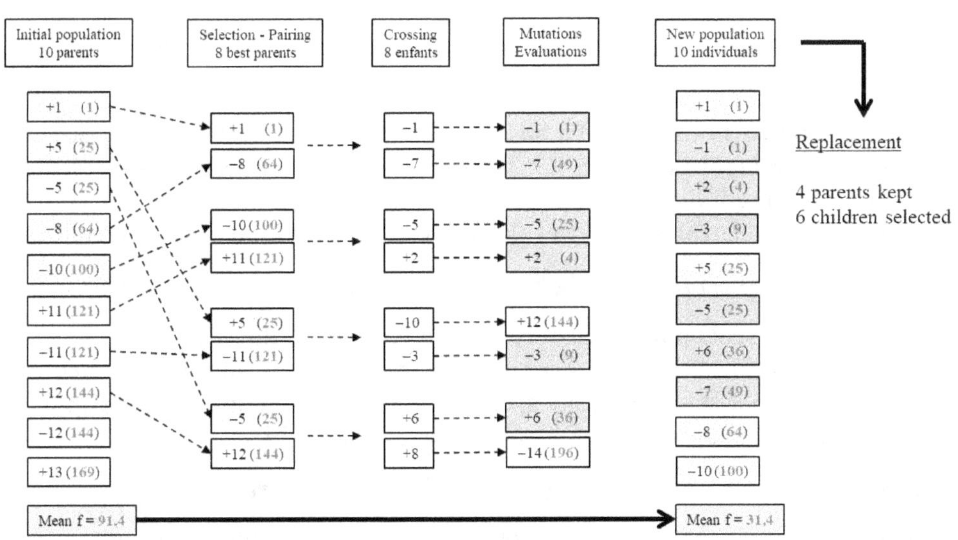

The first iteration significantly reduces the mean of the population (passing from 91.4 to 31.4), but it does not improve the best individual which remains at 1.
It usually takes a large number of iterations and appropriate adjustments to get an accurate solution.

2.12 Conclusion

2.12.1 The key points

- An optimization problem is said difficult when it has integer variables or when it has several local minima;
- gradient-free optimization methods can be local or global, deterministic or random;
- the Nelder-Mead method performs well on a wide variety of real variable problems;
- metaheuristics are capable of finding good solutions to difficult problems, but without guaranteeing convergence or global optimality;
- the quality of the solution depends on the chosen method, its settings and the allocated calculation time.

2.12.2 To go further

- *Métaheuristiques pour l'optimisation difficile (J. Dréo, A. Pétrowski, P. Siarry, E. Taillard, Eyrolles 2003)*

  This reference book presents in detail simulated annealing, tabu search, evolutionary algorithms and ant colonies. Each metaheuristic is presented with its settings, the different possible variants and numerous illustrative examples. The last part proposes the analysis of three industrial case studies.

- *Stochastic optimization (J.J. Schneider, S. Kirkpatrick, Springer 2006)*

  This book presents numerous heuristics (a term the author uses in preference to "metaheuristics") and analyses more particularly the simulated annealing method of which the authors are specialists. The different methods are applied and compared in detail on the travelling salesman problem.

- *Ant colony optimization (M. Rodrigo, T. Stützle, MIT Press 2004)*

  This book is devoted to ant colony algorithms, of which the authors are specialists. It presents the general principles, the application to the travelling salesman problem and other combinatorial problems.

- *Global Optimization by Particle Swarm Method: a Fortran Program (S.K. Mishra, MPRA Paper No. 874, 2006)*

  An article describing the swarm method and its settings. A very powerful Fortran program is provided in the appendix with many difficult test cases.

- *A Simplex Method for Function Minimization (J. Nelder, R. Mead, Computer journal vol. 7 n° 4, 1965)*

  The seminal paper on the Nelder-Mead method.

- *Lipschitzian optimization without the Lipschitz constant (D.R. Jones, C.D. Perttunen, B.E. Stuckman, Journal of Optimization Theory and applications 79, 1993)*

  The seminal article on the DIRECT global optimization method.

3. Unconstrained optimization

This chapter is devoted to continuous unconstrained optimization methods. The function of several real variables is assumed to be differentiable, which allows the use of the gradient and possibly the Hessian. The algorithms presented here are deterministic, which means that their behavior is strictly identical when applied twice in a row to the same problem (as opposed to the metaheuristics presented in chapter 2, that involve a random part).

Section 1 introduces Newton's method first for solving equations and then for minimization. This very powerful method is the core of most optimization algorithms. However, it has two major drawbacks (computation of the Hessian and uncertain robustness), which limit its direct application.

Section 2 presents quasi-Newton methods whose idea is to apply Newton' method without calculating the Hessian. Their principle is to use the sequence of points computed during the iterations to progressively build an approximation of the Hessian.

Section 3 presents line search descent algorithms. Their principle is to search for a new point reducing the function in a given direction. The direction of descent can be constructed in different ways, using the gradient and the quasi-Newton approximation of the Hessian.

Section 4 presents trust region descent algorithms. Their principle is to minimize a quadratic model of the function within a region of given radius. The quadratic model is constructed using the gradient and quasi-Newton approximation of the Hessian and the radius of the region is adapted to obtain a new point reducing the function.

Section 5 presents proximal methods, which generalize descent methods in the case of convex non-differentiable functions. These methods for non-smooth optimization are especially used for image processing and statistical learning.

Section 6 establishes some convergence properties of descent methods. These properties concern in particular the speed of convergence and the achievable accuracy of the solution.

3.1 Newton's method

Newton's method is used to numerically solve a system of nonlinear equations. It is the basis of most optimization algorithms. This section presents Newton's method for solving equations, homotopy methods for overcoming initialization difficulties and application to minimization problems.

3.1.1 System of equations

Consider a system of n nonlinear equations with n unknowns.

$$g(x)=0 \quad \text{with} \quad g: x \in \mathbb{R}^n \mapsto g(x) \in \mathbb{R}^n \tag{3.1}$$

The function g is assumed to be differentiable. Let us write its Taylor expansion to order 1 in a point x_0.

$$g(x)=g(x_0)+\nabla g(x_0)^T(x-x_0)+o\left(\|x-x_0\|\right) \tag{3.2}$$

Approximating in the vicinity of x_0 the function g by the linear function $\hat{g}_0$

$$\hat{g}_0(x) \underset{\text{def}}{=} g(x_0)+G_0(x-x_0) \quad \text{with} \quad G_0=\nabla g(x_0)^T \tag{3.3}$$

and solving the linear system

$$\hat{g}_0(x)=0 \tag{3.4}$$

we obtain the point x_1

$$x_1=x_0-G_0^{-1}g(x_0) \tag{3.5}$$

If the linear function $\hat{g}_0$ is close to g, one can expect that the point x_1 is close to the solution of the system (3.1), which is checked by calculating $g(x_1)$.

Newton's method consists of iterating the process (3.3–3.5) until a satisfactory solution is obtained. Newton's iterations are defined by

$$x_{k+1}=x_k-G_k^{-1}g(x_k) \quad \text{with} \quad G_k=\nabla g(x_k)^T \tag{3.6}$$

Figure 3-1 shows the general principle of Newton's method.

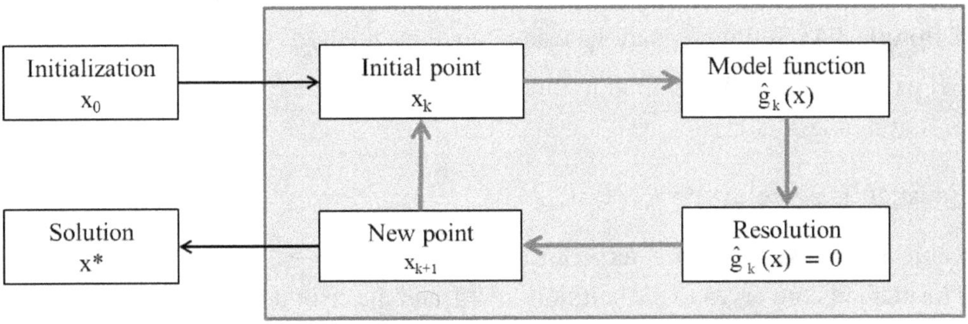

Figure 3-1: Principle of Newton's method.

In dimension 1, the matrix G is the derivative of g and equation (3.3) is the equation of the tangent in x_0 to the curve $y = g(x)$. Newton's method is also called the tangent method and the Newton iteration in dimension 1 is given by

$$x_{k+1} = x_k - \frac{g(x_k)}{g'(x_k)} \tag{3.7}$$

Figure 3-2 shows the iterations associated with successive tangents.

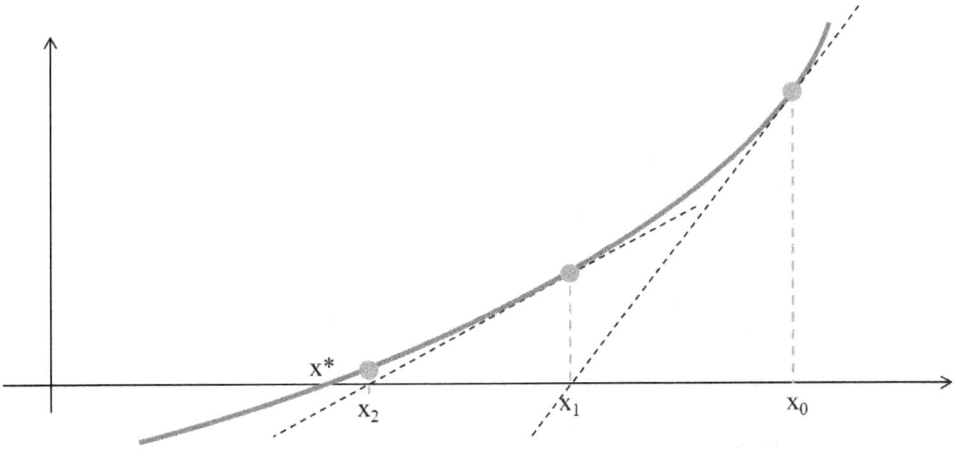

Figure 3-2: Tangent method.

Examples 3.1 and 3.2 show the possible behaviors of Newton's method when solving equations with one unknown.

Example 3-1: Solving equations using Newton's method

Let us apply Newton's method to three different equations.

Equation 1: Solve $g(x) = x^2 - 1 = 0$.

Table 3-1 shows Newton's iterations starting from $x_0 = 4$.
The method converges to the solution $x^* = 1$ and the error decreases quadratically.

Iteration	x_k	$g(x_k)$	$g'(x_k)$	Error
0	4.00000000	1.5E+01	8.0000	3.0E+00
1	2.12500000	3.5E+00	4.2500	1.1E+00
2	1.29779412	6.8E-01	2.5956	3.0E-01
3	1.03416618	6.9E-02	2.0683	3.4E-02
4	1.00056438	1.1E-03	2.0011	5.6E-04
5	1.00000016	3.2E-07	2.0000	1.6E-07
6	**1.00000000**	**2.5E-14**	**2.0000**	**1.3E-14**

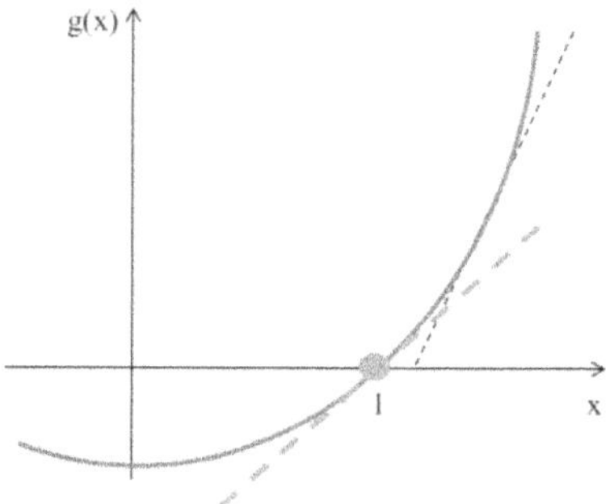

Table 3-1: Quadratic convergence of Newton's method.

Equation 2: Solve $g(x) = (x-1)^2 = 0$.

Table 3-2 shows Newton's iterations starting from $x_0 = 4$.
The method converges to the solution $x^* = 1$ and the error decreases linearly.

Iteration	x_k	$g(x_k)$	$g'(x_k)$	Error
0	4.00000000	9.0E+00	6.0000	3.0E+00
1	2.50000000	2.3E+00	3.0000	1.5E+00
2	1.75000000	5.6E-01	1.5000	7.5E-01
3	1.37500000	1.4E-01	0.7500	3.8E-01
4	1.18750000	3.5E-02	0.3750	1.9E-01
5	1.09375000	8.8E-03	0.1875	9.4E-02
6	1.04687500	2.2E-03	0.0938	4.7E-02
7	1.02343750	5.5E-04	0.0469	2.3E-02
8	1.01171875	1.4E-04	0.0234	1.2E-02
9	1.00585938	3.4E-05	0.0117	5.9E-03
10	1.00292969	8.6E-06	0.0059	2.9E-03
15	1.00009155	8.4E-09	0.0002	9.2E-05
20	**1.00000286**	**8.2E-12**	**0.0000**	**2.9E-06**

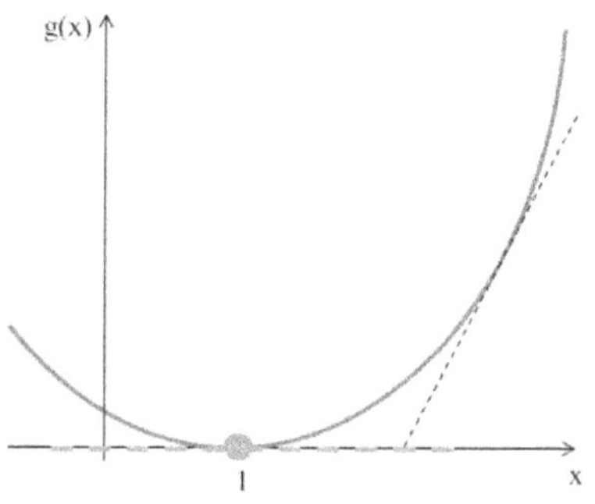

Table 3-2: Linear convergence of Newton's method.

Equation 3: Solve $g(x) = \text{Arctan}\, x = 0$.
Table 3-3 shows Newton's iterations starting from $x_0 = 1.3$ or $x_0 = 1.5$.
The method converges quadratically for $x_0 = 1.3$ and diverges for $x_0 = 1.5$.

Iteration	x_k	$g(x_k)$	$g'(x_k)$	Error
0	**1.300**	**0.915**	**0.372**	**1.3E+00**
1	-1.162	-0.860	0.426	-1.2E+00
2	0.859	0.710	0.575	8.6E-01
3	-0.374	-0.358	0.877	-3.7E-01
4	0.034	0.034	0.999	3.4E-02
5	0.000	0.000	1.000	-2.6E-05
6	**0.000**	**0.000**	**1.000**	**1.2E-14**

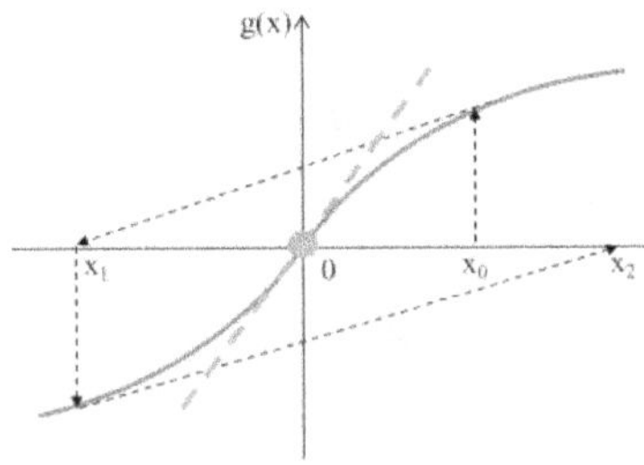

Iteration	x_k	$g(x_k)$	$g'(x_k)$	Error
0	**1.500**	**0.983**	**0.308**	**1.5E+00**
1	-1.694	-1.038	0.258	-1.7E+00
2	2.321	1.164	0.157	2.3E+00
3	-5.114	-1.378	0.037	-5.1E+00
4	32.296	1.540	0.001	3.2E+01
5	-1575.317	-1.570	0.000	-1.6E+03
6	**3894976.008**	**1.571**	**0.000**	**3.9E+06**

Table 3-3: Divergence of Newton's method depending on the initial point.

These examples in dimension 1 show that the convergence of Newton's method is not guaranteed, and that the speed of convergence is variable.
To explain this behavior, let us write Newton's iteration in terms of the deviation from the solution by posing $x_k = x^* + \delta_k$. Formula (3.7) becomes

$$\delta_{k+1} = \delta_k - \frac{g(x^* + \delta_k)}{g'(x^* + \delta_k)} \tag{3.8}$$

Let us expand the function to order 3 and its derivative to order 1 in the neighborhood of x*.

$$\delta_{k+1} = \delta_k - \frac{g(x^*) + g'(x^*)\delta_k + \frac{1}{2}g''(x^*)\delta_k^2 + \frac{1}{6}g'''(x^*)\delta_k^3}{g'(x^*) + g''(x^*)\delta_k} \tag{3.9}$$

After reduction and using $g(x^*) = 0$ (because x* is the solution), we get

$$\delta_{k+1} = \frac{\frac{1}{2}g''(x^*)\delta_k^2 - \frac{1}{6}g'''(x^*)\delta_k^3}{g'(x^*) + g''(x^*)\delta_k} \tag{3.10}$$

In the neighborhood of the solution x*, the deviations δ_k are small. The behavior of Newton's method depends on the values of $g'(x^*)$ and $g''(x^*)$.

- If $g'(x^*) \neq 0$ and $g''(x^*) \neq 0$, and for small δ_k, formula (3.10) reduces to

  $$\delta_{k+1} \approx \frac{1}{2}\frac{g''(x^*)}{g'(x^*)}\delta_k^2 \tag{3.11}$$

  The deviation is squared at each iteration and the convergence is **quadratic**. This behavior can be observed in table 3-1.

- If $g'(x^*) = 0$ and $g''(x^*) \neq 0$, and for small δ_k, formula (3.10) reduces to

  $$\delta_{k+1} \approx \frac{1}{2}\delta_k \tag{3.12}$$

  The deviation is divided by 2 at each iteration and the convergence is **linear**. This behavior can be observed in table 3-2.

- If $g'(x^*) \neq 0$ and $g''(x^*) = 0$, and for small δ_k, formula (3.10) reduces to

  $$\delta_{k+1} = -\frac{1}{6}\frac{g'''(x^*)}{g'(x^*)}\delta_k^3 \tag{3.13}$$

  The deviation is cubed at each iteration and the convergence is **of order 3**. This behavior can be observed in the last iterations of table 3-3 initialized with $x_0 = 1.3$.

These analyses are valid in the vicinity of the solution for $\delta_k \ll 1$. When the initial point is far from the solution, there is no guarantee of convergence (except for particular properties of the function) as shown in table 3-3 starting from $x_0 = 1.5$. In some cases, the radius of convergence of Newton's method is very small and makes the initialization problematic. Moreover, the solution obtained depends on the initialization and can sometimes be very far from it. The following example shows the chaotic behavior of Newton's method depending on the starting point, also called **initial guess**.

Example 3-2: Newton's method areas of attraction

Consider the equation in complex numbers: $z^3 = 1$.

The solutions are the 3 roots of unity:

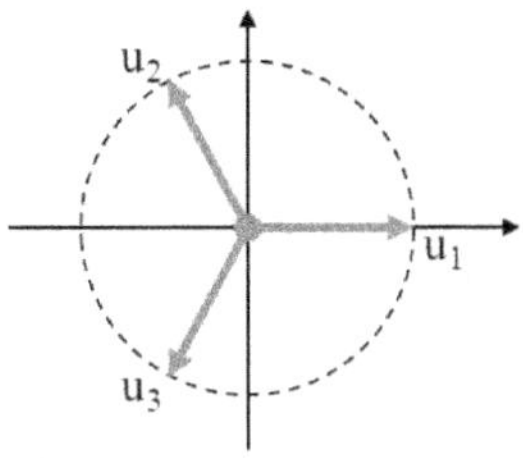

$$u_1 = 1 \ , \ u_2 = \frac{-1+i\sqrt{3}}{2} \ , \ u_3 = \frac{-1-i\sqrt{3}}{2}$$

Newton's iteration is: $z_{k+1} = z_k - \frac{z_k^3 - 1}{3z_k^2} = \frac{2z_k^3 + 1}{3z_k^2}$.

Depending on the initial point z_0, Newton's method converges to one of the three roots u_1, u_2 or u_3. Different starting point are chosen in the unit disk: $|z_0| \leq 1$.

Figure 3-3 is plotted in the complex plane. In each point z_0, the color code indicates to which of the 3 roots Newton's method has converged. There are areas of attraction associated with the 3 roots and "fractal" boundaries appears where the distinction is not clear and may lead to one of the 3 roots. A small variation of the initialization in these areas changes the solution.

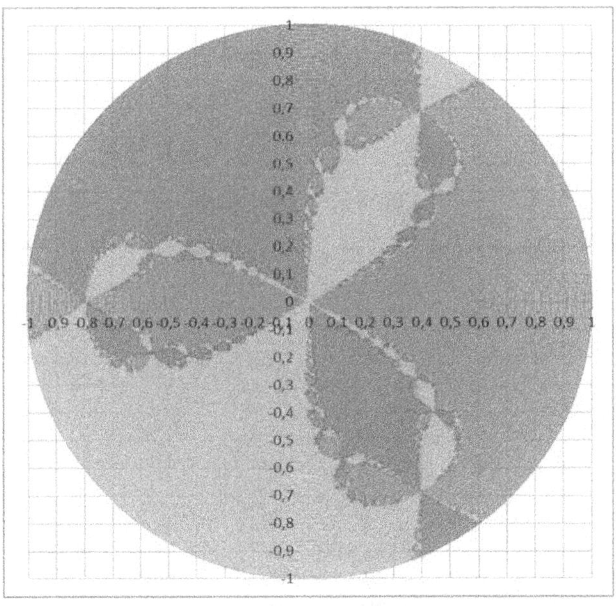

Figure 3-3: Newton's method areas of attraction.

3.1.2 Homotopy method

Consider a system of n nonlinear equations with n unknowns.

$$g_1(x) = 0 \tag{3.14}$$

Suppose it turns out to be very difficult to find a reliable starting point that allows Newton's method to converge. The principle of a homotopy (or continuation) method is to build a good initial guess by solving a sequence of parameterized problems.

For that purpose, we consider another system of n nonlinear equations called P_0.

$$g_0(x) = 0 \tag{3.15}$$

The function g_0 is chosen arbitrarily and does not necessarily have any "resemblance" to the function g_1 in (3.14). This function should be chosen so that the system (3.15) is easy to solve and it yields a solution denoted by x_0.

We then define the sequence of parameterized problems P_α with $0 \le \alpha \le 1$.

$$(P_\alpha) : g_\alpha(x) = 0 \quad \text{with} \quad g_\alpha(x) = \alpha g_1(x) + (1-\alpha) g_0(x) \tag{3.16}$$

- when $\alpha = 0$, we retrieve the "easy" problem P_0 (3.15);
- when $\alpha = 1$, we retrieve the "hard" problem P_1 (3.14).

The homotopy method depicted in figure 3-4 consists in varying α from 0 to 1 by solving each time the problem P_α and using its solution x_α to initialize the next problem. If the variation $\delta\alpha$ is small, the solution $x_{\alpha+\delta\alpha}$ of the problem $P_{\alpha+\delta\alpha}$ should be close to the solution x_α of the problem P_α and the initialization x_α could allow the convergence of Newton's method on the problem $P_{\alpha+\delta\alpha}$.

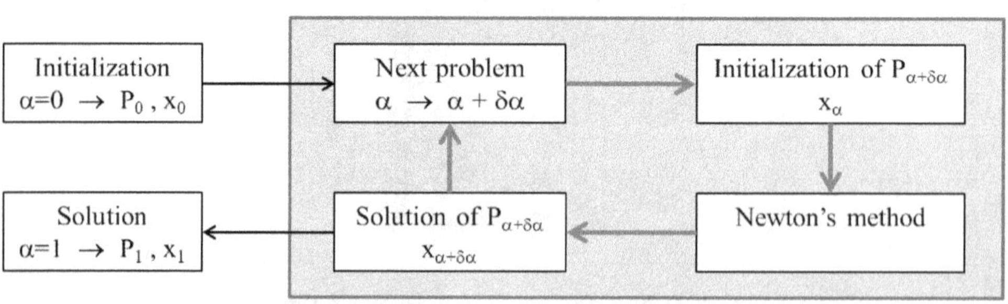

Figure 3-4: Homotopy method.

Example 3-3: Solving equations by homotopy method

Let us try to solve the equation: $g_1(x) = \text{Arctan}\, x = 0$.

The example 3-1 has shown that Newton's method converges from $x_0 = 1.3$ and diverges from $x_0 = 1.5$.

Let us apply a homotopy method to overcome this initialization issue.

For that purpose, we arbitrarily define as "easy" problem P_0 : $g_0(x) = x - 10 = 0$.

The family of parameterized problems P_α is defined by

$$g_\alpha(x) = (1-\alpha)(x-10) + \alpha \text{Arctan}(x) = 0$$

We solve the sequence of problems P_α from $\alpha = 0$ to $\alpha = 1$. The homotopy step is set first to $\delta\alpha = 0.1$ from $\alpha = 0$ to $\alpha = 0.8$, then to $\delta\alpha = 0.05$ from $\alpha = 0.8$ to $\alpha = 1$.

There is a total of 13 successive problems to solve. The problem P_0 is initialized with $x_0 = 100$. Each problem is then initialized with the previous solution.

Table 3-4 shows the iterations of Newton's method for the 13 successive problems. Each column corresponds to a problem associated with a particular value of α indicated in the first row. The solution of each problem is used to initialize the next one. The sequence of solutions x_α is plotted in figure 3-5.

This process makes it possible to obtain the correct solution $(x = 0)$ of the equation: $g_1(x) = \text{Arctan}\, x = 0$ without having to look for a suitable initialization.

α	0	0.1	0.2	0.3	0.4	0.5	0.6	0.7	0.8	0.85	0.9	0.95	1
x_0	100.0	10.0000	9.8367	9.6332	9.3724	9.0264	8.5457	7.8342	6.6815	4.5772	2.9505	1.4112	0.5433
x_1	10.0	9.8367	9.6331	9.3723	9.0263	8.5454	7.8330	6.6744	4.5021	2.7813	0.7989	0.1131	-0.1013
x_2	**10.0**	**9.8367**	9.6332	9.3724	9.0264	8.5457	7.8342	6.6815	4.5770	2.9471	1.2816	0.5052	0.0007
x_3			**9.6332**	**9.3724**	**9.0264**	**8.5457**	**7.8342**	**6.6815**	4.5772	2.9505	1.4052	0.5428	0.0000
x_4									**4.5772**	**2.9505**	1.4112	0.5433	**0.0000**
x_5											**1.4112**	**0.5433**	

Table 3-4: Newton's iterations at each homotopy step.

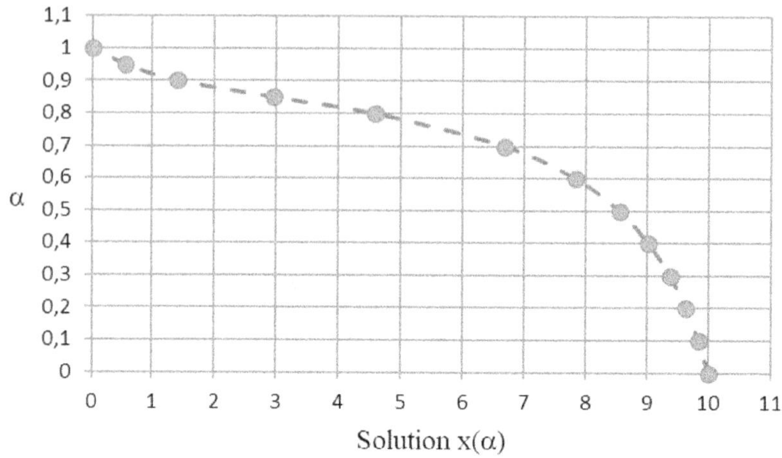

Figure 3-5: Homotopy solution sequence.

The set of solutions $\{x_\alpha, 0 \le \alpha \le 1\}$ defines a curve C in $\mathbb{R}^n$ called the **path of of zeros** of the problem family $(P_\alpha)_{0\le\alpha\le1}$. The homotopy method seeks to follow this path of zeros to arrive at the solution of the problem P_1.

The success of the method relies on the existence of a continuous path of zeros from problem P_0 to problem P_1. Whether or not this path exists depends on the choice of problem P_0 and it is not possible to ensure this in advance. Even when the problem P_0 "resembles" the problem P_1, there is no guarantee of continuity.

Figure 3-6 shows various situations that may be encountered in practice.

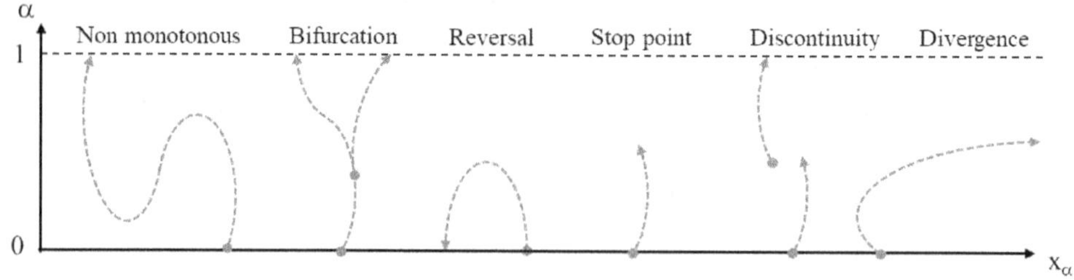

Figure 3-6: Possible forms of the path of zeros.

The example 3-4 shows that difficulties can arise even in very simple cases.

Example 3-4: Path of zeros with disjoint branches

Let us try to solve the equation: $g_1(x) = x^2 - 1 = 0$.

This problem noted P_1 has two solutions: $x_1 = \pm 1$.

Let us start with the problem P_0 : $g_0(x) = x + 2 = 0$ whose solution is $x_0 = -2$.

The family of parameterized problems P_α is defined by

$$g_\alpha(x) = (1-\alpha)(x+2) + \alpha(x^2 - 1) = \alpha x^2 + (1-\alpha)x + (2-3\alpha) = 0$$

The solution of the problem P_α is calculated analytically.

$$x_\alpha = \frac{-(1-\alpha) \pm \sqrt{(1-\alpha)^2 - 4\alpha(2-3\alpha)}}{2\alpha}$$

This solution only exists if

$$(1-\alpha)^2 - 4\alpha(2-3\alpha) \geq 0 \Rightarrow \alpha \leq \frac{5-2\sqrt{3}}{13} \quad \text{or} \quad \frac{5+2\sqrt{3}}{13} \leq \alpha$$

The path of zeros shown in figure 3-7 is therefore not continuous. It consists of two disjoint branches:

- the first branch covers the values $0 \leq \alpha \leq 0.11814$ and it exhibits a reversal for $\alpha = 0.11814$.
  It starts from the solution $x_0 = -2$ and diverges to $-\infty$ when α decreases to zero;
- the second branch covers the values $0.65109 \leq \alpha \leq 1$ and it exhibits a reversal for $\alpha = 0.65109$.
  Its endpoints go from the solution $x_1 = -1$ to the solution $x_1 = 1$.

There exists therefore no path going continuously from the solution of P_0 to that of P_1, although both problems are very simple and their respective solutions very close.

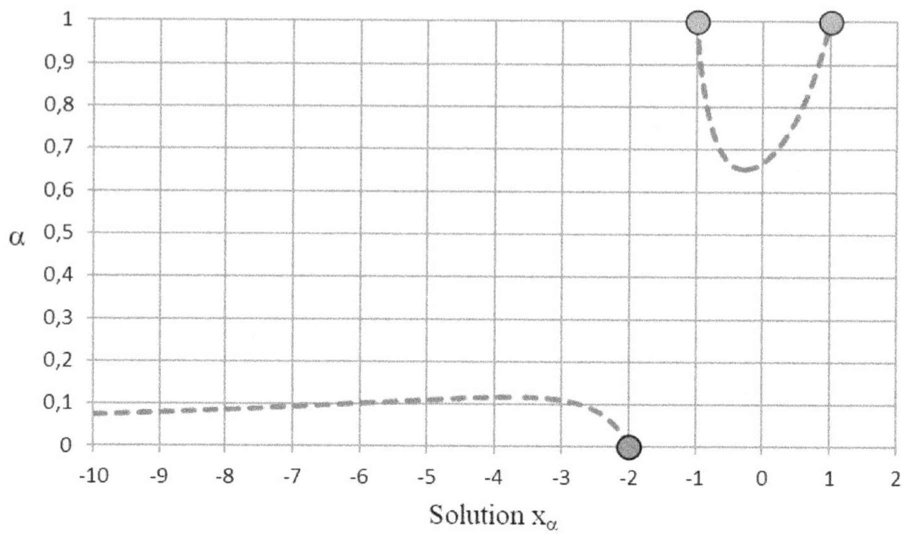

Figure 3-7: Path of zeros with two disjoint branches.

Even if a continuous path of zeros exists, it is not necessarily monotonous (as in figure 3-6). In this case, a discrete homotopy based on a monotonous variation of α will get stuck at the reversal. This problem can be circumvented by using the tangent vector to the path of zeros.

The path of zeros is the set of points $x_\alpha \in \mathbb{R}^n$ solutions of: $g_\alpha(x) = 0$.

Let us place ourselves in $\mathbb{R}^{n+1}$ and consider α as a variable.

The path of zeros is the curve C_g of $\mathbb{R}^{n+1}$ with equation: $g(\alpha, x) = 0$, where the function $g : \mathbb{R}^{n+1} \to \mathbb{R}^n$ is defined by: $g(\alpha, x) \underset{\text{def}}{=} g_\alpha(x)$.

The tangent vector to the curve C_g has the components

$$u = \begin{pmatrix} u_\alpha \\ u_x \end{pmatrix} \begin{matrix} \to \in \mathbb{R} \\ \to \in \mathbb{R}^n \end{matrix} \tag{3.17}$$

This vector is normal to the gradient of the function g.

$$\nabla g^T u = 0 \iff Gu = 0 \tag{3.18}$$

where $G \in \mathbb{R}^{n\times(n+1)}$ is the Jacobian matrix of g.

The tangent vector u therefore lies in the null space of G which is of dimension 1. If Z is a basis of the null space of G (GZ = 0), the tangent vector is expressed as $u = Zu_z$ with $u_z \in \mathbb{R}$, and it depends only on the real u_z (section 1.3.1).

Figure 3-8 shows the curve C_g and the components of the tangent vector u.

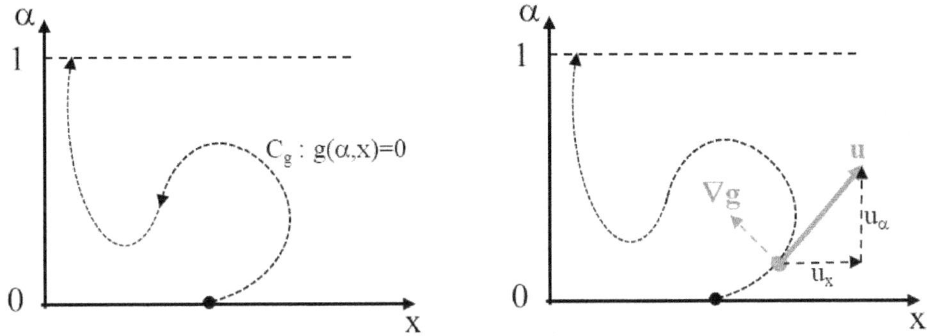

Figure 3-8: Vector tangent to the path of zeros.

The progression along the C_g curve is achieved by prediction-correction.

Prediction stage

A move p is made along the tangent vector from the point $(\alpha_0; x_0)$.

The direction (sign) is chosen with respect to the previous iteration: $u_{k+1}^T u_k > 0$.

The new point is no longer on the curve C_g.

$$\begin{pmatrix} \alpha_0 \\ x_0 \end{pmatrix} \rightarrow \begin{pmatrix} \alpha_p = \alpha_0 + pu_\alpha \\ x_p = x_0 + pu_x \end{pmatrix} \rightarrow g(\alpha_p, x_p) \neq 0 \tag{3.19}$$

Correction stage

A move $v \in \mathbb{R}^{n+1}$ is sought to return to the curve C_g and restore the relation: $g(\alpha, x) = 0$.

$$\begin{pmatrix} \alpha_p \\ x_p \end{pmatrix} \rightarrow \begin{pmatrix} \alpha_c = \alpha_p + v_\alpha \\ x_c = x_p + v_x \end{pmatrix} \rightarrow g(\alpha_c, x_c) = 0 \tag{3.20}$$

The vector $v \in \mathbb{R}^{n+1}$ must satisfy n equations (3.20). This nonlinear system is underdetermined. One can fix one of the components of v and solve the remaining system by Newton's method, or look for the minimum norm restoration (section 1.3.2). Figure 3-9 illustrates the prediction-correction method.

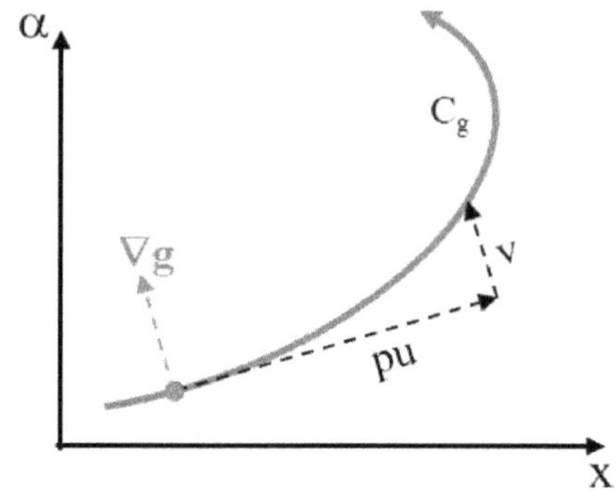

Figure 3-9: Prediction-correction method.

Example 3-5: Tracking of the path of zeros by prediction-correction

Consider the equation: $g_1(x) = 3x^3 - x - \frac{1}{2} = 0$ of solution $x_1 = +0,7461$.

Start with the problem P_0 : $g_0(x) = 3x^3 - x + \frac{1}{2} = 0$ of solution $x_0 = -0,7461$.

The problems P_α are defined by: $g(\alpha,x) = 3x^3 - x + \frac{1}{2} - \alpha = 0$.

The path of zeros has the equation: $\alpha = 3x^3 - x + \frac{1}{2}$. It is not monotonous.

The change direction of α reverses when: $\frac{d\alpha}{dx} = 9x^2 - 1 = 0 \Rightarrow \alpha = \frac{5}{18}$ or $\frac{13}{18}$.

Let us apply a prediction-correction method.

<u>Prediction stage</u>

The tangent vector u is normal to $\nabla g(\alpha,x)$.

$$\nabla g(\alpha,x) = \begin{pmatrix} -1 \\ 9x^2 - 1 \end{pmatrix} \Rightarrow u = \frac{\pm 1}{\sqrt{(9x^2-1)^2 + 1}} \begin{pmatrix} 9x^2 - 1 \\ 1 \end{pmatrix}$$

The vector u is oriented by the direction of the previous step (toward x > 0).

The step length p is fixed: $p = 0.1$.

The predicted point is given by $\begin{pmatrix} \alpha_k \\ x_k \end{pmatrix} \to \begin{pmatrix} \alpha_p = \alpha_k + pu_\alpha \\ x_p = x_k + pu_x \end{pmatrix}$.

This point is not on the path of zeros.

Correction stage

We fix $x_{k+1} = x_p$ and solve the equation with one unknown: $g(\alpha, x_{k+1}) = 0$ by Newton's method. The solution gives the value of α_{k+1} .

The corrected point is given by $\begin{pmatrix} \alpha_p \\ x_p \end{pmatrix} \to \begin{pmatrix} \alpha_{k+1} = \alpha_c = \alpha_p + v_\alpha \\ x_{k+1} = x_c = x_p \end{pmatrix}$.

This point is on the path of zeros.

Figure 3-10 shows the progression of the prediction-correction method with a fixed step length: $p = 0.1$.

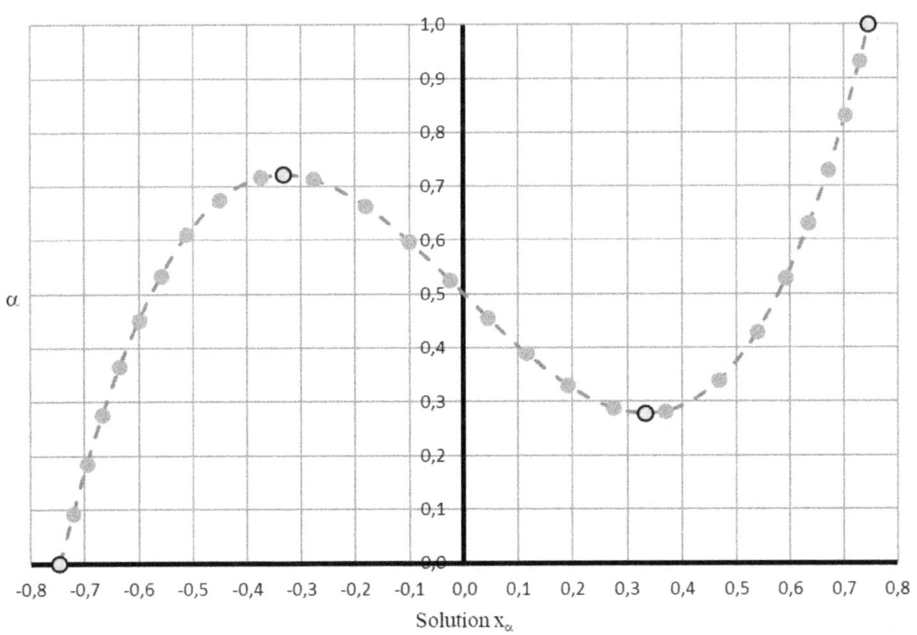

Figure 3-10: Tracking a non-monotonous path of zeros.

The efficiency of the homotopy method depends on the step length p used in the prediction stage. A large step length accelerates the progress along the path of zeros, but with the risk of a failure of Newton's method in the correction step. A small step increases the chances of a fast correction, but requires solving a larger number of successive problems. A strategy with an adaptive step (step increase in case of success, step decrease in case of failure) makes it possible to obtain the best speed/robustness compromise.

3.1.3 Minimization

Let us now consider an unconstrained minimization problem.

$$\min_{x \in \mathbb{R}^n} f(x) \tag{3.21}$$

A necessary condition for x* to be a minimum is

$$\nabla f(x^*) = 0 \tag{3.22}$$

This is a system of n equations with n unknowns to which we can apply Newton's method. The function g is in this case the gradient of f: $g(x) = \nabla f(x)$.

Newton's iteration (3.6) takes the form

$$\begin{aligned} & x_{k+1} = x_k - \nabla g(x_k)^{-T} g(x_k) \quad \text{with} \quad g = \nabla f \\ \Rightarrow \ & x_{k+1} = x_k - \nabla^2 f(x_k)^{-1} \nabla f(x_k) \end{aligned} \tag{3.23}$$

This formula can also be obtained from the second order Taylor expansion of the function f in the point x_k.

$$f(x) = f(x_k) + \nabla f(x_k)^T (x - x_k) + \frac{1}{2}(x - x_k)^T \nabla^2 f(x_k)(x - x_k) + o\left(\|x - x_0\|^2\right) \tag{3.24}$$

If we approximate the function f by the function quadratic function $\hat{f}_k$

$$\hat{f}_k(x) = f(x_k) + \nabla f(x_k)^T (x - x_k) + \frac{1}{2}(x - x_k)^T \nabla^2 f(x_k)(x - x_k) \tag{3.25}$$

and minimize this quadratic function:

$$\nabla \hat{f}_k(x) = 0 \ \Rightarrow \ \nabla f(x_k) + \nabla^2 f(x_k)(x - x_k) = 0 \tag{3.26}$$

we retrieve the point x_{k+1} given by Newton's iteration (3.23).

Newton's method applied to the minimization of a function is thus equivalent to approximating the function by a quadratic model (3.25) and solving the first order optimality condition (3.26).

The following examples of minimization of a function of one variable show the possible behavior of Newton's method.

Example 3-6: Minimization using Newton's method

Consider the minimization problem with one variable

$$\min_{x} f(x) = -x^4 + 12x^3 - 47x^2 + 60x$$

This polynomial of degree 4 is plotted in figure 3-11. It has a minimum in the interval [3; 4] and a maximum in the interval [4; 5].

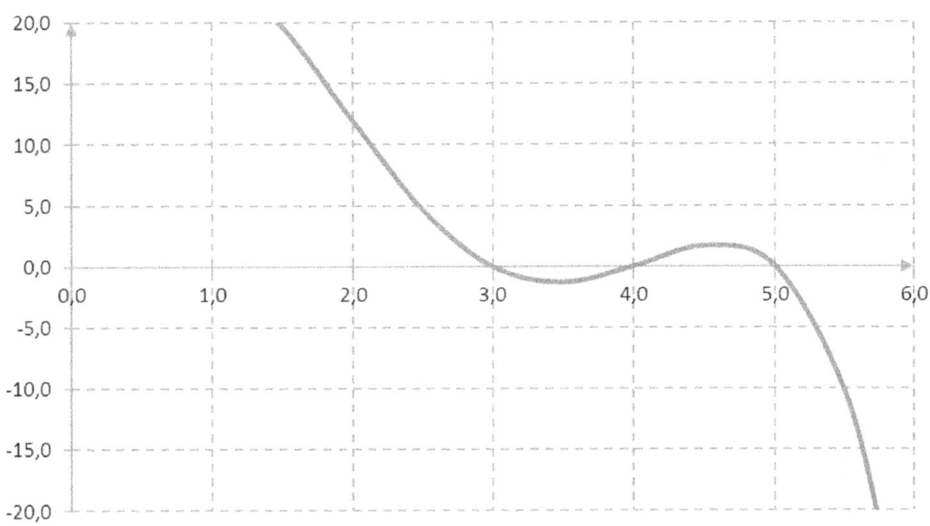

Figure 3-11: Polynomial function of degree 4.

Let us apply Newton's iteration starting from the three different points:

$x_0 = 3$, or $x_0 = 4$, or $x_0 = 5$

Newton's iteration amounts to minimizing a quadratic model of f in the point x_0.

Initial point: $x_0 = 3$

The quadratic model in this point is: $\hat{f}_0(x) = 7x^2 - 48x + 81$.
It is shown in the dotted line opposite.

$\hat{f}_0$ derivative cancels in $x_1 = \dfrac{24}{7}$.

This point minimizes $\hat{f}_0$ and it comes close to the true minimum of f $(x^* \approx 3.5)$.

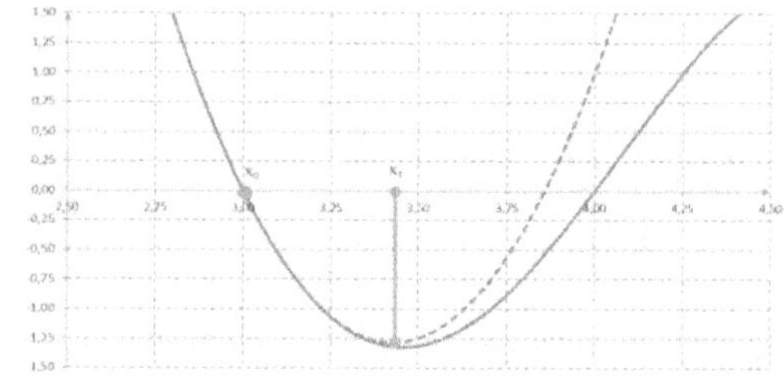

Figure 3-12: Newton's iteration from $x_0 = 3$.

Initial point: $x_0 = 4$

The quadratic model in this point is: $\hat{f}_0(x) = x^2 - 4x$.
It is shown in the dotted line opposite.

$\hat{f}_0$ derivative cancels in $x_1 = 2$.

This point minimizes $\hat{f}_0$, but it moves away from of the true minimum of f $(x^* \approx 3.5)$.

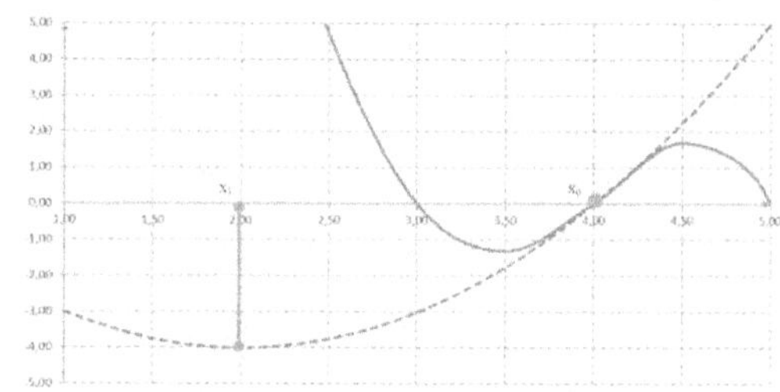

Figure 3-13: Newton's iteration from $x_0 = 4$.

Initial point: $x_0 = 5$

The quadratic model in this point is: $\hat{f}_0(x) = -17x^2 + 160x - 375$.
It is shown in the dotted line opposite.

$\hat{f}_0$ derivative cancels in $x_1 = \dfrac{80}{17}$.

This point maximizes $\hat{f}_0$ and it diverges of the true minimum of f $(x^* \approx 3.5)$.

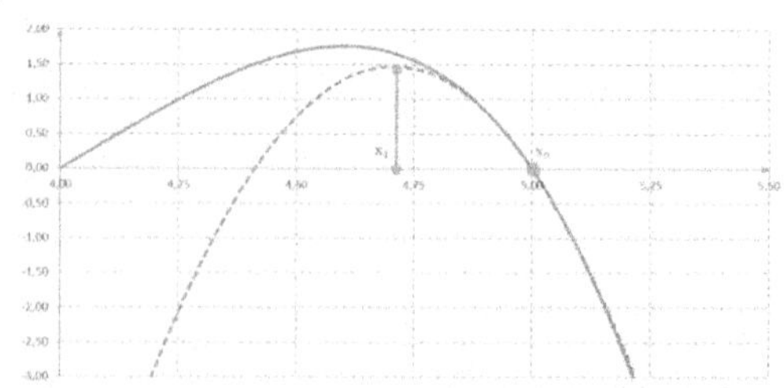

Figure 3-14: Newton's iteration from $x_0 = 5$.

These examples show that Newton's method cannot be applied without care. Indeed, the formula (3.23) amounts to solving the equation $\nabla f(x)=0$, which corresponds to a stationary point of the function f. The other necessary condition for a minimum would be $\nabla^2 f(x) \geq 0$. It is therefore necessary to ensure that the Hessian of f in the point x_k is positive before applying Newton's iteration (3.23).

The features of Newton's method observed for the solution of equations are found again for the minimization of a function:
- the speed of convergence is quadratic near the solution;
- the convergence is not guaranteed and the initialization can be problematic.

Two additional difficulties appear:
- the Hessian of the function must be calculated at each iteration;
- Newton's formula is not applicable if the Hessian is not positive.

Sections 3.2–3.4 show how to circumvent these difficulties by quasi-Newton methods (for the calculation of the Hessian) and globalization techniques (to control the progression to a minimum).

3.1.4 Least squares

A particular class of optimization problems concerns the **identification** of the parameters of a model. Let us assume that we have m measured values $s_1, s_2, \cdots, s_m$ on a physical system or process. In addition, a numerical model is available to simulate the behavior of the system. This model depends on n parameters $x_1, x_2, \cdots, x_n$ whose values are not known. For a given set of parameters $x \in \mathbb{R}^n$, the simulation yields the output values $\sigma_1(x), \sigma_2(x), \cdots, \sigma_m(x)$. These simulated outputs are called "**pseudo-measurements**".

By comparing them to the real measurements $s_1, s_2, \cdots, s_m$, the deviation between the model and the real system can be quantified. The objective is then to find the parameter values $x_1, x_2, \cdots, x_n$ so that the simulated measurements coincide as closely as possible with the real measurements. This will result in a reliable model of the system.

The difference between a simulated measurement σ_i and an actual measurement s_i is called the **residual** noted r_i.

This difference may be weighted by a factor w_i depending on the accuracy or reliability of the measurement s_i.

The vector of residuals is defined by

$$r(x)=\begin{pmatrix} r_1(x) \\ r_2(x) \\ \vdots \\ r_m(x) \end{pmatrix} \quad \text{with} \quad r_i(x)=(\sigma_i(x)-s_i)w_i \tag{3.27}$$

The method of **least squares** consists in searching for the values of the parameters minimizing the norm of the residual vector.

$$\min_{x\in\mathbb{R}^n} f(x)=\frac{1}{2}\|r(x)\|^2=\frac{1}{2}\sum_{i=1}^{m} r_i(x)^2 \tag{3.28}$$

Let us apply Newton's method to this minimization problem. Newton's iterations (3.23) are defined by

$$x_{k+1}=x_k-\nabla^2 f(x_k)^{-1}\nabla f(x_k)$$
$$\text{with} \begin{cases} \nabla f(x)=\sum_{i=1}^{m}\nabla r_i(x) r_i(x) \\ \nabla^2 f(x)=\sum_{i=1}^{m}\left(\nabla r_i(x)\nabla r_i(x)^T+\nabla^2 r_i(x) r_i(x)\right) \end{cases} \tag{3.29}$$

When the number m of measurements is large, the calculation of the second derivatives $\nabla^2 r_i(x)$ of each residual is expensive.

The **Gauss-Newton** method consists in omitting these second derivatives in the calculation of the Hessian, which amounts to linearizing the residuals $r_i(x)$ in the vicinity of x_k.

The gradient and the approximate Hessian can be written in matrix form

$$\begin{cases} \nabla f(x) = \nabla r(x) r(x) \\ \nabla^2 f(x) \approx \nabla r(x)\nabla r(x)^T \end{cases} \tag{3.30}$$

The Gauss-Newton iteration reduces to

$$x_{k+1}=x_k-\left[\nabla r(x_k)\nabla r(x_k)^T\right]^{-1}\nabla r(x_k) r(x_k) \tag{3.31}$$

The approximate Hessian (3.30) is always positive semidefinite, since we have: $u^T\nabla^2 f(x)u=\left\|\nabla r(x)^T u\right\|^2 \geq 0$. This ensures the convergence of iterations (3.31). The Gauss-Newton method specific to the least squares problem (3.28) is thus more robust and economical than the general Newton method.

Linear case

When the outputs $\sigma_i(x)$ are linear functions of the model parameters, the residual vector is of the form

$$r(x) = Ax - b \quad \text{with } A \in \mathbb{R}^{m\times n},\ b \in \mathbb{R}^m \tag{3.32}$$

The solution of the quadratic problem (3.28) is obtained by cancelling the gradient $\nabla f(x) = \nabla r(x) r(x) = A^T(Ax - b) = 0$, which gives the **normal equations**.

$$A^T A x = A^T b \tag{3.33}$$

Solving this linear system yields the solution x directly.

Example 3-7: Least squares estimation of gravity

The aim is to estimate the gravity (assumed to be constant) from 21 measurements of the height of an object dropped in free fall. The measurements are given in Table 3-5 and they are plotted against time in the figure on the right.

Time (s)	Height (m)
0	0.90
1	5.40
2	20.81
3	45.73
4	78.56
5	124.10
6	175.75
7	241.41
8	315.08
9	397.36
10	488.25
11	595.35
12	707.26
13	829.98
14	961.20
15	1103.14
16	1252.89
17	1415.55
18	1586.62
19	1770.20
20	1964.29

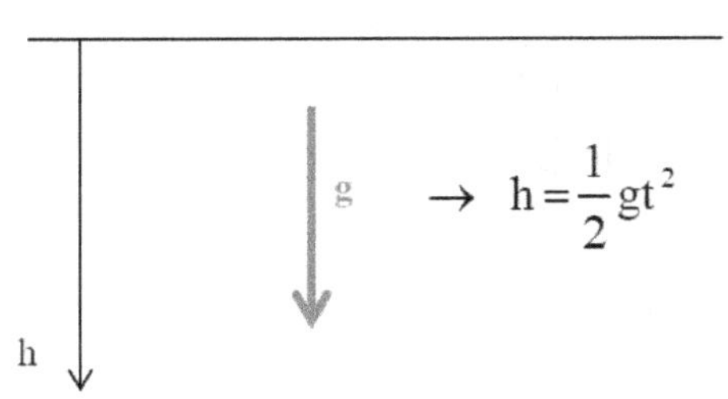

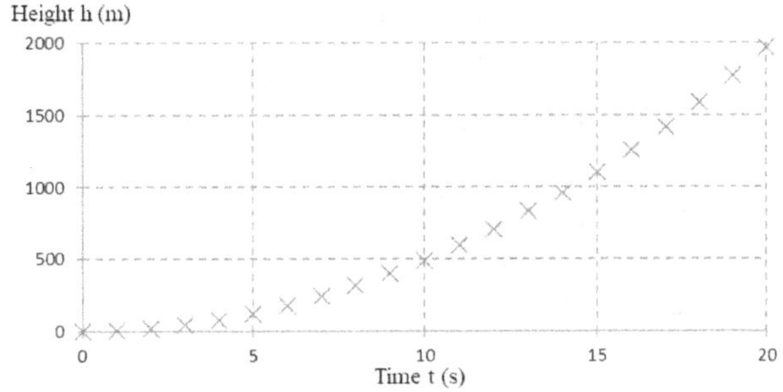

Table 3-5: Height measurements as a function of time.

The model parameter is the gravity g. The observed output is the height.

The measurement h is modelled by the equation: $h_{model} = \frac{1}{2}gt^2$.

This measurement model is <u>linear</u> in g. The residuals are the differences between the model and the measurement: $r = h_{model} - h_{measure}$.

The matrix A is a column whose 21 components are $(t_i^2 / 2)_{i=0 \text{ to } 20}$.

The vector b has the measured values as its components $(h_{i\ measure})_{i=0 \text{ to } 20}$.

The solution of the normal equations (3.33) yields: $g = 9.8070\ m \cdot s^{-2}$ with a residual $f(g) = 21.668$. The average deviation over the 21 measurements is about 1 m, for measured values of the order of a few hundred meters.
The residuals are given in table 3-6. The figure on the right compares the linear model with the measurements represented by the crosses.

Time (s)	h measure (m)	hmodel (m)	Residual (m)
0	0.90	0.00	0.90
1	5.40	4.90	0.49
2	20.81	19.61	1.19
3	45.73	44.13	1.60
4	78.56	78.46	0.10
5	124.10	122.59	1.51
6	175.75	176.53	-0.78
7	241.41	240.27	1.13
8	315.08	313.82	1.25
9	397.36	397.18	0.17
10	488.25	490.35	-2.10
11	595.35	593.32	2.02
12	707.26	706.10	1.15
13	829.98	828.69	1.28
14	961.20	961.09	0.12
15	1103.14	1103.29	-0.14
16	1252.89	1255.30	-2.40
17	1415.55	1417.11	-1.56
18	1586.62	1588.73	-2.11
19	1770.20	1770.16	0.04
20	1964.29	1961.40	2.89

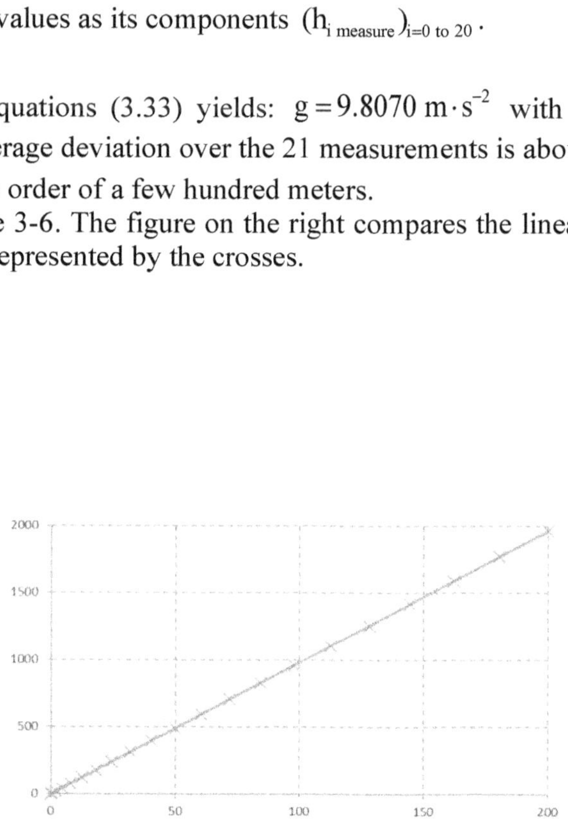

Table 3-6: Residuals of the least squares problem.

Sequential (or recursive) solution

Let us assume that we partition the measurements into two series with m_1 and m_2 measurements respectively. The residual vector (3.32) is of the form

$$r(x) = Ax - b = \begin{pmatrix} A_1 \\ A_2 \end{pmatrix} x - \begin{pmatrix} b_1 \\ b_2 \end{pmatrix} \tag{3.34}$$

The matrices $A_1 \in \mathbb{R}^{m_1 \times n}$, $b_1 \in \mathbb{R}^{m_1}$ are associated with the first measurement set.

The matrices $A_2 \in \mathbb{R}^{m_2 \times n}$, $b_2 \in \mathbb{R}^{m_2}$ are associated with the second measurement set. The least squares problem associated with the first set of measurements is called the initial problem. Its solution x_1 satisfies the normal equations

$$A_1^T A_1 x_1 = A_1^T b_1 \tag{3.35}$$

The least squares problem associated with the whole set of measurements is called the complete problem. Its solution x_2 satisfies the normal equations

$$A^T A x_2 = A^T b \;\Rightarrow\; (A_1^T A_1 + A_2^T A_2) x_2 = A_1^T b_1 + A_2^T b_2 \tag{3.36}$$

Using equality (3.35), we can rewrite (3.36) as

$$\begin{aligned}(A_1^T A_1 + A_2^T A_2) x_2 &= A_1^T b_1 + A_2^T b_2 = A_1^T A_1 x_1 + A_2^T b_2 \\ &= (A_1^T A_1 + A_2^T A_2) x_1 + A_2^T (b_2 - A_2 x_1)\end{aligned} \tag{3.37}$$

which gives the expression x_2 as a function of x_1

$$x_2 = x_1 + (A_1^T A_1 + A_2^T A_2)^{-1} A_2^T (b_2 - A_2 x_1) \tag{3.38}$$

Note that this formula is that of Gauss-Newton (3.31) with $x_k = x_1$ and $x_{k+1} = x_2$. For a quadratic problem, the Gauss-Newton iteration indeed yields the solution (here x_2) in a single iteration from any initial point (here x_1).

Let us assume that the original problem has been solved. Its solution x_1 is known, as well as the matrix $A_1^T A_1$ calculated to solve the normal equations (3.35). The recursive formula (3.38) then gives the solution x_2 of the complete problem by calculating "only" the matrix $A_2^T A_2$ associated with the new measurements. The savings in matrix computations are noticeable when $m_2 << m_1$, because the matrix A_2 is much smaller than the matrix A_1. It is thus possible to continuously update the solution with new measurements without solving the complete problem.

In practice, the solution is updated by the following formulas.

$$\begin{cases} H_k = \rho H_{k-1} + A_k^T A_k \\ x_k = x_{k-1} + H_k^{-1} A_k^T (b_k - A_k x_{k-1}) \end{cases} \quad (3.39)$$

where the matrix $H_k \in \mathbb{R}^{n \times n}$ accumulates the products $A_k^T A_k$ and the **fading coefficient** $\rho \le 1$ favors the latest measurements ($\rho = 1$ to assign the same weight to all measurements). Updating can be systematic with the arrival of each new measurement, or periodic by grouping the latest measurements.

Example 3-8: Recursive least squares estimation of gravity

Let us go back to example 3-7 for gravity estimation. Here, the least squares solution is updated with each new measurement using equation (3.38).
Table 3-7 and the figure on the right show the evolution of the solution as the measurements arrive. The last solution is identical to that found in example 3-7 where all 21 measurements were processed simultaneously.

Time (s)	Matrix H	Estimated g (m/s^2)
0	0.0	0.0000
1	0.3	10.7937
2	4.3	10.4263
3	24.5	10.2074
4	88.5	9.9269
5	244.8	9.9274
6	568.8	9.8341
7	1169.0	9.8440
8	2193.0	9.8450
9	3833.3	9.8305
10	6333.3	9.8046
11	9993.5	9.8177
12	15177.5	9.8195
13	22317.8	9.8204
14	31921.8	9.8167
15	44578.0	9.8136
16	60962.0	9.8068
17	81842.3	9.8041
18	108086.3	9.8016
19	140666.5	9.8029
20	180666.5	**9.8070**

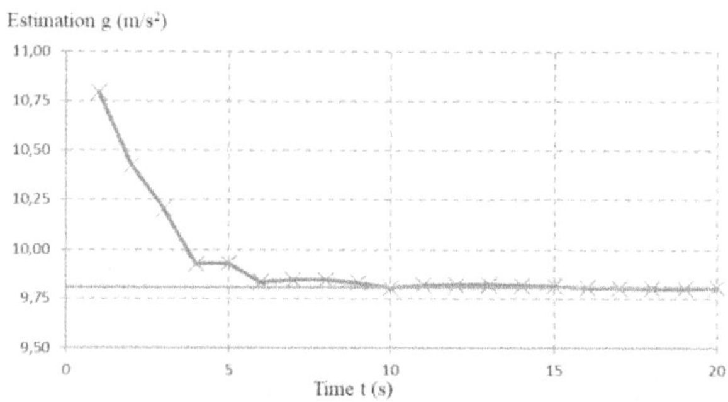

Table 3-7: Recursive least squares solution.

3.2 Quasi-Newton methods

The first shortcoming of Newton's method mentioned in section 3.1.3 is that it requires the calculation of the gradient (for a system of equations) or the Hessian (for a minimization). This calculation, which has to be carried out at each iteration, is very expensive.

Quasi-Newton methods aim to avoid the explicit computation of the gradient or the Hessian by replacing them with approximations built up through iterations. This section presents the Broyden methods (for solving a system of equations) and the DFP, BFGS and SR1 methods (for minimizing a function).

3.2.1 Broyden's method

Let us consider a system of n nonlinear equations $g(x)=0$.
Newton's iterations are defined by (3.6).

$$x_{k+1}=x_k-G_k^{-1}g(x_k) \quad \text{with} \quad G_k=\nabla g(x_k)^T \tag{3.40}$$

The gradient of g in the point x_k can be estimated by finite differences or by a method specific to the system to be solved. Whichever method is used, the computation time of the gradient is of the same order of magnitude as that of g, or even greater. The objective of Broyden's method is to avoid the calculation of the Jacobian matrix $\nabla g(x_k)$ by replacing it with a matrix G_k "close" to $\nabla g(x_k)$.

Let us write a linear model of the function g in the point x_k as in (3.3).

$$\hat{g}_k(x)=g(x_k)+G_k(x-x_k) \tag{3.41}$$

A first condition imposed on the matrix G_k is that the linear model must pass through the point of the previous iteration x_{k-1}.

$$\hat{g}_k(x_{k-1})=g(x_{k-1}) \tag{3.42}$$

This condition called the **secant equation** can be written as

$$y_k=G_kd_k \quad \text{with} \quad \begin{cases} d_k=x_k-x_{k-1} \\ y_k=g(x_k)-g(x_{k-1}) \end{cases} \tag{3.43}$$

The vector $d_k\in\mathbb{R}^n$ is the move between the points x_{k-1} and x_k.
The vector $y_k\in\mathbb{R}^n$ is the variation of the function g between x_{k-1} and x_k.

In dimension 1, equation (3.43) completely determines the real G_k which is the slope of the line passing through the two points.

$$G_k = \frac{g(x_k) - g(x_{k-1})}{x_k - x_{k-1}} \tag{3.44}$$

Replacing into (3.40), we obtain the quasi-Newton iteration in dimension 1:

$$x_{k+1} = x_k - \frac{x_k - x_{k-1}}{g(x_k) - g(x_{k-1})} g(x_k) \tag{3.45}$$

Compared to Newton's iteration (3.7), this formula uses only known values and it avoids the calculation of $g'(x_k)$. The quasi-Newton method based on the formula (3.45) in dimension 1 is also called the secant method.

Figure 3-15 shows the iterations associated with successive secants.

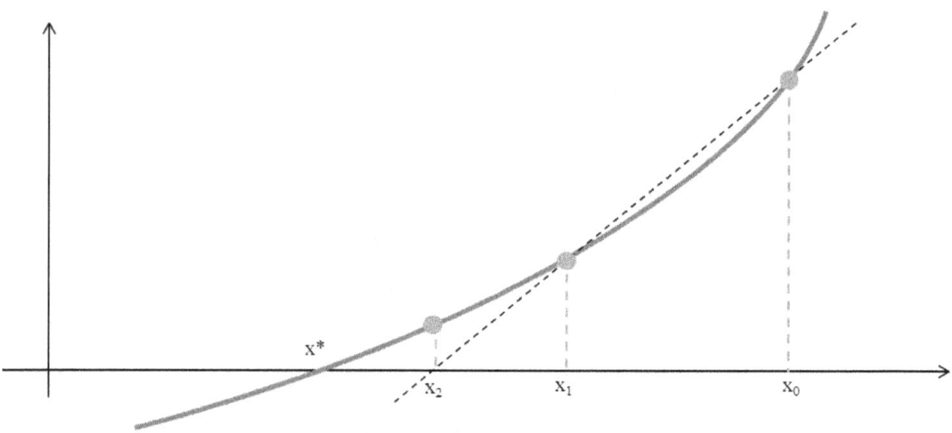

Figure 3-15: Secant method.

The example 3-9 compares the convergence of the quasi-Newton method and the Newton method.

Example 3-9: Solving an equation using the quasi-Newton method

Let us take the equation from example 3-1: $g(x) = x^2 - 1 = 0$.

The quasi-Newton method needs two initial points for a first estimate of the slope. These points are set here in $x_0 = 5$ and $x_0 = 4$.
Table 3-8 compares the iterations of the secant method (quasi-Newton) and the tangent method (Newton) initialized in $x_0 = 4$. The derivative is correctly estimated by the quasi-Newton method, which allows a quadratic convergence to the solution $x^* = 1$.

Secant method (quasi-Newton)

Iteration	x_k	$g(x_k)$	$g'(x_k)$	Error
	5.00000000	2.4E+01		4.0E+00
0	4.00000000	1.5E+01	9.0000	3.0E+00
1	2.33333333	4.4E+00	6.3333	1.3E+00
2	1.63157895	1.7E+00	3.9649	6.3E-01
3	1.21238938	4.7E-01	2.8440	2.1E-01
4	1.04716672	9.7E-02	2.2596	4.7E-02
5	1.00443349	8.9E-03	2.0516	4.4E-03
6	1.00010193	2.0E-04	2.0045	1.0E-04
7	1.00000023	4.5E-07	2.0001	2.3E-07
8	**1.00000000**	**2.3E-11**	**2.0000**	**1.1E-11**

Tangent method (Newton)

Iteration	x_k	$g(x_k)$	$g'(x_k)$	Error
0	4.00000000	1.5E+01	8.0000	3.0E+00
1	2.12500000	3.5E+00	4.2500	1.1E+00
2	1.29779412	6.8E-01	2.5956	3.0E-01
3	1.03416618	6.9E-02	2.0683	3.4E-02
4	1.00056438	1.1E-03	2.0011	5.6E-04
5	1.00000016	3.2E-07	2.0000	1.6E-07
6	**1.00000000**	**2.5E-14**	**2.0000**	**1.3E-14**

Table 3-8: Equation solving by quasi-Newton method.

In dimension n > 1, the secant equation (3.43) imposes only n conditions on the matrix $G_k \in \mathbb{R}^{n\times n}$ and it is therefore not sufficient to determine it completely.

The approach proposed by Broyden in 1965 is to choose the matrix G_k "as close as possible" to that of the previous iteration G_{k-1}.

The linear models of the function g in x_{k-1} and x_k are respectively

$$\begin{cases} \hat{g}_{k-1}(x) = g(x_{k-1}) + G_{k-1}(x - x_{k-1}) \\ \hat{g}_k(x) \ = g(x_k) \ + G_k \ (x - x_k) \end{cases} \tag{3.46}$$

The matrix G_k satisfies the secant equation (3.43).

$$g(x_k) - g(x_{k-1}) = G_k (x_k - x_{k-1}) \tag{3.47}$$

By subtracting member by member the relations (3.46) and using (3.47), the difference between the linear models $\hat{g}_{k-1}$ and $\hat{g}_k$ is given by

$$\hat{g}_k(x) - \hat{g}_{k-1}(x) = (G_k - G_{k-1})(x - x_{k-1}) \tag{3.48}$$

Let us express the vector $x - x_{k-1}$ as a linear combination of the vector d_k and a vector z orthogonal to d_k.

$$x - x_{k-1} = \alpha d_k + z \quad \text{with} \quad d_k^T z = 0 \tag{3.49}$$

Replacing into (3.48) and using the secant equation (3.43), we obtain

$$\begin{aligned} \hat{g}_k(x) - \hat{g}_{k-1}(x) &= \alpha G_k d_k - \alpha G_{k-1} d_k + (G_k - G_{k-1}) z \\ &= \alpha y_k \quad - \alpha G_{k-1} d_k + (G_k - G_{k-1}) z \end{aligned} \tag{3.50}$$

The matrix G_k is only involved in the last term where z is orthogonal to d_k.
To minimize the deviation of the models $\hat{g}_{k-1}$ and $\hat{g}_k$, we choose G_k such that

$$G_k - G_{k-1} = u d_k^T \quad \text{with} \quad u \in \mathbb{R}^n \tag{3.51}$$

which cancels the last term of (3.50). The vector u is calculated by taking the dot product of (3.51) with d_k and using the secant equation $G_k d_k = y_k$.

$$u = \frac{(G_k - G_{k-1}) d_k}{d_k^T d_k} = \frac{y_k - G_{k-1} d_k}{d_k^T d_k} \tag{3.52}$$

Replacing into (3.51), we obtain the **Broyden formula**, which expresses G_k as a function of G_{k-1} and of the variations $d_k = x_k - x_{k-1}$ and $y_k = g(x_k) - g(x_{k-1})$ observed over the last move.

$$\boxed{G_k = G_{k-1} + \frac{(y_k - G_{k-1} d_k) d_k^T}{d_k^T d_k}} \tag{3.53}$$

Let us now show an alternative approach to defining the matrix G_k.
We can check that the matrix G_k defined by (3.53) is a solution to the problem

$$\min_{G \in R^{n \times n}} \| G - G_{k-1} \| \quad \text{with} \quad y_k = G d_k \tag{3.54}$$

where the matrix norm is defined from the vector norm by

$$\|A\| = \max_{x \neq 0} \frac{\|Ax\|}{\|x\|} \tag{3.55}$$

Indeed, for any matrix G satisfying the secant equation $y_k = Gd_k$, we have by using the inequalities on the norms

$$\begin{aligned} \|G_k - G_{k-1}\| &= \left\| \frac{(y_k - G_{k-1}d_k)d_k^T}{d_k^T d_k} \right\| = \left\| \frac{(Gd_k - G_{k-1}d_k)d_k^T}{d_k^T d_k} \right\| = \left\| \frac{(G - G_{k-1})d_k d_k^T}{d_k^T d_k} \right\| \\ &\leq \|G - G_{k-1}\| \left\| \frac{d_k d_k^T}{d_k^T d_k} \right\| \leq \|G - G_{k-1}\| \end{aligned} \tag{3.56}$$

Broyden's formula (3.53) allows to solve the system of equations $g(x) = 0$ by Newton's method (3.40) without calculating the gradient $\nabla g(x_k)$ at each iteration.

For a minimization problem, the system to be solved becomes: $\nabla f(x) = 0$ and the matrix G_k would replace the Hessian $\nabla^2 f(x_k)$, which is supposed to be a positive definite symmetric matrix. Broyden's formula is not suitable, as it does not have this property. The main quasi-Newton methods for minimization are presented in section 3.2.2.

3.2.2 DFP, BFGS and SR1 methods

Newton's iterations for a minimization are defined by (3.23).

$$x_{k+1} = x_k - \nabla^2 f(x_k)^{-1} \nabla f(x_k) \tag{3.57}$$

The objective here is to avoid computing the Hessian $\nabla^2 f(x_k)$ at each iteration by replacing it with some matrix H_k. By denoting $g_k = \nabla f(x_k)$ the gradient of f in x_k, the quasi-Newton iteration for a minimization takes the form

$$x_{k+1} = x_k - H_k^{-1} g_k \tag{3.58}$$

There are many methods for defining the matrix H_k, the most widely used being DFP, BFGS and SR1. These methods require the symmetric matrix H_k to satisfy the secant equation.

$$y_k = H_k d_k \quad \text{with} \quad \begin{cases} d_k = x_k - x_{k-1} \\ y_k = g_k - g_{k-1} \end{cases} \tag{3.59}$$

DFP method

The DFP method (proposed by Davidon, Fletcher and Powell in the 1950s) is historically the first quasi-Newton type method. The useful matrix for the quasi-Newton iteration (3.57) is the inverse of H_k noted: $B_k = H_k^{-1}$.

This inverse matrix must satisfy the secant equation (3.59) of the form

$$B_k y_k = d_k \quad \text{with} \quad \begin{cases} d_k = x_k - x_{k-1} \\ y_k = g_k - g_{k-1} \end{cases} \tag{3.60}$$

The DFP formula updates B_k using B_{k-1} and the variations of x and $g = \nabla f$.

$$\boxed{B_k = B_{k-1} + \frac{d_k d_k^T}{y_k^T d_k} - \frac{B_{k-1} y_k y_k^T B_{k-1}}{y_k^T B_{k-1} y_k} \quad \text{with} \quad \begin{cases} d_k = x_k - x_{k-1} \\ y_k = g_k - g_{k-1} \end{cases}} \tag{3.61}$$

This formula only applies if the condition: $y_k^T d_k > 0$ is satisfied.

Construction of the DFP formula

The DFP formula is built by seeking B_k under the symmetric form: $B_k = AA^T$. The matrix B_{k-1} is assumed to be symmetric positive definite. It therefore admits a Cholesky factorization with a lower triangular matrix L: $B_{k-1} = LL^T$.

The matrix B_k must satisfy the secant equation: $B_k y_k = d_k$ with $\begin{cases} d_k = x_k - x_{k-1} \\ y_k = g_k - g_{k-1} \end{cases}$

This equation is decomposed into a system.

$$B_k y_k = d_k \iff AA^T y_k = d_k \iff \begin{cases} z = A^T y_k \\ Az = d_k \end{cases}$$

For $z \in \mathbb{R}^n$, the second equation imposes n conditions on the matrix A. We are looking for the matrix A satisfying the equation $Az = d_k$ and "as close as possible" to the matrix L associated with the Cholesky factorization of B_{k-1}. This problem is strictly similar to the one treated in section 3.2.1 where the constraint was the secant equation (3.43). The solution is given by Broyden's formula (3.53):

$$A = L + \frac{(d_k - Lz)z^T}{z^T z}$$

The matrix A is replaced in the first equation.

$$z = A^T y_k = L^T y_k + \frac{(d_k - Lz)^T y_k}{z^T z} z$$

This equality requires that $L^T y_k$ and z are collinear: $z = \alpha L^T y_k$.

Replacing z in the previous equation, we find: $\alpha^2 = \dfrac{y_k^T d_k}{y_k^T B_{k-1} y_k}$.

For a solution α to exist, we must have: $\boxed{y_k^T d_k > 0}$. This is consistent with assumptions about B_k, which is positive definite and satisfies $d_k = B_k y_k$, from which we deduce: $y_k^T d_k = y_k^T B_k y_k > 0$ for $y_k \neq 0$.

This yields the expression for A: $A = L + \dfrac{1}{d_k^T y_k}\left(\alpha d_k y_k^T L - \dfrac{d_k^T y_k}{y_k^T B_{k-1} y_k} B_{k-1} y_k y_k^T L\right)$

and then the one for $B_k = AA^T$ giving the DFP formula (3.61).

The condition $y_k^T d_k > 0$ appeared in the construction of the matrix B_k.

Let us show that this **curvature condition** is always satisfied if the point x_k is defined from x_{k-1} by minimization along the quasi-Newton direction $-B_{k-1} g_{k-1}$. The iteration (3.58) is of the form

$$x_k = x_{k-1} - sB_{k-1} g_{k-1} = x_{k-1} + su_{k-1} \tag{3.62}$$

where the step $s > 0$ along the direction $u_{k-1} = -B_{k-1} g_{k-1}$ is obtained by

$$\min_{s>0} \varphi(s) = f(x_{k-1} + su_{k-1}) \tag{3.63}$$

The minimization of the directional function φ gives

$$\varphi'(s) = 0 \quad \Rightarrow \quad \nabla f(x_{k-1} + su_{k-1})^T u_{k-1} = \nabla f(x_k)^T u_{k-1} = g_k^T u_{k-1} = 0 \tag{3.64}$$

This relation is used to simplify the scalar product $y_k^T d_k$.

$$\begin{aligned} y_k^T d_k &= (g_k - g_{k-1})^T (x_k - x_{k-1}) = (g_k - g_{k-1})^T (su_{k-1}) \\ &= -sg_{k-1}^T u_{k-1} = sg_{k-1}^T B_{k-1} g_{k-1} \end{aligned} \tag{3.65}$$

Since the matrix B_{k-1} is positive and $s > 0$, we obtain $y_k^T d_k > 0$, which is the condition required to apply the DFP formula.

The DFP method is usually initialized with the identity matrix: $B_0 = I$.

The example 3-10 shows the evolution of the matrix B_k during the iterations.

Example 3-10: Minimization by DFP method

Consider the minimization problem with 2 variables noted ξ_1 and ξ_2:

$$\min_{\xi_1,\xi_2} f(\xi_1,\xi_2)=\xi_1-\xi_2+2\xi_1^2+2\xi_1\xi_2+\xi_2^2$$

The gradient and the Hessian are given by: $g=\begin{pmatrix}1+4\xi_1+2\xi_2\\-1+2\xi_1+2\xi_2\end{pmatrix}$, $H=\begin{pmatrix}4&2\\2&2\end{pmatrix}$.

Let us perform two iterations of the DFP method from $x_0=\begin{pmatrix}0\\0\end{pmatrix}$, $B_0=\begin{pmatrix}1&0\\0&1\end{pmatrix}$.

At each iteration, the point x_{k+1} is defined as the minimum along the direction: $u_k=-B_k g_k$.

<u>Iteration 1</u>

Move

$$x_0=\begin{pmatrix}0\\0\end{pmatrix},\ B_0=\begin{pmatrix}1&0\\0&1\end{pmatrix},\ g_0=\begin{pmatrix}1\\-1\end{pmatrix}\ \to\ u_0=-B_0g_0=\begin{pmatrix}-1\\1\end{pmatrix}\ \to\ x_1=x_0+su_0=\begin{pmatrix}-s\\s\end{pmatrix}$$

Minimization

$$\min_s \varphi(s)=s^2-2s\ \to\ s=1\ \to\ x_1=\begin{pmatrix}-1\\1\end{pmatrix},\ g_1=\begin{pmatrix}-1\\-1\end{pmatrix}$$

DFP formula

$$d_1=\begin{pmatrix}-1\\1\end{pmatrix},\ y_1=\begin{pmatrix}-2\\0\end{pmatrix}\ \to\ B_1=\begin{pmatrix}1&0\\0&1\end{pmatrix}+\frac{1}{2}\begin{pmatrix}1&-1\\-1&1\end{pmatrix}-\frac{1}{4}\begin{pmatrix}4&0\\0&0\end{pmatrix}=\frac{1}{2}\begin{pmatrix}1&-1\\-1&3\end{pmatrix}$$

<u>Iteration 2</u>

Move

$$x_1=\begin{pmatrix}-1\\1\end{pmatrix},\ B_1=\frac{1}{2}\begin{pmatrix}1&-1\\-1&3\end{pmatrix},\ g_1=\begin{pmatrix}-1\\-1\end{pmatrix}\ \to\ u_1=-B_1g_1=\begin{pmatrix}0\\1\end{pmatrix}\ \to\ x_2=x_1+su_1=\begin{pmatrix}-1\\1+s\end{pmatrix}$$

Minimization

$$\min_s \varphi(s)=1-3(1+s)+(1+s)^2\ \to\ s=\frac{1}{2}\ \to\ x_2=\begin{pmatrix}-1\\1,5\end{pmatrix}\ \to\ g_2=\begin{pmatrix}0\\0\end{pmatrix}$$

DFP formula

$$d_2=\begin{pmatrix}0\\0,5\end{pmatrix},\ y_2=\begin{pmatrix}1\\1\end{pmatrix}\ \to\ B_2=\frac{1}{2}\begin{pmatrix}1&-1\\-1&3\end{pmatrix}+\frac{1}{2}\begin{pmatrix}0&0\\0&1\end{pmatrix}-\begin{pmatrix}0&0\\0&1\end{pmatrix}=\frac{1}{2}\begin{pmatrix}1&-1\\-1&2\end{pmatrix}$$

Solution

The minimum is obtained in 2 iterations: $x^* = \begin{pmatrix} -1 \\ 1{,}5 \end{pmatrix}$ with a zero gradient.

The matrix $B_2 = \frac{1}{2}\begin{pmatrix} 1 & -1 \\ -1 & 2 \end{pmatrix}$ is equal to the inverse of the Hessian: $H = \begin{pmatrix} 4 & 2 \\ 2 & 2 \end{pmatrix}$.

On this example in dimension 2, the DFP method leads to the exact minimum in 2 iterations and perfectly retrieve the Hessian of the function. This behavior occurs for any convex quadratic function. This follows from property 3-1 and 3.2.

Property 3-1: Convergence of DFP for a convex quadratic function

Consider the convex quadratic function on $\mathbb{R}^n$

$$f(x) = \frac{1}{2}x^T Q x + c^T x \quad \text{with} \quad c \in \mathbb{R}^n \ , Q \in \mathbb{R}^{n \times n} \ , Q > 0 \quad \text{and } Q \text{ symmetric}$$

The gradient is: $g(x) = Qx + c$ and the moves (3.43) satisfy: $y_k = Qd_k$.

Let us apply the DFP method to this quadratic function applying an exact one-dimensional minimization at each iteration.

The successive moves $(d_i)_{i=1 \text{ to } k}$ performed over k iterations satisfy then

$$\begin{cases} d_i^T Q d_j = 0 \ , \ 1 \le i < j \le k \\ B_k Q d_i = d_i \ , \ 1 \le i \le k \end{cases} \tag{3.66}$$

Demonstration (see [R11])

The two relations (3.66) are proved by induction on k.

For the value k = 1, the only direction is d_1. The first relation does not apply and the second relation can be checked directly using first: $y_1 = Qd_1$ and then the secant equation (3.60) : $B_1 Q d_1 = B_1 y_1 = d_1$.

Assume both relations (3.66) to be true for k and show them for k + 1.

For the first relation, it must be shown that $d_i^T Q d_{k+1} = 0 \ , \ 1 \le i \le k$.

$$
\begin{aligned}
d_i^T Q d_{k+1} &= d_i^T Q(x_{k+1} - x_k) = d_i^T Q(-sB_k g_k) && \text{because } x_{k+1} = x_k - sB_k g_k \\
&= -s d_i^T Q B_k g_k = -s d_i^T g_k && \text{because } B_k Q d_i = d_i \text{ (induction assumption)} \\
&= -s d_i^T (g_i + y_{i+1} + \cdots + y_k) && \text{because } y_i = g_i - g_{i-1} \\
&= -s d_i^T g_i - s d_i^T (Q d_{i+1} + \cdots + Q d_k) && \text{because } g = Qx + c \Rightarrow y = Qd
\end{aligned}
$$

The first term $d_i^T g_i$ is zero, because an exact one-dimensional minimization is performed at each iteration. Indeed:

$$\min_s \varphi(s) = f(x_{i-1} - s u_{i-1}) \ \rightarrow \ \nabla f(x_{i-1} - s u_{i-1})^T u_{i-1} = 0$$

$$\rightarrow \ g_i^T d_i = 0 \quad \text{since} \ \begin{cases} x_i = x_{i-1} - s u_{i-1} \\ d_i = x_i - x_{i-1} = -s u_{i-1} \\ g_i = \nabla f(x_i) \end{cases}$$

The following terms are zero (induction assumption): $d_i^T Q d_j = 0 \ , 1 < i < j \le k$.
The first relation for k + 1 has been proved.

For the second relations, we calculate $B_{k+1} Q d_i \ , \ 1 \le i \le k+1$ using the DFP formula (3.61).

$$B_{k+1} Q d_i = B_k Q d_i + \frac{d_{k+1} d_{k+1}^T Q d_i}{y_{k+1}^T d_{k+1}} - \frac{B_k y_{k+1} y_{k+1}^T B_k}{y_{k+1}^T B_k y_{k+1}} Q d_i$$

If $1 \le i \le k$, we have: $\begin{cases} B_k Q d_i = d_i & \text{(induction assumption)} \\ d_{k+1}^T Q d_i = 0 & \text{(demontrated just below)} \\ y_{k+1}^T B_k Q d_i = y_{k+1}^T d_i = d_{k+1}^T Q d_i & \text{because } y_{k+1} = Q d_{k+1} \end{cases}$

which yields: $B_{k+1} Q d_i = d_i$

If $i = k+1$, we have directly: $B_{k+1} Q d_{k+1} = B_{k+1} y_{k+1} = d_{k+1}$

using: $y_{k+1} = Q d_{k+1}$ and the secant equation (3.60).

The second relation is thus satisfied for all values of i from 1 to k + 1.

Let us interpret the two relations (3.66).

The first relation (3.66) indicates that the directions are **conjugate** with respect to the Q matrix. The minimum is then reached in n one-dimensional minimizations. This will be shown by property 3-2.

The second relation (3.66) applied with $k = n$ yields: $B_n Q d_i = d_i \ , \ 1 \le i \le n$.

The n vectors $(d_i)_{i=1 \text{ to } n}$ are independent (because they are conjugate with respect to Q) and form a basis of $\mathbb{R}^n$. The matrix $B_n Q$ is therefore the identity and the matrix B_n obtained at the last DFP iteration is the inverse of the Hessian Q of the quadratic function.

Let us now return to the implementation of the DFP method. The DFP formula (3.61) updates the inverse approximation B_k of the Hessian. This matrix is then used to perform the iteration: $x_{k+1} = x_k - sB_k g_k$.

An alternative approach is to invert the formula (3.61) to update the approximation of the Hessian H_k, and then use the inverse H_k^{-1} to perform the iteration: $x_{k+1} = x_k - sH_k^{-1} g_k$. This approach seems artificial at first, as it forces the matrix H_k to be inverted once it is updated. But, it is not equivalent to DFP and it can give better results as shown by the BFGS method developed below. The inversion of (3.61) is performed by the Sherman-Morrison-Woodbury formula.

<u>Sherman-Morrison-Woodbury formula</u> (also called matrix inversion lemma)

Let be the matrices $A, B \in \mathbb{R}^{n\times n}$, $U, V \in \mathbb{R}^{n\times m}$ $(m \le n)$. We have the relation

$$\boxed{B = A + UV^T \iff B^{-1} = A^{-1} - A^{-1}U\left(I + V^T A^{-1} U\right)^{-1} V^T A^{-1}}$$

This formula is demonstrated by calculating BB^{-1} to check that $BB^{-1} = I$.

$$\begin{aligned} BB^{-1} &= (A + UV^T)(A^{-1} - A^{-1}U(I+V^TA^{-1}U)^{-1}V^TA^{-1}) \\ &= I + UV^TA^{-1} - U(I+V^TA^{-1}U)^{-1}V^TA^{-1} - UV^TA^{-1}U(I+V^TA^{-1}U)^{-1}V^TA^{-1} \\ &= I + UV^TA^{-1} - U(I+V^TA^{-1}U)\,(I+V^TA^{-1}U)^{-1}V^TA^{-1} \\ &= I + UV^TA^{-1} - UV^TA^{-1} = I \end{aligned}$$

After (long) calculations, the inverse DFP formula is obtained, which gives the Hessian update H_k from H_{k-1} and the variation vectors d_k and y_k.

$$H_k = \left(I - \frac{y_k d_k^T}{y_k^T d_k}\right) H_{k-1} \left(I - \frac{d_k y_k^T}{y_k^T d_k}\right) + \frac{y_k y_k^T}{y_k^T d_k} \tag{3.67}$$

BFGS method

The method BFGS (developed by Broyden, Fletcher, Goldfarb and Shanno in 1970) is constructed in the same way as DFP, but considering the matrix H_k instead of the inverse matrix B_k. The secant equation is used in the form (3.59) instead of (3.60). The construction process is then strictly identical to that presented for DFP with the permutations $d_k \leftrightarrow y_k$ and $B_k \leftrightarrow H_k$.

The "direct" BFGS formula is thus deduced from (3.61)

$$H_k = H_{k-1} + \frac{y_k y_k^T}{d_k^T y_k} - \frac{H_{k-1} d_k d_k^T H_{k-1}}{d_k^T H_{k-1} d_k} \quad \text{with} \quad \begin{cases} d_k = x_k - x_{k-1} \\ y_k = g_k - g_{k-1} \end{cases} \tag{3.68}$$

and the "inverse" BFGS formula is deduced from (3.67)

$$B_k = \left(I - \frac{d_k y_k^T}{d_k^T y_k} \right) B_{k-1} \left(I - \frac{y_k d_k^T}{d_k^T y_k} \right) + \frac{d_k d_k^T}{d_k^T y_k} \tag{3.69}$$

As for DFP, this formula is subject to the curvature condition: $y_k^T d_k > 0$.

The BFGS method has the same properties as DFP for minimizing a convex quadratic function (property 3-1). These properties assume an exact one-dimensional minimization at each iteration.

In practical applications, the function to be minimized is arbitrary and often expensive to compute. It is necessary to settle for an approximate one-dimensional minimization giving a satisfactory step (section 3.3.2), and then check the curvature condition $y_k^T d_k > 0$ before updating the matrix H_k.

If this condition is not met, one can either keep the previous matrix $(H_k = H_{k-1})$, or reset to the identity $(H_k = I)$, or apply a damped update (section 3.2.3).

The DFP and BFGS methods assume that the matrices H_k (or B_k) are positive, which is not necessarily representative of the Hessian of the function. Indeed, if the point x_k is far from the optimum or if some constraints are active, the Hessian of the function is not necessarily positive. As the matrices H_k can become quite different from the true Hessian, it is advisable to reset them periodically to the identity, for example every n iterations for a problem in $\mathbb{R}^n$. This avoids generating matrices that bear little relation to the true Hessian of the function. The BFGS method has proven to be more efficient than DFP and has become widely used in continuous optimization algorithms. Its drawback is that it assumes the matrices H_k to be positive, which can sometimes reduce its efficiency. The SR1 method can then constitute an interesting alternative.

SR1 method

The DFP and BFGS methods are rank 2 methods, as their update formula depends on two vectors d_k and y_k. The SR1 method (*System Rank One*) is based on a symmetric update depending on a single vector $u \in \mathbb{R}^n$.

$$H_k = H_{k-1} + uu^T \tag{3.70}$$

The matrix H_k must satisfy the secant equation (3.59). The formula SR1 is then

$$H_k = H_{k-1} + \frac{(y_k - H_{k-1}d_k)(y_k - H_{k-1}d_k)^T}{d_k^T(y_k - H_{k-1}d_k)} \quad \text{with} \quad \begin{cases} d_k = x_k - x_{k-1} \\ y_k = g_k - g_{k-1} \end{cases} \tag{3.71}$$

Construction of the SR1 formula

We are looking for the vector $u \in \mathbb{R}^n$ such that

$$\begin{cases} H_k = H_{k-1} + uu^T \\ H_k d_k = y_k \end{cases} \Rightarrow y_k - H_{k-1}d_k = u^T d_k u$$

By posing: $\frac{1}{\gamma} = u^T d_k$, we have: $y_k - H_{k-1}d_k = \frac{1}{\gamma}u \Rightarrow u = \gamma(y_k - H_{k-1}d_k)$

Let us replace this expression of u into: $\frac{1}{\gamma} = u^T d_k \Rightarrow \frac{1}{\gamma^2} = d_k^T(y_k - H_{k-1}d_k)$

Then replace this expression of γ into: $u = \gamma(y_k - H_{k-1}d_k)$ to calculate uu^T

$$uu^T = \gamma^2(y_k - H_{k-1}d_k)(y_k - H_{k-1}d_k)^T = \frac{(y_k - H_{k-1}d_k)(y_k - H_{k-1}d_k)^T}{d_k^T(y_k - H_{k-1}d_k)}$$

This yields the update formula SR1 (3.71).

Unlike the DFP and BFGS formulas, the SR1 formula does not assume the matrix H_k to be positive definite. It is therefore more likely to approximate the true Hessian of the function, which may be non-positive far from the minimum. However, it will be necessary to check that the direction $-H_k^{-1}g_k$ is still a direction of descent, and if necessary modify the matrix H_k to make it positive definite.

The SR1 update is not subject to the curvature condition $y_k^T d_k > 0$, but one must check that the denominator in formula (3.71) is not too small. Otherwise one can either keep the previous matrix $(H_k = H_{k-1})$, or reset to the identity $(H_k = I)$.

Applied to a quadratic function, the SR1 method converges in n iterations and gives the exact Hessian. Table 3-9 summarizes the properties of the DFP, BFGS and SR1 methods.

	BFGS	DFP	SR1
Updated matrix	Hessian H_k	Inverse hessian H_k^{-1}	Hessian H_k
Equation solved	Secant	Inverse secant	Secante
Construction	Broyden	Broyden	Direct resolution
Matrix properties	Symmetrical AA^T Rang 2 Positive definite	Symmetrical AA^T Rang 2 Positive definite	Symmetrical uu^T Rang 1 Indefinite
Required accuracy for the 1-D minimization	Medium accuracy	High accuracy	Low accuracy
Quadratic function	Exact Hessian: $H_n=Q$ Conjugate directions / Q^{-1}	Exact Hessian: $H_n=Q$ Conjugate directions / Q	Exact Hessian: $H_n=Q$
Limitations	Hessian of f indefinite	Hessian of f indefinite 1-D minimization accuracy	Matrices H_k non positive definite

Table 3-9: Comparison of DFP, BFGS and SR1 methods.

For the minimization of a function of a single variable, d_k, y_k, H_k become real numbers. The formulas DFP, BFGS or SR1 simplify to

$$H_k = \frac{y_k}{d_k} = \frac{g(x_k) - g(x_{k-1})}{x_k - x_{k-1}} \tag{3.72}$$

which is equivalent to the secant method (3.44).

The example 3-11 compares the convergence of the quasi-Newton and Newton methods for minimizing a function of one variable.

Example 3-11: Quasi-Newton minimization

Let us retrieve the problem of example 3-6: $\min_x f(x) = -x^4 + 12x^3 - 47x^2 + 60x$.

The local minimum is sought in the interval [3; 4] starting from $x_0 = 3$.
Table 3-10 compares the quasi-Newton and Newton iterations.

The quasi-Newton method requires 2 additional iterations to reach the minimum, but it does not require any calculation of the second derivative.

Quasi-Newton method

Iter	x_k	$f(x_k)$	$f'(x_k)$	h_k
0	3.00000000	0.00000000	-6.00E+00	1.000
1	2.99900000	0.00600700	-6.01E+00	14.000
2	3.42857155	-1.31945027	-3.15E-01	13.267
3	3.45230465	-1.32362420	-3.79E-02	11.672
4	3.45554876	-1.32368634	-4.68E-04	11.527
5	3.45558934	-1.32368635	-7.27E-07	11.509
6	3.45558940	-1.32368635	-1.40E-11	11.509
7	**3.45558940**	**-1.32368635**	**-5.68E-14**	**11.462**

Newton method

x_k	$f(x_k)$	$f'(x_k)$	$f''(x_k)$
3.00000000	0.00000000	-6.00E+00	14.000
3.42857143	-1.31945023	-3.15E-01	11.796
3.45526446	-1.32368574	-3.74E-03	11.513
3.45558935	-1.32368635	-5.77E-07	11.509
3.45558940	**-1.32368635**	**-5.68E-14**	**11.509**

Table 3-10: Minimization by quasi-Newton method.

3.2.3 BFGS improvements

As the BFGS method has proven to be the most efficient, several improvements have been proposed regarding initialization, damped update and memory storage reduction.

Initialization

The BFGS method is initialized with any positive definite matrix H_0.
The different choices can be, in order of increasing computation time:

- the identity matrix: $H_0 = I$;
- the calculation of diagonal terms by finite differences: $H_0 = \mathrm{diag}\left(\dfrac{\partial^2 f}{\partial x_i^2}\right)$;
- the complete calculation of the Hessian by finite differences: $H_0 = \left(\dfrac{\partial^2 f}{\partial x_i \partial x_j}\right)$.

The latter solution is generally prohibitive in terms of computing time.

During the iterations, periodic resets are advisable to prevent the matrices H_k becoming "too different" from the true Hessian. An alternative to the above initialization options is to use the variations of x and g on the last iteration and choose a reset of the form

$$H_0 = \tau I \quad \text{with} \quad \tau = \frac{y^T y}{y^T d} \ , \quad \begin{cases} d = x_k - x_{k-1} \\ y = g_k - g_{k-1} \end{cases} \tag{3.73}$$

Justification of formula (3.73)

The idea is to approximate the Hessian by a diagonal matrix with its largest eigenvalue on all the diagonal. The Hessian H in the point x_k is positive definite. It therefore admits an orthonormal basis of eigenvectors $(u_i)_{i=1,\cdots,n}$ associated with the eigenvalues $(\lambda_i)_{i=1,\cdots,n}$.

The vector $d = x_k - x_{k-1}$ is expressed on this basis as $d = \sum \alpha_i u_i$. The secant equation that the matrix H must satisfy yields: $y = Hd = H\sum \alpha_i u_i = \sum \alpha_i \lambda_i u_i$.

Keeping only the largest eigenvalue noted λ_m and neglecting the other terms, we are left with: $\begin{cases} d \approx \alpha_m u_m \\ y \approx \alpha_m \lambda_m u_m \end{cases} \Rightarrow \lambda_m \approx \dfrac{y^T y}{y^T d}$.

This value corresponds to the diagonal term τ chosen in formula (3.73).

Damped update

The BFGS formula (3.68) only applies if the curvature condition $y_k^T d_k > 0$ is satisfied. This condition is not necessarily satisfied in the case of an approximate one-dimensional minimization. The damped update consists in replacing in the BFGS formula the vector y_k by the vector z_k with a weight θ

$$z_k = \theta y_k + (1-\theta) H_{k-1} d_k \quad \text{with} \quad 0 \le \theta \le 1 \tag{3.74}$$

The **damped BFGS** formula is then

$$H_k = H_{k-1} + \frac{z_k z_k^T}{d_k^T z_k} - \frac{H_{k-1} d_k d_k^T H_{k-1}}{d_k^T H_{k-1} d_k} \quad \text{with} \begin{cases} d_k = x_k - x_{k-1} \\ z_k = \theta(g_k - g_{k-1}) + (1-\theta) H_{k-1} d_k \end{cases} \tag{3.75}$$

The vector z_k is a compromise between the actual gradient variation $y_k = g_k - g_{k-1}$ and the expected variation $H_{k-1} d_k$ (from the previous matrix H_{k-1}):

- for $\theta = 0$, the formula (3.75) keeps the previous matrix $(H_k = H_{k-1})$;
- for $\theta = 1$, the formula (3.75) gives the usual BFGS update.

The value of θ is set by imposing the "reinforced" curvature condition on z_k.

$$d_k^T z_k \ge \alpha d_k^T H_{k-1} d_k \quad \text{with} \quad \alpha > 0 \tag{3.76}$$

This condition is stronger than the usual one $y_k^T d_k > 0$. In practice, the real α can be set to 0.2. If y_k satisfies the condition (3.76), we fix $\theta = 1$, which gives the usual BFGS formula. Otherwise, the value of θ is calculated to have the equality of the two members in (3.76).

$$\theta = \frac{(1-\alpha)d_k^T H_{k-1} d_k}{d_k^T H_{k-1} d_k - d_k^T y_k} \tag{3.77}$$

The damped BFGS formula allows further updating when the curvature condition is not met. It is to be combined with a periodic reset of the matrix H_k, for example by the formula (3.73).

Limitation of memory storage

For large problems, the storage of the matrix H_k can become prohibitive in memory space. This matrix is used to calculate the move

$$x_{k+1} = x_k - s_k B_k g_k \tag{3.78}$$

where $B_k = H_k^{-1}$ $g_k = \nabla f(x_k)$ is the gradient of f and s_k is the step length.

The L-BFGS (*Limited Memory*) method aims to construct the direction of move $B_k g_k$ without explicitly using the matrix B_k. This method is based on the inverse BFGS formula (3.69) rewritten in compact form:

$$B_k = V_k^T B_{k-1} V_k + \rho_k d_k d_k^T \tag{3.79}$$

by posing

$$\rho_k = \frac{1}{d_k^T y_k} \quad \text{and} \quad V_k = I - \frac{y_k d_k^T}{d_k^T y_k} = I - \rho_k y_k d_k^T \tag{3.80}$$

Consider m successive iterations of the form (3.79) starting with the matrix B_0.

$$\begin{aligned}
B_1 &= V_1^T B_0 V_1 + \rho_1 d_1 d_1^T \\
B_2 &= V_2^T B_1 V_2 + \rho_2 d_2 d_2^T = \left(V_2^T V_1^T\right) B_0 \left(V_1 V_2\right) + \rho_1 V_2^T d_1 d_1^T V_2 + \rho_2 d_2 d_2^T \\
&\vdots \\
B_m &= V_m^T B_{m-1} V_m + \rho_m d_m d_m^T \\
&= \left(V_m^T \cdots V_2^T V_1^T\right) B_0 \left(V_1 V_2 \cdots V_m\right) + \rho_1 \left(V_m^T \cdots V_2^T\right) d_1 d_1^T \left(V_2 \cdots V_m\right) \\
&\qquad + \rho_2 \left(V_m^T \cdots V_3^T\right) d_2 d_2^T \left(V_3 \cdots V_m\right) \\
&\qquad + \cdots + \rho_m d_m d_m^T
\end{aligned} \tag{3.81}$$

It is assumed that only the vectors $(d_j)_{j=1\text{ to }m}$ and $(y_j)_{j=1\text{ to }m}$ have been stored in the previous m iterations. No matrix $(V_j$ or $B_j)$ has been stored.
The direction $B_m g_m$ that we are trying to calculate is expressed using (3.81).

$$\begin{aligned} B_m g_m = \; & \left(V_m^T \cdots V_1^T\right) B_0 \left(V_1 \cdots V_m\right) g_m \\ & + \left(V_m^T \cdots V_2^T\right) d_1 \rho_1 d_1^T \left(V_2 \cdots V_m\right) g_m \\ & + \left(V_m^T \cdots V_3^T\right) d_2 \rho_2 d_2^T \left(V_3 \cdots V_m\right) g_m + \cdots \\ & + V_m^T d_{m-1} \rho_{m-1} d_{m-1}^T V_m g_m \\ & + d_m \rho_m d_m^T g_m \end{aligned} \tag{3.82}$$

Let us define the real series $(q_j)_{j=m+1\text{ to }1}$ and $(\alpha_j)_{j=m\text{ to }1}$ as follows.

$$\begin{cases} q_{m+1} = g_m \\ q_j = (V_j \cdots V_m) g_m \end{cases} \quad \text{and} \quad \alpha_j = \rho_j d_j^T q_{j+1} \tag{3.83}$$

We can then write $B_m g_m$ (3.82) as

$$\begin{aligned} B_m g_m = \; & \left(V_m^T \cdots V_1^T\right) B_0 q_m \\ & + \left(V_m^T \cdots V_2^T\right) d_1 \alpha_1 + \left(V_m^T \cdots V_3^T\right) d_2 \alpha_2 + \cdots + V_m^T d_{m-1} \alpha_{m-1} \\ & + d_m \alpha_m \end{aligned} \tag{3.84}$$

which can be factorized into

$$\begin{aligned} B_m g_m = \; & \alpha_m d_m \\ & + V_m^T (\; \alpha_{m-1} d_{m-1} \\ & \qquad + V_{m-1}^T (\; \alpha_{m-2} d_{m-2} \\ & \qquad\qquad + V_{m-2}^T (\cdots \\ & \qquad\qquad\qquad + V_2^T (\alpha_1 d_1 + V_1^T (B_0 q_1)) \cdots))) \end{aligned} \tag{3.85}$$

This expression lends itself to a recursive calculation in two loops:

- the first loop from j = m to 1 calculates the reals $(q_j, \alpha_j)_{j=m\text{ to }1}$.
  Using the expression of V_j (3.80), we observe that

$$q_{j-1} = V_{j-1} q_j = q_j - \rho_{j-1} y_{j-1} d_{j-1}^T q_j = q_j - \alpha_{j-1} y_{j-1} \tag{3.86}$$

The real numbers $(q_j, \alpha_j)_{j=m\text{ to }1}$ can therefore be calculated from $q_{m+1} = g_m$.

$$\begin{cases} \alpha_j = \rho_j d_j^T q_{j+1} \\ q_{j-1} = q_j - \alpha_j y_j \end{cases} \quad \text{for } j = m \text{ to } 1 \tag{3.87}$$

- the second loop from $j = 1$ to m calculates $B_m g_m$ in the factorized form (3.85). Using the expression of V_j (3.80) and noting r_j the contents of the nested brackets in (3.85), we observe that

$$r_{j+1} = \alpha_j d_j + V_j^T r_j = \alpha_j d_j + r_j - \rho_j d_j y_j^T r_j \tag{3.88}$$

Starting from the inner parenthesis of (3.85): $r_1 = B_0 q_1$ where q_1 is given by (3.87), we calculate the series $(r_j)_{j=1 \text{ to } m+1}$ whose last term gives $r_{m+1} = B_m g_m$.

Formulas (3.87) and (3.88) use only vectors, real numbers and the matrix B_0. This matrix is chosen to be diagonal as in (3.73).

$$B_0 = \tau I \quad \text{with} \quad \tau = \frac{y^T d}{y^T y} \quad , \quad \begin{cases} d = x_k - x_{k-1} \\ y = g_k - g_{k-1} \end{cases} \tag{3.89}$$

The L-BFGS method requires choosing the number m of steps to be stored.

At each iteration, the initial matrix B_0 is calculated by (3.89) based on the first (oldest) stored move. Then the direction $B_m g_m$ is calculated by (3.86 - 3.88) and the one-dimensional search in this direction yields a new point. The list of stored moves $(d_j, y_j)_{j=1 \text{ to } m}$ is updated by deleting the oldest move and adding the new move at the end of the list. The L-BFGS method is identical to the classical BFGS method during the first m iterations (as the matrix B_0 does not change). From iteration m+1 onwards, the generated directions $B_m g_m$ become different due to the updating of the matrix B_0. This method using only vectors allows the quasi-Newton idea to be applied with a significant memory saving, which is essential for large-scale programming problems.

3.3 Line search

The second shortcoming of Newton's method mentioned in section 3.1.3 is its lack of robustness. Globalization consists of checking whether the iteration is satisfactory and correcting it if necessary. The two main globalization strategies are the line search presented in this section and the trust region presented in the next section. These strategies generally use Newton's solution as a starting point.

A line search descent method (figure 3-16) consists at each iteration in:
- choosing a direction of descent u_k;
- finding a step length s_k in this direction.

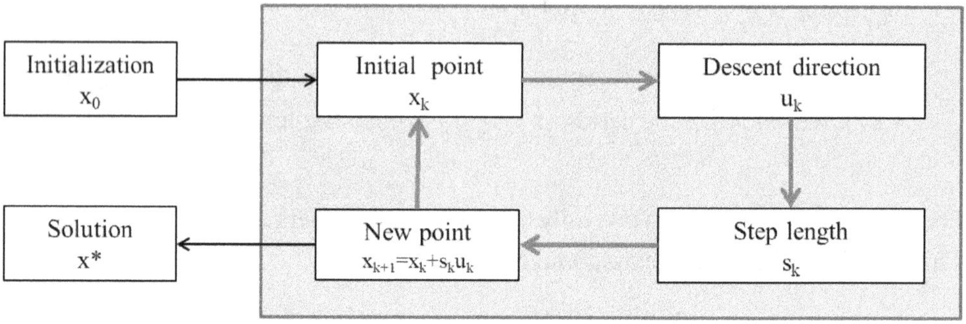

Figure 3-16: Descent by line search.

The possible choices of the direction u_k and the step length s_k are detailed in sections 3.3.1 and 3.3.2 respectively.

3.3.1 Direction of descent

The direction u_k is a descent direction in x_k if

$$g_k^T u_k < 0 \quad \text{with} \quad g_k = \nabla f(x_k) \tag{3.90}$$

Among the various strategies, we examine the steepest descent, the accelerated gradient, the Nesterov method, the conjugate gradient and the preconditioned gradient.

Steepest descent

The method of steepest descent consists in choosing the direction: $u_k = -g_k$. Since the gradient is the direction of steepest ascent in the point x_k, it is natural to look for the new point x_{k+1} opposite to this direction. The point minimizing the function f along u_k is obtained by solving the one-dimensional problem

$$\min_{s>0} \varphi(s) = f(x_k + su_k) \tag{3.91}$$

where the unknown is the step length $s > 0$ along the direction u_k.

Minimizing the directional function φ yields

$$\varphi'(s)=0 \;\Rightarrow\; \nabla f(x_k+su_k)^T u_k = \nabla f(x_{k+1})^T u_k = g_{k+1}^T u_k = 0 \tag{3.92}$$

The gradient in the new point x_{k+1} is orthogonal to the direction of move u_k. This result holds true for any direction u_k. In the case of the steepest descent method: $u_k=-g_k$, the successive moves will be orthogonal as illustrated in figure 3-17.

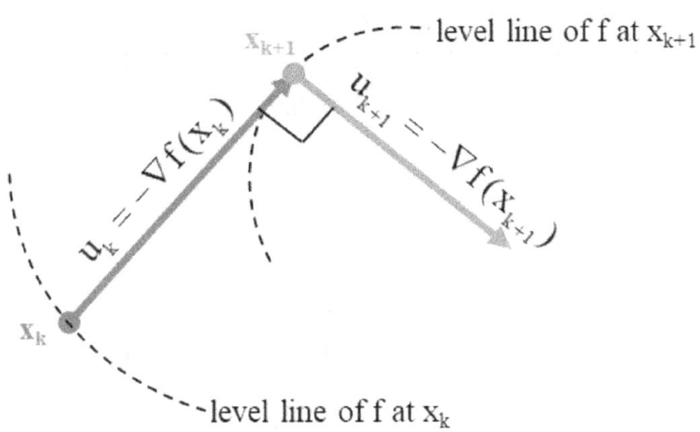

Figure 3-17: Successive directions of steepest descent.

The steepest descent method, also called the simple gradient method, thus leads to a zig-zag progression. Convergence to the minimum can be very slow, as the example 3-12 shows.

Example 3-12: Steepest descent method

Consider the problem with 2 variables: $\min_{x_1,x_2} f(x_1,x_2)=\frac{1}{2}x_1^2+\frac{9}{2}x_2^2$.

The minimum is in (0; 0). Let us apply the steepest descent method starting from the point (9; 1). The direction of descent from any point $(x_1;x_2)$ is given by

$$u=-\nabla f(x_1,x_2) \;\Rightarrow\; \begin{pmatrix}u_1\\u_2\end{pmatrix}=\begin{pmatrix}-x_1\\-9x_2\end{pmatrix}$$

The optimal step length in this direction is calculated analytically.

$$\min_s \varphi(s)=f(x_1+su_1,x_2+su_2)=\frac{1}{2}(x_1+su_1)^2+\frac{9}{2}(x_2+su_2)^2 \;\rightarrow\; s=\frac{x_1^2+81x_2^2}{x_1^2+729x_2^2}$$

Table 3-11 shows the components of the point, the direction and the value of the optimal step during the first 5 iterations and then every 10 iterations. The iterations are plotted in the plane (x_1, x_2) in figure 3-18. The characteristic zig-zag progression of the steepest descent method can be seen.

Fifty iterations are required to obtain the solution to within 10^{-4}.

Iteration	x_1	x_2	f	u_1	u_2	s	Error
0	9.000	1.000	45.000	-9.000	-9.000	0.2	9.055
1	7.200	-0.800	28.800	-7.200	7.200	0.2	7.244
2	5.760	0.640	18.432	-5.760	-5.760	0.2	5.795
3	4.608	-0.512	11.796	-4.608	4.608	0.2	4.636
4	3.686	0.410	7.550	-3.686	-3.686	0.2	3.709
5	2.949	-0.328	4.832	-2.949	2.949	0.2	2.967
10	0.966	0.107	0.519	-0.966	-0.966	0.2	0.972
20	0.104	0.012	0.006	-0.104	-0.104	0.2	0.104
30	1.11E-02	1.24E-03	6.90E-05	-1.11E-02	-1.11E-02	0.2	1.12E-02
40	1.20E-03	1.33E-04	7.95E-07	-1.20E-03	-1.20E-03	0.2	1.20E-03
50	1.28E-04	1.43E-05	9.17E-09	-1.28E-04	-1.28E-04	0.2	1.29E-04

Table 3-11: Iterations of the steepest descent method.

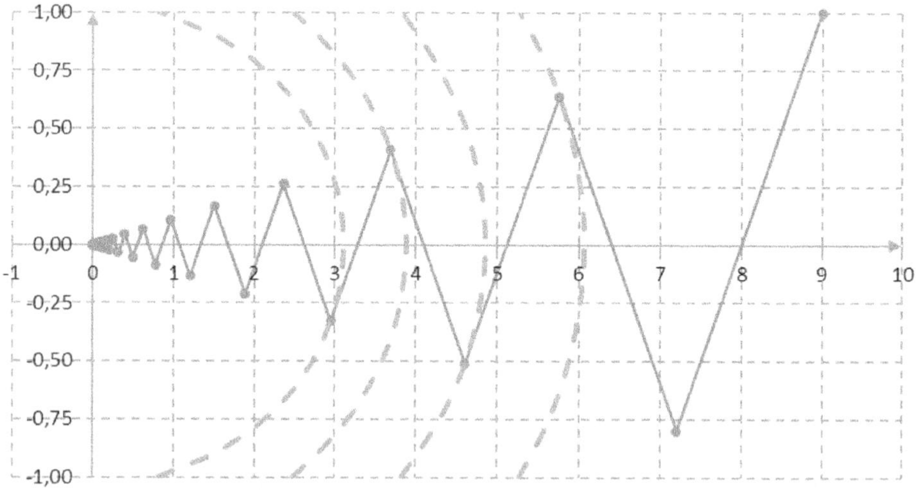

Figure 3-18: Iterations of the steepest descent method.

Accelerated gradient

The steepest descent method generates orthogonal directions, which leads to a slow zig-zag progression. Suppose that p consecutive iterations of steepest slope have been performed, starting from the point x_0 and ending in the point x_p.

$$x_0 \xrightarrow{u_0} x_1 \xrightarrow{u_1} x_2 \xrightarrow{u_2} \cdots\cdots \xrightarrow{u_{p-1}} x_p$$

Since the consecutive directions are orthogonal, one can expect the "average" direction from x_0 to x_p to be a good descent direction. A line search along this direction: $u_p = x_p - x_0$ will yield a new point x_{p+1}.

$$x_p \xrightarrow{u_p} x_{p+1}$$

Starting from x_{p+1}, we again perform p iterations of steepest descent, then a new line search along the direction: $u_{2p+1} = x_{2p+1} - x_{p+1}$ and so on.

The **accelerated gradient** method thus consists in completing the steepest gradient method by periodically inserting a search along the average direction of the last p iterations. The periodicity can be set to p = n, where n is the dimension of the problem. This process is illustrated in figure 3-19 .

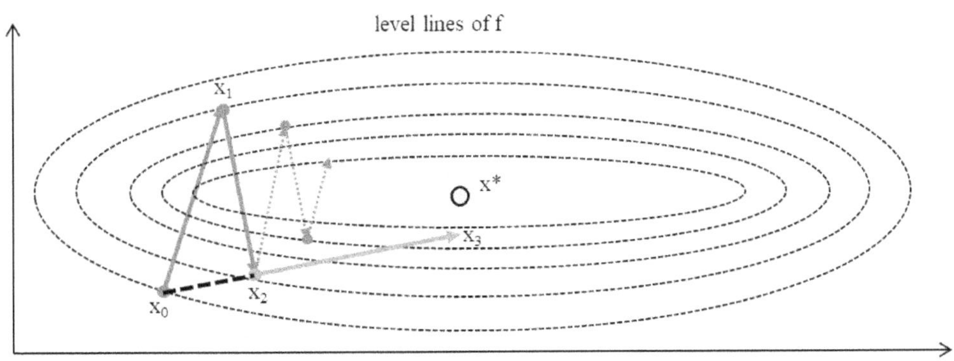

Figure 3-19: Accelerated gradient method.

Let us examine the behavior of this method in the special case of a convex quadratic function in dimension 2.

$$\min_{x \in \mathbb{R}^2} f(x) = \frac{1}{2} x^T Q x + c^T x \tag{3.93}$$

The gradient of the function is given by: $g(x) = \nabla f(x) = Qx + c$.

Since the matrix Q is positive, the minimum is at: $x^* = -Q^{-1}c$.

Consider two consecutive steepest descent iterations from the point x_0 and the gradient directions g_0 and g_2 in the points x_0 and x_2 respectively (figure 3-20).

$$\begin{cases} g_0 = Qx_0 + c \\ g_2 = Qx_2 + c \\ 0 = Qx^* + c \end{cases} \Rightarrow \begin{cases} x_2 - x_0 \underset{\text{noted}}{=} u_A = Q^{-1}(g_2 - g_0) \\ x^* - x_2 \underset{\text{noted}}{=} u^* = -Q^{-1}g_2 \end{cases} \tag{3.94}$$

Since the successive directions are orthogonal: $g_0 \perp g_1$, $g_1 \perp g_2$, the direction g_2 is parallel to the direction g_0 (<u>this is only true in dimension n = 2</u>).

We deduce from (3.94) that the average direction u_A (from x_0 to x_2) is parallel to the direction u^* (from x_2 to the minimum x^*). The line search in this direction u_A will therefore make it possible to reach directly the minimum x^* from the third iteration.

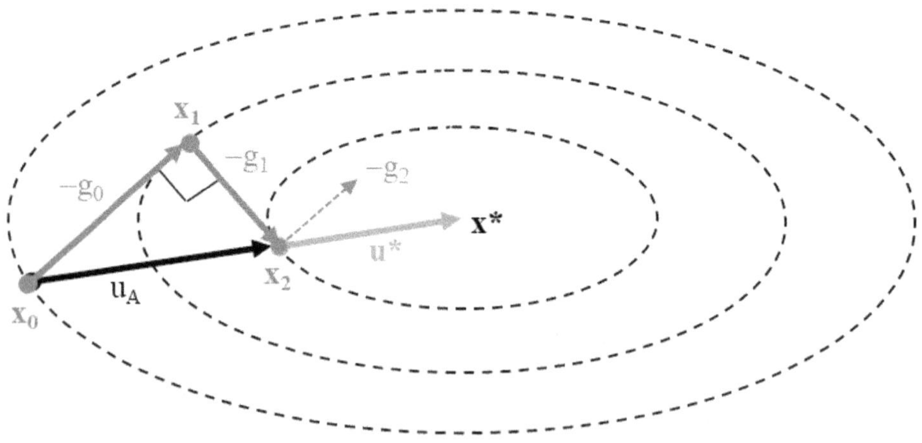

Figure 3-20: Accelerated gradient on a quadratic function in dimension 2.

Example 3-13: Accelerated gradient method

Let us retrieve the problem of example 3-12: $\min_{x_1,x_2} f(x_1,x_2)=\frac{1}{2}x_1^2+\frac{9}{2}x_2^2$.

The initial point is in (9; 1), the minimum is in (0; 0).

The accelerated gradient method consists of two iterations of steepest descent, followed by one iteration in the mean direction.

The optimal step along the direction of components (u_1,u_2) is given by

$$\min_s \varphi(s)=f(x_1+su_1,x_2+su_2)=\frac{1}{2}(x_1+su_1)^2+\frac{9}{2}(x_2+su_2)^2 \quad \rightarrow \quad s=-\frac{u_1x_1+9u_2x_2}{u_1^2+9u_2^2}$$

Table 3-12 shows the components of the point, the direction and the optimal step. The minimum is reached exactly at the third iteration. For comparison, the simple gradient method (steepest descent) reaches an accuracy of 10^{-4} after 50 iterations (table 3-11).

Iteration	x_1	x_2	f	u_1	u_2	s	Error
0	9.0000	1.0000	45.0000	-9.0000	-9.0000	0.2	9.0554
1	7.2000	-0.8000	28.8000	-7.2000	7.2000	0.2	7.2443
2	5.7600	0.6400	18.4320	-3.2400	-0.3600	1.8	5.7954
3	0.0000	0.0000	0.0000				0.0000

Table 3-12: Iterations of the accelerated gradient method.

In the general case (any function f, dimension n > 2), the average direction does not necessarily point toward the minimum x*. Nevertheless, this average direction allows a significant progression, which compensates for the defect of the steepest descent method. The principle of periodically inserting a line search along an average direction applies to any descent method and often allows for faster convergence.

Nesterov's method

Similarly to the accelerated gradient, Nesterov's method is an improvement of the steepest descent method by additional moves.
The principle is to systematically increase the steepest descent move to avoid generating orthogonal directions.

The first two iterations are identical to the steepest slope method.

From the third iteration onwards (k > 2), the move is carried out in two stages:

- the first stage consists in "stretching" the previous move from x_{k-1} to x_k.

  We obtain an intermediate point y_k defined by

$$y_k = x_k + \theta_k (x_k - x_{k-1}) \quad \text{with} \quad \theta_k = \frac{k-1}{k+2} \tag{3.95}$$

  The coefficient θ_k increases during the iterations while remaining below 1;

- the second stage is a classical steepest iteration starting from the intermediate point y_k. The new point x_{k+1} is defined by

$$x_{k+1} = x_k + s_k u_k \quad \text{with} \quad \begin{cases} u_k = -\nabla f(y_k) \\ s_k \ \to \ \min\limits_s f(x_k + s_k u_k) \end{cases} \tag{3.96}$$

These two stages are illustrated in figure 3-21. The idea of Nesterov's method is that the direction of steepest slope $-g(y_k)$ from the offset point y_k will be more favorable than from the point x_k.

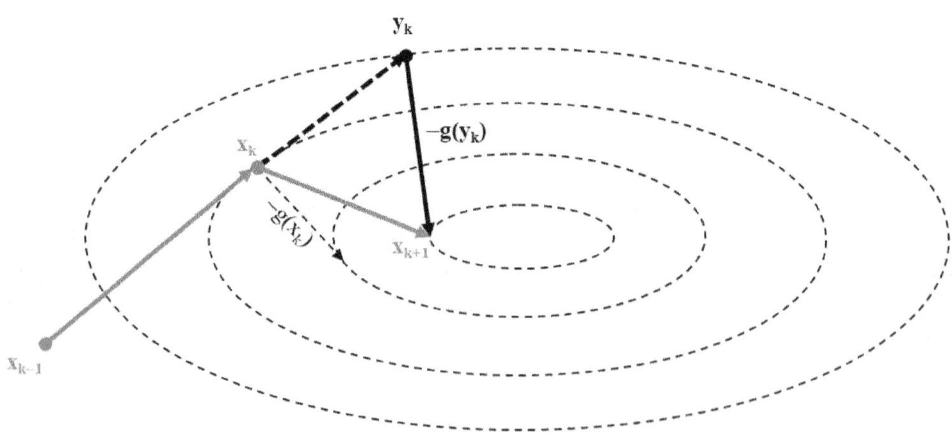

Figure 3-21: Nesterov's method.

Example 3-14: Nesterov's method

Let us retrieve the problem of example 3-12 : $\min_{x_1,x_2} f(x_1,x_2)=\frac{1}{2}x_1^2+\frac{9}{2}x_2^2$.

The initial point is in (9; 1), the minimum is in (0; 0).

The first two iterations follow the steepest descent method.

The next iterations are performed by

$$\begin{cases} y_k = x_k + \theta_k(x_k - x_{k-1}) & \text{with } \theta_k = \dfrac{k-1}{k+2} \\ x_{k+1} = y_k + s_k u_k & \text{with } u_k = -\nabla f(y_k) = \begin{pmatrix} -y_1 \\ -9y_2 \end{pmatrix} \end{cases}$$

where the optimal step s_k is given by

$$\min_s \varphi(s) = f(x_k + su_k) \quad \rightarrow \quad s_k = -\frac{u_{k,1}x_{k,1} + 9u_{k,2}x_{k,2}}{u_{k,1}^2 + 9u_{k,2}^2}$$

Table 3-13 compares Nesterov's method with the steepest descent method. After two identical iterations, a faster decrease of the function is observed.

Figure 3-22 shows the progression of the two methods towards the minimum located in (0; 0).

	Steepest descent			Nesterov		
Iteration	x_1	x_2	f	x_1	x_2	f
0	9.000	1.000	45.0000	9.0000	1.0000	45.0000
1	7.200	-0.800	28.8000	7.2000	-0.8000	28.8000
2	5.760	0.640	18.4320	5.7600	0.6400	18.4320
3	4.608	-0.512	11.7965	4.6154	-0.3077	11.0769
4	3.686	0.410	7.5497	3.5187	0.2630	6.5018
5	2.949	-0.328	4.8318	2.5381	-0.1697	3.3507
6	2.359	0.262	3.0924	1.7054	0.0999	1.4991
7	1.887	-0.210	1.9791	1.0259	-0.0559	0.5403
8	1.510	0.168	1.2666	0.5004	0.0221	0.1274
9	1.208	-0.134	0.8106	0.1173	-0.0025	0.0069
10	0.966	0.107	0.5188	-0.1320	0.0129	0.0095

Table 3-13: Iterations of Nesterov's method.

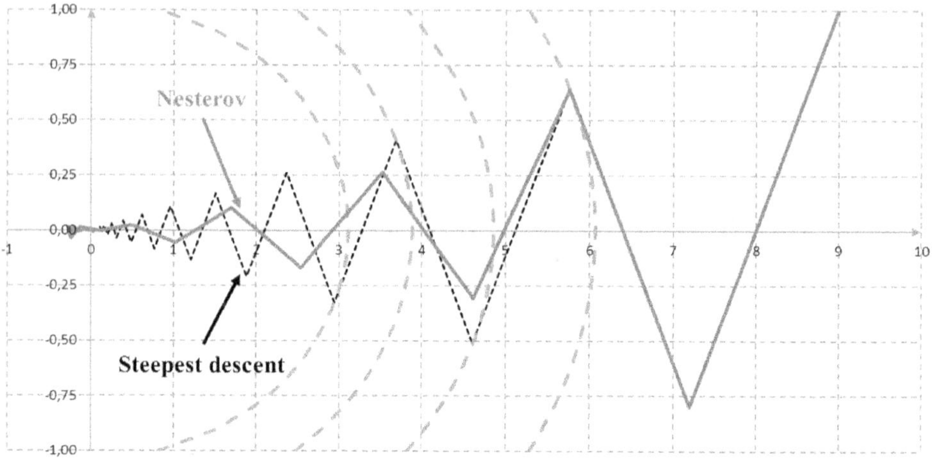

Figure 3-22: Iterations of Nesterov's method.

The lengthening of the steepest descent move, with the coefficient θ_k, produces increasingly favorable directions, allowing a faster approach of the minimum.

Conjugate gradient

Consider the minimization of a convex quadratic function.

$$\min_{x\in\mathbb{R}^n} f(x) = \frac{1}{2}x^T Qx + c^T x \quad , \quad c\in\mathbb{R}^n \, , Q\in\mathbb{R}^{n\times n} \, , Q>0 \tag{3.97}$$

Two directions u_i and u_j are said to be **conjugate with** respect to the Q matrix if they satisfy

$$u_i^T Q u_j = 0 \tag{3.98}$$

The interest of these combined directions comes from property 3-2.

Property 3-2: Minimization by conjugate directions

Let us define from any point x_0 the series of points $(x_k)_{k=1 \text{ to } n}$ by

$$x_{k+1} = x_k + s_k u_k \tag{3.99}$$

where the directions $(u_k)_{k=0 \text{ to } n-1}$ are mutually conjugate and the step s_k is chosen to minimize f along the direction u_k.

$$\min_s f(x_k + su_k) \;\Rightarrow\; u_k^T \nabla f(x_k + s_k u_k) = u_k^T \left[Q(x_k + s_k u_k) + c\right] = 0$$
$$\Rightarrow\; s_k = -\frac{u_k^T (Qx_k + c)}{u_k^T Q u_k} \tag{3.100}$$

Then the point x_k is the minimum of the quadratic function f on the subspace generated by the directions $(u_0, \cdots, u_{k-1})$ and passing through x_0.

In particular, the point x_n is the solution of the quadratic problem (3.97).

Demonstration of property 3-2 (see [R11])

The conjugate directions $(u_i)_{i=0 \text{ to } n-1}$ are independent.

Indeed, if a linear combination of (u_i) is zero, we have

$\sum_{i=0}^{n-1} \alpha_i u_i = 0 \;\Rightarrow u_j^T Q \sum_{i=0}^{n-1} \alpha_i u_i = 0 \;\Rightarrow \alpha_j u_j^T Q u_j = 0 ,\forall j$ because $u_j^T Q u_i = 0$ for $i \neq j$

Since the matrix Q is positive, the coefficient α_j is zero.

Consider the subspace U_k of $\mathbb{R}^n$ generated by the first k directions $(u_0, \cdots, u_{k-1})$.

The gradient of f in the point $x_k = x_0 + \sum_{i=0}^{k-1} s_i u_i$ is given by: $g_k = \nabla f(x_k) = Qx_k + c$.

Let us calculate the dot product of the gradient g_k with the directions $(u_j)_{j=0 \text{ to } k-1}$.

$$u_j^T g_k = u_j^T (Qx_k + c) = u_j^T Q x_0 + \sum_{i=0}^{k-1} s_i u_j^T Q u_i + u_j^T c = u_j^T Q x_0 + s_j u_j^T Q u_j + u_j^T c$$

because $u_j^T Q u_i = 0$ for $i \neq j$.

Using the expression (3.100) for the step s_j and with $x_{j-1} = x_0 + \sum_{i=0}^{j-2} s_i u_i$, we get

$$u_j^T g_k = u_j^T Q x_0 - u_j^T (Q x_j + c) + u_j^T c = u_j^T Q(x_0 - x_j) = -u_j^T \sum_{i=0}^{j-1} s_i Q u_i \quad \Rightarrow \quad u_j^T g_k = 0 \,,\ \forall j < k$$

The gradient of f in x_k is orthogonal to the basis $(u_j)_{j=0 \text{ to } k-1}$, and hence to the subspace U_k. The point x_k is therefore the minimum of f on U_k and the point x_n is the minimum of f on $U_n = \mathbb{R}^n$.

Property 3-2 shows that the minimum of a convex quadratic function is reached in n one-dimensional minimizations along conjugate directions. To exploit this property, one must know how to construct these conjugate directions.

The **conjugate gradient** method defines the new direction u_k as a linear combination of the previous direction u_{k-1} and of the gradient g_k in x_k.

$$\begin{aligned} u_0 &= -g_0 \\ u_k &= -g_k + \beta_k u_{k-1} \end{aligned} \quad \text{with} \quad \begin{cases} g_k = \nabla f(x_k) = Q x_k + c \\ \beta_k = \dfrac{g_k^T Q u_{k-1}}{u_{k-1}^T Q u_{k-1}} \end{cases} \tag{3.101}$$

Justification

We check by induction that the directions defined by (3.101) are conjugate.

Let us assume that the directions $(u_i)_{i=0 \text{ to } k-1}$ are conjugate and show that the new direction u_k is conjugate to all previous directions $(u_i)_{i=0 \text{ to } k-1}$.

For that purpose, we calculate: $u_k^T Q u_i = -g_k^T Q u_i + \beta_k u_{k-1}^T Q u_i$ **(equality E)**

For $i = k-1$: $u_k^T Q u_{k-1} = -g_k^T Q u_{k-1} + \beta_k u_{k-1}^T Q u_{k-1} = 0$

by definition of β_k (3.101)

For $i < k-1$: $u_k^T Q u_i = -g_k^T Q u_i$

because the directions $(u_i)_{i=0 \text{ to } k-1}$ are conjugate by assumption.

Using the relations $\begin{cases} x_{i+1} = x_i + s_i u_i \\ u_{i+1} = -g_{i+1} + \beta_{i+1} u_i \\ u_i = -g_i + \beta_i u_{i-1} \end{cases}$, we put the term Qu_i into the form

$$Qu_i = Q\frac{x_{i+1} - x_i}{s_i} = \frac{1}{s_i}(Qx_{i+1} - Qx_i) = \frac{1}{s_i}(g_{i+1} - g_i) = \frac{1}{s_i}(-u_{i+1} + \beta_{i+1}u_i + u_i - \beta_i u_{i-1})$$

This term is a linear combination of the directions u_{i-1}, u_i, u_{i+1} which belong to the subspace U_k. Since the gradient of f in x_k is orthogonal to U_k (as was demonstrated for property 3-2), we have: $g_k^T Qu_i = 0$.

From equality (E), we can deduce that: $u_k^T Qu_i = 0$, $\forall i < k$.

The formulas (3.101) allow us to construct at each iteration a new conjugate direction u_k from the previous direction u_{k-1} by calculating only the gradient in the new point x_k, hence the name of the conjugate gradient method. The optimal step s_k along this direction is given by (3.100).

Example 3-15: Minimization of a quadratic function by conjugate gradient

Apply the conjugate gradient method to: $\min\limits_{\xi_1,\xi_2} f(\xi_1,\xi_2) = \frac{1}{2}\xi_1^2 + \xi_1\xi_2 + \xi_2^2$.

The gradient and the Hessian are: $g(\xi_1,\xi_2) = \nabla f(\xi_1,\xi_2) = \begin{pmatrix} \xi_1 + \xi_2 \\ \xi_1 + 2\xi_2 \end{pmatrix}, Q = \begin{pmatrix} 1 & 1 \\ 1 & 2 \end{pmatrix}$

The initial point is in (10;– 5), the minimum is in (0; 0).

Iteration 1

The 1st direction is: $u_0 = -g_0 \rightarrow u_0 = \begin{pmatrix} -5 \\ 0 \end{pmatrix}$.

The new point is: $x_1 = x_0 + su_0 = \begin{pmatrix} 10 - 5s \\ -5 \end{pmatrix} = 5\begin{pmatrix} 2 - s \\ -1 \end{pmatrix}$.

Step calculation: $\min\limits_s f(x_0 + su_0) = \frac{25}{2}(2-s)^2 - 25(2-s) + 25 \rightarrow s_0 = 1$

$$\rightarrow x_1 = \begin{pmatrix} 5 \\ -5 \end{pmatrix}, \quad g_1 = \begin{pmatrix} 0 \\ -5 \end{pmatrix}$$

Iteration 2

The 2nd direction is: $u_1 = -g_1 + \beta_1 u_0$ with $\beta_1 = \frac{g_1^T Q u_0}{u_0^T Q u_0} = \frac{25}{25} = 1 \rightarrow u_1 = \begin{pmatrix} -5 \\ 5 \end{pmatrix}$.

The new point is: $x_2 = x_1 + su_1 = \begin{pmatrix} 5-5s \\ -5+5s \end{pmatrix} = 5(1-s)\begin{pmatrix} 1 \\ -1 \end{pmatrix}$.

Step calculation: $\min_s f(x_1 + su_1) = \frac{25}{2}(1-s)^2 - 25(1-s)^2 + 25(1-s)^2 \rightarrow s_1 = 1$

$$\rightarrow x_2 = \begin{pmatrix} 0 \\ 0 \end{pmatrix} , \; g_2 = \begin{pmatrix} 0 \\ 0 \end{pmatrix}$$

The minimum is obtained in 2 iterations with the conjugate directions u_0 and u_1.

The conjugate gradient method reaches the minimum of a quadratic function in n iterations. This method can be applied to the minimization of any function. To do this, the coefficient β_k must be expressed without using the Q matrix (associated with a quadratic function). The most commonly used formulas are:

- the formula of Fletcher-Reeves (1964):

$$\beta_k = \frac{\|g_k)\|^2}{\|g_{k-1}\|^2} \tag{3.102}$$

- the formula of Polak-Ribière (1971):

$$\beta_k = \frac{g_k^T (g_k - g_{k-1})}{\|g_{k-1}\|^2} \tag{3.103}$$

These formulas are equivalent to (3.101) in the case of a quadratic function.

Verification of formulae (3.102) and (3.103) for a quadratic function

The coefficient β_k is given by (3.101) : $\beta_k = \frac{g_k^T Q u_{k-1}}{u_{k-1}^T Q u_{k-1}}$.

With the relations: $\begin{cases} x_k &= x_{k-1} + s_{k-1} u_{k-1} \\ g_k &= Q x_k + c \\ g_{k-1} &= Q x_{k-1} + c \end{cases}$, we have: $Qu_{k-1} = Q\dfrac{x_k - x_{k-1}}{s_{k-1}} = \dfrac{Qx_k - Qx_{k-1}}{s_{k-1}} = \dfrac{g_k - g_{k-1}}{s_{k-1}}$

which we replace in the expression of β_k : $\beta_k = \dfrac{g_k^T(g_k - g_{k-1})}{u_{k-1}^T(g_k - g_{k-1})}$.

The gradient g_k is orthogonal to the subspace of $(u_i)_{i=0 \text{ to } k-1}$ $\rightarrow u_{k-1}^T g_k = 0$

Using the relation: $u_{k-1} = -g_{k-1} + \beta_{k-1}u_{k-2}$, we have

- by scalar product with g_{k-1} : $u_{k-1}^T g_{k-1} = -g_{k-1}^T g_{k-1}$ because $g_{k-1} \perp u_{k-2}$

$$\rightarrow \beta_k = \frac{g_k^T(g_k - g_{k-1})}{g_{k-1}^T g_{k-1}}$$

- by scalar product with g_k : $0 = -g_{k-1}^T g_k$ because $g_k \perp u_{k-1}$ and $g_k \perp u_{k-2}$

$$\rightarrow \beta_k = \frac{g_k^T g_k}{g_{k-1}^T g_{k-1}}$$

We obtain the formulas of Fletcher-Reeves (3.102) and Polak-Ribière (3.103), which are equivalent to (3.101) in the case of a quadratic function.

The conjugate gradient method with Fletcher-Reeves or Polak-Ribière directions is a so-called first order method: it uses only the calculation of the gradient at each iteration, followed by a one-dimensional minimization in the new direction. It is therefore no more expensive than the previous methods (accelerated gradient, Nesterov) and it generally converges faster.

Preconditioned gradient

Preconditioning methods aim to accelerate convergence by approaching Newton's method. On the other hand, these so-called second order methods require more calculations and storage at each iteration.

Their principle is to multiply the gradient by a matrix B_k providing curvature information. Any direction of the form $u_k = -B_k g_k$ is a direction of descent if the matrix B_k is positive (because $g_k^T u_k = -g_k B_k g_k < 0$).

The gradient g_k is then said to be **preconditioned** by the matrix B_k .
Preconditioning is equivalent to applying the steepest descent method after a linear variable change.

Interpretation of preconditioning

Consider the change of variable $\tilde{x} = L^T x$, where the lower triangular matrix L comes from the Cholesky factorization: $B_k^{-1} = H_k = LL^T$.
This factorization exists because B_k is positive definite.

Let us note $\tilde{f}(\tilde{x}) \underset{def}{=} f(x)$ the function of the new variable $\tilde{x}$. Its gradient is: $\nabla\tilde{f}(\tilde{x}) = \nabla f(L^{-T}x) = L^{-1}\nabla f(x)$.

The steepest iteration in variable $\tilde{x}$ is: $\tilde{x}_{k+1} = \tilde{x}_k - s_k \nabla\tilde{f}(\tilde{x}_k)$.

Going back to the variable x, the iteration becomes: $L^T x_{k+1} = L^T x_k - s_k L^{-1}\nabla f(x_k)$.

Pre-multiplying by L^{-T}, we obtain

$x_{k+1} = x_k - s_k L^{-T}L^{-1}\nabla f(x_k) = x_k - s_k (LL^T)^{-1} g_k = x_k - s_k B_k g_k$

The iteration in variable x with the preconditioned direction $B_k g_k$ is therefore identical to the steepest descent iteration in variable $\tilde{x} = L^T x$.

The "optimal" preconditioning is associated with the Hessian of the function to be minimized $B_k = H_k^{-1}$ with $H_k = \nabla^2 f(x_k)$. This would be equivalent to a Newton iteration (3.23). In practice, the preconditioning matrix is obtained by a quasi-Newton method such as DFP, BFGS or SR1 (section 3.3). In the case of SR1, the matrix H_k must be made positive. This is achieved by adding a diagonal matrix: $H'_k = H_k + \tau I$ with a sufficiently large positive real τ, as shown below.

Adding a diagonal matrix

The matrix H is symmetric and therefore has an orthonormal basis of eigenvectors. By noting P the orthogonal transfer matrix in this basis and D the diagonal matrix formed by the eigenvalues of H, we can express H in the form $H = P^{-1}DP$.

Then we have: $H' = H + \tau I = P^{-1}DP + \tau PP^{-1} = P^{-1}(D + \tau I)P$.

The symmetric matrix H' of the form $H' = P^{-1}D'P$ thus has the same eigenvector basis as H (same matrix P). Its eigenvalues are those of H increased by τ. By noting λ_m the most negative eigenvalue of H, it is sufficient to take $\tau = |\lambda_m| + \varepsilon$ (with $\varepsilon > 0$) to obtain a positive definite matrix H'.

An alternative method for obtaining a positive definite matrix from H_k is the modified Cholesky factorization. This method has the advantage of modifying the initial matrix as little as possible.

Modified Cholesky factorization

The Cholesky factorization puts a symmetric positive definite matrix A in the form $A = LDL^T$, with a lower triangular matrix L and a diagonal matrix D formed by the eigenvalues of A.

Cholesky's algorithm can be adapted to a non-positive definite matrix A. The modification is that when a negative value is encountered on the diagonal, it is replaced by a small positive value. This results in the factorization of a positive definite matrix A' "close" to A.

Figure 3-23 depicts the factorization algorithm with modification of the diagonal terms (when necessary).

The matrices are noted: $A = (a_{ij})$, $L = (l_{ij})$, $D = \mathrm{diag}(d_j)$.

For j = 1,2,…,n

$c_{jj} = a_{jj} - \sum_{s=1}^{j-1} d_s l_{js}^2$

$d_j = c_{jj}$ — If $d_j \le \delta$ →

For i = j+1,…,n

$c_{ij} = a_{ij} - \sum_{s=1}^{j-1} d_s l_{is} l_{js}$

$l_{ij} = \dfrac{c_{ij}}{d_j}$

End loop i

End loop j

Modification to get $d_j > 0$
(if A is not definite positive)

$$d_j = \max\left(|c_{jj}|, \left(\frac{\theta_j}{\beta}\right)^2, \delta \right) \text{ with } \theta_j = \max_{j<i\le n} |c_{ij}|$$

$$\Rightarrow \begin{cases} d_j \ge \delta \\ m_{ij} = l_{ij}\sqrt{d_j} \le \beta, \quad i = j+1,\dots,n \end{cases}$$

Parameters : $\delta > 0$ small $(\delta = 10^{-3})$
$\beta > 0$ $(\beta = 1)$

Figure 3-23: Modified Cholesky factorization.

3.3.2 Step length

For a given descent direction u_k, a step length $s_k > 0$ along u_k is sought to find a point x_{k+1} that is better than the initial point x_k.

This is a one-variable minimization problem which is formulated as

$$\min_{s>0} \varphi(s) = f(x_k + su_k) \tag{3.104}$$

where the directional function φ gives the variation of f along the direction u_k.

In a line search algorithm, the problem (3.104) has to be solved at each iteration from a new point x_k and along a new direction u_k (figure 3-16).
Obtaining a precise solution by a classical one-dimensional minimization method such as the golden ratio (section 2.2.3) would be costly in terms of function calls. We can settle for an approximate solution respecting the following two conditions.

Sufficient decrease condition

This condition relates to the function decrease between the initial point x_k and the new point x_{k+1}. The directional derivative of f along the direction u_k is given by: $\varphi'(0) = \nabla f(x_k)^T u_k < 0$. Moving in this direction, one can expect a decrease of f proportional to $\varphi'(0)$.

Armijo's condition, also called Goldstein's or Wolfe's first condition, requires that the step s satisfies

$$\varphi(s) < \varphi(0) + c_1 s\varphi'(0) \tag{3.105}$$

or equivalently

$$f(x_k + su_k) < f(x_k) + c_1 s\nabla f(x_k)^T u_k \tag{3.106}$$

The coefficient c_1 on the directional derivative is usually set to $c_1 = 0{,}1$.

The line with slope $c_1\varphi'(0) < 0$ from the origin $(s = 0)$ is called the Armijo line or the first Goldstein line.

Figure 3-24 shows the variation of the function f along the direction u_k with a solid line and the Armijo line with a dashed line. The origin is in the point x_k.

The condition (3.106) imposes a step s such that the point on the curve of f is below the Armijo line.

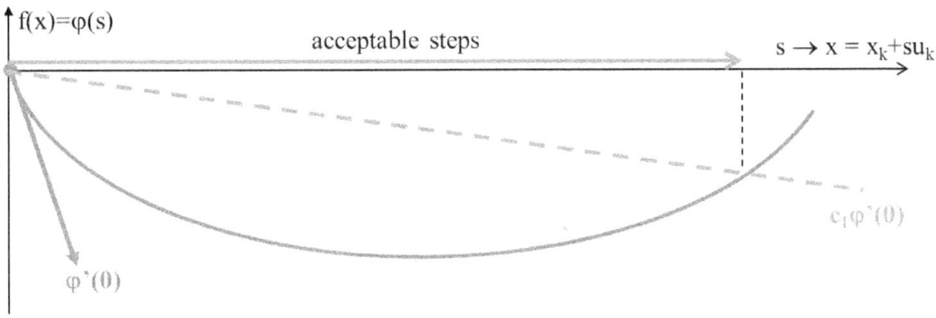

Figure 3-24: Sufficient decrease condition (Armijo condition).

Sufficient move condition

The decrease condition (3.106) is satisfied for very small steps s. To avoid a stagnation in the vicinity of the initial point x_k, a gap is imposed between the initial point x_k and the new point x_{k+1}. This condition is expressed from the directional derivative $\varphi'(0)$ in one of the following two forms:

- the second Goldstein condition concerns the value of the function φ. It requires that the step s satisfies

$$\varphi(s) > \varphi(0) + c_2 s\varphi'(0) \tag{3.107}$$

or equivalently

$$f(x_k + su_k) > f(x_k) + c_2 s \nabla f(x_k)^T u_k \tag{3.108}$$

- the second Wolfe condition concerns the derivative of the function φ. It requires that the step s satisfies

$$\varphi'(s) > c_2\, \varphi'(0) \tag{3.109}$$

or equivalently

$$\nabla f(x_k + su_k)^T u_k > c_2 \nabla f(x_k)^T u_k \tag{3.110}$$

For these two conditions, the coefficient c_2 on the directional derivative is usually set to $c_2 = 0.9$. The line with slope $c_2\varphi'(0) < 0$ from the origin $(s = 0)$ is called the second Goldstein line.

Figure 3-25 shows the variation of the function f along the direction u_k. Goldstein's condition (3.108) is shown in figure 3-25a. Wolfe's condition (3.110) is shown in figure 3-25b.

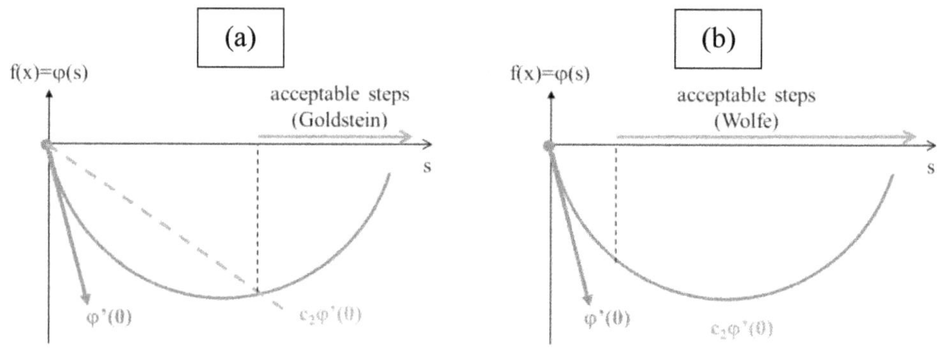

Figure 3-25: Sufficient move condition.

The Wolfe condition has the advantage of ensuring that the curvature condition required by the DFP and BFGS methods is satisfied in $x_{k+1} = x_k + su_k$.

Indeed, the product: $y_{k+1}^T d_{k+1} = \left(\nabla f(x_{k+1}) - \nabla f(x_k)\right)^T (x_{k+1} - x_k)$ is positive if the point x_{k+1} satisfies the condition (3.110). However, Wolfe's condition is expensive to check, because it requires for each value of s the calculation of the gradient of f in the point $x_k + su_k$. It is therefore rarely used in practice. Goldstein's condition (3.106) does not require any additional calculation, as it uses the value of the function already calculated for Armijo's condition (3.106).

In practice, the conditions for accepting the step are therefore

$$\begin{cases} \varphi(s) < \varphi(0) + c_1 s\varphi'(0) \\ \varphi(s) > \varphi(0) + c_2 s\varphi'(0) \end{cases} \Leftrightarrow \begin{cases} f(x_k + su_k) < f(x_k) + c_1 s\nabla f(x_k)^T u_k \\ f(x_k + su_k) > f(x_k) + c_2 s\nabla f(x_k)^T u_k \end{cases} \tag{3.111}$$

The usual values $c_1 = 0.1$ and $c_2 = 0.9$ can be changed, but they have no significant influence on the behavior of a line search algorithm.

The search for a step satisfying (3.111) is achieved by a dichotomy in the interval $0 \le s \le s_{max}$, the upper bound being for example $s_{max} = 1$ (which gives the minimum if the function is quadratic of Hessian H_k).

Figure 3-26 illustrates the search process for an acceptable step.

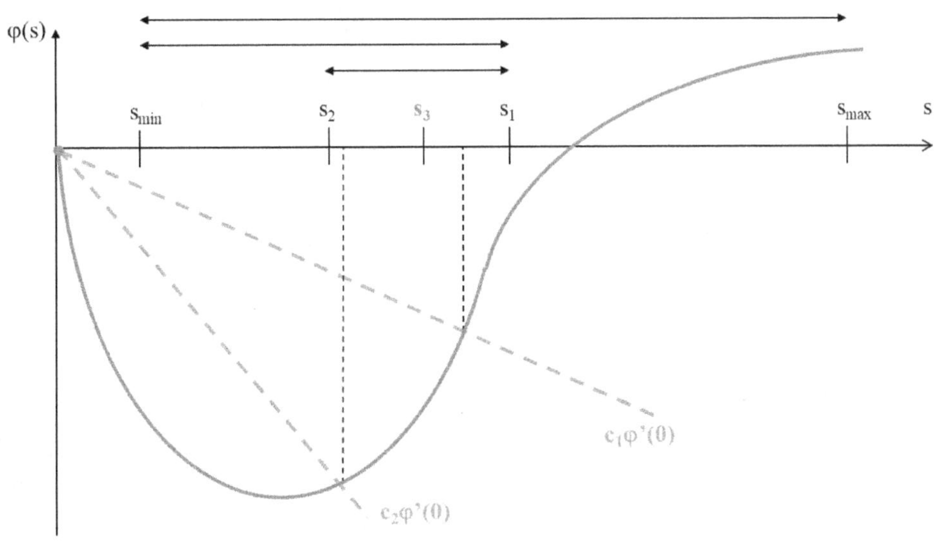

Figure 3-26: Step adjustment by dichotomy.

3.3.3 Algorithm

Figure 3-27 depicts the stages of a line search minimization algorithm using a quasi-Newton method.

We give next some indications on the initialization, the iterations and the stop conditions.

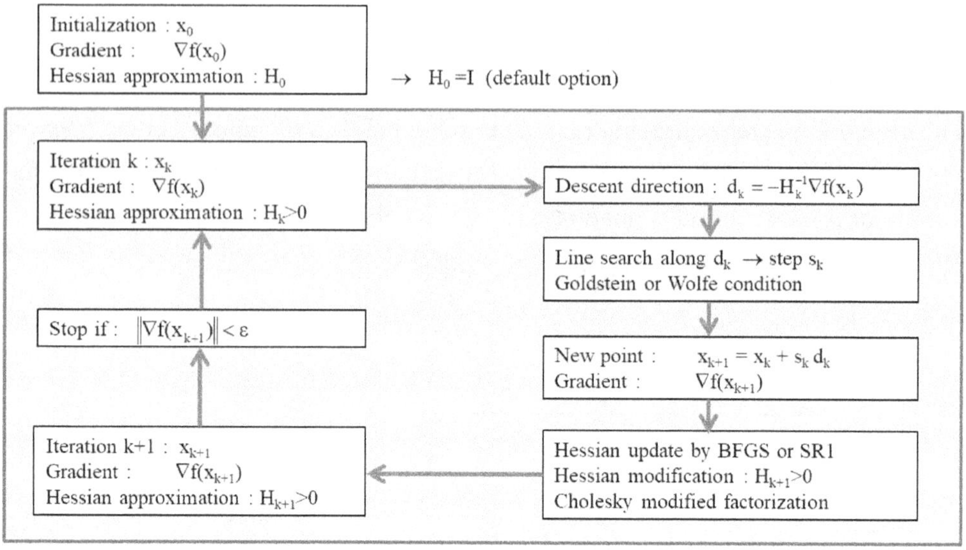

Figure 3-27: Minimization by line search with quasi-Newton.

Initialization

The choice of the initial point x_0 is important if the function has several local minima (which is usually not known). Since the line search method is local, the algorithm will converge to a local minimum "close" to the initial point. In case of doubt, it is necessary to restart the optimization from different initial points to ensure that the minimum obtained is the best. In the general case, there is no guarantee that the global minimum has been found.

The simplest choice for the matrix H_0 is the identity. Other choices are possible (section 3.2.3) at the cost of additional evaluations of the function.

Iterations

Each iteration involves of the calculation of the gradient of f (by n finite difference evaluations) followed by a line search along the preconditioned direction $u_k = -H_k^{-1} g_k$. To limit the number of evaluations of the function, the maximum step length $s_{max} = 1$ can be set, which corresponds to the Newton iteration.

To be accepted, the step must satisfy at least the sufficient decrease condition (Armijo's condition or first Goldstein's condition) and if possible the sufficient move condition (second Goldstein's condition). In case of failure, the matrix H_k can be reset to the identity, which is equivalent to performing a steepest descent iteration. The steepest descent direction ensures that a step can be found that at least satisfies the Armijo condition.

Stopping conditions

The algorithm stops as soon as at least one of the following conditions is met:

- insufficient decrease: $|f(x_k) - f(x_{k-1})| < \varepsilon_f$

  The threshold ε_f depends on the estimated accuracy of the numerical calculation and the desired accuracy on the minimum value of f;

- insufficient move: $\|x_k - x_{k-1}\| < \varepsilon_x$

  The threshold ε_x depends on the desired precision on the variables. This threshold can possibly be different for each component of x;

- sufficiently small gradient: $\|\nabla f(x_k)\| < \varepsilon_g$

  This condition ensures that a local minimum has been reached. In practical applications, it is usually not possible to rigorously cancel the gradient due to numerical inaccuracies in the function and/or the finite difference derivatives. For a scaled problem, one can take $\varepsilon_g \approx \sqrt{\varepsilon_f}$;

- number of evaluations of the function: $N_{fonc} < N_{fonc\ max}$

  This condition is necessary to control the computation time. In practical applications, the function is usually expensive to evaluate and the computation time is proportional to the number of evaluations of the function;

- number of iterations: $N_{iter} < N_{iter\ max}$

  This condition is usually provided in addition to the number of evaluations of the function. It is indeed easier to think in terms of iterations with respect to convergence.

Example 3-16: Minimization by line search with BFGS or SR1

The Rosenbrock is defined as: $f(x_1,x_2)=100\left(x_2-x_1^2\right)^2+\left(1-x_1\right)^2$.

Figure 3-28 illustrates the application of a line search algorithm with BFGS or SR1 starting from the initial point (−1,2 ; 1). The progression along the valley until the minimum in (1 ; 1) is observed.

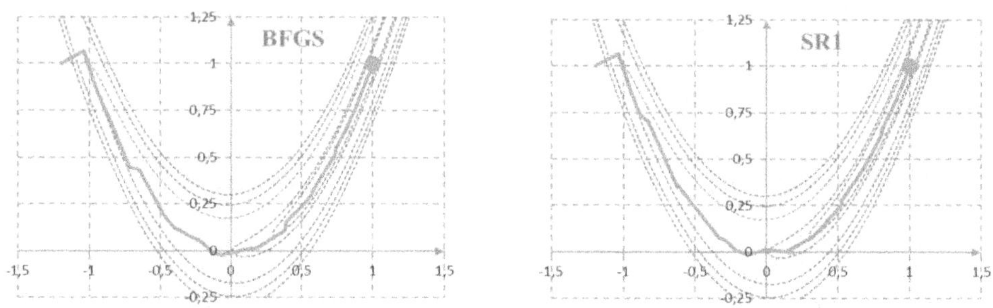

Figure 3-28: Minimization of the Rosenbrock function with BFGS or SR1.

Figure 3-29 shows the decrease of the function during the iterations. In this example, the BFGS method progresses faster towards the minimum, as it benefits from the fact that the "true" Hessian is naturally positive.

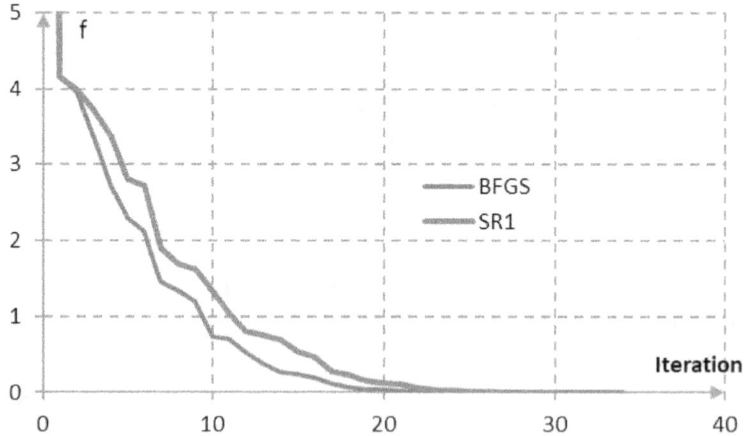

Figure 3-29: BFGS/SR1 comparison on the Rosenbrock function.

Non monotonous descent

Armijo's condition (3.106) imposes a strict decrease of the function at each iteration. This monotonous decrease requirement can slow down progress when the function is ill-conditioned. An alternative strategy is to accept temporary increases over a number m of iterations.

The decrease condition (3.106) is modified by taking the reference value $f(x_{k-m})$ instead of $f(x_k)$.

$$f(x_k + su_k) < f(x_{k-m}) + c_1 s \nabla f(x_k)^T u_k \tag{3.112}$$

The function can thus increase during the first m iterations, but the sub-sequences of period m must then be decreasing.

$$\begin{cases} f(x_0) > f(x_{0+m+1}) & > f(x_{0+2(m+1)}) & > \cdots > f(x_{0+k(m+1)}) \\ f(x_1) > f(x_{1+m+1}) & > f(x_{1+2(m+1)}) & > \cdots > f(x_{1+k(m+1)}) \\ \vdots & & \\ f(x_m) > f(x_{m+m+1}) & > f(x_{m+2(m+1)}) & > \cdots > f(x_{m+k(m+1)}) \end{cases} \tag{3.113}$$

This non-monotonous descent strategy (called "watchdog") allows a faster progression when the function is steep. The period m for the decrease check can be taken equal to the dimension of the problem $(m = n)$.

Example 3-17: Non-monotonous descent strategy

Let us take the Rosenbrock function again: $f(x_1, x_2) = 100\left(x_2 - x_1^2\right)^2 + \left(1 - x_1\right)^2$.

Figure 3-30 compares the progression of a monotonous and a non-monotonous descent of period m = 2.

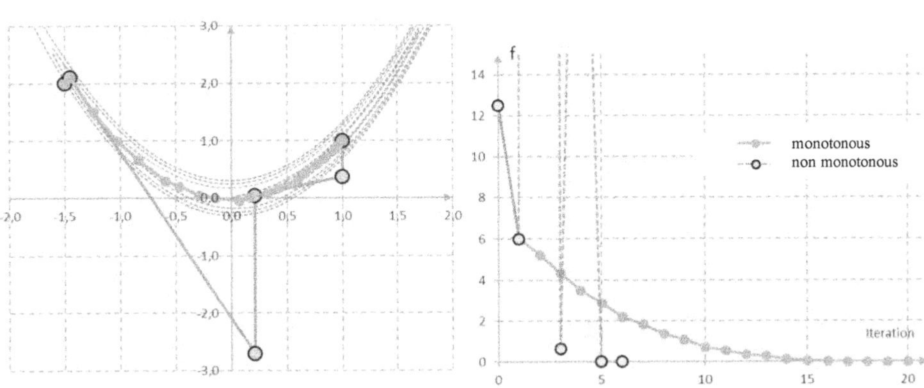

Figure 3-30: Non-monotonous descent on the Rosenbrock function.

The non-monotonous strategy escapes the "valley" at iteration 1 and again at iteration 3. The function is strongly increased during these iterations, but the convergence is globally much faster. The minimum is reached in 6 iterations against 21 iterations for the monotonous descent.

3.4 Trust region

The line search method evaluates points only in a given direction. This strategy can run into difficulties when the function varies greatly along the search direction. The idea of the trust region method is to extend the search area while considering on a local quadratic model of the function. This process is illustrated in figure 3-31 and depicted in figure 3-32. For each quadratic model, the level line through x_k is dotted.

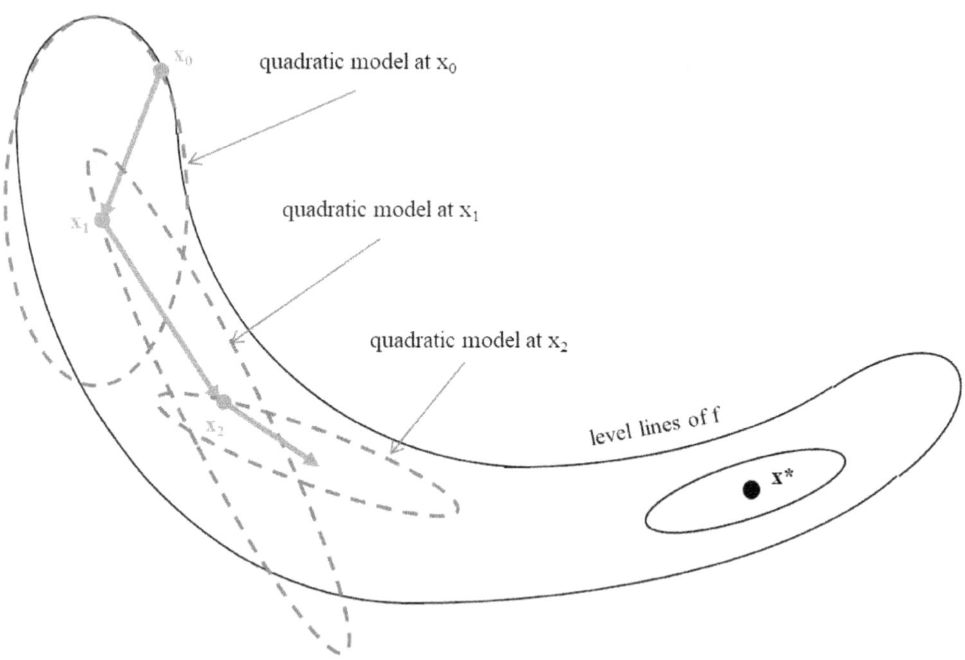

Figure 3-31: Successive quadratic models.

The quadratic model only represents the function correctly in the vicinity of the current point. We define a **trust region** centered on the current point x_k and bounded by a radius r_k.

A trust region descent method consists at each iteration in:

- minimizing the quadratic model within the region of radius r_k;
- adapting the radius r_k to obtain a new satisfactory point $x_{k+1} = x_k + d_k$.

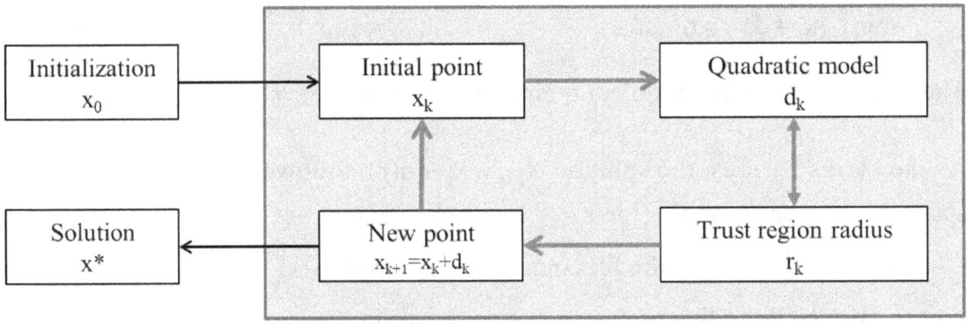

Figure 3-32: Descent by trust region.

The formulation of the quadratic problem is presented in section 3.4.1 and the solution methods are discussed in sections 3.4.2 and 3.4.3.

3.4.1 Quadratic model

The quadratic model of the function f in x_k depends on the function value f_k, the gradient g_k and the approximation of the Hessian H_k (usually obtained by quasi-Newton). This function noted $\hat{f}_k$ has the expression

$$\hat{f}_k(x_k + d) = f_k + g_k^T d + \frac{1}{2} d^T H_k d \tag{3.114}$$

where $d \in \mathbb{R}^n$ represents a move from x_k.

It is assumed that the quadratic model is a correct approximation of the function within a region of center x_k and of radius r_k.

This trust region can be defined as either:

- with the L2 norm (circular region) : $\|d\|_2 = \sqrt{\sum_{i=1}^{n} d_i^2} \le r_k$;
- with the L∞ norm (rectangular region) : $\|d\|_\infty = \max(d_i)_{i=1 \text{ to } n} \le r_k$.

The **trust region problem** consists in minimizing the quadratic model $\hat{f}_k$ while remaining within the trust region.

$$\min_{d \in \mathbb{R}^n} \hat{f}_k(x_k + d) \quad \text{s.t.} \quad \|d\| \le r_k \tag{3.115}$$

The solution d_k of this problem defines a new point $x_{k+1} = x_k + d_k$.

As the radius r varies, the solution $x_{k+1} = x_k + d(r)$ follows a curve (figure 3-33) starting from the point x_k (for $r = 0$) with the tangent $-g_k$ and going:
- to the minimum x_N of the function $\hat{f}_k$ (in the case of H_k positive);
- or to infinity if the function $\hat{f}_k$ is not bounded (in the case of H_k indefinite).

The notation x_N refers to Newton's point (unconstrained minimum).

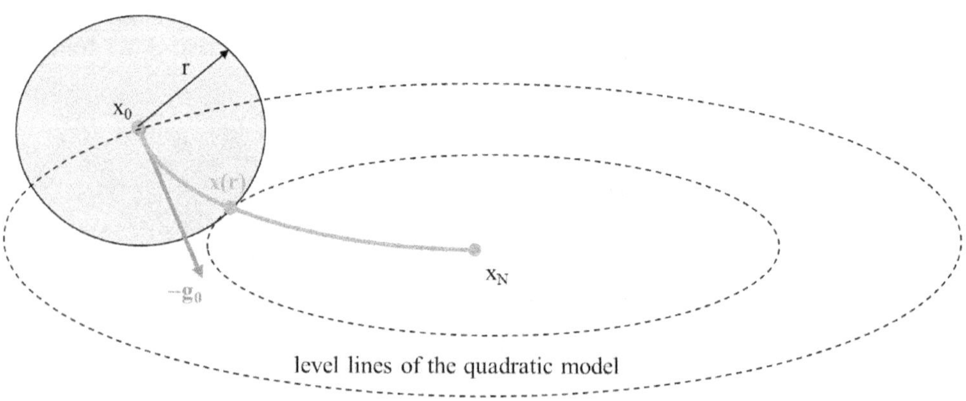

Figure 3-33: Solution as a function of the trust radius.

For a given radius, the point $x_{k+1} = x_k + d(r)$ minimizes the quadratic model $\hat{f}_k$, but not necessarily the "true" function f.

The reduction ratio ρ compares the actual and planned improvements.

$$\rho = \frac{f(x_k) - f(x_{k+1})}{\hat{f}(x_k) - \hat{f}(x_{k+1})} \tag{3.116}$$

A ratio greater than or equal to 1 indicates that the quadratic model represents the function well in the region of radius r. In this case, the radius can be increased in the next iteration to allow a larger move.

Conversely, a small or negative ratio indicates that the quadratic approximation is not good. In this case, the iteration must be retrieved with a reduced trust radius to obtain an acceptable point x_{k+1}.

The strategy for setting the radius is as follows:

- if $\rho > \rho_{max}$ (with $\rho_{max} = 0.9$), the point x_{k+1} is accepted and the radius is doubled for the next iteration: $r_{k+1} = 2r_k$;
- if $\rho < \rho_{min}$ (with $\rho_{min} = 0.01$), the point x_{k+1} is rejected and the iteration is repeated with a halved radius: $r_k \to r_k / 2$;
- if $\rho_{min} < \rho < \rho_{max}$, the point x_{k+1} is accepted and the radius is kept unchanged for the next iteration: $r_{k+1} = r_k$.

The threshold values ρ_{min} and ρ_{max} here above are indicative and are not essential for the operation of the algorithm.

The trust region problem (3.115) has to be solved at each iteration, possibly several times if the solution is rejected and the radius is reduced.
The solution of this quadratic problem must be fast and robust. The two main solution methods are presented in sections 3.4.2 and 3.4.3.

3.4.2 Direct solution

The direct solution is to solve problem (3.115) with the trust region constraint formulated in L_2 norm.

$$\min_{d \in \mathbb{R}^n} g_k^T d + \frac{1}{2} d^T H_k d \quad \text{s.t.} \quad \frac{1}{2}(d^T d - r^2) \le 0 \tag{3.117}$$

The Lagrangian is expressed with a multiplier $\mu \ge 0$ of the inequality constraint.

$$L(d, \mu) = d^T g_k + \frac{1}{2} d^T H_k d + \frac{1}{2} \mu (d^T d - r^2) \tag{3.118}$$

The KKT conditions of order 1 are

$$\begin{cases}(H_k+\mu I)d=-g_k \\ \|d\|\le r \ , \ \mu\ge 0 \\ \mu(\|d\|-r)=0\end{cases} \tag{3.119}$$

If the constraint is inactive: $\mu=0$, the solution is directly: $d_N=-H_k^{-1}g_k$.

If the constraint is active: $\|d\|=r$, the system becomes: $\begin{cases}(H_k+\mu I)d=-g_k \\ \|d\|=r \ , \ \mu\ge 0\end{cases}$.

A simple way to solve the KKT conditions (3.119) is to consider the solution parameterized by μ: $d(\mu)=-(H_k+\mu I)^{-1}g_k$. As μ decreases, the solution point $x_{k+1}=x_k+d(\mu)$ follows a curve similar to the one in figure 3-33 starting for $\mu=+\infty$ from the point x_k ($d=0$) and going for $\mu=0$ either to the minimum $x_N=x_k+d_N$ of the function $\hat{f}_k$ (if H_k is positive definite) or to infinity if the function $\hat{f}_k$ is not bounded (if H_k is indefinite). We can thus search by dichotomy the value of μ satisfying the condition $\|d(\mu)\|=r$ if the constraint is active.

This method presents a difficulty when the matrix H_k is indefinite.
If $D=\text{diag}(\sigma_1,\cdots,\sigma_n)$ is the diagonal matrix of eigenvalues of H_k and P is the transfer matrix in an orthogonal basis of eigenvectors, we have

$$H_k=PDP^T \Rightarrow P(D+\mu I)P^T d=-g_k \tag{3.120}$$

For $\mu=-\sigma_i$, where σ_i is a negative eigenvalue, the matrix $D+\mu I$ is no longer invertible and $d(\mu)\to\infty$. The function $\|d(\mu)\|$ is thus not continuous and the solution of the equation $\|d(\mu)\|=r$ (if it exists) must be sought in each interval between two consecutive negative eigenvalues. Moreover, the function $\|d(\mu)\|$ is not monotonous in these intervals, which greatly complicates the solution.

It is in fact not useful to satisfy precisely the constraint $\|d(\mu)\|=r$, which only serves to bound the solution of the quadratic problem. It is more straightforward to check the reduction ratio ρ (3.116) and adjust μ to obtain a satisfactory solution. However, this does not eliminate the issues mentioned above when the matrix H_k is indefinite (non-continuous and non-monotonous function $\|d(\mu)\|$). Furthermore, the multiplier is difficult to estimate *a priori*, which can make this method costly in evaluations. The following section presents a simpler and more robust method.

3.4.3 Dogleg solution

The dogleg method aims to avoid too many function evaluations, especially when the matrix H_k is not positive definite. For that purpose, the problem (3.115) is simplified by using Newton and Cauchy points.

The **Newton point** $x_N = x_k + d_N$ is the minimum of the quadratic function $\hat{f}_k$ if this one is convex (matrix H_k positive definite).

$$d_N = -H_k^{-1} g_k \tag{3.121}$$

The **Cauchy point** $x_C = x_k + d_C$ is the minimum of $\hat{f}_k$ along the direction $-g_k$.

$$\begin{aligned} &\min_{s>0} \hat{f}(-sg_k) = f_k - s g_k^T g_k + \frac{1}{2} s^2 g_k^T H_k g_k \\ &\rightarrow\ s = \frac{g_k^T g_k}{g_k^T H_k g_k} \quad \rightarrow\ d_C = -\frac{g_k^T g_k}{g_k^T H_k g_k} g_k \end{aligned} \tag{3.122}$$

The Newton and Cauchy points form a two-segment path:
- the first segment goes from the initial point x_k to the Cauchy point x_C;
- the second segment goes from the Cauchy point x_C to the Newton point x_N.

This path shown in figure 3-34 is called the **dogleg path**. It is used as an approximation to the exact solution which follows the dotted curve.

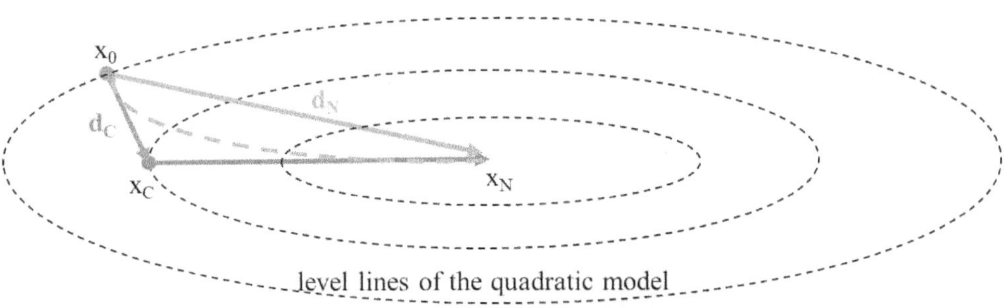

Figure 3-34: Newton and Cauchy points.

The dogleg path is parameterized by the step s varying from 0 to 2.

$$\begin{cases} d(s) = s d_C & \text{if } 0 \le s \le 1 \\ d(s) = d_C + (s-1)(d_N - d_C) & \text{if } 1 \le s \le 2 \end{cases} \tag{3.123}$$

The trust region problem reduced to the dogleg path is formulated as

$$\min_{0\le s\le 2} g_k^T d(s)+\frac{1}{2}d(s)^T H_k d(s) \quad \text{s.t.} \quad \|d(s)\|\le r \tag{3.124}$$

When the matrix H_k is positive definite, the quadratic function (3.124) decreases with s while the distance $\|d(s)\|$ increases. It is therefore sufficient to walk the dogleg path backwards from the Newton point ($s=2$) until entering the trust region $\|d(s)\|\le r$. The solution is:

- in the Newton point if this one lies within the trust region: $\|d_N\|<r$;
- on the second segment if the Cauchy point is within the trust region whereas the Newton point is outside: $\|d_C\|<r<\|d_N\|$.

  The step s satisfies then: $\|d_C+(s-1)(d_N-d_C)\|=r$ which leads to an equation of second degree with only one positive root (negative root product).

$$s=1+\frac{-b+\sqrt{b^2-4ac}}{2a} \quad \text{with} \quad \begin{cases} a=\|d_N-d_C\|^2 \\ b=2d_C^T\left(d_N-d_C\right) \\ c=\|d_C\|^2-r^2 \end{cases} \tag{3.125}$$

- on the first segment if the Cauchy point is outside the trust region: $r<\|d_C\|$.

  The step s is then directly given by

$$s=\frac{r}{\|d_C\|} \Rightarrow d=r\frac{d_C}{\|d_C\|}=-r\frac{g_k}{\|g_k\|} \tag{3.126}$$

These three situations are illustrated in figure 3-35.

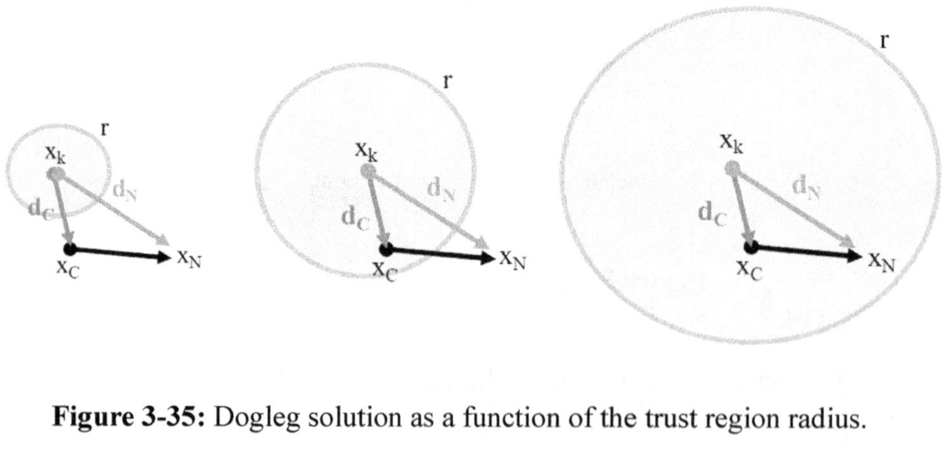

Figure 3-35: Dogleg solution as a function of the trust region radius.

Example 3-18: Iteration of the dogleg method

Let us retrieve the function of example 3-12: $\min_{x_1,x_2} f(x_1,x_2) = \frac{1}{2}x_1^2 + \frac{9}{2}x_2^2$.

and illustrate an iteration of the dogleg method from the initial point: $x_0 = \begin{pmatrix} 9 \\ 1 \end{pmatrix}$.

The gradient and Hessian in the initial point are

$$g_0 = \nabla f(x_0) = \begin{pmatrix} 9 \\ 9 \end{pmatrix} \ , \ H_0 = \nabla^2 f(x_0) = \begin{pmatrix} 1 & 0 \\ 0 & 9 \end{pmatrix}$$

The Cauchy point is obtained by

$$\min_s \hat{f}(x_0 - sg_0) = 45 - 18(9s) + 5(9s)^2 \rightarrow s = \frac{1}{5} \quad \rightarrow \quad x_C = \begin{pmatrix} 7.2 \\ -0.8 \end{pmatrix}$$

The Newton point is obtained by

$$x_N = x_0 - H_0^{-1}g_0 = \begin{pmatrix} 9 \\ 1 \end{pmatrix} - \begin{pmatrix} 1 & 0 \\ 0 & 1/9 \end{pmatrix}\begin{pmatrix} 9 \\ 9 \end{pmatrix} \qquad \rightarrow \quad x_N = \begin{pmatrix} 0 \\ 0 \end{pmatrix}$$

The first dogleg segment has the equation

$$x_{d1}(s) = x_0 + s(x_C - x_0) = \begin{pmatrix} 9 - 1.8s \\ 1 - 1.8s \end{pmatrix} \qquad \text{for} \quad 0 \le s \le 1$$

The second dogleg segment has the equation

$$x_{d2}(s) = x_C + s(x_N - x_C) = \begin{pmatrix} 7.2 - 7.2s \\ -0.8 + 0.8s \end{pmatrix} \quad \text{for} \quad 0 \le s \le 1$$

(we have chosen here to parameterize each segment with a step s from 0 to 1).

Figure 3-36 shows two trust regions of respective radii $r = 1$ and $r = 4$.

For $r = 1$, the dogleg solution is on the first segment: $\quad x_d \approx \begin{pmatrix} 8.293 \\ 0.293 \end{pmatrix}$.

For $r = 4$, the dogleg solution is on the second segment: $x_d \approx \begin{pmatrix} 5.331 \\ -0.592 \end{pmatrix}$.

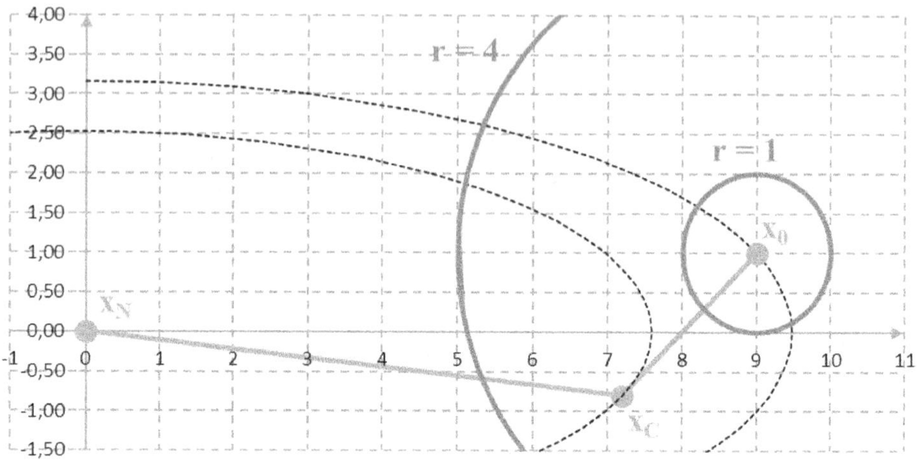

Figure 3-36: Dogleg solution for two values of the trust region radius.

When the matrix H_k is not positive definite, the quadratic function is not lower bounded and there are potentially the same difficulties as mentioned in section 3.4.2. The evaluation of the Newton and Cauchy points allows to detect these difficulties related to negative eigenvalues. Indeed, if the direction $H_k^{-1}g_k$ is not a descent direction, the Cauchy step (3.122) becomes negative. In this case, it is sufficient to reduce the radius until an acceptable reduction ratio ρ (3.116) is obtained. By restricting the search space to two segments, the dogleg method avoids solving the system (3.119), which makes it more robust to the case of indefinite matrices.

The dogleg method restricts the solution to the two segments defined by the vectors d_C and $d_N - d_C$. A simple improvement is to extend the search space to the whole subspace generated by the vectors d_C and d_N. The move from the point x_k is then of the form $d = \alpha_C d_C + \alpha_N d_N$ and depends on the two real coefficients α_C, α_N. The dogleg problem in dimension 2 is formulated as

$$\min_{\alpha_C,\alpha_N} g_k^T d(\alpha_C,\alpha_N) + \frac{1}{2} d(\alpha_C,\alpha_N)^T H_k d(\alpha_C,\alpha_N) \quad \text{s.t.} \quad \|d(\alpha_C,\alpha_N)\| \le r \qquad (3.127)$$

If the Newton point is not in the trust region, it is sufficient to solve the problem (3.127) with two unknowns by considering the constraint as active.

Squaring this, and with a multiplier λ, the Lagrangian is written as

$$\begin{aligned} L(\alpha_C,\alpha_N,\lambda) = g_k^T(\alpha_C d_C + \alpha_N d_N) + \frac{1}{2}(\alpha_C d_C + \alpha_N d_N)^T H_k (\alpha_C d_C + \alpha_N d_N) \\ + \lambda\left[(\alpha_C d_C + \alpha_N d_N)^2 - r^2\right] \end{aligned} \tag{3.128}$$

The KKT conditions give a linear system in α_C, α_N, parameterized by λ.

By expressing α_C, α_N as a function of λ, we are reduced to an equation of fourth degree in λ which can be solved analytically. This dogleg solution in two dimensions allows, at the cost of additional calculations, to improve the classical dogleg solution.

3.4.4 Algorithm

Figure 3-37 depicts the stages of a trust region minimization algorithm using a quasi-Newton method.

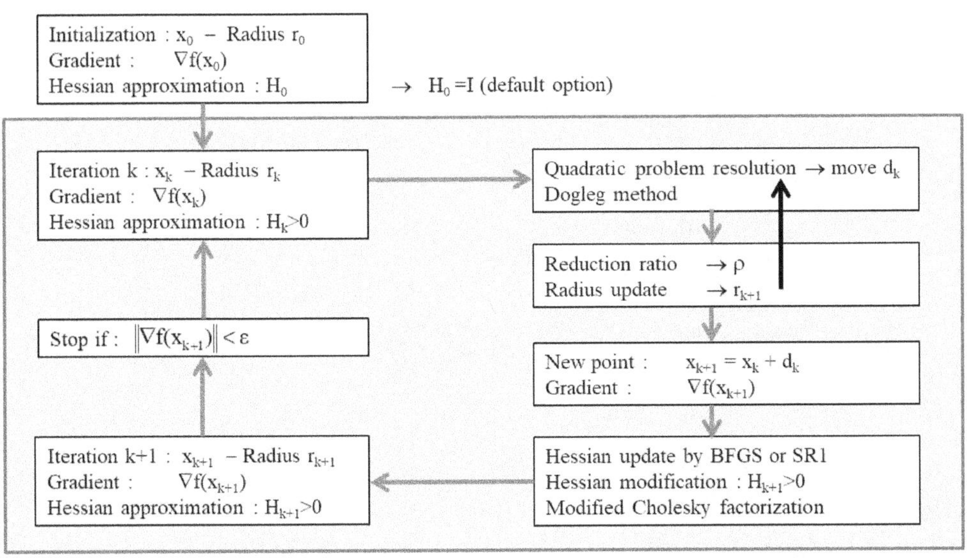

Figure 3-37: Trust region minimization with quasi-Newton.

The indications given in section 3.3.3 for the line search algorithm remain valid for the trust region algorithm. The dogleg solution can be seen as a compromise between the steepest descent method (Cauchy point) and the quasi-Newton method (Newton point). It is thus likely to make better progress than the line search at each iteration, but at the cost of a higher number of evaluations. Both approaches (line search or trust region) should be considered for each problem, neither being systematically superior to the other.

The following example (to be compared to example 3-16) illustrates the behavior of the trust region algorithm on the Rosenbrock function.

The algorithm is applied by considering four options:

- circular or rectangular trust region;
- quasi-Newton BFGS or SR1 method.

Example 3-19: Trust region minimization with BFGS or SR1

The Rosenbrock function with 2 variables is defined by

$$f(x_1,x_2)=100\left(x_2-x_1^2\right)^2+\left(1-x_1\right)^2$$

Its minimum is in (1 ; 1).

Figure 3-38 illustrates the application of a trust region algorithm with BFGS starting from the initial point (−1.2 ; 1).

The trust region is circular on the left, rectangular on the right.

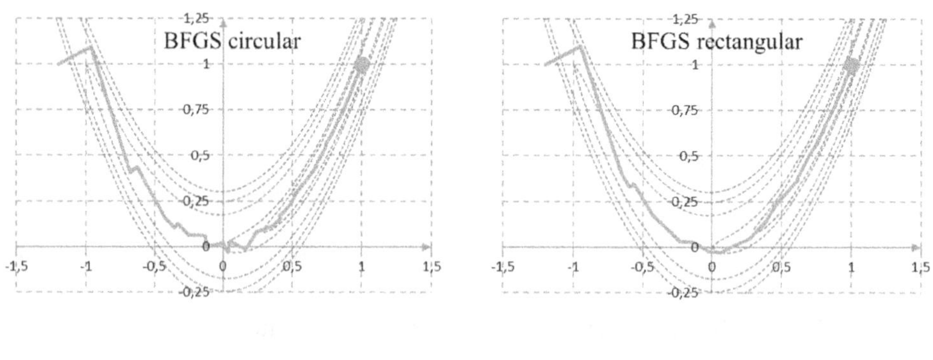

Figure 3-38: BFGS with circular/rectangular trust region.

Figure 3-39 illustrates the application of a trust region algorithm with SR1 starting from the initial point (−1.2 ; 1).

The trust region is circular in figure 3-39a, rectangular in figure 3-39b.

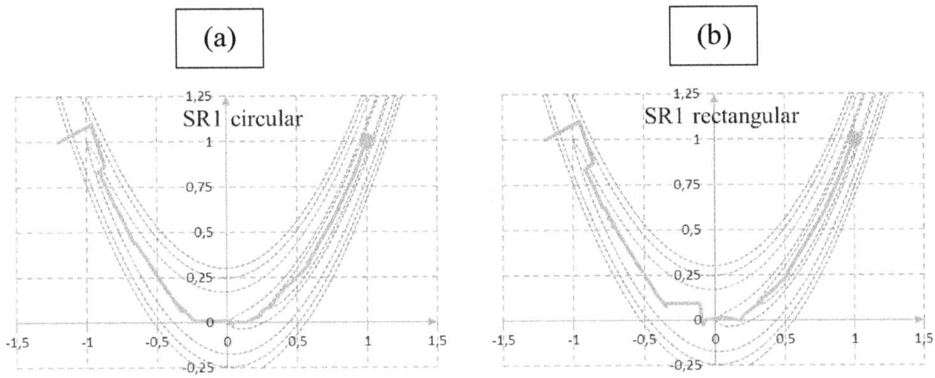

Figure 3-39: SR1 with circular/rectangular trust region.

Figure 3-40 shows the decrease of the cost function according to the option chosen. The trust region is circular in figure 3-40a, rectangular in figure 3-40b.

In this example, the progression with BFGS is faster than with SR1, and the circular trust region gives a faster progression than the rectangular trust region.

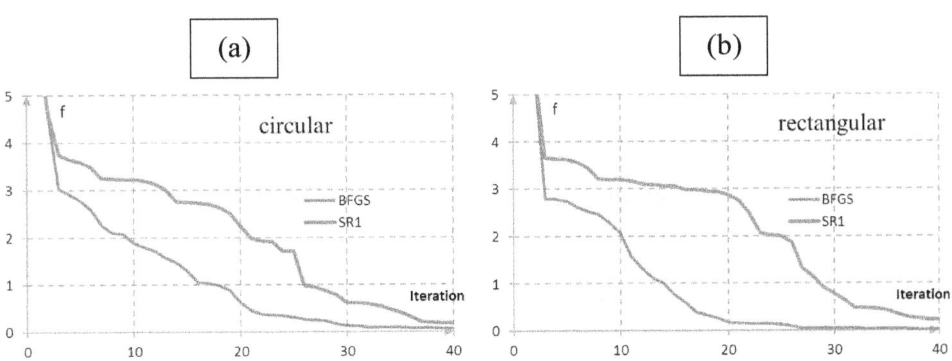

Figure 3-40: Cost function decrease.

3.5 Proximal methods

Proximal methods generalize descent methods in the case of non-differentiable convex functions. These non-smooth optimization methods are used in image processing, robust or stochastic optimization, or in learning processes (deep-learning, machine-learning).

3.5.1 Proximal operator

Consider a convex function f that is not necessarily differentiable and can take infinite values: $f:\mathbb{R}^n \to \mathbb{R}\cup\{+\infty\}$.

The domain of f is the set of points for which f remains finite.

$$\text{dom}(f)=\left\{x\in\mathbb{R}^n \ / \ f(x)<+\infty\right\} \tag{3.129}$$

A **subgradient** of f in the point x_0 is a vector $g\in\mathbb{R}^n$ such that

$$\forall x\in\text{dom}(f)\,,\ f(x)\geq f(x_0)+g^T(x-x_0) \tag{3.130}$$

The **subdifferential** of f in the point x_0 is the set of subgradients in that point. This subset of $\mathbb{R}^n$ is denoted $\partial f(x_0)$.

$$\partial f(x_0)=\left\{g\in\mathbb{R}^n \ / \ \forall x\in\text{dom}(f)\,,\ f(x)\geq f(x_0)+g^T(x-x_0)\right\} \tag{3.131}$$

Example 3-20: Sub-differential

The absolute value function:

$f(x)=|x|$ has the subdifferential:

$$\partial f(x)=\begin{cases}\{-1\} & \text{if } x<0\\ [-1\,;+1] & \text{if } x=0\\ \{+1\} & \text{if } x>0\end{cases}$$

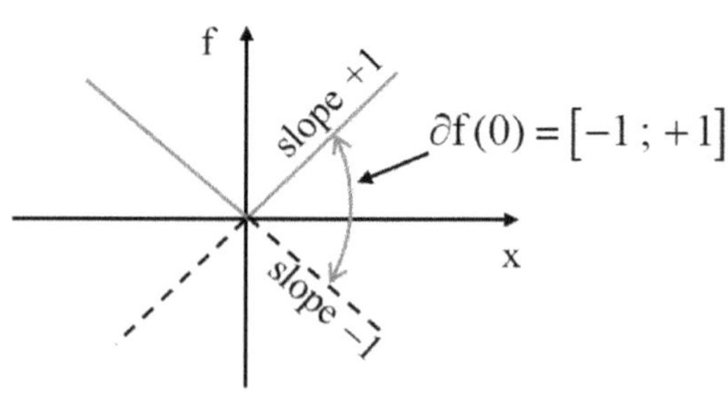

The subdifferential has the following properties, which are easily proved from its definition (3.131).

- if f is differentiable in x_0, it has only one subgradient which is the gradient of f in x_0. The subdifferential reduces to: $\partial f(x_0) = \{\nabla f(x_0)\}$;
- the point x* is a minimum of f if and only if the null vector belongs to the subdifferential of f in x* : $0 \in \partial f(x^*)$.

Let us now turn to the definition of the proximal operator. Consider a convex function f, a point $v \in \mathbb{R}^n$ and the optimization problem

$$\min_x \left(f(x) + \frac{1}{2}\|x - v\|_2^2 \right) \tag{3.132}$$

This problem consists in minimizing the function f while remaining close to v. The point x_p solution of this problem is the **proximal point** of v with respect to f.

$$x_p = \arg\min_x \left(f(x) + \frac{1}{2}\|x - v\|_2^2 \right) \tag{3.133}$$

The **proximal operator** of the function f associates the point $v \in \mathbb{R}^n$ with the point $x_p \in \mathbb{R}^n$. This operator is a function from $\mathbb{R}^n$ in $\mathbb{R}^n$ noted prox_f.

$$\text{prox}_f(v) = \arg\min_x \left(f(x) + \frac{1}{2}\|x - v\|_2^2 \right) \tag{3.134}$$

Using the characterization of the minimum by the subdifferential, we have

$$x_p = \text{prox}_f(v) \iff 0 \in \partial f(x_p) + x_p - v \tag{3.135}$$

Let us apply the definition (3.134) to the function ρf where ρ is a strictly positive real. Dividing the function to be minimized by $\rho > 0$, we obtain

$$\text{prox}_{\rho f}(v) = \arg\min_x \left(f(x) + \frac{1}{2\rho}\|x - v\|_2^2 \right) \tag{3.136}$$

The coefficient ρ introduces a weighting between the minimization of f and the distance to the point v. The operator $\text{prox}_{\rho f}$ is the proximal operator of f with the parameter ρ. Using the sub-differential of f, the operator $\text{prox}_{\rho f}$ is expressed as

$$\text{prox}_{\rho f} = (I + \rho \partial f)^{-1} \tag{3.137}$$

This formula, called the resolvent of ∂f, is demonstrated below.

The subdifferential ∂f represents here the operator associating to the point x the set of subgradients of f in this point.

Demonstration

The proximal point $x_p = \text{prox}_{\rho f}(v)$ minimizes $F(x) = f(x) + \frac{1}{2\rho}\|x - v\|_2^2$.

According to (3.135), the null vector must belong to the subdifferential of F in x_p

$$0 \in \partial F(x_p) = \partial f(x_p) + \frac{x_p - v}{\rho}.$$

Multiplying by $\rho > 0$ and passing v to the 1st member, we get: $v \in (I + \rho\partial f)(x_p)$

Applying the inverse operator, we obtain: $(I + \rho\partial f)^{-1}(v) \in \{x_p\}$.

The proximal point is unique, because the function F is convex. This membership relation is therefore an equality: $(I + \rho\partial f)^{-1}(v) = x_p$, hence the formula (3.137).

The proximal operator has the following properties, which can be proved from its definition (3.134).

- Separable function in sum

  $$f(x,y) = \varphi(x) + \psi(y) \rightarrow \text{prox}_f(v,w) = \left(\text{prox}_\varphi(v)\,,\,\text{prox}_\psi(w)\right)$$

  This property, which generalizes to any number of variables, makes it possible to parallelize the calculation of the proximal operators of each component.

- Composition of functions

  $$f(x) = a\varphi(x) + b\,,\ a > 0 \rightarrow \text{prox}_{\rho f}(v) = \text{prox}_{a\rho\varphi}(v)$$

  $$f(x) = \varphi(ax + b)\,,\ a \neq 0 \rightarrow \text{prox}_{\rho f}(v) = \frac{1}{a}\left(\text{prox}_{a^2\rho\varphi}(av + b) - b\right)$$

  $$f(x) = \varphi(x) + c^T x + b \rightarrow \text{prox}_{\rho f}(v) = \text{prox}_{\rho\varphi}(v - \rho c)$$

  $$f(x) = \varphi(Qx) \rightarrow \text{prox}_{\rho f}(v) = Q^T \text{prox}_{\rho\varphi}(Qv)$$

  with Q orthogonal

  $$f(x) = \varphi(x) + \frac{\alpha}{2}\|x - c\|_2^2 \rightarrow \text{prox}_{\rho f}(v) = \text{prox}_{\rho'\varphi}\left(\frac{\rho'}{\rho}v + \alpha\rho' c\right)$$

  $$\text{with } \rho' = \frac{\rho}{1 + \alpha\rho}$$

For simple functions such as those below, the proximal operator is calculated analytically. More complex examples are discussed in section 3.5.5.

Example 3-21: Analytical expressions of the proximal operator

- Indicator function of the convex set C

$$f(x) = I_C(x) = \begin{cases} 0 & \text{if } x \in C \\ +\infty & \text{if } x \notin C \end{cases} \rightarrow \text{prox}_f(v) = P_C(v) = \arg\min_{x \in C} \|x - v\|_2^2$$

This proximal operator is the Euclidean projection P_C on the set C.

- Quadratic function with a positive semidefinite Q matrix

$$f(x) = \frac{1}{2} x^T Q x + c^T x + b \rightarrow \text{prox}_{\rho f}(v) = (I + \rho Q)^{-1}(v - \rho c)$$

In the case $Q = I$, $c = 0$, $b = 0$ which corresponds to the L_2 norm, the proximal operator called the contraction operator is defined by

$$f(x) = \frac{1}{2} x^T x = \frac{1}{2}\|x\|_2^2 \rightarrow \text{prox}_{\rho f}(v) = \frac{1}{1+\rho} v$$

- Absolute value

$$f(x) = |x| \rightarrow \text{prox}_{\rho f}(v) = \begin{cases} v + \rho & \text{if } v \le -\rho \\ 0 & \text{if } -\rho \le v \le +\rho \\ v - \rho & \text{if } +\rho \le v \end{cases}$$

This proximal operator is called the thresholding operator.

The interest of the proximal operator comes mainly from the following property.

Property 3-3: Fixed point

A point x* is a minimum of f if and only if x* is a fixed point of prox_f.

$$x^* \text{ minimum of } f \Leftrightarrow \text{prox}_f(x^*) = x^* \tag{3.138}$$

Demonstration

First, assume that x* is a minimum of f: $\forall x\ ,\ f(x) \ge f(x^*)$.

Add $\frac{1}{2}\|x - x^*\|_2^2 \ge 0$ to the 1st member and $\frac{1}{2}\|x^* - x^*\|_2^2 = 0$ to the 2nd member.

The result is: $\forall x\ ,\ f(x) + \frac{1}{2}\|x - x^*\|_2^2 \ge f(x^*) + \frac{1}{2}\|x^* - x^*\|_2^2$.

This inequality shows that x* minimizes the function: $F(x)=f(x)+\frac{1}{2}\|x-x^*\|_2^2$.
By definition of the proximal point (3.133), we have then: $x^*=\text{prox}_f(x^*)$ and x^* is a fixed point of the proximal operator.

Assume conversely that x* is a fixed point of the proximal operator.
Consider the function: $F(x)=f(x)+\frac{1}{2}\|x-x^*\|_2^2$. Its subdifferential in point x is the subset of $\mathbb{R}^n$ defined by $\partial f(x)+x-x^*$.
Let us place ourselves in the minimum of F corresponding to: $x_p=\text{prox}_f(x^*)$.
The subdifferential of F in x_p contains 0 (characterization of the minimum of F).
Since by assumption: $\text{prox}_f(x^*)=x^*$, we deduce
$0\in\partial F(x_p)=\partial f(x_p)+x_p-x^*=\partial f(x^*)$ which shows that x* is a minimum of f.

Property 3-3 suggests that a minimum of f can be found by a fixed point algorithm. The convergence relies on the following properties:

- the operator prox_f is strongly non-expansive, which means that
  $\forall x_1,x_2,\ \|\text{prox}_f(x_1)-\text{prox}_f(x_2)\|_2^2\le(\text{prox}_f(x_1)-\text{prox}_f(x_2))^T(x_1-x_2)$;
- the operator prox_f is Lipschitzian of constant 1, which means that
  $\forall x_1,x_2,\ \|\text{prox}_f(x_1)-\text{prox}_f(x_2)\|_2\le\|x_1-x_2\|_2$

3.5.2 Interpretations

By iterating the proximal operator until convergence to a fixed point, we obtain a minimum of f. The proximal iteration is defined by

$$x_{k+1}=\text{prox}_{\rho f}(x_k) \tag{3.139}$$

The parameter $\rho>0$ can be constant or variable. The calculation of $\text{prox}_{\rho f}(x_k)$ involves minimizing the function: $F(x)=f(x)+\frac{1}{2\rho}\|x-x_k\|_2^2$ at each iteration.

The quadratic term added to the function f is called the Tikhonov regularization. It convexifies the function in the vicinity of the current iterate x_k and forces the new iterate x_{k+1} to stay in the vicinity of x_k. This procedure can speed up the convergence to the minimum of f. The minimization of the function $F(x)$ can also be interpreted as a trust region method (section 3.4) treated by penalization. The coefficient $\rho > 0$ then represents the trust radius.

The proximal algorithm is not especially interesting for any function f, since minimizing F is not simpler than minimizing f. However, it is interesting for the function forms presented in section 3.5.3.

Let us give some interpretations of this algorithm in relation to classical descent methods and to methods for solving differential equations.

<u>Connection to descent methods</u>

For a differentiable function, formula (3.137) gives: $\text{prox}_{\rho f} = (I + \rho\nabla f)^{-1}$ and the proximal iteration is of the form

$$x_{k+1} = (I + \rho\nabla f)^{-1} x_k \tag{3.140}$$

- Connection to a gradient method

For ρ small, the first-order expansion of formula (3.140) gives

$$x_{k+1} = x_k - \rho\nabla f(x_k) + o(\rho) \tag{3.141}$$

The proximal iteration is similar to a simple gradient method with a step length ρ.

- Connection to a Newtonian method

Let us approximate the function f by its second-order expansion in the vicinity of the point x_k with the notations: $f_k = f(x_k)$, $g_k = \nabla f(x_k)$, $H_k = \nabla^2 f(x_k)$.

$$f(x) \approx f_k + g_k^T (x - x_k) + \frac{1}{2}(x - x_k)^T H_k (x - x_k) \tag{3.142}$$

By applying the formula for a quadratic function (example 3-21) with here: $Q = H_k$, $c = g_k - Qx_k$, the proximal iteration (3.140) is of the form

$$x_{k+1} = x_k - \left(H_k + \frac{1}{\rho} I \right)^{-1} g_k \tag{3.143}$$

We recognize a modified Newton iteration (with the Tikhonov regularization) also called Levenberg-Marquardt method as in (3.119). The matrix H_k can be an approximation of the Hessian obtained by quasi-Newton.

Connection to differential equations

Consider the following differential equation called the gradient flow of f.

$$\frac{dx(t)}{dt} = -\nabla f(x(t)) \quad , \quad x(t) \in \mathbb{R}^n \tag{3.144}$$

This equation admits as equilibrium points the stationary points of f.
Let us look at the forward and backward Euler methods in turn.

- Forwards Euler method

The forwards Euler method consists in discretizing the equation into the form

$$\frac{x(t_{k+1}) - x(t_k)}{t_{k+1} - t_k} = -\nabla f(x(t_k)) \tag{3.145}$$

By noting $x_k = x(t_k)$, $h_k = t_{k+1} - t_k$, we obtain x_{k+1} in the explicit form

$$x_{k+1} = x_k - h_k \nabla f(x_k) \tag{3.146}$$

We retrieve the iteration of a simple gradient method with a step length h_k.

- Backwards Euler method

The backwards Euler method consists in discretizing the equation into the form

$$\frac{x(t_{k+1}) - x(t_k)}{t_{k+1} - t_k} = -\nabla f(x(t_{k+1})) \tag{3.147}$$

We obtain x_{k+1} in the implicit form

$$x_{k+1} + h_k \nabla f(x_{k+1}) = x_k \quad \Leftrightarrow \quad (I + h_k \nabla f) x_{k+1} = x_k \tag{3.148}$$

We retrieve the iteration (3.140) of a proximal method with a coefficient h_k.

The convergence properties of the gradient and proximal methods are thus closely related to the properties of the forwards and backwards Euler methods.

These interpretations generalize to the non-differentiable case by considering the Moreau envelope defined as: $M_{\rho f}(v) = \inf_x \left(f(x) + \frac{1}{2\rho} \|x - v\|_2^2 \right)$.

This function is a differentiable regularization of f having the same minima as f.
Its gradient is given by: $\nabla M_{\rho f}(v) = \left(v - \text{prox}_{\rho f}(v)\right) / \rho$.

These more complex developments are detailed in reference [R13].

3.5.3 Proximal gradient

The proximal gradient method applies to problems of the form

$$\min_{x\in\mathbb{R}^n} f(x)+c(x) \tag{3.149}$$

The function $f:\mathbb{R}^n \to \mathbb{R}$ is assumed convex and differentiable, while the function $c:\mathbb{R}^n \to \mathbb{R}\cup\{+\infty\}$ is simply assumed convex. This function c can express a set of constraints and take the value $+\infty$ for non-feasible points.

The **proximal gradient** method consists in performing the iterations

$$x_{k+1} = \text{prox}_{s_k c}(x_k - s_k g_k) \tag{3.150}$$

where $g_k = \nabla f(x_k)$ and $s_k > 0$ is the step length.

The idea behind this method is as follows:

- the point $\xi = x_k - s_k g_k$ reduces the function f by a simple gradient step;
- the point $x_{k+1} = \text{prox}_{s_k c}(\xi)$ reduces the function c by staying close to ξ.

In the case where $c(x) = I_C(x)$ (indicator function of the convex set C defined in example 3-21), we obtain the projected gradient method.
The convergence of this method is established by the following property.

Property 3-4: Fixed point of the proximal gradient

x* is a minimum of (3.149) if and only if x* is a fixed point of (3.150).

Demonstration

Assume x* is a minimum of f + c. We have the following equivalences.

x* minimum of f + c $\Leftrightarrow\ 0 \in \nabla f(x^*) + \partial c(x^*)$

$$\Leftrightarrow\ 0 \in s\nabla f(x^*) - x^* + x^* + s\partial c(x^*)$$

$$\Leftrightarrow\ (I - s\nabla f)(x^*) \in (I + s\partial c)(x^*)$$

$$\Leftrightarrow\ (I + s\partial c)^{-1}(I - s\nabla f)(x^*) = x^*$$

with equality because the right-hand side is reduced to the single element x*.

$$\Leftrightarrow\ \text{prox}_{sc}(I - s\nabla f)(x^*) = x^*$$

using the expression for the proximal operator (3.137).
The last line means that x* is a fixed point of the iterations (3.150).

Step adjustment

The step length s_k is set by line search. Among the possible methods, Beck and Teboulle method is based on the function

$$\hat{f}_s(x, x_k) \underset{def}{=} f(x_k) + g_k^T(x - x_k) + \frac{1}{2s}\|x - x_k\|_2^2 \tag{3.151}$$

This function $\hat{f}_s(x, x_k)$ is an upper bound of $f(x)$. It satisfies

$$\forall s \in \left]0, \frac{1}{L}\right], \forall x \in \mathbb{R}^n, \begin{cases} f(x) \le \hat{f}_s(x, x_k) \\ \hat{f}_s(x, x) = f(x) \end{cases} \tag{3.152}$$

where L is the Lipschitz constant of f.

Demonstration

The equality $\hat{f}_s(x, x) = f(x)$ is obvious.

To show the inequality $f(x) \le \hat{f}_s(x, x_k)$, we calculate $\hat{f}_s(x, x_k) - f(x)$.

$\hat{f}_s(x, x_k) - f(x) = f(x_k) - f(x) + g_k^T(x - x_k) + \frac{1}{2s}\|x - x_k\|_2^2$ **(equality E)**

If f is Lipschitzian of constant L, we have

$\forall x,\ f(x) \le f(x_k) + g_k^T(x - x_k) + \frac{1}{2}L\|x - x_k\|_2^2$ which is replaced into (E).

$\hat{f}_s(x, x_k) - f(x) \ge -\frac{1}{2}L\|x - x_k\|_2^2 + \frac{1}{2s}\|x - x_k\|_2^2$

For any step s smaller than $1/L$, we obtain the inequality (3.152).

The Lipschitz constant is generally not known. The Beck and Teboulle method consists in starting from an initial value (for example $s = 1$) and halving it until the following condition is satisfied.

$$f(\xi) \le \hat{f}_s(\xi, x_k) \quad \text{with} \quad \xi = x_k - s g_k \tag{3.153}$$

It is shown that this step setting ensures convergence of the proximal gradient.

The proximal gradient method actually amounts to minimizing at each iteration the upper bounding function $\hat{f}_s(x, x_k)$ which represents a first-order expansion of f in x_k with a quadratic penalty term of the trust region type.

Accelerated proximal gradient

The accelerated proximal gradient method takes up Nesterov's idea presented in section 3.3.1 by inserting a move before iteration (3.150).
This move is an extrapolation of the previous step with a coefficient θ_k.
The gradient assessed in the extrapolated point y_k allows for faster progress.
The iteration of the accelerated proximal gradient is defined by

$$\begin{cases} y_k = x_k + \theta_k (x_k - x_{k-1}) & \text{with } \theta_k = \dfrac{k}{k+3} \\ x_{k+1} = \text{prox}_{s_k c}(y_k - s_k g_k) & \text{with } g_k = \nabla f(y_k) \end{cases} \tag{3.154}$$

Different settings of the coefficient $\theta_k \in]0\,;1[$ are possible.

Connection to differential equations

Consider the differential equation

$$\frac{dx(t)}{dt} = -\nabla f\big(x(t)\big) - \nabla c\big(x(t)\big) \ , \ x(t) \in \mathbb{R}^n \tag{3.155}$$

By applying an Euler method in the so-called forward-backward form:

$$\frac{x_{k+1} - x_k}{h_k} = -\nabla f(x_k) - \nabla c(x_{k+1}) \tag{3.156}$$

we get

$$\big(I + h_k \nabla c\big)(x_{k+1}) = \big(I - h_k \nabla f\big)(x_k) \tag{3.157}$$

This formula is the iteration of the proximal gradient (3.150) written as

$$x_{k+1} = \big(I + s_k \nabla c\big)^{-1} \big(I - s_k \nabla f\big)(x_k) \tag{3.158}$$

with a step $s_k = h_k$.

The proximal gradient method thus amounts to integrating the differential equation (3.155) by a forwards Euler method for the differentiable part f and a backwards Euler method for the non-differentiable part c.

The proximal gradient algorithm is for this reason also called forward-backward. Versions specialized to the frequent case where $c(x) = \|x\|_1$ are ISTA (*Iterative Shrinkage-Thresholding Algorithm*) or FISTA (*Fast ISTA*) which applies acceleration techniques of the type (3.154).

3.5.4 Primal-dual method

Let us retrieve the problem (3.149) with functions $f,c:\mathbb{R}^n \to \mathbb{R}\cup\{+\infty\}$. These functions are convex and the function f can now be non-differentiable.

$$\min_{x\in\mathbb{R}^n} f(x)+c(x) \tag{3.159}$$

The **primal-dual proximal method** consists in performing the iterations

$$\begin{cases} x_{k+1} = \text{prox}_{sf}(y_k - u_k) \\ y_{k+1} = \text{prox}_{sc}(x_{k+1} + u_k) \\ u_{k+1} = u_k + x_{k+1} - y_{k+1} \end{cases} \tag{3.160}$$

This method called ADMM (*Alternating Direction Method of Multipliers*) or Douglas-Rachford separation updates 3 vectors x_k, y_k, u_k at each iteration. It separates the proximal operators of f and c, which is interesting when prox_f and prox_c are simple to compute while prox_{f+c} is not. The vectors x_k and y_k are associated respectively with the functions f and c and the vector u_k aims at ensuring their convergence (or consensus) towards the solution of the problem (3.159). This convergence is established by the following property.

Property 3-5: Fixed point of the primal-dual method

The iterations (3.160) converge to a minimum of (3.159).

Demonstration

Let x*, y*, u* be a fixed point of (3.160). We have: $\begin{cases} x^* = \text{prox}_{sf}(y^*-u^*) \\ y^* = \text{prox}_{sc}(x^*+u^*) \\ u^* = u^*+x^*-y^* \end{cases}$.

The 3rd line gives $y^* = x^*$, which is replaced above: $\begin{cases} x^* = \text{prox}_{sf}(x^*-u^*) \\ x^* = \text{prox}_{sc}(x^*+u^*) \end{cases}$.

Using the expression for the proximal operator (3.137), this system is written as

$$\begin{cases} x^* \in (I+s\partial f)^{-1}(x^*-u^*) \\ x^* \in (I+s\partial c)^{-1}(x^*+u^*) \end{cases} \Leftrightarrow \begin{cases} x^*-u^* \in (I+s\partial f)(x^*) = x^*+s\partial f(x^*) \\ x^*+u^* \in (I+s\partial c)(x^*) = x^*+s\partial c(x^*) \end{cases}$$

Summing member by member, we obtain: $0 \in (\partial f + \partial c)(x^*)$
which shows that x* is a minimum of f + c.

Connection to augmented Lagrangian methods

Let us rewrite problem (3.159) in the so-called consensus form:

$$\min_{x,y\in\mathbb{R}^n} f(x)+c(y) \quad \text{s.t.} \quad x-y=0 \tag{3.161}$$

This formulation separates the minimizations of f and c while imposing the consensus constraint on the solution. The augmented Lagrangian (section 4.6) of this problem is defined with a multiplier $\lambda \in \mathbb{R}^n$ associated with the equality constraint and a penalty coefficient $\rho > 0$.

$$L_\rho(x,y,\lambda) = f(x)+c(y)+\lambda^T(x-y)+\frac{1}{2}\rho\|x-y\|_2^2 \tag{3.162}$$

The augmented Lagrangian algorithm (section 4.6.4) for solving problem (3.161) consists in alternating minimization in (x,y) and updating the multipliers λ according to the value of the constraints (here $x-y=0$).
Since the variables (x,y) are separable, we can put the iterations in the form

$$\begin{cases} \min_x L_\rho(x,y_k,\lambda_k) & \rightarrow x_{k+1} \\ \min_y L_\rho(x_{k+1},y,\lambda_k) & \rightarrow y_{k+1} \\ \lambda_{k+1} = \lambda_k + \rho(x_{k+1}-y_{k+1}) \end{cases} \tag{3.163}$$

Let us explain these iterations for the augmented Lagrangian (3.162) by entering the linear terms in $\lambda^T(x-y)$ into the norm.

$$\begin{cases} x_{k+1} = \arg\min_x f(x)+\frac{1}{2}\rho\left\|x-y_k+\frac{1}{\rho}\lambda_k\right\|_2^2 \\ y_{k+1} = \arg\min_y c(y)+\frac{1}{2}\rho\left\|x_{k+1}-y-\frac{1}{\rho}\lambda_k\right\|_2^2 \\ \lambda_{k+1} = \lambda_k + \rho(x_{k+1}-y_{k+1}) \end{cases} \tag{3.164}$$

By setting $u_k = \frac{1}{\rho}\lambda_k$ and $s = \frac{1}{\rho}$, we retrieve the ADMM iterations (3.160).

The primal-dual method (3.160) is thus equivalent to an augmented Lagrangian algorithm applied to problem (3.161). It forms the basis of many variants by combining it with the proximal gradient (3.150) and/or by using the specificities of the f and c functions to accelerate convergence. We can mention especially the Chambolle-Pock or Tseng algorithms. The applications concern problems of image processing, point-saddle (robust min-max optimization), stochastic or multi-objective optimization.

Linearized version

The linearized version of the primal-dual method applies to the problem

$$\min_{x\in\mathbb{R}^n} f(x)+c(Ax) \tag{3.165}$$

with the functions $f:\mathbb{R}^n \to \mathbb{R}\cup\{+\infty\}$ and $c:\mathbb{R}^m \to \mathbb{R}\cup\{+\infty\}$ being convex but not necessarily differentiable and a matrix $A\in\mathbb{R}^{m\times n}$. The presence of the matrix A can complicate the calculation of the proximal operator of $c(Ax)$.

The linearized version avoids this computation by modifying the iterations (3.160) as follows:

$$\begin{cases} x_{k+1} = \text{prox}_{tf}\left(x_k - \frac{t}{s}A^T(Ax_k - y_k + u_k)\right) \\ y_{k+1} = \text{prox}_{sc}(Ax_{k+1} + u_k) \\ u_{k+1} = u_k + Ax_{k+1} - y_{k+1} \end{cases} \tag{3.166}$$

This algorithm uses only the proximal operators of f and c, with respective coefficients t and s. For $t=s$ and $A=I$, we retrieve the formulas (3.160).

3.5.5 Calculation of the proximal operator

This section introduces the notions of conjugate convex and Moreau identity which facilitate the calculation of commonly used proximal operators.

Consider a function $f:\mathbb{R}^n \to \mathbb{R}\cup\{+\infty\}$ which is not necessarily convex. The **convex conjugate function** of f (also called the Legendre-Fenchel transform of f) is defined by

$$f^*(v) = \sup_{x\in\mathbb{R}^n}\left(v^Tx - f(x)\right) \tag{3.167}$$

The objective of this convex function f* is to approximate f by the upper envelope of all the affine minoring functions. For $v\in\mathbb{R}^n$ we look for the largest $\alpha\in\mathbb{R}$ such that: $\forall x\in\mathbb{R}^n$, $v^Tx+\alpha\le f(x)$, which leads to

$$-\alpha = \sup_{x\in\mathbb{R}^n}\left(v^Tx - f(x)\right) = f^*(v)$$

Taking the conjugate of f*, we define the bi-conjugate function f**.

$$f^{**}(x) = \sup_{v\in\mathbb{R}^n}\left(x^Tv - f^*(v)\right) \tag{3.168}$$

The bi-conjugate function f** has the following two properties:
- it is the largest closed convex function lower than f;
- it is identical to f if and only if f is closed convex.

These properties can be demonstrated using the characterization of a convex function as the upper envelope of affine minoring functions.

The subdifferentials of f and f* are related by the following property.

Property 3-6: Sub-differential of the conjugate function

If f is closed convex, then: $\forall(x,v)\in\mathbb{R}^n\times\mathbb{R}^n\ ,\ v\in\partial f(x)\Leftrightarrow x\in\partial f^*(v)$

Demonstration

By definition of the subdifferential: $v\in\partial f(x)\Leftrightarrow\forall y\ ,\ f(y)\geq f(x)+v^T(y-x)$

which is rewritten as: $\forall y\ ,\ v^Tx-f(x)\geq v^Ty-f(y)$

Taking the upper bound, we have: $v^Tx-f(x)\geq\sup\limits_y\left(v^Ty-f(y)\right)\underset{def}{=}f^*(v)$

with equality for $y=x$, hence: $v^Tx-f(x)=f^*(v)$

The convex function f is identical to its bi-conjugate.

Thus, we have: $v^Tx-f^*(v)=f^{**}(x)=\sup\limits_y\left(x^Ty-f^*(y)\right)$,

which leads to: $\forall y\ ,\ v^Tx-f^*(v)\geq x^Ty-f^*(y)$

$\Leftrightarrow\ \forall y\ ,\ f^*(y)\geq f^*(v)+x^T(y-v)\ \Leftrightarrow\ x\in\partial f^*(v)$

The main interest of the conjugate function comes from the **Moreau identity**.

$$\forall v\in\mathbb{R}^n\ ,\ \text{prox}_f(v)+\text{prox}_{f^*}(v)=v \tag{3.169}$$

Demonstration

Note $x_p=\text{prox}_f(v)$ and $y_p=v-x_p$.

The proximal point x_p satisfies: $0\in\partial f(x_p)+x_p-v\ \Rightarrow\ y_p\in\partial f(x_p)$.

Using property 3-6, we have: $y_p \in \partial f(x_p) \Leftrightarrow x_p \in \partial f^*(y_p)$

Let us replace $x_p = v - y_p$: $v - y_p \in \partial f^*(y_p) \Rightarrow 0 \in \partial f^*(y_p) + y_p - v$

which means that y_p is the proximal point of v for f*: $y_p = \text{prox}_{f^*}(v)$.

Therefore, we have: $\text{prox}_{f^*}(v) = y_p = v - x_p = v - \text{prox}_f(v)$.

Moreau's identity generalizes the decomposition theorem of a vector on a subspace and its orthogonal complement, as shown below.

Example 3-22: Orthogonal decomposition

Consider a subspace L of $\mathbb{R}^n$ and the indicator function of L defined by

$$I_L(x) = \begin{cases} 0 & \text{if } x \in L \\ +\infty & \text{if } x \notin L \end{cases}$$

The conjugate convex function of I_L is defined by

$$I_L^*(v) = \sup_{x \in \mathbb{R}^n}\left(v^T x - I_L(x)\right) = \sup_{x \in L} v^T x \quad \textbf{(equation E)} \text{ since } I_L(x) = +\infty \text{ if } x \notin L.$$

Let $L^\perp$ be the orthogonal complement of L: $L^\perp = \left\{y \in \mathbb{R}^n \;/\; \forall x \in L,\; y^T x = 0\right\}$.

- if $v \in L^\perp$, then $\forall x \in L,\; v^T x = 0$ and equation (E) gives: $I_L^*(v) = 0$;
- if $v \notin L^\perp$, there exists $x \in L, v^T x > 0$ and $\sup\limits_{x \in L} v^T x$ is unbounded: $I_L^*(v) = +\infty$.

The conjugate convex function of I_L is therefore the indicator of $L^\perp$: $I_L^* = I_{L^\perp}$.

Let us apply Moreau's identity to the function I_L, recalling that its proximal operator is the Euclidean projection on L: $\text{prox}_{I_L}(v) = P_L(v)$ (example 3-21).

$$\begin{cases} \text{prox}_{I_L}(v) = P_L(v) \\ \text{prox}_{I_L^*}(v) = P_{L^\perp}(v) \end{cases} \Rightarrow \forall v \in \mathbb{R}^n,\; P_L(v) + P_{L^\perp}(v) = v$$

The vector v is thus expressed as the sum of its projections on L and $L^\perp$.

Moreau's identity is useful when the proximal operator of f* is simpler to calculate than that of f. In many applications (inverse problems, identification, learning), the function f is a sum of norms. The following example shows the calculation of the proximal operator of a norm.

Example 3-23: Proximal operator associated with a norm

Consider the function: $f(x)=\|x\|$ where $\|.\|$ is any norm on $\mathbb{R}^n$.

Let us first calculate the convex conjugate function of f: $f^*(v)=\sup_{x\in\mathbb{R}^n}\left(v^Tx-\|x\|\right)$

Any vector $x\in\mathbb{R}^n$ can be put into the form $x=\alpha u$ where:
- the real $\alpha\geq 0$ is the norm of x;
- the unit vector $u\in\mathbb{R}^n$, $\|u\|=1$ is the direction of x.

The function f * is expressed as: $f^*(v)=\sup_{\alpha\geq 0\,,\,\|u\|=1}\alpha(v^Tu-1)$ **(equation E)**

Let $B=\left\{v\ /\ \|v\|_*\leq 1\right\}$ be the unit ball associated with the dual norm of $\|.\|$.

This dual standard noted $\|.\|_*$ is defined by: $\|v\|_*=\sup_{\|u\|\leq 1}v^Tu$.

- if $v\in B$, then for any u of norm 1, we have: $v^Tu\leq 1$.
  Equation (E) gives in this case: $f^*(v)=0$ (sup reached for $\alpha=0$).
- if $v\notin B$, $\sup_{\|u\|=1}v^Tu$ is positive. Equation (E) gives: $f^*(v)=+\infty$ for $\alpha=\infty$.

The conjugate convex function of $f(x)=\|x\|$ is therefore the indicator function of the unit ball B associated to the dual norm: $f^*(v)=I_B(v)$.

Let us apply Moreau's identity to the function f by recalling that the proximal operator of the indicator function of B is the Euclidean projection on B.

$\forall v\in\mathbb{R}^n$, $\text{prox}_f(v)+\text{prox}_{f^*}(v)=v \quad\Rightarrow\quad \text{prox}_f(v)=v-P_B(v)$

By introducing the coefficient $\rho>0$, we obtain the general formula

$$f(x)=\|x\| \;\rightarrow\; \text{prox}_{\rho f}(v)=v-\rho P_B\left(\frac{v}{\rho}\right)$$

Reference [R13] details the computation of many usual proximal operators (norms, projections, etc.). An important application concerns the LASSO (*Least Absolute Shrinkage and Selection Operator*) form problems:

$$\min_{x\in\mathbb{R}^n}\frac{1}{2}\|Ax-b\|_2^2+\gamma\|x\|_1 \tag{3.170}$$

By using the explicit expression of the proximal operator, as well as the decomposition property when the variables are separable, one can solve high-dimensional problems of the type (3.170) very quickly.

3.6 Convergence

This section presents some convergence results for descent methods, in particular on the speed of convergence and on the accuracy of the solution.

3.6.1 Global convergence

A descent algorithm is said to be **globally convergent** if the iterate limit point $\lim_{k\to\infty} x_k = x^*$ is a stationary point: $\nabla f(x^*) = 0$. If we ensure that the function decreases at each iteration, the limit point is a local minimum.

Global convergence does not mean convergence to a global optimum, but simply that the algorithm converges to a stationary point whatever the initial point. The global convergence property is demonstrated with mildly restrictive assumptions.

Property 3-7: Global convergence

Consider a line search descent method with iterations of the form: $x_{k+1} = x_k + s_k u_k$ where u_k is the direction and $s_k > 0$ is the step length.
Note θ_k the angle between u_k and the direction of steepest descent $-\nabla f(x_k)$.
Assume that the directions u_k satisfy the condition: $\cos\theta_k \geq \delta > 0$ and that the step lengths s_k satisfy the Wolfe conditions (3.110).

Then the descent method converges to a stationary point: $\nabla f(x^*) = 0$.

Demonstration (see [R11], [R12])

It is assumed that the gradient of the cost function is Lipschitzian (an assumption that is not very restrictive and is often satisfied in practice).
Therefore, there is a constant L such that: $\forall x, x', \; \|\nabla f(x') - \nabla f(x)\| < L\|x' - x\|$

The successive iterates: $x_{k+1} = x_k + s_k u_k$ satisfy Wolfe's conditions.

$$\begin{cases} f(x_{k+1}) < f(x_k) + c_1 s_k \nabla f(x_k)^T u_k \\ \nabla f(x_{k+1})^T u_k > c_2 \nabla f(x_k)^T u_k \end{cases} \quad \text{with } 0 < c_1 < c_2 < 1$$

The 2nd Wolfe condition yields: $\left[\nabla f(x_{k+1}) - \nabla f(x_k)\right]^T u_k > (c_2 - 1)\nabla f(x_k)^T u_k$.

Using the Lipschitz condition on the gradient in this inequality

$$\left[\nabla f(x_{k+1})-\nabla f(x_k)\right]^T u_k \le \left\|\nabla f(x_{k+1})-\nabla f(x_k)\right\|\left\|u_k\right\| < L s_k \left\|u_k\right\|^2$$

we deduce an inequality on the step s_k

$$L s_k \left\|u_k\right\|^2 > (c_2-1)\nabla f(x_k)^T u_k \quad \Rightarrow s_k > \frac{(c_2-1)\nabla f(x_k)^T u_k}{L\left\|u_k\right\|^2}$$

which is reported in the 1st Wolfe condition (with $\nabla f(x_k)^T u_k < 0$)

$$f(x_{k+1}) < f(x_k)+c_1 s_k \nabla f(x_k)^T u_k < f(x_k)+\frac{c_1(c_2-1)}{L}\frac{\left\|\nabla f(x_k)^T u_k\right\|^2}{\left\|u_k\right\|^2}$$

Noting: $c_L = \dfrac{c_1(1-c_2)}{L} > 0$ and with: $\nabla f(x_k)^T u_k = -\left\|\nabla f(x_k)\right\| \left\|u_k\right\| \cos\theta_k$,

we obtain the inequality: $f(x_{k+1}) < f(x_k) - c_L \left\|\nabla f(x_k)\right\|^2 \cos^2\theta_k$.

Let us sum up these inequalities member by member from $k=0$ to $k=n-1$.

$$f(x_n) < f(x_0) - c_L \sum_{k=0}^{n-1} \left\|\nabla f(x_k)\right\|^2 \cos^2\theta_k$$

Assuming that f is lower bounded (otherwise the optimization problem has no solution), and passing to the limit $n \to \infty$, we obtain the **Zoutendijk inequality**.

$$\sum_{k=0}^{\infty} \left\|\nabla f(x_k)\right\|^2 \cos^2\theta_k < \infty \quad \text{which implies: } \lim_{k\to\infty} \left\|\nabla f(x_k)\right\|^2 \cos^2\theta_k = 0.$$

Since by assumption: $\cos\theta_k \ge \delta > 0$, we obtain: $\lim_{k\to\infty} \left\|\nabla f(x_k)\right\| = \left\|\nabla f(x^*)\right\| = 0$.

It is therefore sufficient that the successive directions of descent do not become "too orthogonal" to the gradient of f to ensure global convergence. The steepest descent, Newton and quasi-Newton methods satisfy this condition.

Property 3-7 is also demonstrated for the Goldstein conditions (3.108).

Global convergence is in fact guaranteed for any algorithm in which a steepest descent iteration is periodically inserted. Although this property is reassuring on the asymptotic behavior, it does not guarantee anything on the speed of convergence which is an essential element for applications.

3.6.2 Speed of convergence

The speed of convergence is measured on the deviation from the optimum. We can consider the deviations on the variables $\|x_k - x^*\|$ or the deviations on the function values $f(x_k) - f(x^*)$ (which are positive because x^* is a minimum).

The convergence is said to be **linear** if there exists a real $\delta < 1$ such that

$$\lim_{k\to\infty} \frac{\|x_{k+1} - x^*\|}{\|x_k - x^*\|} < \delta \quad \text{or} \quad \lim_{k\to\infty} \frac{f(x_{k+1}) - f(x^*)}{f(x_k) - f(x^*)} < \delta \tag{3.171}$$

The convergence is said to be **superlinear** of order $\gamma > 1$ if

$$\lim_{k\to\infty} \frac{\|x_{k+1} - x^*\|}{\|x_k - x^*\|^\gamma} < \infty \quad \text{or} \quad \lim_{k\to\infty} \frac{f(x_{k+1}) - f(x^*)}{[f(x_k) - f(x^*)]^\gamma} < \infty \tag{3.172}$$

The quadratic convergence corresponds to $\gamma = 2$.

The following property concerns the method of steepest descent.

Property 3-8: Speed of convergence of the steepest descent method

The steepest descent method applied to the minimization of a convex quadratic function f has a linear convergence with

$$\frac{f(x_{k+1}) - f(x^*)}{f(x_k) - f(x^*)} \le \left(\frac{\sigma_n - \sigma_1}{\sigma_n + \sigma_1}\right)^2 \tag{3.173}$$

where σ_1 and σ_n are the smallest and largest eigenvalue of the Hessian of f.

Demonstration elements

There are two stages in the demonstration: expressing of the improvement ratio and then bounding it.

<u>Expression of the improvement ratio</u>
The quadratic function to be minimized is of the form

$f(x) = \frac{1}{2}x^T Q x + c^T x$ with a symmetric positive definite matrix Q.

We can always reduce to a symmetric matrix by posing: $Q' = \frac{1}{2}(Q + Q^T)$.

The minimum of the quadratic function is: $x^* = -Q^{-1}c \;\rightarrow\; f(x^*) = -\frac{1}{2}c^T Q^{-1} c$.

The steepest descent method performs a move from the point x_k in the opposite direction to the gradient: $u_k = -\nabla f(x_k) = -Qx_k - c$ with a step s.
The value of the function in the point $x_{k+1} = x_k + su_k$ is (using $Qx_k = -u_k - c$)

$$f(x_k + su_k) = \frac{1}{2}(x_k + su_k)^T Q(x_k + su_k) + c^T(x_k + su_k) = f(x_k) + \frac{1}{2}s^2 u_k^T Q u_k - s u_k^T u_k$$

The optimal step s_k is given by

$$\min_s f(x_k + su_k) \;\rightarrow\; s_k = \frac{u_k^T u_k}{u_k^T Q u_k} \;\rightarrow\; f(x_{k+1}) = f(x_k) - \frac{1}{2}\frac{(u_k^T u_k)^2}{u_k^T Q u_k}$$

The improvement ratio of the function is

$$\frac{f(x_{k+1}) - f(x^*)}{f(x_k) - f(x^*)} = 1 - \frac{(u_k^T u_k)^2}{2u_k^T Q u_k}\frac{1}{f(x_k) - f(x^*)}$$

Using $x_k = -Q^{-1}(u_k + c)$, we have

$$f(x_k) - f(x^*) = \frac{1}{2}x_k^T Q x_k + c^T x_k + \frac{1}{2}c^T Q^{-1} c$$

$$= \frac{1}{2}(u_k + c)^T Q^{-1}(u_k + c)^T - c^T Q^{-1}(u_k + c) + \frac{1}{2}c^T Q^{-1} c = \frac{1}{2}u_k^T Q^{-1} u_k$$

which gives the following equality E_1 for the improvement ratio of the function.

$$\frac{f(x_{k+1}) - f(x^*)}{f(x_k) - f(x^*)} = 1 - \frac{1}{q} \quad \text{with } q = \frac{(u_k^T Q u_k)(u_k^T Q^{-1} u_k)}{(u_k^T u_k)^2} \qquad \textbf{(equality E}_1\textbf{)}$$

The symmetric positive definite matrix Q admits an orthonormal basis of eigenvectors. Let P be the orthogonal transfer matrix in this basis $(P^{-1} = P^T)$ and D the diagonal eigenvalue matrix of Q.

$$D = \mathrm{diag}(\sigma_i) \text{ with } 0 < \sigma_1 \le \sigma_2 \le \cdots \le \sigma_{n-1} \le \sigma_n .$$

Expressing the matrix Q as $Q = PDP^T$ and introducing the unit vector $v = P^T \frac{u_k}{\|u_k\|}$

we obtain for q

$$q = \frac{(u_k^T PDP^T u_k)(u_k^T PD^{-1}P^T u_k)}{(u_k^T u_k)^2} = \frac{\left(\|u_k\| v^T D v \|u_k\|\right)\left(\|u_k\| v^T D^{-1} v \|u_k\|\right)}{\|u_k\|^4}$$

$$= (v^T D v)(v^T D^{-1} v)$$

The unit vector v has components $(v_1, v_2, \cdots, v_n)$ in the eigenvector basis of Q.

By noting $\alpha_i = v_i^2$, q is expressed as

$$q = \left(\sum_{i=1}^{n} \alpha_i \sigma_i\right)\left(\sum_{i=1}^{n} \alpha_i \frac{1}{\sigma_i}\right) \quad \text{with} \quad \begin{cases} \alpha_1 + \alpha_2 + \cdots + \alpha_n = 1 \\ 0 \le \alpha_i \le 1 \end{cases} \qquad \textbf{(equality E}_2\textbf{)}$$

We will now upper bound the value of q as a function of $(\alpha_1, \cdots, \alpha_n)$.

Upper bound on the improvement ratio

Consider the random variable S taking the values $(\sigma_1, \cdots, \sigma_n)$ with the respective probabilities $(\alpha_1, \cdots, \alpha_n)$ which satisfy: $\begin{cases} \alpha_1 + \alpha_2 + \cdots + \alpha_n = 1 \\ 0 \le \alpha_i \le 1 \end{cases}$.

The expectation of S is: $E[S] = \sum_{i=1}^{n} \alpha_i \sigma_i$.

The variable $\frac{1}{S}$ takes the values $\left(\frac{1}{\sigma_1}, \cdots, \frac{1}{\sigma_n}\right)$ with the probabilities $(\alpha_1, \cdots, \alpha_n)$.

Its expectation is: $E\left[\frac{1}{S}\right] = \sum_{i=1}^{n} \alpha_i \frac{1}{\sigma_i}$.

With this random variable S, the equality E2 is expressed as: $q = E[S]E\left[\frac{1}{S}\right]$.

Since the variable S takes values between σ_1 and σ_n (which are both positive), we have

$$(S - \sigma_1)(\sigma_n - S) \ge 0 \;\Rightarrow\; S(\sigma_1 + \sigma_n - S) \ge \sigma_1 \sigma_n \;\Rightarrow\; \frac{1}{S} \le \frac{\sigma_1 + \sigma_n - S}{\sigma_1 \sigma_n}$$

$$\Rightarrow\; E\left[\frac{1}{S}\right] \le \frac{\sigma_1 + \sigma_n - E[S]}{\sigma_1 \sigma_n}$$

Let us multiply by $E[S] > 0$ to make q appear.

$$q = E[S]E\left[\frac{1}{S}\right] \le \frac{(\sigma_1 + \sigma_n)E[S] - E[S]^2}{\sigma_1 \sigma_n}$$

By writing the numerator as

$$E[S]^2 - (\sigma_1 + \sigma_n)E[S] = \left(E[S] - \frac{\sigma_1 + \sigma_n}{2}\right)^2 - \frac{(\sigma_1 + \sigma_n)^2}{4}$$

we obtain the inequality

$$q \le \frac{(\sigma_1 + \sigma_n)^2}{4\sigma_1 \sigma_n} - \frac{1}{\sigma_1 \sigma_n}\left(E[S] - \frac{\sigma_1 + \sigma_n}{2}\right)^2 \le \frac{(\sigma_1 + \sigma_n)^2}{4\sigma_1 \sigma_n}$$

Taking the expression of q as a function of u, this inequality is written as

$$\forall u\ ,\ q=\frac{(u^TQu)(u^TQ^{-1}u)}{(u^Tu)^2}\le\frac{(\sigma_1+\sigma_n)^2}{4\sigma_1\sigma_n}\qquad \textbf{(inequality } \mathbf{I_1})$$

This result is called the **Kantorovich inequality**.

Replacing the Kantorovich inequality I_1 into the equality E_1, we obtain the relation (3.173) bounding the improvement ratio of the function.

Property 3-8 has been proved for a quadratic function. It extends to any function f in the neighborhood of its minimum x*, by considering the eigenvalues of the Hessian $\nabla^2 f(x^*)$.

The inequality (3.173) can be rewritten by introducing the conditioning of the Hessian $\kappa=\frac{\sigma_n}{\sigma_1}\ge 1$.

$$\frac{f(x_{k+1})-f(x^*)}{f(x_k)-f(x^*)}\le\left(\frac{\kappa-1}{\kappa+1}\right)^2 \qquad (3.174)$$

If $\kappa=1$ (perfect conditioning), the solution is obtained in one iteration, as the level lines are circular and the gradient points towards the center. If κ is large (poor conditioning), the second member is close to 1, which means that the improvement might be small at each iteration. Although this bound is only a worst case, it is found in practice that the steepest descent method converges very slowly on ill-conditioned problems (example 3-12).

Conjugate gradient methods have superlinear convergence, but require n one-dimensional minimizations per iteration. Quasi-Newton methods have superlinear (or quadratic under certain assumptions) convergence. Newton's method has quadratic convergence. These methods are theoretically not sensitive to Hessian conditioning, but their behavior can be affected by numerical accuracy errors.

3.6.3 Numerical accuracy

Numerical errors due to the finite precision calculation affect both the solution obtained and the directions of a descent method.

Accuracy of the solution

Let us apply from the minimum x* a move of step s along the unit direction u. The variation of f to order 2 is given by

$$f(x^*+su) - f(x^*) = su^T g^* + \frac{1}{2}s^2u^TH^*u = \frac{1}{2}s^2u^TH^*u \tag{3.175}$$

with $g^* = \nabla f(x^*) = 0$ and $H^* = \nabla^2 f(x^*) > 0$.

If there is a numerical error ε_f in the evaluation of f, the algorithm can stop in any point $x = x^* + su$ such that

$$\varepsilon_f = \frac{1}{2}s^2u^TH^*u \tag{3.176}$$

If the direction u is an eigenvector of H* associated with the eigenvalue σ, then $u^TH^*u = \sigma$ and the error on the solution is

$$\varepsilon_x = \|x - x^*\| = s = \sqrt{\frac{2\varepsilon_f}{\sigma}} \tag{3.177}$$

The error is maximum along the direction of the smallest eigenvalue of H*.

When the matrix H is close to singular, or if it has a bad conditioning ($\kappa = \frac{\sigma_n}{\sigma_1} \gg 1$) it is difficult to obtain an accurate x* solution.

Accuracy of the direction

The directions of a descent method are obtained by solving at each iteration a linear system of the form

$$Hd = -g \tag{3.178}$$

where g is the gradient of the function and H is an approximation of the Hessian.

The positive definite matrix H has an orthonormal basis of eigenvectors. In this basis, the system (3.178) gives n independent linear equations:

$$\sigma_i d_i = -g_i \tag{3.179}$$

where d_i, g_i are the components of d and g in the eigenvector basis.

Let us examine the effect of numerical errors on a simple linear equation.

$$ax = b \Rightarrow x^* = \frac{b}{a} \tag{3.180}$$

The errors ε_a and ε_b on the numbers a and b cause an error ε_x on the solution x*. Summing up the absolute errors, we obtain at order 1

$$\frac{\varepsilon_x}{x^*} = \frac{\varepsilon_a}{a} + \frac{\varepsilon_b}{b} \tag{3.181}$$

For a numerical calculation with a relative machine accuracy ε_m (usually equal to 10^{-16}), the errors ε_a and ε_b on the real numbers a and b are

$$\begin{cases} \varepsilon_a = \max(a\varepsilon_m, \varepsilon_m) \\ \varepsilon_b = \max(b\varepsilon_m, \varepsilon_m) \end{cases} \tag{3.182}$$

Assume that b is not small, and that the error on b is: $\varepsilon_b = b\varepsilon_m$.

- if a is large, $\varepsilon_a = a\varepsilon_m$ and the relative error on x* is of the order of: $\frac{\varepsilon_x}{x^*} \sim \varepsilon_m$;
- if a is small, $\varepsilon_a = \varepsilon_m$ and the relative error on x* is of the order of: $\frac{\varepsilon_x}{x^*} \sim \frac{\varepsilon_m}{a}$.

The relative error in this case may be much higher than ε_m.

Returning to (3.179), this means that the direction components along eigenvectors associated to small eigenvalues can be imprecise, which is detrimental to a line search algorithm.

This is especially the case for a penalization method. The penalized function $f_p = f + \rho c^2$ has a gradient given by: $\nabla f_p = \nabla f + 2\rho \nabla c.c$ of the order of ρ and a Hessian with some eigenvalues of the order of ρ (section 4.2.4).

Passing into the eigenvector basis of the Hessian, the system (3.178) giving the direction takes the form

$$\begin{pmatrix} D_1 & 0 \\ 0 & \rho D_2 \end{pmatrix} \begin{pmatrix} d_1 \\ d_2 \end{pmatrix} = -\begin{pmatrix} \rho g_1 \\ \rho g_2 \end{pmatrix} \Leftrightarrow \begin{cases} \frac{1}{\rho} D_1 d_1 = -g_1 \\ D_2 d_2 = -g_2 \end{cases} \tag{3.183}$$

For a large penalty coefficient ρ, the error is larger on the components d_1. The direction of descent becomes imprecise, which can block the algorithm.

3.7 Conclusion

3.7.1 The key points

- Newton's method is fast and accurate, but it requires the calculation of the gradient (equations) or the Hessian (minimization) and is not robust;
- quasi-Newton methods (Broyden, BFGS, SR1) avoid the computation of the gradient/Hessian by having performances close to Newton's method;
- globalization aims to control the progression by homotopy (equations), line search or trust region (minimization) methods;
- a line search method looks for the minimum along a descent direction. The Armijo condition defines an acceptable move;
- a trust region method searches for the minimum of a quadratic model in a region of given radius. The dogleg method gives an approximate solution.

3.7.2 To go further

- *Programmation mathématique (M. Minoux, Lavoisier 2008, 2e édition)*

  Chapter 4 presents the quasi-Newton, conjugate direction and line search descent methods. Theoretical results are given on the convergence of the algorithms.

- *Introduction à l'optimisation différentiable*
  (M. Bierlaire, Presses polytechniques et universitaires normandes 2006)

  Newton's method is presented in chapters 7 and 10. Quasi-Newton methods are presented in chapters 8 (Broyden) and 13 (BFGS, SR1). Chapter 11 presents the line search descent. Chapter 12 presents the trust region descent. Chapter 6 presents least squares problems. Numerous detailed examples illustrate how the algorithms work.

- *Numerical optimization (J. Nocedal, S.J. Wright, Springer 2006)*

  Quasi-Newton methods are presented in chapters 11 (Broyden), 6 (BFGS, SR1) and 7 (BGFS extensions). Chapter 3 presents the line search descent. Chapter 4 presents the trust region descent. Chapter 5 presents the conjugate direction methods. Chapter 10 presents least squares problems. A lot of practical advice is given on the implementation of the algorithms.

- *Practical methods of optimization (R. Fletcher, Wiley 1987, 2nd edition)*

  Chapter 3 presents the Newton and quasi-Newton methods. Chapter 2 presents the line search descent. Chapter 5 presents the trust region descent. Chapter 4 presents the conjugate direction methods. Chapter 6 presents least squares problems. Many convergence proofs are given.

- *Numerical methods for unconstrained optimization and nonlinear equations (J.E. Dennis, R.B. Schnabel, Siam 1996)*

  Newton's method is presented in chapter 5. Quasi-Newton methods are presented in chapters 8 (Broyden) and 9 (BFGS, SR1). Chapter 6 presents line search and trust region descent. Chapter 10 presents least squares problems. The algorithms are detailed and illustrated with examples. Numerous demonstrations are also given on the convergence of the algorithms.

- *Numerical optimization (J.F. Bonnans, J.C. Gilbert, C. Lemaréchal, C.A. Sagastizabal, Springer 2003)*

  Chapter 4 presents the Newton and quasi-Newton methods. Chapter 3 presents the line search descent. Chapter 6 presents the trust region descent. Chapter 5 presents the conjugate direction methods. Many demonstrations are given on the convergence of the algorithms.

- *Analyse numérique (Collectif direction J. Baranger, Hermann 1991)*

  Chapters 1 and 6 present the quasi-Newton methods in detail with many demonstrations.

- *Practical mathematical optimization (J.A. Snyman, Springer 2005)*

  The book offers variations on the classic descent methods with many illustrative examples.

- *Introduction to numerical continuation methods (E.L. Allgower, K. Georg, Siam 2003)*

  The book is devoted to homotopy (or continuation) methods for solving systems of equations.

4. Constrained optimization

This chapter is devoted to continuous optimization methods with constraints, referred to as nonlinear programming. Linear programming (linear cost function and constraints) is the subject of chapter 5.

Section 1 presents the categories of methods: primal, primal-dual or dual, depending on the formulation chosen for the optimization problem. These methods measure the iterate improvement by a merit function or a filter taking into account the reductions of the cost function and the violation of the constraints.

Section 2 presents the penalization methods. Their principle is to reduce to an unconstrained problem by concatenating the cost function and the constraint violation into an augmented function. Depending on the form of penalization, it is possible to approximate or obtain the exact solution of the constrained problem.

Section 3 presents reduced gradient methods. Their principle is to accept only feasible solutions. The search for a new point is based on a linearization of the constraints and the restoration of a feasible point. These methods have the advantage of providing a feasible solution, even when the optimum is not reached.

Section 4 introduces sequential quadratic programming methods. Their principle is to solve the optimality conditions by a Newton method, which is equivalent to solving a sequence of quadratic-linear problems. These methods converge faster than the reduced gradient, but intermediate solutions are not feasible.

Section 5 presents interior point methods. Their principle is to solve the optimality conditions by a Newton method, but treating the inequality constraints by barrier penalization. These methods were initially developed for linear problems and are interesting in the presence of a large number of inequality constraints.

Section 6 presents the augmented Lagrangian methods. Their principle is to solve the dual problem with a quadratic penalty. This eliminates the intrinsic difficulties of the dual problem and allows one to reduce to a sequence of unconstrained problem. The exact solution can be obtained without indefinitely increasing the penalty coefficient.

4.1 Classification of methods

The three main categories of constrained optimization methods are primal, primal-dual and dual methods. These methods differ in the formulation of the problem to be solved.

4.1.1 Problem formulations

This section recalls the formulations for tackling an optimization problem with constraints and for dealing with inequality constraints.

Primal formulation

The standard form of an optimization problem with n variables x, p equality constraints c_E and q inequality constraints c_I is as follows:

$$\min_{x \in \mathbb{R}^n} f(x) \quad \text{s.t.} \quad \begin{cases} c_E(x) = 0 \\ c_I(x) \le 0 \end{cases} \tag{4.1}$$

This formulation defines the **primal** problem. The variables x are called the **primal variables**.

KKT conditions

The Lagrangian of the problem (4.1) is the function of $\mathbb{R}^{n+p+q}$ in $\mathbb{R}$ defined by

$$\begin{aligned} L(x,\lambda,\mu) &= f(x) + \lambda^T c_E(x) + \mu^T c_I(x) \\ &= f(x) + \sum_{j=1}^{p} \lambda_j c_{Ej}(x) + \sum_{j=1}^{q} \mu_j c_{Ij}(x) \end{aligned} \tag{4.2}$$

The multipliers $\lambda \in \mathbb{R}^p$ are associated with the p equality constraints c_E.
The multipliers $\mu \in \mathbb{R}^q$ are associated with the q inequality constraints c_I.
The multipliers λ, μ are called **dual variables**. A local solution of problem (4.1) must satisfy the KKT optimality conditions of order 1.

$$\begin{cases} \nabla_x L(x,\lambda,\mu) = 0 \\ c_E(x) = 0 \\ c_I(x) \le 0 \\ \mu \ge 0 \\ \mu_j c_{Ij}(x) = 0 \quad , \quad j=1 \text{ to } q \end{cases} \tag{4.3}$$

Dual formulation

The dual function ψ is defined as the minimum of the Lagrangian with respect to the x variables.

$$\psi(\lambda,\mu) = \min_{x\in D} L(x,\lambda,\mu) \tag{4.4}$$

The domain of definition of the dual function is: $D = \{\lambda,\mu \geq 0 \;/\; \psi(\lambda,\mu) > -\infty\}$.
The **dual problem** of problem (4.1) consists in maximizing the dual function.

$$\max_{\lambda,\mu\geq 0} \psi(\lambda,\mu) \tag{4.5}$$

The solution of the dual problem satisfies the KKT conditions for a local minimum (4.3) and also yields a lower bound on the global minimum of the problem (4.1).

Inequality constraints

Some inequality constraints are inactive on the solution of problem (4.1) and have zero multipliers: $c_{Ij}(x) < 0$ and $\mu_j = 0$. If these inactive constraints were known in advance, we could ignore them and transform the remaining active inequality constraints into equalities. Problem (4.1) could then be reformulated into a problem with only equality constraints.

$$\min_{x\in\mathbb{R}^n} f(x) \;\text{ s.t. }\; \begin{cases} c_E(x) = 0 \\ c_I(x) \leq 0 \end{cases} \quad\Leftrightarrow\quad \min_{x\in\mathbb{R}^n} f(x) \;\text{ s.t. }\; c_{act}(x) = 0 \tag{4.6}$$

This problem formulation with only equality constraints forms the basis of many constrained optimization algorithms. Two methods can be used to get back to it.

The first method consists in transforming inequalities into equalities by introducing q positive **slack variables** denoted y.

$$\min_{x\in\mathbb{R}^n} f(x) \;\text{ s.t. }\; \begin{cases} c_E(x) = 0 \\ c_I(x) \leq 0 \end{cases} \quad\Leftrightarrow\quad \min_{\substack{x\in\mathbb{R}^n \\ y\in\mathbb{R}^q}} f(x) \;\text{ s.t. }\; \begin{cases} c_E(x) = 0 \\ c_I(x) + y = 0 \\ y \geq 0 \end{cases} \tag{4.7}$$

This new problem has $n+q$ variables and $p+q$ equality constraints. We can apply an optimization algorithm with equality constraints, taking care that the slack variables remain positive at each iteration.

The second method, known as **active constraint** method, consists in selecting at each iteration the set A_k of constraints active in the current point x_k. The inactive constraints in x_k $(c_{Ij}(x_k)<0)$ are ignored to construct a new point x_{k+1}.

We then examine the set A_{k+1} of active constraints in the new point x_{k+1}.
If $A_{k+1} = A_k$, then the initial selection of active constraints was correct and the new point is accepted. Otherwise, the iteration is resumed from the point x_k by modifying the selection of active constraints. For example, one can consider directly the expected set A_{k+1}, or trials can be carried out by progressively including/excluding inequalities according to their respective active/inactive status in x_k and x_{k+1}.

4.1.2 Primal, primal-dual and dual methods

The numerical methods are based either on the primal formulation (4.1), or on the KKT conditions (4.3) or on the dual formulation (4.5).

Primal methods

The primal methods tackle the optimization problem in its direct formulation (4.1) by considering only the primal variables x. Multipliers are not used. Constraints are treated either by penalization or by reduction.

The **penalization** approach consists in transforming the constrained problem (4.1) into an unconstrained problem of the form

$$\min_{x \in \mathbb{R}^n} f_\rho(x) \underset{def}{=} f(x) + \rho C(x) \tag{4.8}$$

The function $f_\rho : \mathbb{R}^n \to \mathbb{R}$ is called penalized cost or **augmented cost**. The positive real $\rho > 0$ is the penalty coefficient. The penalty function $C : \mathbb{R}^n \to \mathbb{R}$ measures the violation of the constraints. It can be defined in different ways, the most common being the following quadratic form.

$$C(x) = \frac{1}{2}\|c_E(x)\|_2^2 + \frac{1}{2}\|\max(0, c_I(x))\|_2^2 \tag{4.9}$$

The penalization approach allows the application of an unconstrained optimization algorithm, which greatly simplifies the solution. The drawbacks are an approximate satisfaction of the constraints (depending on the form of penalty chosen) and the difficulties related to the conditioning of the augmented cost (4.8). The penalization methods are presented in section 4.2.

For some applications, a strict compliance with the constraints is imperative and takes precedence over the solution optimality. The **reduction** approach consists in generating a series of feasible points decreasing the cost function. Even if the optimum is not reached, intermediate solutions are acceptable.

These methods require an initial feasible point x_0, which can be constructed by minimizing the penalty function (4.9) until it is cancelled. At each iteration, the active constraints are linearized in the current point x_k, and a minimization is performed in the space tangent to the active constraints (reduced space) in order to determine a new point x_d. If the constraints are nonlinear, a feasible point x_{k+1} must be restored from x_d and then checked whether it is better than x_k. The reduced/projected gradient methods are presented in section 4.3.

Primal-dual methods

The primal-dual methods are based on the KKT (4.3). These conditions form a system of equations and inequations whose unknowns are the primal variables x and the dual variables λ,μ. The inequations arising from the inequality constraints do not allow a Newton method to be applied directly to the solution of the KKT system. The inequality constraints are treated either by slack variables or by active constraints.

The first method is to introduce q positive slack variables y as in (4.7). The KKT conditions (4.3) take the form

$$\begin{cases} \nabla f(x)+\nabla c_E(x)\lambda+\nabla c_I(x)\mu=0 \\ c_E(x)=0 \\ c_I(x)+y=0 \\ \mu_j y_j=0 \quad , \quad j=1 \text{ to } q \\ y\geq 0\,,\, \mu\geq 0 \end{cases} \tag{4.10}$$

The first 4 lines in (4.10) form a nonlinear system that can be solved by a Newton method. However, the complementarity conditions (involving the cancellation of entire μ_j or y_j) introduce a combinatorial aspect that can block the iterations if one variable approaches zero too quickly.

The **interior point** methods introduce a barrier parameter $h\geq 0$ and replace the complementarity conditions by: $\mu_j y_j = h$. The perturbed KKT system is solved for decreasing values of h, so as to approximate the solution of (4.10) while remaining within the domain $\{y\geq 0\,;\,\mu\geq 0\}$. These interior point methods are presented in section 4.5.

The second method consists in considering only the active constraints noted c(x) at each iteration as in (4.6). The KKT system takes the form

$$\nabla L(x,\lambda)=0 \quad \text{with } L(x,\lambda)=f(x)+\lambda^T c(x) \tag{4.11}$$

This system of nonlinear equations is of dimension n + m (where m is the number of active constraints). Newton's iteration in the point x_k is given by

$$\nabla^2 L(x_k,\lambda_k)\begin{pmatrix} d_x \\ d_\lambda \end{pmatrix}=-\nabla L(x_k,\lambda_k) \tag{4.12}$$

where (d_x, d_λ) are the components of the move on the variables (x,λ). Detailing (4.12) in terms of the variables x and λ, we obtain

$$\begin{cases} \nabla^2_{xx} L(x_k,\lambda_k)d_x+\nabla c(x_k)d_\lambda=-\nabla_x L(x_k,\lambda_k) \\ \nabla c^T(x_k)d_x=-c(x_k) \end{cases} \tag{4.13}$$

Note that this linear system in (d_x, d_λ) corresponds to the KKT conditions of the following local quadratic-linear problem, obtained by expanding the Lagrangian to order 2 and the constraints to order 1 in (x_k,λ_k).

$$\min_{d_x\in R^n} \nabla_x L(x_k,\lambda_k)^T d_x+\frac{1}{2}d_x^T\nabla^2_{xx}L(x_k,\lambda_k)d_x \;\text{ s.t. }\; \nabla c(x_k)^T d_x+c(x_k)=0 \tag{4.14}$$

The **sequential quadratic programming** methods presented in section 4.4 use this equivalence by solving a sequence of local quadratic problems. These methods based on Newton iterations require a globalization process. Moreover, the successive points are not feasible. Their quality is evaluated by a merit function or a filter (section 4.1.3) taking into account the double objective of minimizing the cost function and satisfying the constraints.

Dual methods

Dual methods approach the optimization problem in the formulation (4.5), which is a max-min problem:

$$\max_{\lambda,\mu\geq 0}\left[\min_x L(x,\lambda,\mu)\right] \quad \text{with } L(x,\lambda,\mu)=f(x)+\lambda^T c_E(x)+\mu^T c_I(x) \tag{4.15}$$

To simplify the notations, the constraints and multipliers are grouped.

$$c \underset{def}{=} \begin{pmatrix} c_E \\ c_I \end{pmatrix}, \quad \nu \underset{def}{=} \begin{pmatrix} \lambda \\ \mu \end{pmatrix} \tag{4.16}$$

Problem (4.15) becomes

$$\max_{\nu}\left[\min_{x} L(x,\nu)\right] \quad \text{with} \quad L(x,\nu)=f(x)+\nu^T c(x) \tag{4.17}$$

The dual variables ν are the first level variables. The primal variables x are second level variables obtained by minimizing the Lagrangian for fixed ν . The advantage of this formulation is that the constraints are not applied explicitly. One can thus use unconstrained algorithms for the first level problem as well as for the second level problem.

The dual function ψ is defined by

$$\psi(\nu)=\min_{x\in D} L(x,\nu) \tag{4.18}$$

and its domain of definition is: $D=\{\nu \,/\, \psi(\nu)>-\infty\}$.
Ideally, one should find the global minimum of the Lagrangian with respect to x. We would then converge to a saddle point (if such a point exists), which would ensure that we have found the global minimum (theorem 1-7). In practice, we have to be satisfied with a local minimum of the Lagrangian, and therefore end up with a point satisfying the KKT conditions of local minimum. Uzawa method consists in solving the problem (4.18) by an algorithm of steepest descent.

The formulation (4.18) presents a difficulty linked to the domain of definition D which is not known in advance. Iterations can go outside the domain and the Lagrangian is no longer lower bounded. **Augmented Lagrangian** methods avoid this difficulty by adding penalty terms to the classical Lagrangian in order to guarantee the existence of a minimum. The augmented Lagrangian methods are presented in section 4.6.

4.1.3 Measuring improvement

An optimization algorithm is an iterative method generating a sequence of points. These points must improve the satisfaction of the constraints on the one hand, and decrease the cost function on the other. The question then arises of comparing different points, taking into account these two objectives.
Let us assume that we want to minimize several functions simultaneously.

$$\begin{cases} \min\limits_{x\in\mathbb{R}^n} c_1(x) \\ \vdots \\ \min\limits_{x\in\mathbb{R}^n} c_m(x) \end{cases} \tag{4.19}$$

To compare two points, a dominance relation is defined. This relation can be defined by components or by norm.

Dominance by components

The functions to be minimized are compared separately. The point $x \in \mathbb{R}^n$ dominates the point $y \in \mathbb{R}^n$ if it is better for each of the functions to be minimized.

$$x \ll y \quad \Leftrightarrow \begin{cases} c_1(x) \leq c_1(y) \\ \vdots \\ c_m(x) \leq c_m(y) \end{cases} \tag{4.20}$$

Figure 4-1 shows the dominated and non-dominated regions associated with a point $x \in \mathbb{R}^n$ in the case of two functions $c_1(x)$ and $c_2(x)$ to be minimized.

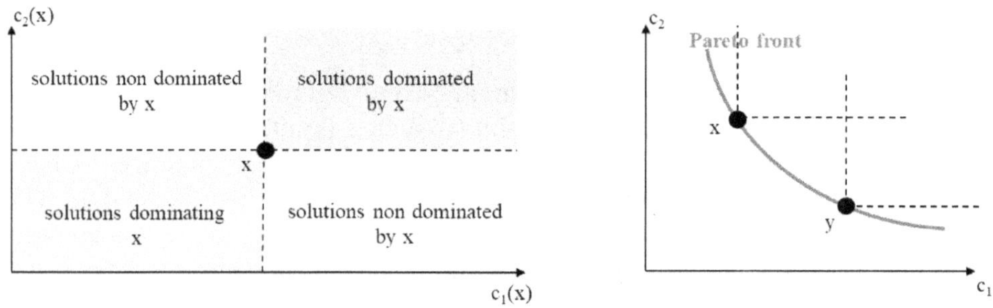

Figure 4-1: Dominance relation and Pareto front.

The x-axis represents the value of the function $c_1(x)$, the y-axis represents the value of the function $c_2(x)$. The **Pareto front** in these axes is the set of non-dominated solutions for which there is no point that simultaneously improves all the functions to be minimized.

When an algorithm produces a sequence of points, the associated **filter** stores the non-dominated solutions. Each new point evaluated is compared to the filter solutions and replaces the solutions it dominates. During the iterations, the filter thus allows to store the set of the best solutions encountered with respect to the m functions to be minimized. In the ideal case, the points of the filter at the end of the algorithm are points of the Pareto front.

Figure 4-2 below shows a filter comprising four solutions $\{x_1; x_2; x_3; x_4\}$ for two functions $c_1(x)$ and $c_2(x)$ to be minimized.

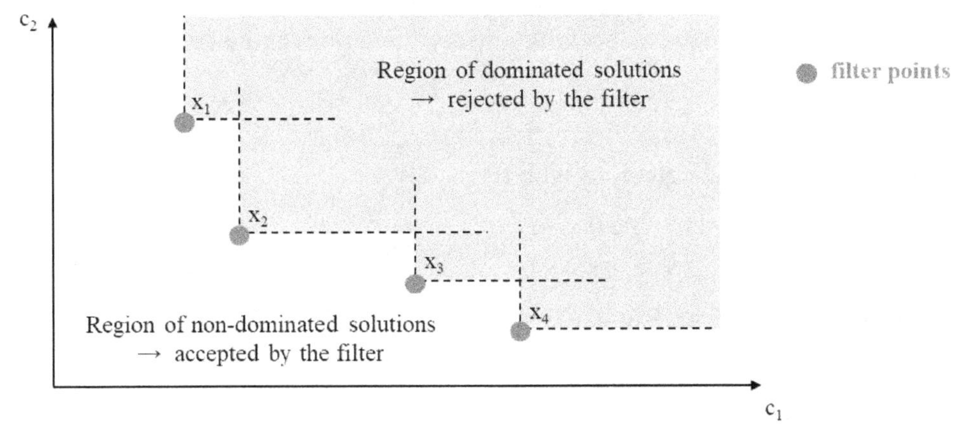

Figure 4-2: Filter points.

Component dominance leads to storing many points in the filter, in proportion to the number of functions to be minimized. A more economical approach is based on norm dominance.

Dominance by norm

The m functions to be minimized in (4.19) form a vector $c \in \mathbb{R}^m$. The norm dominance relation is based on the norm of this vector.

$$x \ll y \quad \Leftrightarrow \quad \|c(x)\| \le \|c(y)\| \tag{4.21}$$

Any norm can be used, in particular:

- the L_1 norm: $\|c\|_1 = \sum_k |c_k|$;
- the L_2 norm: $\|c\|_2 = \sqrt{\sum_k c_k^2}$;
- the L_∞ norm: $\|c\|_\infty = \max_k |c_k|$.

Relation (4.21) is an order relation that ranks the set of points on the basis of the single function $f_m(x) = \|c(x)\|$, called **merit function**. The filter is then reduced to the best point. Norm dominance is simpler to implement, but it is less "accurate" than component dominance. One function may deteriorate while others improve. Weights can be assigned to the functions so as to favor the most important ones.

The concepts of dominance and filter apply to the solution of a system of equations, which amounts to minimizing the norms of the functions.
The example 4-1 compares a norm filter (merit function) and a component filter.

Example 4-1: Solving equations with filter

Consider the system: $\begin{cases} 1-x_1=0 \\ 10(x_2-x_1^2)=0 \end{cases}$ whose solution is (1; 1).

Let us try to solve this system by Newton's method with a line search method and starting from the initial point (0; 0).

Noting $c(x_1,x_2)=\begin{pmatrix} 1-x_1 \\ 10(x_2-x_1^2) \end{pmatrix}$, Newton's direction is given by

$$d_N=-\nabla c^{-T}c=-\begin{pmatrix} -1 & -20x_1 \\ 0 & 10 \end{pmatrix}^{-T}\begin{pmatrix} 1-x_1 \\ 10(x_2-x_1^2) \end{pmatrix}=\begin{pmatrix} 1-x_1 \\ x_1(2-x_1)-x_2 \end{pmatrix}$$

The new point is defined by a line search along Newton's direction.

$$x'=x+sd_N \quad \rightarrow \quad \begin{pmatrix} x'_1 \\ x'_2 \end{pmatrix}=\begin{pmatrix} x_1 \\ x_2 \end{pmatrix}+s\begin{pmatrix} 1-x_1 \\ x_1(2-x_1)-x_2 \end{pmatrix}$$

Newton's point $s=1$ is tried first. If it is not accepted, the step s is halved until an acceptable point is produced. The acceptance criterion depends on the filter chosen. Let us look at a norm filter, then a component filter.

Norm filter

The merit function is defined here as the L₁ norm of the vector $c(x_1,x_2)$.

$$f_m(x_1,x_2)=\|c(x_1,x_2)\|_1=|1-x_1|+|10(x_2-x_1^2)|$$

The filter contains only the minimum merit function point. A new point is accepted if the merit function decreases. The initial value being $f_m(0,0)=1$, the second term of the merit function must remain small (this term is preponderant because of the factor 10). The new points will therefore remain in the vicinity of the curve of equation: $x_2=x_1^2$. Table 4-1 below shows the successive points accepted on the basis of the merit function. It can be seen that the step length s (of the form $s=1/2^n$) increases and that the last two iterations accept the Newton point $s=1$. The figure on the right shows the position of the points that remain in the vicinity of the curve of equation: $x_2=x_1^2$. The exact solution (1; 1) is obtained with 11 iterations and 85 evaluations of the function $c(x_1,x_2)$.

Iter	Step s	x_1	x_2	$f_m(x_1,x_2)$	c_1	c_2
0	0.0000	**0.0000**	**0.0000**	1.0000	1.0000	0.0000
1	0.0625	0.0625	0.0000	0.9766	0.9375	-0.0391
2	0.0625	0.1211	0.0076	0.9499	0.8789	-0.0710
3	0.0625	0.1760	0.0213	0.9207	0.8240	-0.0967
4	0.1250	0.2790	0.0588	0.9117	0.7210	-0.1907
5	0.1250	0.3691	0.1115	0.8789	0.6309	-0.2481
6	0.1250	0.4480	0.1728	0.8312	0.5520	-0.2792
7	0.2500	0.5860	0.3034	0.8139	0.4140	-0.3999
8	0.2500	0.6895	0.4347	0.7175	0.3105	-0.4070
9	0.5000	0.8448	0.6691	0.5998	0.1552	-0.4445
10	1.0000	1.0000	0.9759	0.2410	0.0000	-0.2410
11	1.0000	**1.0000**	**1.0000**	0.0000	0.0000	0.0000

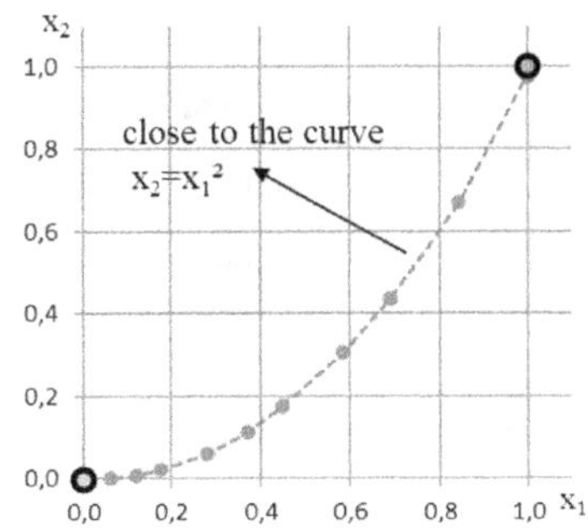

Table 4-1: Iterations with filter in norm.

<u>Component filter</u>

The filter compares each component of $c(x_1, x_2)$. The new point is accepted by the filter if it is not dominated, that is if it decreases at least one of the two components.

Table 4-2 shows the successive points accepted on the basis of the filter. It can be seen that Newton's point $s = 1$ is accepted from the first iteration, as the component x_1 decreases. This point would have been rejected on the basis of a merit function f_m, as the component x_2 increases very strongly. This acceptance allows to move away from the curve of equation: $x_2 = x_1^2$, and then to converge to the exact solution (1; 1) from the second iteration. The solution is obtained with only 2 evaluations of the function $c(x_1, x_2)$.

In this example, the component filter is more effective, as it allows more freedom of move. However, this is an illustrative example and the conclusion cannot be generalized.

Iter	Step s	x_1	x_2	$f_m(x_1,x_2)$	c_1	c_2
0	0	**0.0000**	**0.0000**	1.0000	1.0000	0.0000
1	1	1.0000	0.0000	10.0000	0.0000	-10.000
2	1	**1.0000**	**1.0000**	0.0000	0.0000	0.0000

Table 4-2: Iterations with component filter.

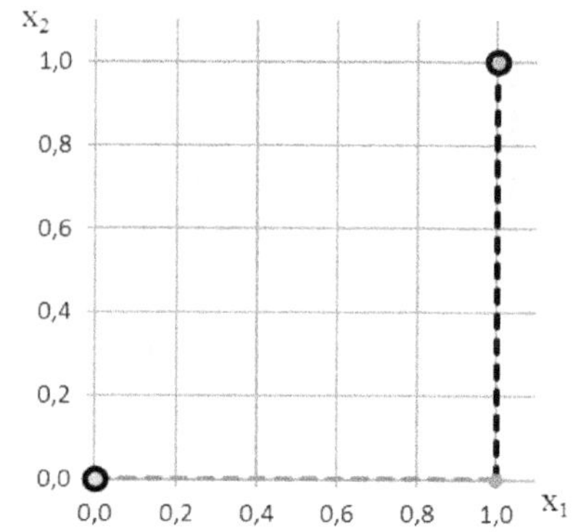

The filter concept applies in a similar way to an optimization problem in standard form with p equality constraints c_E and q inequality constraints c_I.

$$\min_{x} f(x) \ \text{ s.t. } \begin{cases} c_E(x) = 0 \\ c_I(x) \le 0 \end{cases} \tag{4.22}$$

For this problem, three forms of dominance can be considered, depending on whether the functions f, c_E and c_I are compared separately or whether they are grouped together to reduce the size of the filter.

Dominance by separate constraints

The dominance relation is defined by

$$x \ll y \iff \begin{cases} f(x) \le f(y) \\ |c_{Ej}(x)| \le |c_{Ej}(y)| \ , \ j=1 \text{ to } p \\ c_{Ij}(x) \le c_{Ij}^{+}(y) \ , \ j=1 \text{ to } q \end{cases} \tag{4.23}$$

with the notation $c_I^{+}(y) \underset{\text{def}}{=} \max\left(0, c_I(y)\right)$.

The condition on inequality constraints requires that:

- inactive inequalities in y remain satisfied in x: $c_{Ij}(x) \le c_{Ij}^{+}(y) = 0$;
- unsatisfied inequalities in y decrease: $c_{Ij}(x) \le c_{Ij}^{+}(y) = c_{Ij}(y)$.

Each constraint is examined separately, as well as the cost function. A point x is better than a point y if the cost function is lower, if each equality constraint is closer to 0 and if each violated (*i.e.* positive) inequality constraint is closer to 0. This dominance with separate constraints leads to many points being retained in the filter.

Dominance by constraint norm

The dominance relation is defined by

$$x \ll y \quad \Leftrightarrow \quad \begin{cases} f(x) \le f(y) \\ \|c(x)\| \le \|c(y)\| \end{cases} \tag{4.24}$$

The vector $c \in \mathbb{R}^{p+q}$ is defined by $c = (c_E, c_I^+)$ with $c_I^+ \underset{def}{=} \max(0, c_I)$.

The constraint violation is measured by the norm of the vector c. We can use:

- the L_1 norm: $\|c(x)\|_1 = \|c_E(x)\|_1 + \|c_I^+(x)\|_1$ with $\|c\|_1 = \sum_k |c_k|$;
- the L_2 norm: $\|c(x)\|_2 = \|c_E(x)\|_2 + \|c_I^+(x)\|_2$ with $\|c\|_2 = \sqrt{\sum_k c_k^2}$;
- the L_∞ norm: $\|c(x)\|_\infty = \|c_E(x)\|_\infty + \|c_I^+(x)\|_\infty$ with $\|c\|_\infty = \max_k |c_k|$.

The dominance relation (4.24) is less strict than the previous one (4.23). It is indeed possible for one constraint to deteriorate while others improve. Weights can be assigned to the constraints in order to favor the most important ones. This dominance relation leads to fewer points being retained in the filter.

Dominance by merit function

The two objectives of minimizing the cost function and the constraint violation are "concatenated" into a **merit function**. The merit function is used to compare two solutions, but it can also be seen as a new cost function to be minimized. In this case, it is called an **augmented cost** or **penalized function**.

The merit function $f_\rho : \mathbb{R}^n \to \mathbb{R}$ is of the form

$$f_\rho(x) = f(x) + \rho C(x) \tag{4.25}$$

The penalty function $C : \mathbb{R}^n \to \mathbb{R}^+$ measuring the constraint violation is weighted by the coefficient $\rho > 0$. A penalty in norm is usually chosen as in (4.24). The dominance relation is then defined by

$$x \ll y \quad \Leftrightarrow \quad f_\rho(x) \le f_\rho(y) \quad \text{with} \quad C(x) \underset{def}{=} \|c(x)\| \tag{4.26}$$

This order relation reduces the filter to a single point, which simplifies its implementation. It is used in most optimization algorithms.

When the filter has several points, it is necessary at each iteration to choose which point(s) of the filter the algorithm will try to improve. For example, the last point entered in the filter can be selected or the merit function (4.25) can be used to rank the filter points and choose the best one for the current iteration.

Another strategy is to try to systematically improve several points in the filter. Although more costly at each iteration, this strategy can lead to faster progress by expanding the search area.

4.2 Penalization

4.2.1 Penalized problem

The penalization method consists in replacing the problem with constraints

$$\min_x f(x) \text{ s.t. } \begin{cases} c_E(x) = 0 \\ c_I(x) \le 0 \end{cases} \tag{4.27}$$

by the "equivalent" problem without constraints

$$\min_x f_\rho(x) \underset{def}{=} f(x) + \rho C(x) \tag{4.28}$$

The function to be minimized is the **augmented cost** $f_\rho : \mathbb{R}^n \to \mathbb{R}$ which depends on the **penalty coefficient** $\rho > 0$ and on the **penalty function** $C : \mathbb{R}^n \to \mathbb{R}^+$. This function measuring the constraint violation must be zero in a feasible point and strictly positive in a non-feasible point.

The penalty function is generally chosen of the form

$$C(x) = \|c_E(x)\| + \|c_I^+(x)\| \quad \text{with} \quad c_I^+(x) = \max(0\,,\, c_I(x)) \tag{4.29}$$

The vector $c_I^+(x)$ has the same dimension as $c_I(x)$ by replacing satisfied inequalities with zero. This vector only account for unsatisfied inequalities.

Any vector norm can be used to define the function C (4.29).
Let us recall the definition of the L_p norm of a vector $y \in \mathbb{R}^n$.

$$\|y\|_p = \left(\sum_{i=1}^{n} |y_i|^p \right)^{\frac{1}{p}} \tag{4.30}$$

The norms commonly used are:
- the L_1 norm (sum of the absolute values of the components);
- the L_2 norm (usual Euclidean norm);
- the L_∞ norm (largest absolute component) : $\|y\|_\infty = \max(|y_i|)$ which is the limit of the L_p norm for $p \to \infty$.

The solution of the penalized problem (4.28) depends on the form of penalization chosen and it does not necessarily coincide with the exact solution of problem (4.27). The gap between the penalized solution and the exact solution depends on the chosen norm and the value of the penalization coefficient ρ. Increasing ρ reduces this gap and can even cancel it if one chooses a L_1 penalty (property 4-2).

However, the penalization degrades the conditioning of the Hessian and makes the minimization of f_ρ more difficult. It is generally not effective to set a very large value of ρ at the beginning and a progressive strategy is preferable.

We start by solving problem (4.28) with a low penalty (for example $\rho_1 = 1$) to obtain a local minimum $x_{\rho 1}$. Since this point does not perfectly satisfy the constraints of problem (4.27), we give ourselves some tolerances $(\varepsilon_{Ej})_{j=1 \text{ to } p}$ on the equality constraints and $(\varepsilon_{Ij})_{j=1 \text{ to } q}$ on the inequality constraints.
If the tolerances are met in $x_{\rho 1}$

$$\begin{cases} |c_{Ej}(x_{\rho 1})| < \varepsilon_{Ej} & , \ j = 1 \text{ to } p \\ c_{Ij}(x_{\rho 1}) \ < \varepsilon_{Ij} & , \ j = 1 \text{ to } q \end{cases} \tag{4.31}$$

this point is an acceptable approximation to the solution of problem (4.27).

Otherwise, the penalty must be increased to force better compliance with the constraints. The penalty is multiplied by a factor (*e.g.* $\rho_2 = 10\rho_1$) and the problem (4.28) is solved again, taking $x_{\rho 1}$ as starting point.

The following example shows that even a moderate penalty can be sufficient to obtain a good quality solution.

Example 4-2: Box problem

Let us go back to the box problem (example 1-7):

$$\min_{h,r} f(h,r) = r^2 + rh \quad \text{s.t.} \quad c(h,r) = r^2 h - 2v_0 = 0$$

The solution for a volume v_0 is: $r = v_0^{\frac{1}{3}}$, $h = 2v_0^{\frac{1}{3}}$ giving an area: $f = 3v_0^{\frac{2}{3}}$.

<u>Approximate solution by penalization</u>

The problem penalized in L_2 norm is: $\min_{h,r} f_\rho(h,r) = r^2 + rh + \frac{1}{2}\rho(r^2 h - 2v_0)^2$.

Deriving with respect to r and h

$$\begin{cases} \dfrac{\partial f_\rho}{\partial h} = r + \rho r^2 (r^2 h - 2v_0) = 0 & \rightarrow \ \rho r(r^2 h - 2v_0) = -1 \\ \dfrac{\partial f_\rho}{\partial r} = 2r + h + 2\rho r h (r^2 h - 2v_0) = 0 & \rightarrow \ h = 2r \end{cases}$$

we obtain an equation in r: $2\rho r(r^3 - v_0) + 1 = 0$.

This equation is solved numerically by varying the penalty ρ from 10^{-3} to 10^2. The results are given in table 4-3.

The volume is set to $v_0 = 1\,000$. The exact solution is: $r = 10$, $h = 20$, $f = 300$. It can be seen that a moderate penalty $(\rho = 1)$ gives a fairly accurate solution for a tolerance on the constraint of 0.2.

ρ	r	h	f	c
0.001	9.641582	19.28316	278.881	792.565
0.01	9.966442	19.93288	297.990	979.933
0.1	9.996664	19.99333	299.800	997.999
1	9.999667	19.99933	299.980	999.800
10	9.999967	19.99993	299.998	999.980
100	9.999997	19.99999	300.000	999.998

Table 4-3: Solution of the box problem by penalization.

Increasing ρ improves the quality of the solution, but makes solving the penalized problem more difficult. If it is not possible to obtain a sufficiently accurate solution, a more refined strategy is to penalize the constraints individually with separate coefficients. Each coefficient is increased or decreased according to the accuracy obtained on the constraint and the tolerance targeted.

Let us now look at the different forms of penalization and their properties.

4.2.2 Differentiable penalization

The unconstrained problem (4.28) can be solved by the descent methods discussed in chapter 3. To apply a gradient-based method, it is natural to choose a differentiable penalty function C.

Let us assume that the function C is differentiable. Its gradient in a feasible point x_a (which is by construction a local minimum of C) is zero and the gradient of the merit function f_ρ in x_a reduces to

$$\nabla f_\rho(x_a) = \nabla f(x_a) \tag{4.32}$$

This applies in particular to the point x* solution of problem (4.27).

The point x* also satisfies the KKT condition.

$$\nabla L(x^*, \lambda^*, \mu^*) = 0 \quad \Rightarrow \quad \nabla f(x^*) = -\nabla c_E(x^*)\lambda^T - \nabla c_I(x^*)\mu^T \tag{4.33}$$

The gradient of f in x* given by (4.33) is in general not zero. According to (4.32), the same will be true for the gradient of f_ρ. The point x* is then not a minimum of the merit function, hence the following property.

Property 4-1: Differentiable penalization

The minimum of a differentiable penalized function does usually not coincide with the minimum x* of the optimization problem (4.27).

Such a penalization is said to be inexact. Despite this unfavorable fact, using a differentiable penalization (especially a quadratic one) has interesting properties which are discussed below. Moreover, if we use the Lagrangian L instead of the cost function f to define the merit function (which we then note L_ρ)

$$L_\rho(x) = L(x, \lambda, \mu) + \rho C(x) \tag{4.34}$$

formula (4.32) becomes

$$\nabla L_\rho(x_a) = \nabla_x L(x_a, \lambda, \mu) \tag{4.35}$$

and the minimization of L_ρ allows to satisfy the KKT condition (4.33).

A differentiable penalization on the Lagrangian no longer has the defect of being inexact. These **augmented Lagrangian** methods are presented in section 4.6.

4.2.3 Exact penalization

A differentiable penalization does not allow us to obtain the exact solution of the problem (4.27). Let us then consider a penalty function C defined with the L_1 norm (sum of the absolute values of the components).

$$C(x) = \|c_E(x)\|_1 + \|c_I^+(x)\|_1 \quad \text{with} \quad \|c\|_1 = \sum_k |c_k| \quad \text{and} \quad c_I^+(x) = \max(0, c_I(x)) \tag{4.36}$$

This function, called the L_1 penalty, is not differentiable in a feasible point, because of the function c_I^+. Its interest comes from the following property.

Property 4-2: Exact penalty

Assume that x* is a local minimum of the problem (4.27) associated with the multipliers λ^* and μ^*.

If we choose a penalty coefficient ρ that is larger in absolute value than the multipliers λ^* and μ^*, then x* is a local minimum of the penalized problem in L_1 norm (4.36).

Demonstration (see [R3], [R11], [R12])

The triplet (x^*, λ^*, μ^*) satisfies the KKT conditions of problem (4.27).

$$\begin{cases} \nabla_x L(x^*, \lambda^*, \mu^*) = 0 \;\Rightarrow\; \nabla f(x^*) = -\nabla c_E(x^*)\lambda^* - \nabla c_I(x^*)\mu^* \\ c_E(x^*) = 0 \;,\; c_I(x^*) \le 0 \\ \mu_j^* = 0 \;\text{ or }\; c_{Ij}(x^*) = 0 \;\text{ for }\; j = 1 \text{ to } q \end{cases}$$

The merit function is defined as: $f_\rho(x) = f(x) + \rho\|c_E(x)\|_1 + \rho\|c_I^+(x)\|_1$.

Its value in $x^* + d$, where d is a small move from x^* is

$$f_\rho(x^* + d) = f(x^* + d) + \rho\|c_E(x^* + d)\|_1 + \rho\|c_I^+(x^* + d)\|_1$$

Let us evaluate each of the terms to give an expansion to order 1.

- Term $f(x^* + d)$

The function f is differentiable and its gradient appears in the KKT condition.

$$f(x^* + d) = f(x^*) + \nabla f(x^*)^T d = f(x^*) - \lambda^{*T} \nabla c_E(x^*)^T d - \mu^{*T} \nabla c_I(x^*)^T d$$

By introducing the vectors $d_E \in \mathbb{R}^p$ and $d_I \in \mathbb{R}^q$ defined by: $\begin{cases} d_E = \nabla c_E(x^*)^T d \\ d_I = \nabla c_I(x^*)^T d \end{cases}$

and noting that: $f_\rho(x^*) = f(x^*) + \rho\|c_E(x^*)\|_1 + \rho\|c_I^+(x^*)\|_1 = f(x^*)$, we obtain

$$f(x^* + d) = f_\rho(x^*) - \lambda^{*T} d_E - \mu^{*T} d_I = f(x^*) - \sum_{j=1}^{p} \lambda_j^* d_{Ej} - \sum_{j=1}^{q} \mu_j^* d_{Ij}$$

- Term $\|c_E(x^* + d)\|_1$

The function c_E is zero in x*. It is differentiable everywhere and can be expanded to order 1. Using the vector d_E defined above, we obtain

$$\|c_E(x^*+d)\|_1 = \|c_E(x^*) + \nabla c_E(x^*)^T d\|_1 = \|\nabla c_E(x^*)^T d\|_1 = \|d_E\|_1 = \sum_{j=1}^{p} |d_{Ej}|$$

- Term $\|c_I^+(x^* + d)\|_1$

The function c_I^+ is not differentiable in a point where an inequality constraint c_{Ij} cancels. It is then necessary to examine each component separately, depending on whether the constraint is active or not. With the vector d_I defined above, we have

$$\|c_I^+(x^*+d)\|_1 = \sum_{j=1}^{q} |c_{Ij}^+(x^*+d)| \quad \text{with} \quad \begin{cases} c_{Ij}^+(x^*+d) = \max\left(0, c_{Ij}(x^*+d)\right) \\ c_{Ij}(x^*+d) = c_{Ij}(x^*) + \nabla c_{Ij}(x^*)^T d = c_{Ij}(x^*) + d_{Ij} \end{cases}$$

- if the constraint is active: $c_{Ij}(x^*) = 0 \Rightarrow c_{Ij}^+(x^*+d) = \max(0, d_{Ij})$;
- if the constraint is inactive: $c_{Ij}(x^*) < 0$ and it remains so for d small and we have: $c_{Ij}(x^*+d) < 0 \Rightarrow c_{Ij}^+(x^*+d) = 0$.

Moreover, its multiplier μ_j^* is zero (KKT condition).

The terms of $f_\rho(x^*+d)$ associated with an inactive constraint are therefore all zero. Let us now group the different terms of $f_\rho(x^*+d)$ keeping only the active inequalities (indexes noted from 1 to q_{act}).

$$\begin{aligned} f_\rho(x^* + d) &= f_\rho(x^*) - \sum_{j=1}^{p} \lambda_j^* d_{Ej} - \sum_{j=1}^{q_{act}} \mu_j^* d_{Ij} + \rho \sum_{j=1}^{p} |d_{Ej}| + \rho \sum_{j=1}^{q_{act}} \max(0, d_{Ij}) \\ &= f_\rho(x^*) + \sum_{j=1}^{p} |d_{Ej}| (\rho \pm \lambda_j^*) + \sum_{j=1}^{q_{act}} \left[\rho \max(0, d_{Ij}) - \mu_j^* d_{Ij}\right] \end{aligned}$$

- for equality constraints, the sign $\pm$ depends on the sign of d_{Ej};
- for inequality constraints, the multiplier μ_j^* is positive.

It can be checked that if ρ is greater than all multipliers (in absolute value): $\rho > \|\lambda^*, \mu^*\|_\infty$, then the right terms are all positive. We have for any move d: $f_\rho(x^*+d) \geq f_\rho(x^*)$, meaning that x* is a local minimum of f_ρ.

This property shows that the exact solution x* of problem (4.27) is obtained by a single unconstrained minimization of the L_1 penalized function, with a finite value of the penalty coefficient.

Example 4-3: Exact L_1 penalty

Consider the optimization problem with one variable: $\min_x x$ s.t. $x \geq 0$.

The Lagrangian is $L(x,\mu) = x - \mu x$ where $\mu \geq 0$ is the multiplier of the inequality constraint. The KKT conditions give the local minimum.

$$\begin{cases} \nabla_x L(x^*,\mu^*) = 1 - \mu^* = 0 \\ \mu^*(-x^*) = 0 \end{cases} \Rightarrow \begin{cases} \mu^* = 1 \\ x^* = 0 \end{cases}$$

Now let us look for the minimum of the L_1 penalized problem.

The penalized function has the expression

$$f_\rho(x) = x + \rho|\max(0,-x)| = \begin{cases} x & \text{if } x \geq 0 \\ (1-\rho)x & \text{if } x < 0 \end{cases}$$

If $\rho \leq 1$, the function f_ρ is increasing and has no minimum.

If $\rho > 1$, the function f_ρ has its minimum in $x_\rho = 0$.

We check that if $\rho > \mu^* = 1$, then the minimum x_ρ of the penalized problem yields the local minimum x* of the initial optimization problem.

Property 4-2 remains valid for a penalty:

- in L_∞ norm: $C(x) = \|c_E(x)\|_\infty + \|c_I^+(x)\|_\infty$ with $\|c\|_\infty = \max_k |c_k|$;
- in L_2 norm: $C(x) = \|c_E(x)\|_2 + \|c_I^+(x)\|_2$ with $\|c\|_2 = \sqrt{\sum_k c_k^2}$.

These norms which allow to find the exact minimum are non-differentiable in a feasible point, in accordance with property 4-1. This may pose difficulties for a gradient-based descent algorithm.

There is also the difficulty that the multipliers λ^* and μ^* are not known *a priori*, which makes it impossible to set a suitable penalty value directly. For all these reasons, it is generally preferred to use the quadratic penalization described in the next section.

4.2.4 Quadratic penalization

Let us now consider a very frequently used quadratic penalty function (sum of squares of the components).

$$C(x) = \frac{1}{2}\|c_E(x)\|_2^2 + \frac{1}{2}\|c_I^+(x)\|_2^2 \quad \text{with } \|c\|_2^2 = \sum_k c_k^2 \quad \text{and } c_I^+(x) = \max\left(0, c_I(x)\right) \tag{4.37}$$

This quadratic penalization is differentiable even in points where the inequality constraints cancel. The penalized problem is formulated as

$$\min_x f_\rho(x) = f(x) + \frac{1}{2}\rho\|c_E(x)\|_2^2 + \frac{1}{2}\rho\|c_I^+(x)\|_2^2 \tag{4.38}$$

The interest of this penalization comes from its properties of convergence to a global (property 4-3) or local (property 4-4) minimum of problem (4.27)

Property 4-3: Convergence to a global minimum

If x_ρ is a global minimum of the penalized problem (4.38) and if $\lim_{\rho\to\infty} x_\rho = x^*$, then x^* is a global minimum of the problem (4.27).

Demonstration (see [R12])

Let x_a be a feasible point of problem (4.27):

$$\begin{cases} c_E(x_a) = 0 \\ c_I(x_a) \le 0 \Rightarrow c_I^+(x_a) = \max\left(0, c_I(x_a)\right) = 0 \end{cases}$$

The penalized function value in x_a is

$$f_\rho(x_a) = f(x_a) + \frac{1}{2}\rho\|c_E(x_a)\|_2^2 + \frac{1}{2}\rho\|c_I^+(x_a)\|_2^2 = f(x_a)$$

If x_ρ is a global minimum of f_ρ, we have: $f_\rho(x_\rho) \le f_\rho(x_a) = f(x_a)$.

By replacing $f_\rho(x_\rho)$: $f(x_\rho) + \frac{1}{2}\rho\|c_E(x_\rho)\|_2^2 + \frac{1}{2}\rho\|c_I^+(x_\rho)\|_2^2 \le f(x_a)$ **(inequality I)**

Let us use this inequality (I) to show that the limit (which exists by assumption) $x^* = \lim_{\rho\to\infty} x_\rho$ is a global minimum of (4.27).

It must be shown that x* is feasible and better than any other point.

- Feasibility

Putting inequality (I) into the form: $\|c_E(x_\rho)\|_2^2 + \|c_I^+(x_\rho)\|_2^2 \le \frac{2}{\rho}\left[f(x_a) - f(x_\rho)\right]$

and taking the limit $\rho \to \infty$, we deduce: $\begin{cases} \lim_{\rho\to\infty} c_E(x_\rho) = c_E(x^*) = 0 \\ \lim_{\rho\to\infty} c_I^+(x_\rho) = c_I^+(x^*) = 0 \end{cases}$.

The limit point x* is therefore feasible for the problem (4.27).

- Minimum

The terms $\rho\|c\|$ are positive. We have: $f(x_\rho) \le f(x_\rho) + \frac{1}{2}\rho\|c_E(x_\rho)\|_2^2 + \frac{1}{2}\rho\|c_I^+(x_\rho)\|_2^2$

and by replacing into inequality (I): $f(x_\rho) \le f(x_a)$.

This inequality is true for any feasible point x_a and for any value of ρ.

Taking the limit $\rho \to \infty$, we obtain: $f(x^*) \le f(x_a)$ for any feasible point x_a.

The feasible point x*, which is better than any other feasible point, is therefore a global minimum of the problem (4.27).

This property assumes that one is able to determine a global minimum of f_ρ, which is rarely the case in practice. When only a local minimum of f_ρ can be determined, we have the following convergence result.

Property 4-4: Convergence to a local minimum

If x_ρ is a <u>local</u> minimum of the penalized problem (4.38) and if $\lim_{\rho\to\infty} x_\rho = x^*$, and if the constraints are linearly independent at x*,

then (x^*, λ^*, μ^*) verifies the KKT conditions of local minimum

with the Lagrange multipliers defined by : $\begin{cases} \lim_{\rho\to\infty} \rho c_E(x_\rho) = \lambda^* \\ \lim_{\rho\to\infty} \rho c_I^+(x_\rho) = \mu^* \end{cases}$.

Demonstration (see [R12])

The merit function f_ρ has a minimum in x_ρ. For any small move d, we have: $f_\rho(x_\rho + d) \ge f_\rho(x_\rho)$

Replacing the merit function, we have for all d small the following **inequality (I)**.

$$f(x_\rho + d) + \frac{1}{2}\rho\left\|c_E(x_\rho + d)\right\|_2^2 + \frac{1}{2}\rho\left\|c_I^+(x_\rho + d)\right\|_2^2 \ge f(x_\rho) + \frac{1}{2}\rho\left\|c_E(x_\rho)\right\|_2^2 + \frac{1}{2}\rho\left\|c_I^+(x_\rho)\right\|_2^2$$

Let us expand to order 1 each term in $(x_\rho + d)$ of the first member.

<u>Term</u> $f(x_\rho + d)$

The function f is differentiable: $f(x_\rho + d) = f(x_\rho) + \nabla f(x_\rho)^T d + o(\|d\|)$

<u>Term</u> $\left\|c_E(x_\rho + d)\right\|_2^2 = c_E(x_\rho + d)^T c_E(x_\rho + d)$

The function c_E is differentiable: $c_E(x_\rho + d) = c_E(x_\rho) + \nabla c_E(x_\rho)^T d + o(\|d\|)$

$$\Rightarrow \left\|c_E(x_\rho + d)\right\|_2^2 = c_E(x_\rho)^T c_E(x_\rho) + 2\left[\nabla c_E(x_\rho) c_E(x_\rho)\right]^T d + o(\|d\|)$$

<u>Term</u> $\left\|c_I^+(x_\rho + d)\right\|_2^2 = c_I^+(x_\rho + d)^T c_I^+(x_\rho + d)$

The function c_I^+ is not differentiable in a point where an inequality constraint c_{Ij} cancels. Each inequality c_{Ij} must be examined according to its sign in x_ρ.

- 1st case - a strictly negative constraint in x_ρ remains so in $x_\rho + d$:

  $c_{Ij}^+(x_\rho + d) = 0$

- 2nd case - a strictly positive constraint in x_ρ remains so in $x_\rho + d$:
  $c_{Ij}^+(x_\rho + d) = c_{Ij}(x_\rho + d) > 0$
  The function c_{Ij} is differentiable: $c_{Ij}(x_\rho + d) = c_{Ij}(x_\rho) + \nabla c_{Ij}(x_\rho)^T d + o(\|d\|)$
  $\Rightarrow \|c_{Ij}^+(x_\rho + d)\|_2^2 = c_{Ij}(x_\rho)^T c_{Ij}(x_\rho) + 2\left[\nabla c_{Ij}(x_\rho) c_{Ij}(x_\rho)\right]^T d + o(\|d\|)$
- 3rd case - for a constraint zero at x_ρ, the value in $x_\rho + d$ depends on d:
  - a move d decreasing c_{Ij} yields: $c_{Ij}^+(x_\rho + d) = 0$;
  - a move d increasing c_{Ij} yields: $c_{Ij}^+(x_\rho + d) = c_{Ij}(x_\rho + d)$.

In the latter case, the expansion of $\|c_{Ij}^+(x_\rho + d)\|_2^2$ to order 1 (as in the 2nd case above) has only zero terms because $c_{Ij}(x_\rho) = 0$.
In the end, it can be seen that all 3 cases are covered by the following formula:
$$\|c_I^+(x_\rho + d)\|_2^2 = c_I^+(x_\rho)^T c_I^+(x_\rho) + 2\left[\nabla c_I(x_\rho) c_I^+(x_\rho)\right]^T d + o(\|d\|)$$

Let us now report the different terms in inequality (I) which simplifies to
$$\nabla f(x_\rho)^T d + \rho\left[\nabla c_E(x_\rho) c_E(x_\rho)\right]^T d + \rho\left[\nabla c_I(x_\rho) c_I^+(x_\rho)\right]^T d \geq 0$$
This inequality must hold for any small d, which is only possible if
$$\nabla f(x_\rho) + \rho \nabla c_E(x_\rho) c_E(x_\rho) + \rho \nabla c_I(x_\rho) c_I^+(x_\rho) = 0 \quad \textbf{(equality E)}$$
Let us use this equality (E) to establish that the limit point x* is feasible (under conditions) and to obtain the Lagrange multipliers of problem (4.27).

Feasibility

Taking the limit $\rho \to \infty$, with the assumption $\lim_{\rho\to\infty} x_\rho = x^*$, we deduce from (E)
$$\begin{cases} \lim_{\rho\to\infty} \nabla c_E(x_\rho) c_E(x_\rho) = \nabla c_E(x^*) c_E(x^*) = 0 \\ \lim_{\rho\to\infty} \nabla c_I(x_\rho) c_I^+(x_\rho) = \nabla c_I(x^*) c_I^+(x^*) = 0 \end{cases}$$
If the constraints are linearly independent in x*, then x* is feasible.
$$\begin{cases} c_E(x^*) = 0 \\ c_I^+(x^*) = 0 \end{cases}$$
Else x* may be non-feasible (it then minimizes the norm of the constraints).

Lagrange multipliers

Defining $\lambda_\rho \in \mathbb{R}^p$ and $\mu_\rho \in \mathbb{R}^q$ by: $\begin{cases} \lambda_\rho = \rho c_E(x_\rho) \\ \mu_\rho = \rho c_I^+(x_\rho) \geq 0 \end{cases}$, the equality (E) is written as

$$\nabla f(x_\rho)+\nabla c_E(x_\rho)\lambda_\rho+\nabla c_I(x_\rho)\mu_\rho=0$$

Then taking the limit $\rho\to\infty$, with: $\begin{cases}\lim\limits_{\rho\to\infty}\lambda_\rho \underset{\text{noted}}{=} \lambda^* \\ \lim\limits_{\rho\to\infty}\mu_\rho \underset{\text{noted}}{=} \mu^*\end{cases}$, we obtain

$$\nabla f(x^*)+\nabla c_E(x^*)\lambda^*+\nabla c_I(x^*)\mu^*=0$$

The feasible point x* (if the constraints are linearly independent) thus satisfies all KKT conditions of order 1, taking as multipliers the limits λ^* and μ^*.

In particular, if an inequality constraint c_{Ij} is inactive in x*, then for ρ large enough we have

$$c_{Ij}(x_\rho)<0 \;\Rightarrow\; \mu_{\rho j}=\rho c_{Ij}^+(x_\rho)=0 \;\Rightarrow\; \lim_{\rho\to\infty}\mu_{\rho j}=\mu_j^*=0$$

The multipliers of the inactive inequality constraints in x* are zero and we retrieve the complementary condition.

A local minimization of the function with quadratic penalization f_ρ thus allows to obtain simultaneously an estimate x_ρ of a local minimum of the problem (4.27) and estimates λ_ρ and μ_ρ of the Lagrange multipliers in this point.

The increase of the penalty ρ allows to approach the exact solution, but at the expense of the conditioning of the penalized problem. Consider the problem (4.38) with only the active constraints noted $(c_i(x))_{i=1\text{ to }m}$.

$$\min_x f_\rho(x)=f(x)+\frac{1}{2}\rho\|c(x)\|_2^2=f(x)+\frac{1}{2}\rho\sum_{i=1}^m c_i(x)^2 \tag{4.39}$$

Denoting $\rho c=\lambda_\rho$, the Hessian of f_ρ is given by

$$\begin{aligned}\nabla^2 f_\rho&=\nabla^2 f+\rho\sum_{i=1}^m c_i\nabla^2 c_i+\rho\sum_{i=1}^m \nabla c_i\nabla c_i^T\\&=\nabla^2 f+\sum_{i=1}^m \lambda_{\rho i}\nabla^2 c_i+\rho\sum_{i=1}^m \nabla c_i\nabla c_i^T\end{aligned} \tag{4.40}$$

When $\rho\to\infty$, the first two terms converge to the Hessian of the Lagrangian $\nabla^2 L$ in $(x^*;\lambda^*)$.

$$\nabla^2 f_\rho \underset{\rho\to\infty}{\approx} \nabla^2 L+\rho\sum_{i=1}^m \nabla c_i\nabla c_i^T \tag{4.41}$$

The Hessian of f_ρ is the sum of $\nabla^2 L$ and a matrix of rank m (if the constraints are linearly independent) affected by the factor ρ. This Hessian has m eigenvalues of the order of ρ (coming from $\rho\nabla c\nabla c^T$) and n−m bounded eigenvalues. Its conditioning (ratio between the largest and the smallest eigenvalue) degrades when ρ increases, which limits the achievable accuracy of the numerical solution (section 3.5.3). The following example shows the effect of the penalization on the level lines of the augmented cost.

Example 4-4: Effect of the penalization on the level lines

Consider the problem: $\min\limits_{x_1,x_2} f(x_1,x_2)=\frac{1}{2}(x_2^2-x_1^2)$ s.t. $x_1=1$

whose solution is: $(x_1;x_2)^*=(1\,;0)$.

- The level lines of the cost function are hyperbolas;
- the 0-level lines of the cost function are the asymptotes: $x_2=\pm x_1$;
- the 0-level line of the constraint is the vertical line of equation: $x_1=1$.

These level lines and the solution are shown in figure 4-3.

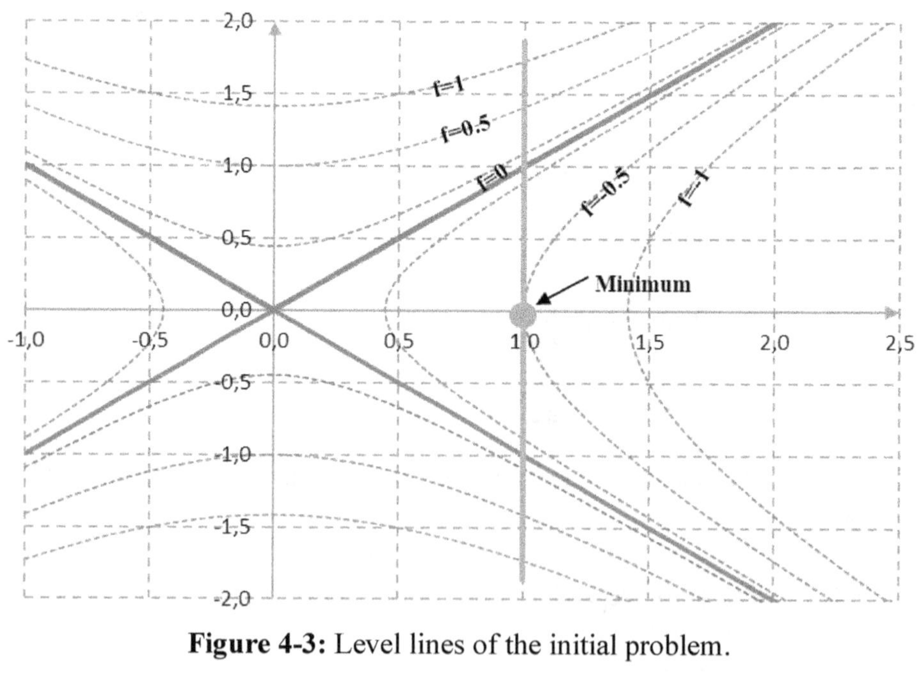

Figure 4-3: Level lines of the initial problem.

The penalized formulation of this problem is

$$\min_{x_1,x_2} f_\rho(x_1,x_2)=\frac{1}{2}(x_2^2-x_1^2)+\frac{1}{2}\rho(x_1-1)^2$$

The minimum of the augmented cost f_ρ is in: $(x_{\rho 1};x_{\rho 2})^* = \left(\frac{\rho}{\rho-1}\,;0\right)$.

As ρ increases, this point approaches the exact minimum $(x_1;x_2)^* = (1\,;0)$.

The level lines of the augmented cost and the solution are shown in figure 4-4 for four different penalty values: $\rho=1\,,2\,,10$ and 100. It can be seen that the penalized problem has no minimum for $\rho=1$, the penalty being then too low. The increase of ρ tightens the level lines, which reflects the ill conditioning of the problem.

The Hessian of the augmented cost: $H_\rho=\begin{pmatrix}\rho-1 & 0\\ 0 & 1\end{pmatrix}$ has eigenvalues 1 and $\rho-1$.

The conditioning $\kappa=\rho-1$ is of the order of ρ as was predicted in (4.41).

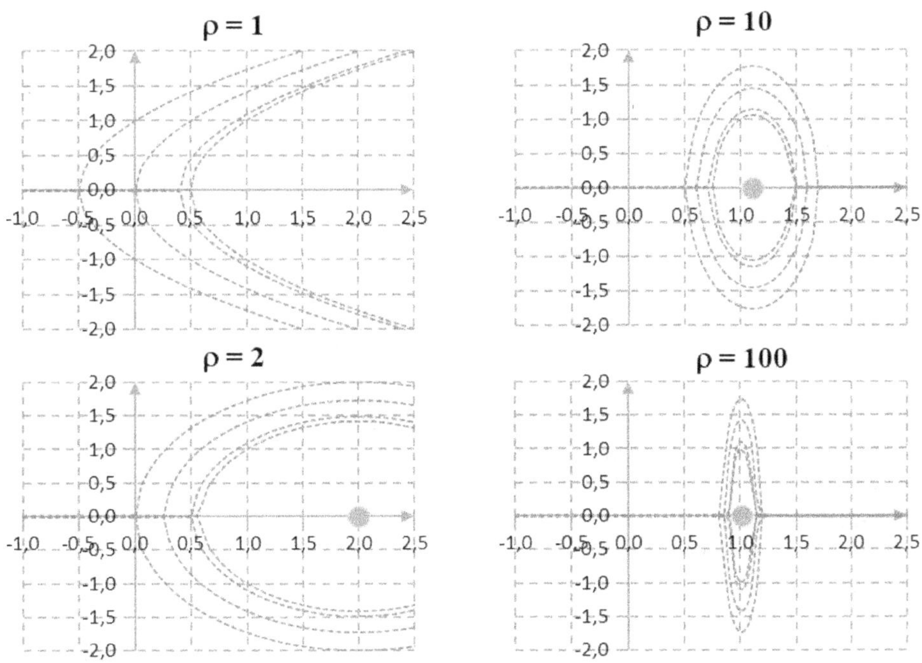

Figure 4-4: Augmented cost level lines.

4.2.5 Barrier penalization

The projection defined in (4.29) : $c_I^+(x) = \max(0, c_I(x))$ makes the merit function non-differentiable in points saturating the constraints. This can cause difficulties for gradient descent algorithms. Furthermore, this form of penalization does not guarantee that the minimum x_ρ will actually satisfy the inequality constraints.

The barrier method consists in penalizing the inequalities by the function

$$B(x) = -\sum_{j=1}^{q} \ln\left[-c_{Ij}(x)\right] \tag{4.42}$$

This function is defined only in the feasible domain $(c_I < 0)$ excluding its boundary. The function B is called either a **barrier function** (because it prevents crossing the line $c_I = 0$) or an **interior penalization** (because it forces to stay within the feasible domain with respect to the inequality constraints).

The merit function with this penalization is defined by

$$\min_x f_\rho(x) = f(x) + \frac{1}{\rho} B(x) \tag{4.43}$$

The penalization has a push-back effect with respect to the limit $c_I = 0$. When $\rho \to \infty$ this effect decreases and the inequality constraints can approach 0.

An alternative form of barrier function is

$$B'(x) = -\sum_{j=1}^{q} \frac{1}{c_{Ij}(x)} \tag{4.44}$$

The function B' has the same push-back effect with respect to the limit $c_I = 0$, but it has the defect that it does not prohibit positive values. The function B (4.42) is more frequently used, especially in interior point methods.

The barrier penalization applies only to inequality constraints. The equality constraints can be kept explicitly in the problem formulation.

$$\min_x f_\rho(x) = f(x) + \frac{1}{\rho} B(x) \quad \text{s.t.} \quad c_E(x) = 0 \tag{4.45}$$

Alternatively, they can be treated by quadratic penalization as in section 4.2.4.

$$\min_x f_\rho(x) = f(x) + \frac{1}{2}\rho \|c_E(x)\|_2^2 + \frac{1}{\rho} B(x) \tag{4.46}$$

The barrier penalization has the same properties of convergence to the minimum as the quadratic penalization (property 4-3).

Denoting x_ρ the minimum of f_ρ, we have: $\lim_{\rho\to\infty} x_\rho = x^*$.

Let us express the gradient of the Lagrangian of problem (4.45) with the multipliers λ_ρ of the equality constraints

$$\nabla_x L_\rho(x_\rho, \lambda_\rho) = \nabla f(x_\rho) - \frac{1}{\rho}\sum_{j=1}^{q}\frac{1}{c_{Ij}(x_\rho)}\nabla c_{Ij}(x_\rho) + \nabla c_E(x_\rho)\lambda_\rho \tag{4.47}$$

and also the gradient of the Lagrangian of problem (4.27) with multipliers λ, μ

$$\nabla_x L(x, \lambda, \mu) = \nabla f(x) + \nabla c_E(x)\lambda + \nabla c_I(x)\mu \tag{4.48}$$

By identifying these gradients when $\rho \to \infty$, we establish the convergence property for the multipliers of the inequality constraints.

$$\lim_{\rho\to\infty} -\frac{1}{\rho c_I(x_\rho)} = \mu^* \tag{4.49}$$

The barrier penalization approach leads to interior point methods (section 4.5).

4.3 Reduced gradient

Reduced gradient methods are primal methods with active constraints. Their principle is to build a sequence of feasible solutions decreasing the cost function. Each iteration consists of a move in the space tangent to the constraints followed by a complementary move to restore a feasible solution. These methods are interesting when satisfying the constraints is imperative and takes precedence over the optimality of the solution.

4.3.1 Move in tangent space

Let us first consider a problem with m linear constraints ($m < n$).

$$\min_{x\in\mathbb{R}^n} f(x) \quad \text{s.t.} \quad Ax = b \quad , \quad A \in \mathbb{R}^{m\times n} \, , \, b \in \mathbb{R}^n \tag{4.50}$$

We apply a move p starting from a feasible point x_0 : $Ax_0 = b$.
The new point $x_0 + p$ is feasible if: $A(x_0 + p) = b$ which leads to

$$Ap = 0 \tag{4.51}$$

The vectors $p \in \mathbb{R}^n$ satisfying this equation form the null space of the constraints, also called the hyperplane of the constraints.

Let us now turn to a problem with m nonlinear constraints.

$$\min_{x\in\mathbb{R}^n} f(x) \quad \text{s.t.} \quad c(x)=0 \quad , \quad c:\mathbb{R}^n \to \mathbb{R}^m \tag{4.52}$$

Let us replace the constraints $c(x)$ by their first-order expansion in the point x_0.

$$\min_{x\in\mathbb{R}^n} f(x) \quad \text{s.t.} \quad c(x_0)+\nabla c(x_0)^T(x-x_0)=0 \tag{4.53}$$

This problem is of the form (4.50) with $A=\nabla c(x_0)^T$ and $b=\nabla c(x_0)^T x_0 - c(x_0)$. The tangent null space or tangent hyperplane of the constraints in x_0 is the set of vectors $p\in\mathbb{R}^n$ satisfying

$$\nabla c(x_0)^T p = 0 \tag{4.54}$$

This tangent space is locally defined in x_0. Starting from a feasible point x_0, a move p in the tangent space gives a non-feasible point due to the non-linearities of the constraint as shown in figure 4-5. If the move p is small, one can hope to return to the level line, $c(x)=0$ by a complementary (so-called restoring) move detailed in section 4.3.2.

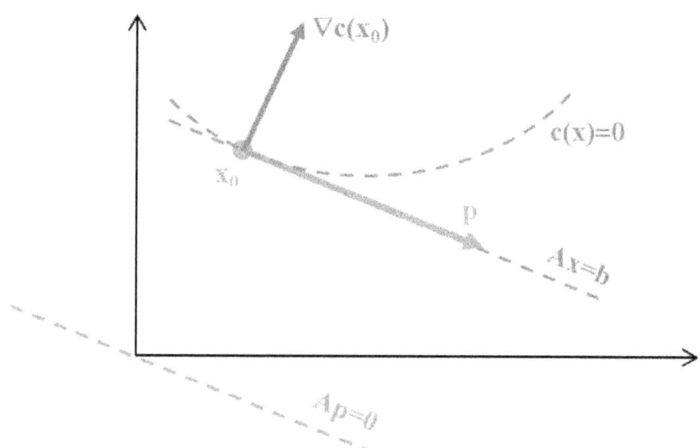

Figure 4-5: Hyperplane tangent to the constraints.

By choosing a basis $Z\in\mathbb{R}^{n\times(n-m)}$ of the tangent space and a complementary basis $Y\in\mathbb{R}^{n\times m}$ in $\mathbb{R}^n$ (section 1.3.1), the move p is decomposed into

$$p = Yp_Y + Zp_Z \quad \text{with} \quad \begin{cases} AZ=0 \\ AY \text{ invertible} \end{cases} \quad \text{and} \quad A=\nabla c(x_0)^T \tag{4.55}$$

Any move in the tangent space satisfies

$$Ap = AYp_Y = 0 \quad \rightarrow \quad \begin{cases} p_Y = 0 \\ p = Zp_Z \end{cases} \tag{4.56}$$

This move depends only on the free components $p_Z \in \mathbb{R}^{n-m}$.
The **reduced function** f_r in the point x_0 is the cost function restricted to moves in the tangent space: $p = Zp_Z$. It is defined by

$$f_r(p_Z) \underset{def}{=} f(x_0 + Zp_Z) \tag{4.57}$$

We are looking for the $n-m$ components p_Z minimizing the reduced function.
The steepest ascent direction is along the gradient of f_r, called **reduced gradient**.

$$\nabla f_r(p_Z) = Z^T \nabla f(x_0 + Zp_Z) \tag{4.58}$$

The reduced gradient g_r is associated with the variables p_Z in x_0.

$$g_r = Z^T g_0 \tag{4.59}$$

This vector is of dimension $n-m$. The direction d in $\mathbb{R}^n$ is given by (4.55).

$$d = Zg_r = ZZ^T g_0 \tag{4.60}$$

The reduced gradient depends on the choice of the basis Z. The usual choices presented in section 1.3.1 are:
- either a basis directly formed from the columns of the matrix A;
- or an orthogonal basis formed from the QR factorization of the matrix A.

The first choice is simpler, the second is better conditioned.

Basis formed by columns of the Jacobian matrix

We choose a basis $B \in \mathbb{R}^{m \times m}$ formed by m independent columns of A.
The basis Z of the tangent space is then defined by

$$A = \begin{pmatrix} \overset{m}{B} & \overset{n-m}{N} \end{pmatrix} \quad \rightarrow \quad Z = \overset{n-m}{\begin{pmatrix} -B^{-1}N \\ I \end{pmatrix}} \begin{matrix} m \\ n-m \end{matrix} \tag{4.61}$$

The gradient of f has components (g_B, g_N) associated with the columns (B, N).
The reduced gradient on the chosen basis B is given by

$$g_r = Z^T g = \begin{pmatrix} -B^{-1}N \\ I \end{pmatrix}^T \begin{pmatrix} g_B \\ g_N \end{pmatrix} = g_N - \left(B^{-1}N\right)^T g_B \tag{4.62}$$

Basis formed by orthogonal columns

The matrix $A^T \in \mathbb{R}^{n\times m}$ $(m \le n)$ admits a QR factorization of the form

$$A^T = QR = {}_n\begin{pmatrix} \overset{m}{Q_1} & \overset{n-m}{Q_2} \end{pmatrix}\begin{pmatrix} \overset{m}{R_1} \\ 0 \end{pmatrix}\begin{matrix} m \\ n-m \end{matrix} \tag{4.63}$$

The matrix Q is orthogonal $(QQ^T = I)$. The matrix R_1 is upper triangular.

The rectangular sub-matrices Q_1 and Q_2 are formed by columns that are orthogonal to each other and satisfy the relations

$$\begin{aligned} QQ^T = I &\Rightarrow \begin{pmatrix} Q_1 & Q_2 \end{pmatrix}\begin{pmatrix} Q_1^T \\ Q_2^T \end{pmatrix} = I \Rightarrow Q_1Q_1^T + Q_2Q_2^T = I \\ Q^TQ = I &\Rightarrow \begin{pmatrix} Q_1^T \\ Q_2^T \end{pmatrix}\begin{pmatrix} Q_1 & Q_2 \end{pmatrix} = I \Rightarrow Q_1^TQ_2 = 0 \end{aligned} \tag{4.64}$$

The basis Z of the tangent space is then defined by

$$Z = Q_2 \quad \to \quad AZ = R_1^TQ_1^TQ_2 = 0 \tag{4.65}$$

The reduced gradient and associated direction (4.55) are given on this basis by

$$\begin{aligned} g_r &= Z^Tg_0 = Q_2^Tg_0 \\ d &= Zg_r \;\; = Q_2Q_2^Tg_0 \end{aligned} \tag{4.66}$$

This direction d is called **projected gradient** because of the following property.

Property 4-5: Projected gradient

The direction (4.66) associated with the reduced gradient on an orthogonal basis is also the projection of the gradient on the tangent hyperplane.

Demonstration

The orthogonal projection of g on the hyperplane of equation $Ap = 0$ is calculated by $g_p = Pg$ with the projection matrix: $P = I - A^T\left(AA^T\right)^{-1}A$ (see example 1-20).

Let us express this matrix P with the QR factorization of A given by (4.63).

$$AA^T = \begin{pmatrix} R_1^T & 0 \end{pmatrix}\begin{pmatrix} Q_1^T \\ Q_2^T \end{pmatrix}\begin{pmatrix} Q_1 & Q_2 \end{pmatrix}\begin{pmatrix} R_1 \\ 0 \end{pmatrix} = \begin{pmatrix} R_1^T & 0 \end{pmatrix}\begin{pmatrix} I & 0 \\ 0 & I \end{pmatrix}\begin{pmatrix} R_1 \\ 0 \end{pmatrix} = R_1^TR_1$$

$\Rightarrow \left(AA^T\right)^{-1} = R_1^{-1}R_1^{-T}$ because the matrix R_1 triangular is invertible.

We obtain : $$P = I - A^T\left(AA^T\right)^{-1}A = I - \begin{pmatrix} Q_1 & Q_2 \end{pmatrix}\begin{pmatrix} R_1 \\ 0 \end{pmatrix}R_1^{-1}R_1^{-T}\begin{pmatrix} R_1^T & 0 \end{pmatrix}\begin{pmatrix} Q_1^T \\ Q_2^T \end{pmatrix}$$

and with (4.64) : $P = I - \begin{pmatrix} Q_1 & Q_2 \end{pmatrix}\begin{pmatrix} I \\ 0 \end{pmatrix}\begin{pmatrix} I & 0 \end{pmatrix}\begin{pmatrix} Q_1^T \\ Q_2^T \end{pmatrix} = I - Q_1Q_1^T = Q_2Q_2^T$

The direction $d = Q_2Q_2^Tg_0$ defined in (4.66) is therefore the projection of the gradient on the tangent hyperplane, since $Q_2Q_2^T = P$.

Property 4-6: Direction of steepest descent

The projected gradient is the direction of steepest descent in the tangent space.

Demonstration

The direction of steepest descent in the tangent space is the unit vector d solution of the problem: $\min\limits_{d\in\mathbb{R}^n} g_0^Td$ s.t. $\begin{cases} Ad = 0 \\ \|d\| = 1 \Leftrightarrow d^Td = 1 \end{cases}$.

The Lagrangian for this problem is: $L(d,\lambda,\mu) = g_0^Td + \lambda^TAd + \mu(d^Td - 1)$.

The KKT conditions are written as

$$\begin{cases} g_0 + A^T\lambda + 2\mu d = 0 & \rightarrow\ d = -(g_0 + A^T\lambda)/(2\mu) \\ Ad = 0 & \rightarrow\ Ag_0 + AA^T\lambda = 0 \quad \rightarrow\ \lambda = -(AA^T)^{-1}Ag_0 \\ \|d\| = 1 & \rightarrow\ 2\mu = \pm\left\|g_0 + A^T\lambda\right\| \end{cases}$$

For the direction d, we get: $d = \dfrac{g_0 - A^T(AA^T)^{-1}Ag_0}{\left\|g_0 - A^T(AA^T)^{-1}Ag_0\right\|} = \dfrac{Pg_0}{\|Pg_0\|}$.

The projected gradient Pg_0 is the steepest descent direction in the tangent space.

Property 4-6 implies that the projected gradient method will have a very slow convergence, as does the steepest descent method (section 3.5.2). It is therefore preferable to combine the reduction technique with a quasi-Newton method. This is done by replacing in formula (4.60) the gradient g_0 by a quasi-Newton direction of the form $H^{-1}g_0$, where H is an approximation of the Hessian of f obtained by a DFP, BFGS or SR1 method (section 3.2.2).

The reduced gradient g_r is then replaced by the reduced direction d_r

$$d_r = Z^T H^{-1} g_0 \tag{4.67}$$

and the direction of descent in $\mathbb{R}^n$ (opposite to the gradient) is given by

$$d = -Zd_r = -ZZ^T H^{-1} g_0 \tag{4.68}$$

The quasi-Newton method can be applied to the full or reduced Hessian of the cost function. These options are discussed in section 4.3.3.

The following example illustrates the calculation of the reduced gradient and the projected gradient by formulae (4.62) and (4.67).

Example 4-5: Reduced and projected gradient

Consider the problem: $\min\limits_{x} f(x) = x_1 + x_2$ s.t. $c(x) = x_1^2 + (x_2 - 1)^2 - 1 = 0$.

The figure on the right shows:

- the 0-level line of the constraint (circle of center (0 ; 1) and radius 1);
- the level lines of the cost function (lines of slope -1);
- the minimum (black dot).

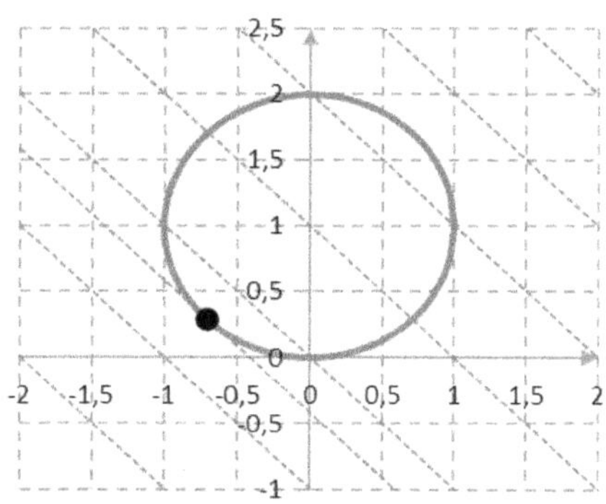

To facilitate the calculations, we switch to polar coordinates.

$$\begin{cases} x_1 = r\cos\theta \\ x_2 = r\sin\theta + 1 \end{cases} \rightarrow \begin{cases} f(r,\theta) = r(\cos\theta + \sin\theta) + 1 \\ c(r,\theta) = r^2 - 1 \end{cases}$$

The constraint imposes $r = 1$, so that the problem reduces to a function (noted F) of the single variable θ : $\min\limits_{\theta} F(\theta) = \cos\theta + \sin\theta + 1$

The minimum is obtained for $F'(\theta) = 0$ and $F''(\theta) > 0$.

$$\left. \begin{cases} F'(\theta) = -\sin\theta + \cos\theta = 0 \rightarrow \tan\theta = 1 & \rightarrow \theta = \dfrac{\pi}{4} \text{ or } \dfrac{5\pi}{4} \\ F''(\theta) = -\cos\theta - \sin\theta \geq 0 \rightarrow \cos\theta(1 + \tan\theta) \leq 0 & \rightarrow \cos\theta \leq 0 \end{cases} \right\} \rightarrow \theta^* = \frac{5\pi}{4}$$

The minimum (black dot on the figure) is at: $\begin{cases} x_1^* = -1/\sqrt{2} \approx -0.70711 \\ x_2^* = 1 - 1/\sqrt{2} \approx 0.29289 \end{cases}$.

Let us place ourselves in a feasible point x_0 with polar coordinates $(r_0 = 1 ; \theta_0)$.
The gradients of f and c are given by

$$\nabla f(x_0) = \begin{pmatrix} 1 \\ 1 \end{pmatrix} \quad , \quad \nabla c(x_0) = \begin{pmatrix} 2x_1 \\ 2(x_2 - 1) \end{pmatrix} = 2r_0 \begin{pmatrix} \cos\theta_0 \\ \sin\theta_0 \end{pmatrix}$$

We calculate the reduced gradient with the two choices of bases (4.61) or (4.65).

<u>Reduced gradient on a column basis of the Jacobian matrix</u>

The Jacobian matrix of constraints has 2 columns.

$$A = \nabla c(x_0)^T = 2r_0 \begin{pmatrix} \cos\theta_0 & \sin\theta_0 \end{pmatrix}$$

Let us choose the first column as the basis B: $B = (\cos\theta_0)$, $N = (\sin\theta_0)$.
We can then calculate with (4.61):

- the matrix Z: $\quad Z = \begin{pmatrix} -B^{-1}N \\ I \end{pmatrix} = \begin{pmatrix} -\tan\theta_0 \\ 1 \end{pmatrix}$;
- the reduced gradient g_r : $g_r = Z^T g_0 = \begin{pmatrix} -\tan\theta_0 \\ 1 \end{pmatrix}^T \begin{pmatrix} 1 \\ 1 \end{pmatrix} = 1 - \tan\theta_0$;
- the associated direction: $d = Zg_r = \begin{pmatrix} -\tan\theta_0 \\ 1 \end{pmatrix} (1 - \tan\theta_0) = \dfrac{\cos\theta_0 - \sin\theta_0}{\cos^2\theta_0} \begin{pmatrix} -\sin\theta_0 \\ \cos\theta_0 \end{pmatrix}$.

We obtain a direction tangent to the circle in x_0.

<u>Reduced gradient on an orthogonal basis</u>

Using an orthogonal basis is equivalent to projecting the gradient on the tangent space. Let us determine the projected gradient directly g_p without going through the QR factorization of the matrix A. The projection matrix P is calculated by

$$P = I - A^T (AA^T)^{-1} A \quad \text{with} \quad A = \nabla c(x_0)^T \quad \rightarrow \; P = \begin{pmatrix} \sin^2\theta_0 & -\sin\theta_0 \cos\theta_0 \\ -\sin\theta_0 \cos\theta_0 & \cos^2\theta_0 \end{pmatrix}$$

$$g_p = Pg_0 \quad \text{with} \quad g_0 = \nabla f(x_0) \qquad \rightarrow \; g_p = (\cos\theta_0 - \sin\theta_0) \begin{pmatrix} -\sin\theta_0 \\ \cos\theta_0 \end{pmatrix}$$

The direction of the projected gradient $d = \dfrac{g_p}{\|g_p\|} = \pm \begin{pmatrix} -\sin\theta_0 \\ \cos\theta_0 \end{pmatrix}$ is a vector tangent to the circle in x_0, shown in figure 4-6.

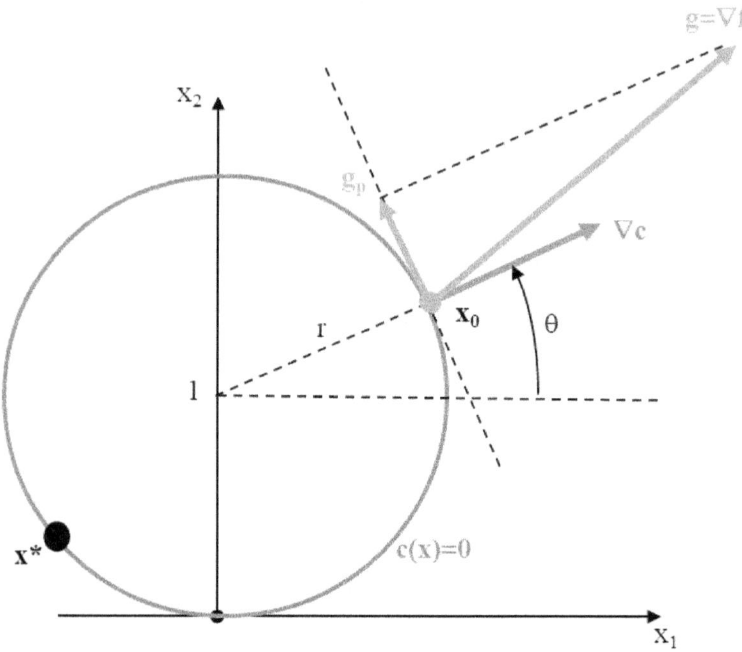

Figure 4-6: Projected gradient.

Both reduction methods give the same tangent direction to the circle in this example. This is because the null space is of dimension 1. In the general case, the two reduction methods give different directions.

4.3.2 Restoration move

The move in the tangent space aims to minimize the cost function, but it does not preserve feasibility when the constraints are not linear. It must be completed by another move to return to a feasible solution.

This so-called restoration move should be inexpensive to calculate and deviate as little as possible from the tangent move. The principles of the different restoration methods discussed in section 1.3.2 are recalled below.

The move p_1 in the tangent space ($Ap_1 = 0$) yields the point $x_1 = x_0 + p_1$.
If the constraints in x_1 are not satisfied: $c(x_1) \neq 0$, then a move p_2 is sought such that the point $x_2 = x_1 + p_2$ is feasible: $c(x_2) = 0$.

This move illustrated in figure 4-7 is based on a first-order expansion of the constraints in x_1 and it reuses the already calculated Jacobian matrix $\nabla c(x_0)$ in order to limit the amount of computation.

$$c(x_1+p_2) \approx c(x_1) + \nabla c(x_0)^T p_2 = 0 \tag{4.69}$$

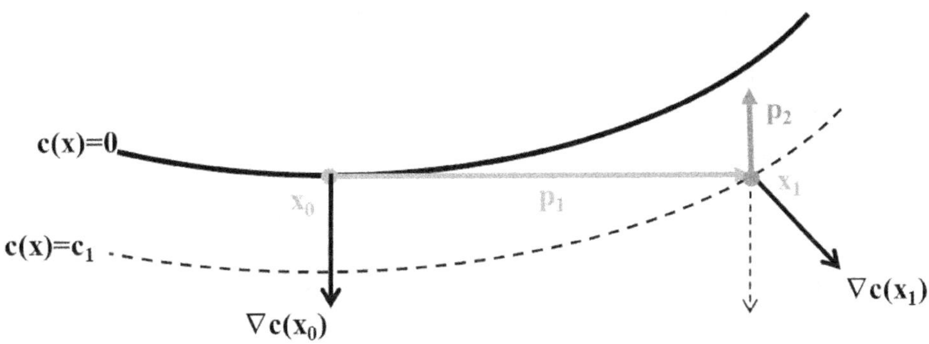

Figure 4-7: Tangent move and restoration.

The linear system (4.69) is underdetermined (n unknowns, m < n equations). It can be solved by one of the following two methods. The objective is to deviate as little as possible from the point x_1 where the cost function was minimized.

Minimum norm restoration

The minimum norm move p_2 is obtained by solving the problem

$$\min_{p_2\in\mathbb{R}^n} \|p_2\| \text{ s.t. } A_1 p_2 = b_1 \text{ with } \begin{cases} A_1 = \nabla c(x_0)^T \\ b_1 = -c(x_1) \end{cases} \tag{4.70}$$

This projection problem has been solved in example 1-20. The solution is

$$p_2 = A_1^T \left(A_1 A_1^T\right)^{-1} b_1 \tag{4.71}$$

Restoration outside the tangent space

The move p_1 in the tangent space is of the form $p_1 = Zp_Z$, where the components p_Z have been chosen to minimize the reduced function (4.57).

In order to degrade the cost function as little as possible, we look for the move p_2 in the complementary basis space Y : $p_2 = Yp_Y$. We solve the system

$$A_1p_2 = b_1 \Rightarrow A_1Yp_Y = b_1 \Rightarrow p_Y = (A_1Y)^{-1}b_1 \tag{4.72}$$

which yields the move

$$p_2 = Y(A_1Y)^{-1}b_1 \tag{4.73}$$

Note that the first formula (4.71) is a special case of the second (4.73) if we choose as complementary basis: $Y = A_1^T$ (theorem 1-5).
As the move p_2 is defined from the linearized constraints in x_1 and without recalculating the Jacobian matrix in x_1, there is no guarantee that the point $x_2 = x_1 + p_2$ is feasible. If $c(x_2) \neq 0$, either an additional move p_3 must be calculated from x_2 using the same approach (4.69), or the initial move p_1 must be reduced in the hope of returning to the "linearity domain" of the constraints (figure 4-8).

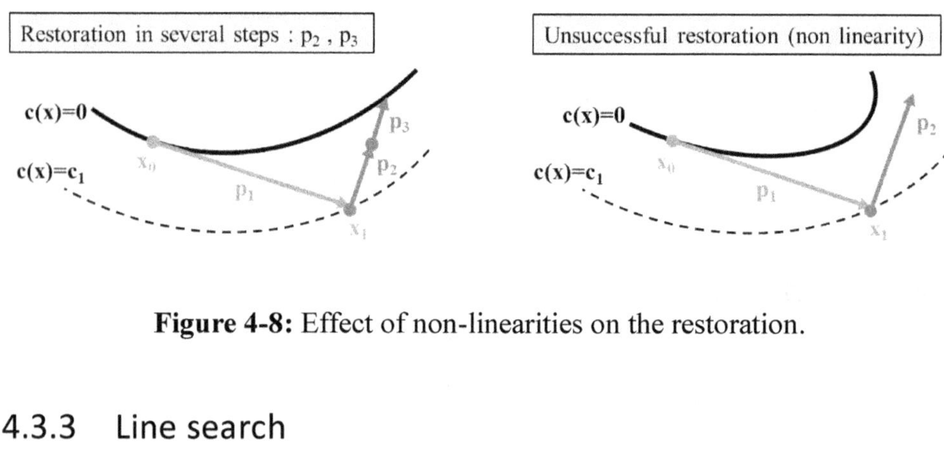

Figure 4-8: Effect of non-linearities on the restoration.

4.3.3 Line search

The goal at each iteration is to find from a feasible point a new feasible point decreasing the cost function. The iterations are performed by a line search method including a restoration stage.

Note x_0 the starting point of the iteration. The elements available in x_0 are:
- the Jacobian matrix of active constraints: $A = \nabla c(x_0)^T$;
- a basis $Z \in \mathbb{R}^{n \times (n-m)}$ of the tangent space and a complementary basis $Y \in \mathbb{R}^{n \times m}$;
- the gradient $g_0 = \nabla f(x_0)$ and an approximation H of the Hessian of f.

The new point is obtained by a step length $s > 0$ along the descent direction d (4.68) in the tangent space

$$x_1 = x_0 + p_1 \quad \text{with} \quad \begin{cases} p_1 = sd \\ d = -ZZ^T H^{-1} g_0 \end{cases} \tag{4.74}$$

followed by one (or more) restoration move defined by (4.73)

$$x_2 = x_1 + p_2 \quad \text{with} \quad p_2 = -Y(AY)^{-1} c(x_1) \tag{4.75}$$

A step s is sought by dichotomy, such that the point x_2 becomes feasible and satisfies the sufficient decrease and move conditions (section 3.3.2). For example, one can impose Goldstein's conditions (3.111) based on the directional derivative $g_0^T d$ along the direction of descent d.

$$\begin{cases} f(x_0 + sd) < f(x_0) + c_1 s g_0^T d \quad , \quad c_1 = 0{,}1 \\ f(x_0 + sd) > f(x_0) + c_2 s g_0^T d \quad , \quad c_2 = 0{,}9 \end{cases} \tag{4.76}$$

The non-linearities of the constraints leads to two difficulties:
- the restoration may be impossible (figure 4-8);
- the restoration may degrade the cost function (figure 4-9).

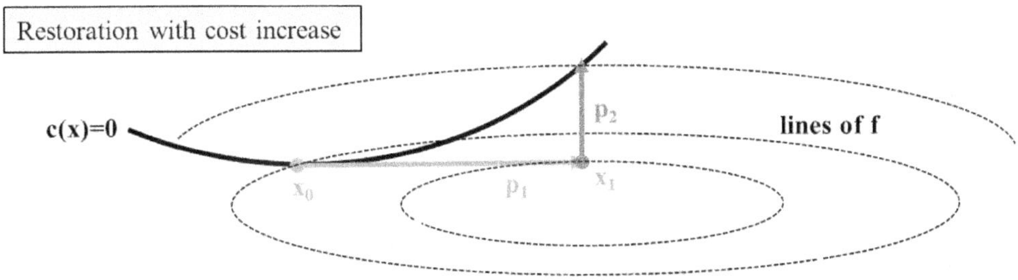

Figure 4-9: Effect of the restoration on the cost function.

In both cases, the step length s must be reduced to find an acceptable point x_2. When the constraints are highly nonlinear, the progress can become very slow and require many iterations. These difficulties could be mitigated by recalculating the Jacobian matrix $\nabla c(x_1)$ to correct the direction of restoration (figure 4-7), but this should be done for each new trial of step s and would be far too costly. It is more efficient to use the second order direction introduced below.

Direction of order 2

Consider the direction $p_c = p_1 + p_2$ where p_1 and p_2 are calculated respectively by (4.74) and (4.75). By expanding the constraints to order 2 in the vicinity of x_0 with their Hessian G

$$\begin{cases} c(x_0 + p_1) = c(x_0) + Ap_1 + \frac{1}{2} p_1^T G p_1 \\ c(x_0 + p_c) = c(x_0) + Ap_c + \frac{1}{2} p_c^T G p_c \end{cases} \tag{4.77}$$

then using the relation imposed on p_1 and p_2

$$\begin{cases} Ap_1 = -c(x_0) \\ Ap_2 = -c(x_0 + p_1) \end{cases} \tag{4.78}$$

we obtain the values of the constraints in $x_0 + p_1$ and $x_0 + p_c$

$$\begin{cases} c(x_0 + p_1) = \frac{1}{2} p_1^T G p_1 \\ c(x_0 + p_c) = p_1^T G p_2 + \frac{1}{2} p_2^T G p_2 \end{cases} \tag{4.79}$$

as well as the following relation between p_2 and p_1

$$Ap_2 = -c(x_0 + p_1) = -\frac{1}{2} p_1^T G p_1 \tag{4.80}$$

These formulae show that p_2 is of order 2 with respect to p_1 (4.80) and that the value of the constraints passes from order 2 in $x_0 + p_1$ to order 3 in $x_0 + p_c$ (4.79). The direction p_c called **second-order direction** takes into account the non-linearities of the constraints and it remains closer to the feasible line better than the tangent direction p_1 (figure 4-10).

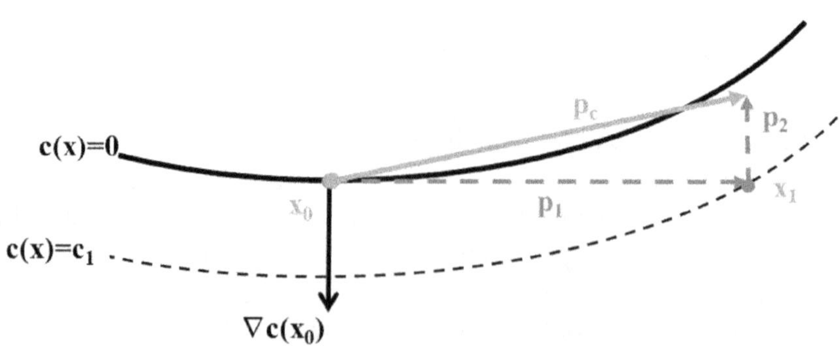

Figure 4-10: Direction of order 2.

A line search along the direction p_c (instead of p_1) yields points that are easier to restore and generally results in larger steps.

The second-order direction can be defined at the beginning of the iteration with a fixed initial step (for example $s=1$ corresponding to Newton's step) and with the associated move p_2 (4.75), even if the point x_2 is not feasible. It can also be used at the end of the iteration to try to improve the point x_2 obtained.

Management of active constraints

The line search with the moves p_1 (4.74) and p_2 (4.75) assumes that the active constraints remain the same in the initial point x_0 and in the point x_2. If the active constraints change, the iteration has to be resumed by updating the set of active constraints and the matrices A, Z, Y. The difficulty is to guess which inequality constraints should be deactivated or activated at the beginning of the line search.

A systematic method is to perform the iteration once with only the equality constraints and then resume the iteration adding the active inequality constraints in the point x_2. This method is simple, but it does not use the information of active inequalities in the initial point x_0. Each iteration will be performed at least twice, which leads to more calculations.

A more adaptive method is to calculate the point $x_1 = x_0 + p_1$ keeping the constraints that were active in x_0. Since the aim of the move p_1 is to minimize the cost function, the point x_1 can be considered to give a good indication of which inequalities should be deactivated or activated. It is then necessary to check that the inactive inequalities in x_1 remain effectively inactive in x_2. This method is faster when the set of active constraints varies little during the iterations.

When the set of active constraints changes, the Z and Y matrices must be recalculated. These matrices result from a factorization of the Jacobian matrix $A = \nabla c(x_0)^T$. They can be updated economically from the previous factorization by taking into account the rows of the constraints removed or added to the A matrix. This avoids a full factorization from scratch which is computationally costly on large problems.

4.3.4 Quasi-Newton method

As indicated in (4.68), the reduction in tangent space can be combined with a quasi-Newton method. The direction of descent in x_k is of the form

$$d_k = -Z_k Z_k^T H_k^{-1} g_k \tag{4.81}$$

In this formula:

- Z_k is a basis of the constraint tangent space in x_k : $\nabla c(x_k)^T Z_k = 0$;
- $g_k = \nabla f(x_k)$ is the gradient of f in x_k, calculated numerically;
- H_k is a quasi-Newton approximation of the Hessian of f in x_k.

The matrix H_k is updated at each iteration by a DFP, BFGS or SR1 formula as for unconstrained optimization (section 3.2.2). The update uses the variations of the point $p_k = x_k - x_{k-1}$ and of the gradient $y_k = g_k - g_{k-1}$ at the previous iteration and it satisfies the secant equation

$$H_k p_k = y_k \tag{4.82}$$

For a problem with constraints, the true Hessian $\nabla^2 f$ has no reason to be positive, even in the vicinity of the solution. However, the matrix H_k must be positive to obtain a descent direction by formula (4.81). This can be enforced by a damped BFGS method (section 3.2.3), or by a SR1 method with Hessian modification (section 3.3.1). The disadvantage is that the corrected H_k matrix deviates from the true Hessian and may lead to an inefficient descent direction.

An alternative is to apply the quasi-Newton method to the reduced Hessian $H_r = Z^T HZ$ instead of the full Hessian H. The KKT conditions of order 2 (1.127) indicate that the reduced Hessian of the Lagrangian is positive in the neighborhood of the solution. Even though we deal here with the reduced Hessian of the cost function (and not of the Lagrangian), we may hope that it is more likely to be positive than the full Hessian.

Let us show how to apply a quasi-Newton method to the reduced Hessian. Consider the first-order expansion of the gradient of f in x_0.

$$\nabla^2 f(x_0)p \approx \nabla f(x_0 + p) - \nabla f(x_0) \tag{4.83}$$

The move p is of the form $p = Yp_Y + Zp_Z$ where the restoration component Yp_Y is much smaller than the minimization component Zp_Z. This statement comes from the formula (4.80) linking the moves p_2 and p_1

Neglecting the component Yp_Y and premultiplying (4.83) by Z^T, we obtain

$$Z^T\nabla^2 f(x_0)Zp_Z = Z^T\left[\nabla f(x_0+p)-\nabla f(x_0)\right] \tag{4.84}$$

where the reduced Hessian $Z^T\nabla^2 f(x_0)Z$ appears.

Equation (4.84) is a secant equation of the form

$$H_r s = y \tag{4.85}$$

where

- H_r is an approximation of the reduced Hessian of f;
- $s = p_Z$ is the variation of the point in tangent space;
- $y = g_r(x_0+p) - g_r(x_0)$ is the variation of the reduced gradient $g_r = Z^T\nabla f$.

A quasi-Newton formula (DFP, BGFS or SR1) based on the secant equation (4.85) can be applied to update the reduced Hessian approximation.
The matrix H_r of dimension $n-m$ is smaller than H, which reduces the amount of algebraic computation, especially for problems with numerous constraints. This approach assumes that the set of active constraints does not change from one iteration to the next. If the active constraints change, the reduced Hessian must be reset to the identity.

Applying a quasi-Newton method to the reduced Hessian, the direction of descent in the tangent space (dimension $n-m$) is defined in x_k by

$$d_{r,k} = -H_{r,k}^{-1} g_{r,k} = -H_{r,k}^{-1} Z_k^T g_k \tag{4.86}$$

and the resulting direction d_k in $\mathbb{R}^n$ is given by

$$d_k = Z_k d_{r,k} = -Z_k H_{r,k}^{-1} Z_k^T g_k \tag{4.87}$$

4.3.5 Algorithm

The reduced gradient method requires starting from a feasible point. Such a point can be obtained by solving the preliminary problem

$$\min_{x\in\mathbb{R}^n} C(x) = \|c_E(x)\|_2^2 + \|\max(0, c_I(x))\|_2^2 \tag{4.88}$$

where the function C measures the violation of the constraints.

This unconstrained problem can be solved by a gradient-free method (chapter 2) or a descent method (chapter 3). A zero value of C corresponds to a feasible point of the constrained problem (4.1).

Each iteration involves the selection of the active constraints, the line search in the tangent space with restoration and the quasi-Newton updating according to the methods described in sections 4.3.1 to 4.3.4. Many variants are possible, especially concerning the restoration strategy (direction, number of trials) and they may be more or less efficient depending on the problem to be treated.

The algorithm stops either on insufficient move of variables, or on insufficient improvement of the cost function, or on maximum number of iterations or function evaluations. The advantage of the reduced gradient method is that the resulting solution is always feasible, even if it is not optimal. This is critical for some applications.

Figure 4-11 depicts the stages of the reduced gradient algorithm.

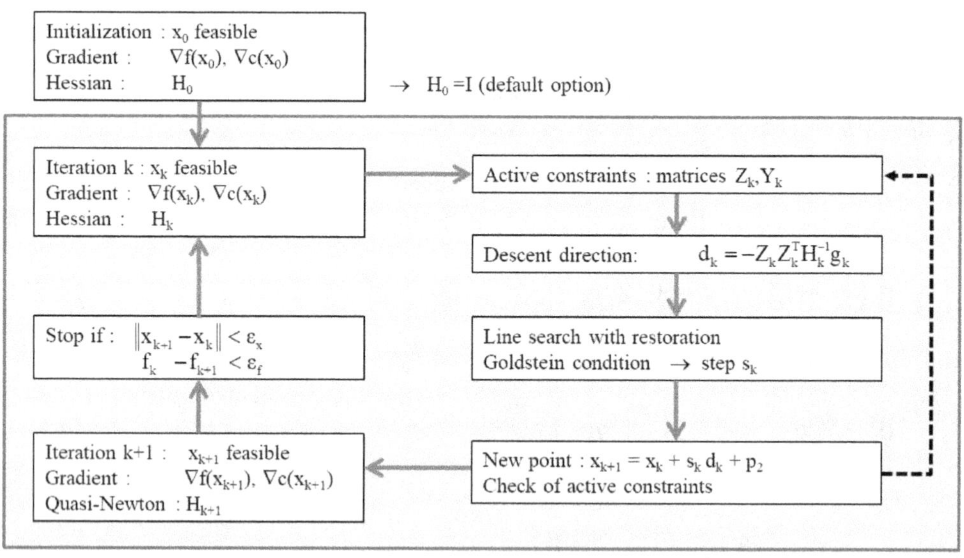

Figure 4-11: Reduced gradient algorithm.

The following example illustrates the iterations of a reduced gradient algorithm.

Example 4-6: Reduced gradient algorithm

We retrieve the problem of example 4-5.

$$\min_x f(x)=x_1+x_2 \quad \text{s.t.} \quad c(x)=x_1^2+(x_2-1)^2-1=0$$

The solution is at: $\begin{cases} x_1^* = \ \ -1/\sqrt{2} \approx -0.70711 \\ x_2^* = 1-1/\sqrt{2} \approx \ \ 0.29289 \end{cases}$.

We pass to polar coordinates to facilitate the calculations.

The gradients of f and c are given by

$$\nabla f(x_0)=\begin{pmatrix}1\\1\end{pmatrix} \ , \ \nabla c(x_0)=\begin{pmatrix}2x_1\\2(x_2-1)\end{pmatrix}=2r_0\begin{pmatrix}\cos\theta_0\\ \sin\theta_0\end{pmatrix}$$

The reduced gradient algorithm is applied from the initial point: $\begin{pmatrix}x_1\\x_2\end{pmatrix}=\begin{pmatrix}0.1\\1\end{pmatrix}$.

- The feasible current point is on the circle: $x=\begin{pmatrix}x_1\\x_2\end{pmatrix}=\begin{pmatrix}r\cos\theta\\ r\sin\theta+1\end{pmatrix}$;
- the projected gradient (example 4-5) is: $g_p=(\cos\theta-\sin\theta)\begin{pmatrix}-\sin\theta\\ \cos\theta\end{pmatrix}$;
- the descent move applies a step length s_1 along the projected gradient.

  $x'=x-s_1d_1 \quad \text{with } d_1 \text{ along } g_p \quad \to d_1=\pm\begin{pmatrix}-\sin\theta\\ \cos\theta\end{pmatrix}$

  The sign is chosen according to $\nabla f(x)$ to have a direction of descent;
- the restorative move applies a step length s_2 along the gradient of c.

  $x''=x'-s_2d_2 \quad \text{with } d_2 \text{ along } \nabla c \quad \to d_2=\pm\begin{pmatrix}\cos\theta\\ \sin\theta\end{pmatrix}$

  The sign is chosen according to $c(x')$ to bring the constraint towards 0.

The step s_2 is adjusted to restore a feasible point: $c(x'')$.
The step s_1 is adjusted to satisfy an Armijo condition (sufficient decrease).

Table 4-4 shows the iterations with the step setting s_1 along the descent direction, the coordinates of the point x' and the restoration step setting s_2 along ∇c.

Iter	x_1	x_2	f(x)	c(x)	Descent step s_1	x'_1	x'_2	c(x')	Restoration step s_2
1	**0.10000**	**1.00000**	**1.10000**	**-0.99000**	0.00000	0.10000	1.00000	-0.99000	4.50000
2	1.00000	1.00000	2.00000	0.00000	1.00000	1.00000	0.00000	1.00000	-0.50000
3	0.00000	0.00000	0.00000	0.00000	0.50000	-0.50000	0.00000	0.25000	-0.06699
4	-0.50000	0.13397	-0.36603	0.00000	0.18301	-0.65849	0.22548	0.03349	-0.00844
5	-0.65005	0.24011	-0.40994	0.00000	5.492E-02	-0.69178	0.27581	3.016E-03	-7.547E-04
6	-0.69080	0.27696	-0.41385	0.00000	1.612E-02	-0.70246	0.28809	2.599E-04	-6.497E-05
7	-0.70237	0.28819	-0.41418	0.00000	4.722E-03	-0.70573	0.29150	2.230E-05	-5.576E-06
8	-0.70572	0.29151	-0.41421	0.00000	1.383E-03	-0.70670	0.29249	1.913E-06	-4.783E-07
9	-0.70670	0.29249	-0.41421	0.00000	4.051E-04	-0.70699	0.29277	1.641E-07	-4.103E-08
10	-0.70699	0.29277	-0.41421	0.00000	1.187E-04	-0.70707	0.29286	1.408E-08	-3.520E-09
11	-0.70707	0.29286	-0.41421	0.00000	3.475E-05	-0.70710	0.29288	1.208E-09	-3.020E-10
12	**-0.70710**	**0.29288**	**-0.41421**	**0.00000**					

Table 4-4: Iterations of the reduced gradient.

Figure 4-12 shows the iterations in the plane (x_1,x_2) with a zoom on the right on the convergence towards the solution.
The tracking of the constraint with alternating descent and restoration steps can be observed.

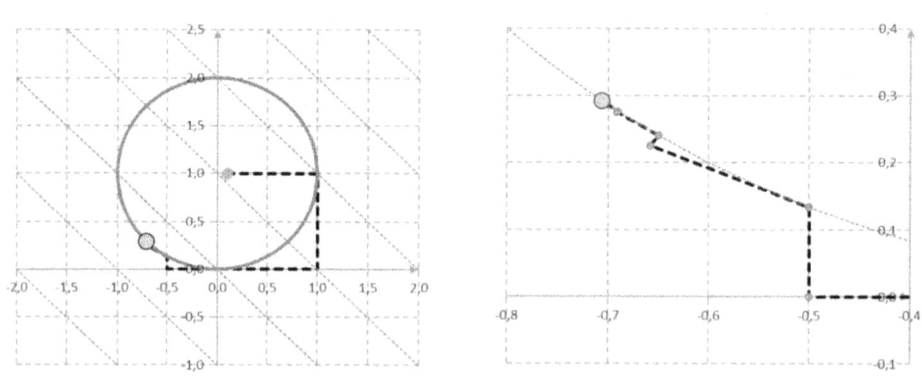

Figure 4-12: Iterations of the reduced gradient.

4.4 Sequential quadratic programming

Sequential Quadratic Programming (SQP) methods are primal-dual methods with active constraints. Their principle is to solve the KKT conditions by Newton's method, which is equivalent to solving a sequence of quadratic-linear problems. These methods are interesting in the presence of strongly nonlinear constraints.

4.4.1 Local quadratic model

Consider an optimization problem in standard form.

$$\min_{x\in\mathbb{R}^n} f(x) \quad \text{s.t.} \begin{cases} c_E(x)=0 \\ c_I(x)\le 0 \end{cases} \tag{4.89}$$

Let $c(x)$ denote the vector of the m constraints active in a point x_0 and suppose that the same constraints are active in the solution x* of the problem (4.89).

It is then equivalent to solve the problem with equality constraints

$$\min_{x\in\mathbb{R}^n} f(x) \quad \text{s.t.} \ c(x)=0 \tag{4.90}$$

The Lagrangian is defined with the multipliers $\lambda\in\mathbb{R}^m$ of the m constraints.

$$L(x,\lambda)=f(x)+\lambda^T c(x) \tag{4.91}$$

The gradient and the Hessian of the Lagrangian are given by

$$\begin{aligned} \nabla L(x,\lambda) &= \begin{pmatrix} \nabla_x L(x,\lambda) \\ \nabla_\lambda L(x,\lambda) \end{pmatrix} = \begin{pmatrix} \nabla f(x)+\nabla c(x)\lambda \\ c(x) \end{pmatrix} \\ \nabla^2 L(x,\lambda) &= \begin{pmatrix} \nabla^2_{xx} L(x,\lambda) & \nabla^2_{x\lambda} L(x,\lambda) \\ \nabla^2_{\lambda x} L(x,\lambda) & \nabla^2_{\lambda\lambda} L(x,\lambda) \end{pmatrix} = \begin{pmatrix} \nabla^2_{xx} L(x,\lambda) & \nabla c(x) \\ \nabla c(x)^T & 0 \end{pmatrix} \end{aligned} \tag{4.92}$$

The KKT conditions of order 1 form a system of dimension n + m.

$$\nabla L(x,\lambda)=0 \iff \begin{cases} \nabla_x L(x,\lambda)=0 \\ \nabla_\lambda L(x,\lambda)=0 \end{cases} \iff \begin{cases} \nabla f(x)+\nabla c(x)\lambda=0 \\ c(x)=0 \end{cases} \tag{4.93}$$

This nonlinear system can be solved by Newton's method. From the point (x_k, λ_k), Newton's iteration gives a move $d=(d_x,d_\lambda)$ such that

$$\nabla^2 L(x_k,\lambda_k)d=-\nabla L(x_k,\lambda_k) \tag{4.94}$$

or by detailing the components in (x, λ) with the formulas (4.92)

$$\begin{cases} \nabla^2_{xx}L(x_k,\lambda_k)d_x + \nabla c(x_k)d_\lambda = -\nabla_x L(x_k,\lambda_k) \\ \nabla c(x_k)^T d_x = -c(x_k) \end{cases} \tag{4.95}$$

Let us show that these equations can be deduced from a local quadratic problem.

Local quadratic problem

Let us take the point (x_k, λ_k) and define the following quadratic-linear problem noted (QP) with variables $d_{QP} \in \mathbb{R}^n$.

$$\min_{d_{QP}\in\mathbb{R}^n} \frac{1}{2} d_{QP}^T Q d_{QP} + g^T d_{QP} \quad \text{s.t.} \quad A d_{QP} = b \tag{4.96}$$

The matrices $Q \in \mathbb{R}^{n\times n}$, $A \in \mathbb{R}^{m\times n}$ and the vectors $g \in \mathbb{R}^n$, $b \in \mathbb{R}^m$ are constant, defined in (x_k, λ_k) by

$$\begin{cases} Q = \nabla^2_{xx}L(x_k,\lambda_k) \\ g = \nabla f(x_k) \end{cases} \text{and} \quad \begin{cases} A = \nabla c(x_k)^T \\ b = -c(x_k) \end{cases} \tag{4.97}$$

The Lagrangian of this QP problem is denoted L_{QP}. It depends on the multipliers $\lambda_{QP} \in \mathbb{R}^m$ associated with the m linear constraints.

$$L_{QP}(d_{QP},\lambda_{QP}) = \frac{1}{2} d_{QP}^T Q d_{QP} + g^T d_{QP} + \lambda_{QP}^T (A d_{QP} - b) \tag{4.98}$$

The KKT conditions of order 1 form a system of dimension $n + m$.

$$\nabla L(d_{QP},\lambda_{QP}) = 0 \Leftrightarrow \begin{cases} \nabla_d L(d_{QP},\lambda_{QP}) = 0 \\ \nabla_\lambda L(d_{QP},\lambda_{QP}) = 0 \end{cases} \Leftrightarrow \begin{cases} Q d_{QP} + g + A^T\lambda_{QP} = 0 \\ A d_{QP} - b = 0 \end{cases} \tag{4.99}$$

Replacing Q, A, g, b with their expressions (4.97) yields

$$\begin{cases} \nabla^2_{xx}L(x_k,\lambda_k)d_{QP} + \nabla c(x_k)\lambda_{QP} = -\nabla f(x_k) \\ \nabla c(x_k)^T d_{QP} = -c(x_k) \end{cases} \tag{4.100}$$

We observe that this system is identical to Newton's iteration (4.95) by posing

$$\begin{cases} d_{QP} = d_x \\ \lambda_{QP} = d_\lambda + \lambda_k \end{cases} \tag{4.101}$$

Performing Newton's iteration on the KKT system (4.93) in the point (x_k, λ_k) is thus equivalent to solving the local quadratic problem (4.96) in the same point.

The following example compares the iterations by Newton's method and the solution of successive quadratic problems.

Example 4-7: Equivalence Newton – Quadratic Problem

Consider the problem

$$\min_x f(x) = 2x_1^2 + 2x_2^2 - 2x_1x_2 - 4x_1 - 6x_2 \quad \text{s.t.} \quad c(x) = 2x_1^2 - x_2 = 0$$

The Lagrangian is: $L(x,\lambda) = 2x_1^2 + 2x_2^2 - 2x_1x_2 - 4x_1 - 6x_2 + \lambda(2x_1^2 - x_2)$.

The KKT conditions $\begin{cases} 4x_1 - 2x_2 - 4 + 4\lambda x_1 = 0 \\ 4x_2 - 2x_1 - 6 - \lambda = 0 \\ 2x_1^2 - x_2 = 0 \end{cases}$ have the solution $\begin{cases} x_1^* \approx 1{,}06902 \\ x_2^* \approx 2{,}28563 \\ \lambda^* \approx 1{,}00446 \end{cases}$

The sequence of Newton's iterations on the KKT system and the solution of successive quadratic problems are compared.

<u>Newton's iterations</u>

Newton's iterations are defined by: $\nabla F(x_1, x_2, \lambda) \begin{pmatrix} d_{x1} \\ d_{x2} \\ d_\lambda \end{pmatrix} = -F(x_1, x_2, \lambda)$

with $F(x_1, x_2, \lambda) = \begin{pmatrix} 4x_1 - 2x_2 - 4 + 4\lambda x_1 \\ 4x_2 - 2x_1 - 6 - \lambda \\ 2x_1^2 - x_2 \end{pmatrix} \rightarrow \nabla F(x_1, x_2, \lambda) = \begin{pmatrix} 4 + 4\lambda & -2 & 4x_1 \\ -2 & 4 & -1 \\ 4x_1 & -1 & 0 \end{pmatrix}$

Let us detail the calculations at the first iteration.

Starting from the initial point:

$$\begin{pmatrix} x_1 \\ x_2 \\ \lambda \end{pmatrix} = \begin{pmatrix} 0 \\ 1 \\ 0 \end{pmatrix} \rightarrow F(x_1, x_2, \lambda) = \begin{pmatrix} -6 \\ -2 \\ -1 \end{pmatrix} \rightarrow \nabla F(x_1, x_2, \lambda) = \begin{pmatrix} 4 & -2 & 0 \\ -2 & 4 & -1 \\ 0 & -1 & 0 \end{pmatrix}$$

the first iteration gives: $\begin{pmatrix} 4 & -2 & 0 \\ -2 & 4 & -1 \\ 0 & -1 & 0 \end{pmatrix} \begin{pmatrix} d_{x1} \\ d_{x2} \\ d_\lambda \end{pmatrix} = \begin{pmatrix} 6 \\ 2 \\ 1 \end{pmatrix} \rightarrow \begin{pmatrix} d_{x1} \\ d_{x2} \\ d_\lambda \end{pmatrix} = \begin{pmatrix} 1 \\ -1 \\ -8 \end{pmatrix}$

and the new point is: $\begin{pmatrix} x_1 \\ x_2 \\ \lambda \end{pmatrix} \rightarrow \begin{pmatrix} x_1 + d_{x1} \\ x_2 + d_{x2} \\ \lambda + d_\lambda \end{pmatrix} = \begin{pmatrix} 1 \\ 0 \\ -8 \end{pmatrix}$

Table 4-5 summarizes the iterations. The solution is reached in 7 iterations.

Iter	x_1	x_2	λ	$f(x_1,x_2)$	$c(x_1,x_2)$	$F(x_1,x_2,\lambda)$	$dF(x_1,x_2,\lambda)$		
						-6.000	**4.000**	**-2.000**	**0.000**
1	**0.00000**	**1.00000**	**0.00000**	**-4.00000**	**-1.00000**	**-2.000**	**-2.000**	**4.000**	**-1.000**
						-1.000	**0.000**	**-1.000**	**0.000**
						-32.000	-28.00	-2.000	4.000
2	1.00000	0.00000	-8.00000	-2.00000	2.00000	0.000	-2.000	4.000	-1.000
						2.000	4.000	-1.000	0.000
						8.640	15.200	-2.000	4.800
3	1.20000	2.80000	2.80000	-9.76000	0.08000	0.000	-2.000	4.000	-1.000
						0.080	4.800	-1.000	0.000
						7.4E-01	8.664	-2.000	4.346
4	1.08639	2.33466	1.16588	-10.16445	0.02582	0.0E+00	-2.000	4.000	-1.000
						2.6E-02	4.346	-1.000	0.000
						1.1E-02	8.027	-2.000	4.277
5	1.06933	2.28636	1.00676	-10.14342	0.00058	0.0E+00	-2.000	4.000	-1.000
						5.8E-04	4.277	-1.000	0.000
						2.8E-06	8.018	-2.000	4.276
6	1.06902	2.28563	1.00446	-10.14283	1.88E-07	0.0E+00	-2.000	4.000	-1.000
						1.9E-07	4.276	-1.000	0.000
						2.1E-13	**8.018**	**-2.000**	**4.276**
7	**1.06902**	**2.28563**	**1.00446**	**-10.14283**	**1.60E-14**	**0.0E+00**	**-2.000**	**4.000**	**-1.000**
						1.6E-14	**4.276**	**-1.000**	**0.000**

Table 4-5: Iterations of Newton's method.

Successive quadratic problems

The quadratic problems are defined by

$$\min_{d_x \in \mathbb{R}^2} \nabla f(x)^T d_x + \frac{1}{2} d_x^T \nabla_{xx}^2 L(x,\lambda) d_x \quad \text{s.t.} \quad \nabla c(x)^T d_x + c(x) = 0$$

with $\nabla f(x) = \begin{pmatrix} 4x_1 - 2x_2 - 4 \\ 4x_2 - 2x_1 - 6 \end{pmatrix}$, $\nabla c(x) = \begin{pmatrix} 4x_1 \\ -1 \end{pmatrix}$, $\nabla_{xx}^2 L(x,\lambda) = \begin{pmatrix} 4+4\lambda & -2 \\ -2 & 4 \end{pmatrix}$

Let us detail the calculations at the first iteration.
Starting from the initial point:

$$\begin{pmatrix} x_1 \\ x_2 \\ \lambda \end{pmatrix} = \begin{pmatrix} 0 \\ 1 \\ 0 \end{pmatrix} \rightarrow \nabla f(x) = \begin{pmatrix} -6 \\ -2 \end{pmatrix} , \nabla c(x) = \begin{pmatrix} 0 \\ -1 \end{pmatrix} , \nabla_{xx}^2 L(x,\lambda) = \begin{pmatrix} 4 & -2 \\ -2 & 4 \end{pmatrix}$$

the QP problem is

$$\min_{d_{x1},d_{x2}} \begin{pmatrix}-6\\-2\end{pmatrix}^T \begin{pmatrix}d_{x1}\\d_{x2}\end{pmatrix} + \frac{1}{2}\begin{pmatrix}d_{x1}\\d_{x2}\end{pmatrix}^T \begin{pmatrix}4 & -2\\-2 & 4\end{pmatrix}\begin{pmatrix}d_{x1}\\d_{x2}\end{pmatrix} \quad \text{s.t.} \quad \begin{pmatrix}0\\-1\end{pmatrix}\begin{pmatrix}d_{x1}\\d_{x2}\end{pmatrix}_x -1=0$$

The solution is given by: $\begin{cases}\min\limits_{d_{x1}} 2d_{x1}^2 - 4d_{x1} \\ d_{x2} = -1\end{cases} \Rightarrow \begin{cases}d_{x1} = 1 \\ d_{x2} = -1\end{cases}$ with $\lambda_{QP} = -8$

and the new point is: $\begin{pmatrix}x_1\\x_2\\\lambda\end{pmatrix} \rightarrow \begin{pmatrix}x_1+d_{x1}\\x_2+d_{x2}\\\lambda_{QP}\end{pmatrix} = \begin{pmatrix}1\\0\\-8\end{pmatrix}$

Table 4-6 summarizes the iterations. The solution is reached in 7 iterations.

Iter	x_1	x_2	λ	$f(x_1,x_2)$	$c(x_1,x_2)$	∇f	∇c	∇L	$\nabla^2 L(x_1,x_2,\lambda)$	
1	**0.00000**	**1.00000**	**0.00000**	**-4.00000**	**-1.00000**	**-6.000**	**0.000**	**-6.000**	**4.000**	**-2.000**
						-2.000	**-1.000**	**-2.000**	**-2.000**	**4.000**
2	1.00000	0.00000	-8.0000	-2.00000	2.00000	0.000	4.000	-32.000	-28.000	-2.000
						-8.000	-1.000	0.000	-2.000	4.000
3	1.20000	2.80000	2.80000	-9.76000	0.08000	-4.800	4.800	8.640	15.200	-2.000
						2.800	-1.000	0.000	-2.000	4.000
4	1.08639	2.33466	1.16588	-10.16445	0.02582	-4.324	4.346	0.74262	8.664	-2.000
						1.166	-1.000	0.00000	-2.000	4.000
5	1.06933	2.28636	1.00676	-10.14342	0.00058	-4.295	4.277	1.1E-02	8.027	-2.000
						1.007	-1.000	0.0E+00	-2.000	4.000
6	1.06902	2.28563	1.00446	-10.14283	1.9E-07	-4.295	4.276	2.8E-06	8.018	-2.000
						1.004	-1.000	0.0E+00	-2.000	4.000
7	**1.06902**	**2.28563**	**1.00446**	**-10.14283**	**1.6E-14**	**-4.295**	**4.276**	**2.1E-13**	**8.018**	**-2.000**
						1.004	**-1.000**	**0.0E+00**	**-2.000**	**4.000**

Table 4-6: Successive quadratic problems.

The iterations are identical to those of Newton's method. At each iteration, we observe that the multiplier λ_k updated by Newton's method is equal to the multiplier λ_{QP} of the quadratic problem.

The interpretation of this Newton/quadratic problem equivalence for constrained optimization is given below.

Equivalence Newton – Quadratic Problem

The KKT equations: $\nabla L(x,\lambda)=0$ are the first order conditions of the dual problem: $\max_{\lambda}\left[\min_{x} L(x,\lambda)\right]$.

It is known from section 3.1.3 that Newton's iteration for solving a system of the form $\nabla F=0$ is equivalent to minimizing or maximizing a local quadratic model of the function F. Newton's iteration on the KKT equations: $\nabla L(x,\lambda)=0$ is here equivalent to minimizing in x/maximizing in λ a local quadratic model of the Lagrangian $L(x,\lambda)$.

Let us write the expansion of the Lagrangian $L(x,\lambda)$ to the second order in the point (x_k, λ_k).

$$\begin{aligned} L(x_k+d_x,\lambda_k+d_\lambda)= \; & L(x_k,\lambda_k)+\nabla_x L(x_k,\lambda_k)^T d_x+\nabla_\lambda L(x_k,\lambda_k)^T d_\lambda \\ & +\frac{1}{2}d_x^T\nabla_{xx}^2 L(x_k,\lambda_k)d_x+\frac{1}{2}d_\lambda^T\nabla_{\lambda\lambda}^2 L(x_k,\lambda_k)d_\lambda+d_\lambda^T\nabla_{\lambda x}^2 L(x_k,\lambda_k)d_x \end{aligned}$$

Let us express the derivatives of the Lagrangian.

$$L(x,\lambda)=f(x)+\lambda^T c(x) \;\Rightarrow\; \begin{cases} \nabla_x L(x_k,\lambda_k) &= \nabla f(x_k)+\nabla c(x_k)\lambda_k \\ \nabla_\lambda L(x_k,\lambda_k) &= c(x_k) \\ \nabla_{\lambda x}^2 L(x_k,\lambda_k) &= \nabla c(x_k)^T \\ \nabla_{\lambda\lambda}^2 L(x_k,\lambda_k) &= 0 \end{cases}$$

Replacing and grouping the terms yields

$$\begin{aligned} L(x_k+d_x,\lambda_k+d_\lambda)= \; & L(x_k,\lambda_k)+\nabla f(x_k)^T d_x+\frac{1}{2}d_x^T\nabla_{xx}^2 L(x_k,\lambda_k)d_x \\ & +(\lambda_k+d_\lambda)^T\left[c(x_k)+\nabla c(x_k)^T d_x\right]-\lambda_k^T c(x_k) \end{aligned}$$

or with the matrices and vectors defined by (4.97)

$$L(x_k+d_x,\lambda_k+d_\lambda)-L(x_k,\lambda_k)-\lambda_k^T b=\frac{1}{2}d_x^T Q d_x+g^T d_x+(d_\lambda+\lambda_k)^T(A d_x-b)$$

The second member is the Lagrangian L_{QP} (4.98) with $\begin{cases} d_{QP}=d_x \\ \lambda_{QP}=d_\lambda+\lambda_k \end{cases}$.

This Lagrangian comes from the quadratic problem (4.96) whose solution leads to minimize L_{QP} in d_{QP} and maximize it in λ_{QP}.

The solution of the quadratic-linear problem (4.96) is therefore equivalent to Newton's iteration on the KKT equations. This quadratic-linear problem consists in minimizing the Lagrangian expanded to order 2 under the constraints expanded to order 1.

The solution of the system (4.100) is expressed analytically by

$$\begin{cases} \lambda_{QP} = -(AQ^{-1}A^T)^{-1}(AQ^{-1}g+b) \\ d_{QP} = -Q^{-1}(A^T\lambda_{QP}+g) \end{cases} \tag{4.102}$$

These formulas requiring matrix inversions are not used in practice. It is more efficient to solve directly the linear system (4.100) which already has a block of zeros. The SQP algorithm consists of solving the local quadratic problem (4.96) at each iteration and then applying the move (d_{QP}, λ_{QP}).

Its implementation requires some adaptations:

- Newton's method is prone to divergence. As with the descent methods, a globalization by line search (section 3.3) or by trust region (section 3.4) allows the Newton solution to be checked and corrected if necessary;
- the Hessian of the Lagrangian $Q = \nabla^2_{xx}L$ is very expensive to compute and it is replaced by an approximation obtained by a quasi-Newton method;
- the solution (d_{QP}, λ_{QP}) is a minimum of the problem (4.96) only if the matrix Q is positive (KKT condition of order 2). It is therefore necessary to ensure at each iteration that this matrix is positive, by modifying it if necessary.

4.4.2 Globalization

Solving the quadratic problem is equivalent to an iteration of Newton. As for an unconstrained minimization, Newton's method is not robust. The following example shows different possible behaviors.

Example 4-8: SQP algorithm without globalization

We retrieve the problem of example 4-5 (presented in [R3]).

$$\min_x f(x) = x_1 + x_2 \quad \text{s.t.} \quad c(x) = x_1^2 + (x_2-1)^2 - 1 = 0$$

The Lagrangian is: $L(x,\lambda) = x_1 + x_2 + \lambda(x_1^2 + (x_2-1)^2 - 1)$.

The KKT conditions $\begin{cases} 1+2\lambda x_1 = 0 \\ 1+2\lambda(x_2-1) = 0 \\ x_1^2 + (x_2-1)^2 - 1 = 0 \end{cases}$ give $\begin{cases} x_1^* = \ -1/\sqrt{2} \approx -0.70711 \\ x_2^* = 1-1/\sqrt{2} \approx \ 0.29289 \\ \lambda^* = \ 1/\sqrt{2} \approx \ 0.70711 \end{cases}$.

The quadratic problem is formulated with the matrices Q, g, A, b (4.97).

$$Q = \nabla^2_{xx}L(x,\lambda) = \begin{pmatrix} 2\lambda & 0 \\ 0 & 2\lambda \end{pmatrix}, \quad g = \nabla f(x) = \begin{pmatrix} 1 \\ 1 \end{pmatrix}, \quad A = \nabla c(x)^T = \begin{pmatrix} 2x_1 \\ 2(x_2-1) \end{pmatrix}^T, \quad b = -c(x)$$

Let us apply the SQP algorithm from different initial points, without move control. For each case, the iterations are given in a table and represented graphically.

Initial point: $\begin{pmatrix} x_1 \\ x_2 \\ \lambda \end{pmatrix} = \begin{pmatrix} 1 \\ -1 \\ 1 \end{pmatrix}$

Iter	x_1	x_2	λ
1	**1.00000**	**-1.00000**	**1.00000**
2	0.00000	-0.50000	0.50000
3	-1.00000	-0.08333	0.47222
4	-0.77401	0.24973	0.60672
5	-0.70743	0.28900	0.69818
6	-0.70714	0.29291	0.70707
7	**-0.70711**	**0.29289**	**0.70711**

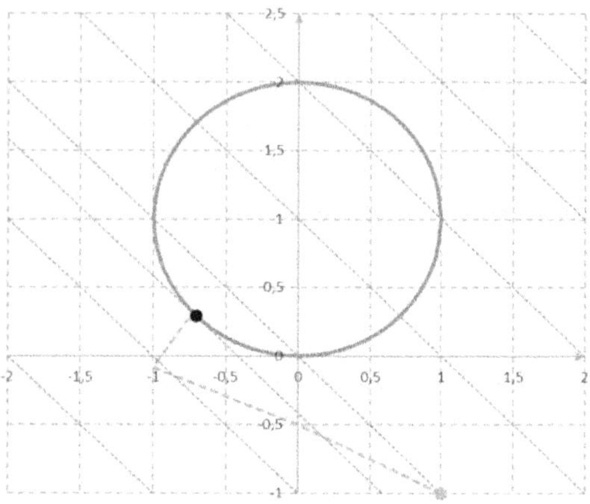

The convergence is fast in 7 iterations.

Initial point: $\begin{pmatrix} x_1 \\ x_2 \\ \lambda \end{pmatrix} = \begin{pmatrix} -0.1 \\ 1 \\ 1 \end{pmatrix}$

Iter	x_1	x_2	λ
1	**-0.10000**	**1.00000**	**1.00000**
2	-5.05000	0.50000	-44.5000
3	-2.62404	0.75032	-21.2782
4	-1.50286	0.87826	-8.90106
5	-1.08612	0.96364	-2.13558
6	-1.01047	1.19247	0.31161
7	-1.33383	-0.65614	0.39510
8	-0.96379	0.10912	0.48447
9	-0.72273	0.25387	0.63996
10	-0.70890	0.29344	0.70407
11	-0.70710	0.29289	0.70710
12	**-0.70711**	**0.29289**	**0.70711**

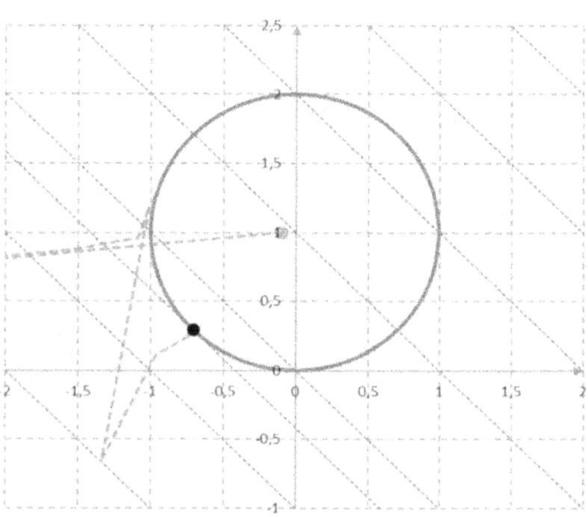

The first iteration is clearly away from the solution and the first 6 iterations are chaotic. Nevertheless, the end point corresponds well to the solution.

Initial point: $\begin{pmatrix} x_1 \\ x_2 \\ \lambda \end{pmatrix} = \begin{pmatrix} 0.1 \\ 1 \\ 1 \end{pmatrix}$

Iter	x_1	x_2	λ
1	**0.10000**	**1.00000**	**1.00000**
2	5.05000	0.50000	-54.5000
3	2.62404	0.75028	-26.2801
4	1.50282	0.87782	-11.4197
5	1.08576	0.95907	-3.50192
6	1.00831	1.11015	-0.71030
7	0.92650	1.72824	-0.55351
8	0.70291	1.74580	-0.67324
9	0.70870	1.70662	-0.70579
10	**0.70711**	**1.70711**	**-0.70710**

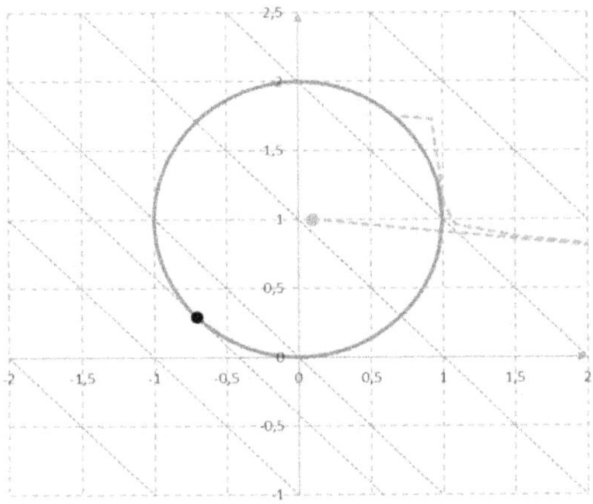

Although the initial point is very close to the previous one, the iterations do not converge to the solution. The final point satisfies the KKT equations, but it is a local maximum, diametrically opposed to the minimum on the circle representing the constraint.

The globalization consists in checking that the new point is closer to the solution of the problem (4.89). For a constrained minimization, the improvement can be measured by a merit function with a L_1 penalization such as

$$f_\rho(x) = f(x) + \rho \|c(x)\|_1 \quad \text{with} \quad \|c(x)\|_1 = \sum_{i=1}^{m} |c_i(x)| \tag{4.103}$$

We know from property 4-2 that the minimum of this function gives the exact solution x^* of the problem (4.89) if the penalty coefficient ρ is large enough. The globalization is done either by line search or by trust region.

Globalization by line search

The solution d_{QP} of the quadratic problem is used as the line search direction. The following property allows us to calculate the directional derivative of the merit function f_ρ along the direction d_{QP} and to ensure that it is a descent direction.

Property 4-7: Directional derivative of the merit function

The directional derivative of the merit function (4.103) in the point x_k along the direction d_{QP} from the quadratic problem (4.96) is given by

$$\left(f_\rho\right)_d'(x_k) = \nabla f(x_k)^T d_{QP} - \rho\left\|c(x_k)\right\|_1 \tag{4.104}$$

It also satisfies the inequality

$$\left(f_\rho\right)_d'(x_k) \le -d_{QP}^T \nabla_{xx}^2 L(x_k,\lambda_k) d_{QP} - \left(\rho - \left\|\lambda_{QP}\right\|_\infty\right)\left\|c(x_k)\right\|_1 \tag{4.105}$$

where λ_{QP} is the multiplier coming from the quadratic problem (4.96).

Demonstration

The directional derivative of f_ρ along the direction d_{QP} is defined by

$$\left(f_\rho\right)_d'(x_k) \underset{def}{=} \lim_{s\to 0} \frac{f_\rho(x_k + s d_{QP}) - f_\rho(x_k)}{s}$$

Let us expand the merit function to order 1.

$$\begin{aligned} f_\rho(x_k + s d_{QP}) - f_\rho(x_k) &= f(x_k + s d_{QP}) + \rho\left\|c(x_k + s d_{QP})\right\|_1 - f(x_k) - \rho\left\|c(x_k)\right\|_1 \\ &= s d_{QP}^T \nabla f(x_k) + \rho\left\|c(x_k) + s d_{QP}^T \nabla c(x_k)\right\|_1 - \rho\left\|c(x_k)\right\|_1 + o(s) \\ &= s d_{QP}^T \nabla f(x_k) + \rho\left\|(1-s)c(x_k)\right\|_1 - \rho\left\|c(x_k)\right\|_1 + o(s) \end{aligned}$$

because the direction d_{QP} satisfies (4.100) : $d_{QP}^T \nabla c(x_k) = -c(x_k)$.

For $0 < s < 1$, we obtain: $\quad f_\rho(x_k + s d_{QP}) - f_\rho(x_k) = s d_{QP}^T \nabla f(x_k) - s\rho\left\|c(x_k)\right\|_1 + o(s)$

hence the formula (4.104) : $\displaystyle\lim_{s\to 0} \frac{f_\rho(x_k + s d_{QP}) - f_\rho(x_k)}{s} = d_{QP}^T \nabla f(x_k) - \rho\left\|c(x_k)\right\|_1$

For the inequality (4.105), we use the equations (4.100) satisfied by the solution of the quadratic problem. These equations allow to express $d_{QP}^T \nabla f(x_k)$.

$$\begin{cases} \nabla_{xx}^2 L(x_k,\lambda_k) d_{QP} + \nabla c(x_k)\lambda_{QP} = -\nabla f(x_k) \\ \nabla c(x_k)^T d_{QP} = -c(x_k) \end{cases}$$

$$\begin{aligned} \Rightarrow\ d_{QP}^T \nabla f(x_k) &= -d_{QP}^T \nabla_{xx}^2 L(x_k,\lambda_k) d_{QP} - d_{QP}^T \nabla c(x_k)\lambda_{QP} \\ &= -d_{QP}^T \nabla_{xx}^2 L(x_k,\lambda_k) d_{QP} + c(x_k)^T \lambda_{QP} \end{aligned}$$

Furthermore, we have

$$c(x_k)^T \lambda_{QP} = \sum_{i=1}^m c_i(x_k)\lambda_{QPi} \le \sum_{i=1}^m \left|c_i(x_k)\right|\left|\lambda_{QPi}\right| \le \sum_{i=1}^m \left|c_i(x_k)\right|\left\|\lambda_{QP}\right\|_\infty = \left\|c(x_k)\right\|_1 \left\|\lambda_{QP}\right\|_\infty$$

From this, we can deduce

$$(f_\rho)'_d(x_k) = d_{QP}^T\nabla f(x_k) - \rho\|c(x_k)\|_1 = -d_{QP}^T\nabla_{xx}^2 L(x_k,\lambda_k)d_{QP} + c(x_k)^T\lambda_{QP} - \rho\|c(x_k)\|_1$$
$$\leq -d_{QP}^T\nabla_{xx}^2 L(x_k,\lambda_k)d_{QP} + \left(\|\lambda_{QP}\|_\infty - \rho\right)\|c(x_k)\|_1$$

Formula (4.104) is useful for expressing the Goldstein conditions (3.111) of sufficient decrease and move. These conditions have the following form:

$$\begin{cases} f_\rho(x_k + sd_{QP}) < f_\rho(x_k) + c_1 s\left(f_\rho\right)'_d(x_k) & , \ c_1 = 0{,}1 \\ f_\rho(x_k + sd_{QP}) > f_\rho(x_k) + c_2 s\left(f_\rho\right)'_d(x_k) & , \ c_2 = 0{,}9 \end{cases} \quad (4.106)$$

The formula (4.105) helps to set the penalty coefficient so that d_{QP} is a descent direction. Let Z_k be a basis of the constraint tangent space associated with the Jacobian matrix $A=\nabla c(x_k)^T$. If the reduced Hessian $Z_k^T\nabla_{xx}^2 L(x_k,\lambda_k)Z_k$ is positive, which is the case in the vicinity of the solution of (4.89), we obtain a descent direction $\left(f_\rho\right)'_d(x_k) < 0$ for a penalty coefficient such that

$$\rho > \|\lambda_{QP}\|_\infty \quad (4.107)$$

In practice, we start from a low penalty coefficient that is gradually increased. Choosing a too high initial penalty would slow down progress by giving too much weight to the constraints in the merit function.

Globalization by trust region

We add to the quadratic problem (4.96) a trust region constraint with radius r_k

$$\min_{d_{QP}\in\mathbb{R}^n} \frac{1}{2} d_{QP}^T Q d_{QP} + g^T d_{QP} \quad \text{s.t.} \begin{cases} Ad_{QP} = b \\ \|d_{QP}\|_2 \leq r_k \end{cases} \quad (4.108)$$

with the matrices and vectors defined in (x_k, λ_k) by

$$\begin{cases} Q = \nabla_{xx}^2 L(x_k,\lambda_k) \\ g = \nabla f(x_k) \end{cases} \text{ and } \begin{cases} A = \nabla c(x_k)^T \\ b = -c(x_k) \end{cases} \quad (4.109)$$

The solution of problem (4.108) gives a new point $x_{k+1} = x_k + d_{QP}$.

The reduction ratio σ compares the actual and expected improvements.

$$\sigma = \frac{f_\rho(x_k) - f_\rho(x_{k+1})}{\hat{f}_\rho(x_k) - \hat{f}_\rho(x_{k+1})} \tag{4.110}$$

The actual improvement is measured by the merit function f_ρ defined as

$$f_\rho(x) = f(x) + \rho\|c(x)\| \tag{4.111}$$

The expected improvement is measured by the model function $\hat{f}_\rho$, which is similar to the merit function (4.111) applied to the quadratic-linear problem (4.108).

$$\hat{f}_\rho(x_k + d_{QP}) = f(x_k) + g^T d_{QP} + \frac{1}{2} d_{QP}^T Q d_{QP} + \rho\|c(x_k) + A d_{QP}\| \tag{4.112}$$

One can choose a L_1 or L_2 penalization to evaluate the ratio σ.

The acceptance of the new point and the setting of the trust radius follow the same principles as for an unconstrained optimization (section 3.4.1):

- a ratio greater than 1 indicates that the quadratic model represents the function well in the region of radius r_k. In this case, the radius is increased for the next iteration to allow for a larger move;
- a small or negative ratio indicates that the quadratic approximation is not good. In this case, the iteration must be repeated with a reduced trust radius to obtain an acceptable point x_{k+1}.

4.4.3 Constraint management

The SQP algorithm requires some precautions regarding the constraints.

<u>Active constraints</u>

The quadratic problem (4.96) is defined by linearizing the active constraints in the point x_k. If the active constraints in the new point x_{k+1} are different, the iteration is to be resumed according to one of the following two approaches.

A first approach is to linearize the set of equality and inequality constraints at each iteration and solve the local quadratic problem:

$$\min_{d_{QP} \in \mathbb{R}^n} \frac{1}{2} d_{QP}^T \nabla^2_{xx} L(x_k, \lambda_k) d_{QP} + \nabla f(x_k)^T d_{QP} \quad \text{s.t.} \begin{cases} c_E(x_k) + \nabla c_E(x_k)^T d_{QP} = 0 \\ c_I(x_k) + \nabla c_I(x_k)^T d_{QP} \le 0 \end{cases} \tag{4.113}$$

This formulation avoids the difficulties of selecting active constraints, but the presence of inequality constraints complicates the solution.

A second approach to guessing the active constraints in x_{k+1} is to solve the following linear problem obtained by ignoring the second order term and adding a trust region constraint.

$$\min_{d_{QP}\in\mathbb{R}^n} \nabla f(x_k)^T d_{QP} \quad \text{s.t.} \begin{cases} c_E(x_k)+\nabla c_E(x_k)^T d_{QP} = 0 \\ c_I(x_k)+\nabla c_I(x_k)^T d_{QP} \le 0 \\ \left\|d_{QP}\right\|_\infty \le r_k \end{cases} \tag{4.114}$$

This preliminary problem can be solved efficiently by a specific linear programming method (chapter 5). Its solution is used to select the active constraints that will be treated as equalities for the current iteration.

Incompatible constraints

Linearization can make the problem quadratic incompatible, especially if one includes inequality constraints as in (4.108) or (4.113). This difficulty can be avoided by relaxation or by penalization.

The relaxation method consists in solving the prior problem:

$$\min_{d'\in\mathbb{R}^n} \left\|Ad'-b\right\|_2 \quad \text{s.t.} \ \left\|\delta\right\|_2 \le 0.8 r_k \tag{4.115}$$

The solution d' of this problem is in the trust region (with a margin due to the coefficient 0.8) and its cost measures the constraint deviation: $\delta c = Ad'-b$.
By adding this deviation to the constraint second member in problem (4.108)

$$\min_{d_{QP}\in\mathbb{R}^n} \frac{1}{2} d_{QP}^T Q d_{QP} + g^T d_{QP} \quad \text{s.t.} \begin{cases} Ad_{QP} = b+\delta c \\ \left\|d_{QP}\right\|_2 \le r_k \end{cases} \tag{4.116}$$

we make sure that the constraints become compatible. The value $d_{QP} = d'$ is then feasible and can initialize the solution of the relaxed quadratic problem (4.116).

The penalty method consists in replacing problem (4.108) by

$$\min_{d_{QP}\in\mathbb{R}^n} \frac{1}{2} d_{QP}^T Q d_{QP} + g^T d_{QP} + \rho\left\|Ad_{QP}-b\right\|_1 \quad \text{s.t.} \ \left\|d_{QP}\right\|_\infty \le r_k \tag{4.117}$$

This problem with a L_1 penalization is always compatible. Inequality constraints can also be incorporated into the cost function. The merit function measuring the improvement (4.111) is in this case defined with the L_1 norm.

Direction of order 2

In the presence of strongly nonlinear constraints, the direction d_{QP} from the quadratic problem may be ineffective in reducing the merit function. In this case, the second-order direction constructed in two steps can be used.

The quadratic problem is first solved in its form (4.96) with the second member of the constraints being: $b = -c(x_k)$.

$$\min_{d_{QP}\in\mathbb{R}^n} \frac{1}{2} d_{QP}^T Q d_{QP} + g^T d_{QP} \quad \text{s.t.} \quad A d_{QP} = b \tag{4.118}$$

A solution d_{QP} is obtained and the constraints are evaluated in $x_k + d_{QP}$.

A high constraint value $c(x_k + d_{QP}) \underset{\text{noted}}{=} \delta c$ indicates that the constraints are strongly nonlinear. We then solve the quadratic problem again by changing the second member of the constraints to: $b' = -c(x_k) - \delta c$.

$$\min_{d_{QP}\in\mathbb{R}^n} \frac{1}{2} d_{QP}^T Q d_{QP} + g^T d_{QP} \quad \text{s.t.} \quad A d_{QP} = b' \tag{4.119}$$

The new solution d'_{QP} takes into account the curvature of the constraints. This direction d'_{QP} is used instead of d_{QP} for the line search. The following example illustrates the use of the second-order direction.

Example 4-9: Direction of order 2

Consider the problem with 2 variables:

$$\min_x f(x) = 2(x_1^2 + x_2^2 - 1) - x_1 \quad \text{s.t.} \quad c(x) = x_1^2 + x_2^2 - 1 = 0$$

The solution is: $x^* = \begin{pmatrix}1\\0\end{pmatrix}$, $\lambda^* = -\dfrac{3}{2}$ $\rightarrow$ $\nabla^2_{xx} L(x^*, \lambda^*) = \begin{pmatrix}1 & 0\\0 & 1\end{pmatrix}$.

Let us place ourselves in a point x_k with polar coordinates $x_k = \begin{pmatrix}\cos\theta\\ \sin\theta\end{pmatrix}$.

The gradients of f and c are: $\nabla f(x_k) = \begin{pmatrix}4\cos\theta - 1\\ 4\sin\theta\end{pmatrix}$, $\nabla c(x_k) = \begin{pmatrix}2\cos\theta\\ 2\sin\theta\end{pmatrix}$.

First resolution of the quadratic problem

Taking $Q = I$, the quadratic problem in x_k is formulated as:

$$\min_{d_1,d_2} (4\cos\theta - 1)d_1 + 4d_2\sin\theta + \frac{1}{2}(d_1^2 + d_2^2) \text{ s.t. } d_1\cos\theta + d_2\sin\theta = 0$$

The solution is obtained by elimination: $d_1 = -d_2\tan\theta$.

$$\min_{d_2} d_2\tan\theta + \frac{1}{2\cos^2\theta}d_2^2 \rightarrow d_2 = -\sin\theta\cos\theta \qquad \rightarrow d_{QP} = \sin\theta\begin{pmatrix}\sin\theta\\ -\cos\theta\end{pmatrix}$$

The new point is: $x_{k+1} = \begin{pmatrix}\cos\theta + \sin^2\theta\\ \sin\theta - \sin\theta\cos\theta\end{pmatrix}$.

Let us calculate the distance from the solution $x^* = \begin{pmatrix}1\\0\end{pmatrix}$ to the points x_k and x_{k+1}

$$\begin{cases} x_k - x^* = \begin{pmatrix}\cos\theta - 1\\ \sin\theta\end{pmatrix} & \Rightarrow \|x_k - x^*\|^2 = 2(1-\cos\theta) \\ x_{k+1} - x^* = (1-\cos\theta)\begin{pmatrix}\cos\theta\\ \sin\theta\end{pmatrix} & \Rightarrow \|x_{k+1} - x^*\|^2 = 1-\cos\theta \end{cases}$$

The new point x_{k+1} is closer to the solution x^*.

Let us now compare the values of cost and constraints in x_k and x_{k+1}:

- cost function: $\begin{cases} f(x_k) = -\cos\theta \\ f(x_{k+1}) = -\cos\theta + \sin^2\theta \end{cases}$;
- constraint: $\begin{cases} c(x_k) = 0 \\ c(x_{k+1}) = \sin^2\theta \end{cases}$.

It is observed that the cost function and the constraint have increased. Although closer to the solution, the point x_{k+1} will be rejected, as it increases the merit function. Let us now consider the second-order direction.

Second resolution of the quadratic problem

The quadratic problem is solved again by modifying the second member of the constraints by the value observed in x_{k+1}. The correction is: $\delta c = c(x_{k+1}) = \sin^2\theta$.

The corrected quadratic problem is formulated as

$$\min_{d_1,d_2} (4\cos\theta - 1)d_1 + 4d_2\sin\theta + \frac{1}{2}(d_1^2 + d_2^2) \text{ s.t. } d_1\cos\theta + d_2\sin\theta = -\sin^2\theta$$

The solution is obtained by elimination: $d_1 = -(d_2 + \sin\theta)\tan\theta$ which leads to

$$\min_{d_2} d_2 \tan\theta + \frac{1}{2} d_2^2 + \frac{1}{2\cos^2\theta}(d_2 + \sin\theta)^2 \quad \rightarrow \; d_2 = -\sin\theta\cos\theta - \sin^3\theta$$

The move corrected to order 2 is: $d'_{QP} = \sin\theta \begin{pmatrix} \sin\theta \\ -\cos\theta \end{pmatrix} - \sin^2\theta \begin{pmatrix} \cos\theta \\ \sin\theta \end{pmatrix}$.

We observe in figure 4-13 that this move d'_{QP} is made of:

- the previous tangential component associated with the move d_{QP};
- an additional radial component from the second order correction.

The second order move corrects the non-linearity of the constraint. The point obtained degrades the constraint (we are no longer on the level line corresponding to the circle), but it decreases the cost function (level lines drawn in dotted lines). This point decreasing the merit function can be accepted for the next iteration.

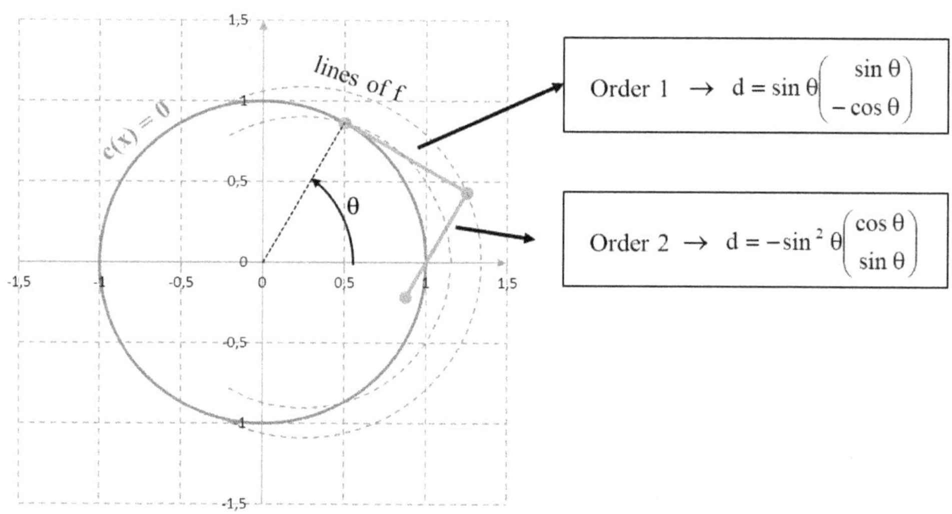

Figure 4-13: Direction of order 2.

4.4.4 Quasi-Newton method

The matrix Q of the quadratic problem (4.96) is the Hessian of the Lagrangian $\nabla^2_{xx}L$ in the point (x_k, λ_k). As this matrix is very expensive to compute numerically, it is replaced by an approximation H_k updated at each iteration by a quasi-Newton method (section 3.2.2).

From the first-order expansion of the Lagrangian gradient in the point (x_k, λ_k):

$$\nabla^2_{xx} L(x_k, \lambda_k)(x_k - x_{k-1}) \approx \nabla_x L(x_k, \lambda_k) - \nabla_x L(x_{k-1}, \lambda_k) \quad (4.120)$$

we obtain the secant equation satisfied by the matrix H_k:

$$H_k p_k = y_k \quad \text{with} \quad \begin{cases} p_k = x_k - x_{k-1} \\ y_k = \nabla_x L(x_k, \lambda_k) - \nabla_x L(x_{k-1}, \lambda_k) \end{cases} \quad (4.121)$$

The matrix H_k is updated at each iteration by a DFP, BFGS or SR1 formula as for unconstrained optimization (section 3.2.2). The update uses the changes in the point and gradient of the Lagrangian during the iteration.

The full Hessian $\nabla^2_{xx} L$ has no reason to be positive definite, even in the neighborhood of the solution. For the quadratic problem to admit a solution, the matrix H_k must be positive. This can be achieved either by a damped BFGS method (3.2.3) or by a SR1 method corrected by adding a diagonal matrix or by a modified Cholesky factorization (section 3.3.1).
The drawback is that the corrected matrix H_k may differ significantly from the true Hessian, leading to a direction d_{QP} that is unfavorable to the progression.

An alternative is to apply the method to the reduced Hessian $H_r = Z^T H Z$ instead of the full Hessian H. The KKT conditions of order 2 (1.127) indicate indeed that the reduced Hessian of the Lagrangian is positive in the neighborhood of the solution. The development is identical to that presented in section 4.3.4 using a basis Z_k of the tangent space to the constraints in x_k.
We obtain the "reduced" secant equation of the form:

$$H_r s_k = y_k \quad (4.122)$$

where
- H_r is an approximation of the Lagrangian reduced Hessian $Z_k^T \nabla^2_{xx} L(x_k, \lambda_k) Z_k$;
- $s_k = p_{Zk}$ is the change of the point in tangent space;
- $y_k = Z_k^T \left[\nabla_x L(x_k, \lambda_k) - \nabla_x L(x_{k-1}, \lambda_k) \right]$ is the change of the reduced gradient of L.

The reduced Hessian approximation is updated by a quasi-Newton formula (DFP, BGFS, SR1) based on the reduced secant equation (4.122). The matrix H_r of dimension $n - m$ is smaller than H, which reduces the algebraic computations for problems with a large number of constraints. This approach assumes that the set of active constraints does not change from one iteration to the next. If the active constraints change, the reduced Hessian must be reset to the identity.

4.4.5 Algorithm

The sequential quadratic programming algorithm starts from a freely chosen point (x_0, λ_0). Since the multipliers have no simple physical meaning, they can either be initialized to zero or a least squares solution can be determined.

The least squares solution consists in choosing for a given x_0 the value λ_0 that minimizes the norm of the gradient of the Lagrangian.

$$\min_{\lambda \in \mathbb{R}^m} \left\| \nabla_x L(x_0, \lambda) \right\|^2 \tag{4.123}$$

This will give us the value λ_0 which "best" approximates the solution of the KKT equations. The derivative of (4.123) with respect to λ is given by

$$\begin{aligned} \frac{d}{d\lambda} \left\| \nabla_x L(x_0, \lambda) \right\|^2 &= \frac{d}{d\lambda} \left\| \nabla f(x_0) + \nabla c(x_0)\lambda \right\|^2 \\ &= 2\nabla c(x_0)^T \left[\nabla f(x_0) + \nabla c(x_0)\lambda \right] \end{aligned} \tag{4.124}$$

We obtain the value of λ_0 cancelling the derivative.

$$\lambda_0 = -\left[\nabla c(x_0)^T \nabla c(x_0) \right]^{-1} \nabla c(x_0)^T \nabla f(x_0) \tag{4.125}$$

This least squares solution can be used to reset the multipliers, for example if the active constraints change during an iteration.

Each iteration involves the selection of active constraints, the globalization by line search or trust region with the merit function in L_1 penalization, and the quasi-Newton update with possible modifications to keep a positive matrix. The initial penalty value is small (for example $\rho = 1$ if the problem is scaled).

When the directional derivative of the merit function is not negative, the Hessian can be reset to the identity. If this is not sufficient, the penalty must be increased. The increase should be gradual (for example by a factor of 2). The non-linearity of the constraints may in some cases block the progression based on the merit function. Possible remedies are the use of the second-order direction, a non-monotonous descent strategy (section 3.3.3) or the acceptance of solutions through a filter (section 4.1.3).

The algorithm stops when the Lagrangian gradient norm is sufficiently small. This criterion may be difficult to achieve in practice due to numerical inaccuracies in the simulation or in the calculation of gradients by finite differences. Other stopping criteria must therefore be provided: insufficient move of the variables, or maximum number of iterations or function evaluations.

Figure 4-14 depicts the stages of the SQP algorithm.

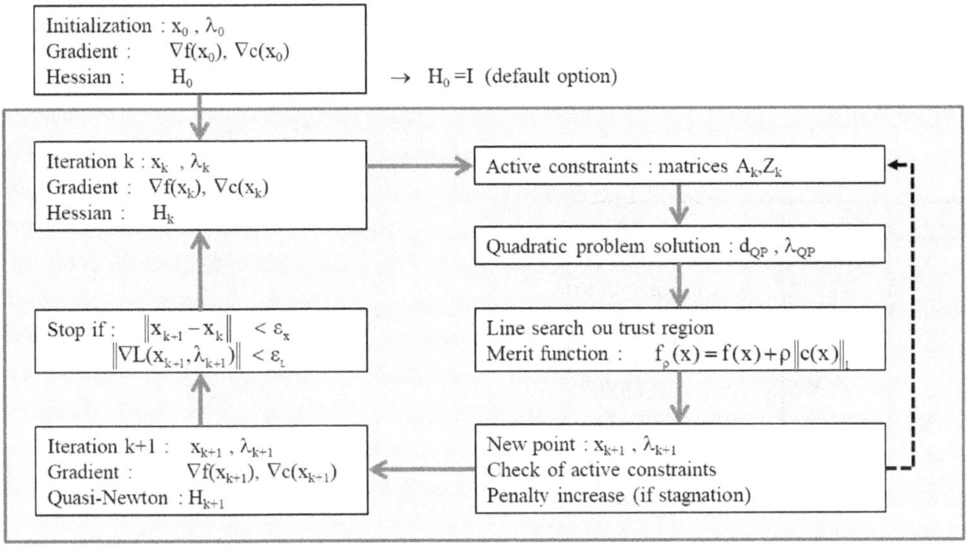

Figure 4-14: Sequential quadratic programming algorithm.

The example 4-10 illustrates the iterations of an SQP algorithm.

Example 4-10: Sequential quadratic programming algorithm

We retrieve the problem of example 4-8 (presented in [R3]).

$$\min_{x} f(x) = x_1 + x_2 \quad \text{s.t.} \quad c(x) = x_1^2 + (x_2 - 1)^2 - 1 = 0$$

In example 4-8, the SQP algorithm was applied without globalization starting from 3 different initial points. Starting from the 2nd point, the convergence was chaotic during the first iterations. Starting from the 3rd point, the algorithm converged to a maximum. A line search globalization along the direction d_{QP} is applied here. The exact Hessian matrix is used at each iteration.

$$\nabla^2_{xx} L(x_k, \lambda_k) = \begin{pmatrix} 2\lambda_k & 0 \\ 0 & 2\lambda_k \end{pmatrix}$$

This matrix is made positive by adding a diagonal matrix with a coefficient τ chosen as a multiple of 10.

$$H_k = \nabla^2_{xx} L(x_k, \lambda_k) + \tau I = (2\lambda_k + \tau)\begin{pmatrix} 1 & 0 \\ 0 & 1 \end{pmatrix} \quad \text{with} \quad 2\lambda_k + \tau > 0$$

Let us apply the globalized SQP algorithm from points 2 and 3 of example 4-8. The iterations are given in a table and represented graphically.

Initial point: $\begin{pmatrix} x_1 \\ x_2 \\ \lambda \end{pmatrix} = \begin{pmatrix} -0.1 \\ 1 \\ 1 \end{pmatrix}$

Iter	x_1	x_2	λ
1	**-0.10000**	**1.00000**	**1.00000**
2	-1.33750	0.87500	-44.5000
3	-1.03171	0.82117	1.63129
4	-0.94371	0.58304	0.62377
5	-0.74975	0.22132	0.65803
6	-0.71035	0.29156	0.70147
7	-0.70710	0.29288	0.70708
8	**-0.70711**	**0.29289**	**0.70711**

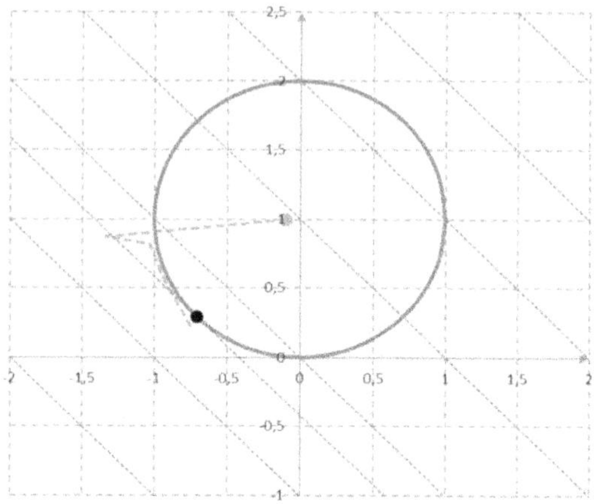

The progress is steady and convergence is achieved in 8 iterations instead of 12. Table 4-7 gives at each iteration the value of τ and the step s of the line search.

Iter	x_1	x_2	λ	∇L	$\nabla^2 L(x_1,x_2,\lambda)$		τ	dx	Step s
1	**-0.10000**	**1.00000**	**1.00000**	**9.90000**	**2.00000**	**0.00000**	**0.0**	**-4.95000**	**0.25**
				1.00000	**0.00000**	**2.00000**	**0.0**	**-0.50000**	**0.25**
2	-1.33750	0.87500	-44.500	-3.36370	-89.000	0.00000	100.0	0.30579	1.00
				0.59218	0.00000	-89.000	100.0	-0.05383	1.00
3	-1.03171	0.82117	1.63129	-0.28710	3.26258	0.00000	0.0	0.08800	1.00
				0.77690	0.00000	3.26258	0.0	-0.23812	1.00
4	-0.94371	0.58304	0.62377	-0.24198	1.24754	0.00000	0.0	0.19396	1.00
				0.45126	0.00000	1.24754	0.0	-0.36172	1.00
5	-0.74975	0.22132	0.65803	-0.05185	1.31605	0.00000	0.0	0.03940	1.00
				-0.09244	0.00000	1.31605	0.0	0.07024	1.00
6	-0.71035	0.29156	0.70147	-4.6E-03	1.40294	0.00000	0.0	3.2E-03	1.00
				-1.9E-03	0.00000	1.40294	0.0	1.3E-03	1.00
7	-0.70710	0.29288	0.70708	4.7E-06	1.41417	0.00000	0.0	-3.3E-06	1.00
				-1.7E-05	0.00000	1.41417	0.0	1.2E-05	1.00
8	**-0.70711**	**0.29289**	**0.70711**	**-4.2E-10**	**1.41421**	**0.00000**	**0.0**	**3.0E-10**	**1.00**
				2.7E-10	**0.00000**	**1.41421**	**0.0**	**-1.9E-10**	**1.00**

Table 4-7: SQP iterations with globalization (starting point 2).

Initial point: $\begin{pmatrix} x_1 \\ x_2 \\ \lambda \end{pmatrix} = \begin{pmatrix} 0.1 \\ 1 \\ 1 \end{pmatrix}$

Iter	x_1	x_2	λ
1	**0.10000**	**1.00000**	**1.00000**
2	1.33750	0.87500	-54.5000
3	1.03687	0.87643	4.23389
4	0.97837	0.75123	-0.24333
5	0.94133	0.60794	-0.35556
6	0.50173	-0.43482	-0.26135
7	-0.82925	-0.67191	0.26961
8	-1.05655	-0.18790	0.45516
9	-0.80511	0.23137	0.58156
10	-0.70800	0.28512	0.69118
11	-0.70721	0.29295	0.70699
12	**-0.70711**	**0.29289**	**0.70711**

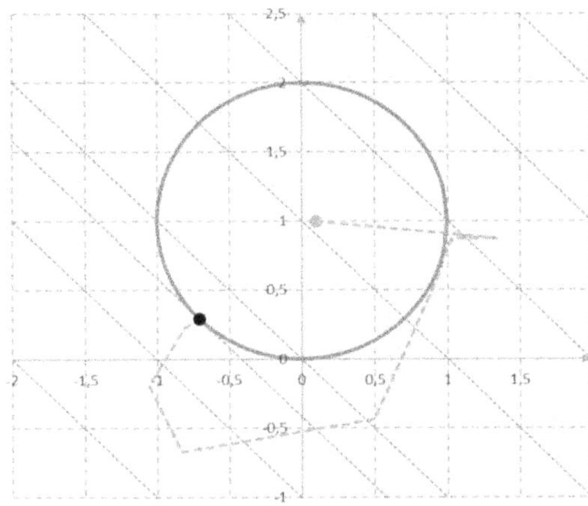

We converge well towards the minimum (instead of the maximum) with more globalization corrections (τ and s in table 4-8) than in the previous case.

Iter	x_1	x_2	λ	∇L	$\nabla^2 L(x_1,x_2,\lambda)$		τ	dx	Step s
1	**0.10000**	**1.00000**	**1.00000**	**-9.90000**	**2.00000**	**0.00000**	**0.0**	**4.95000**	**0.25**
				1.00000	**0.00000**	**2.00000**	**0.0**	**-0.50000**	**0.25**
2	1.33750	0.87500	-54.5000	12.32567	-109.000	0.00000	150.0	-0.30063	1.00
				-0.05847	0.00000	-109.000	150.0	0.00143	1.00
3	1.03687	0.87643	4.23389	0.49539	8.46779	0.00000	0.0	-0.05850	1.00
				1.06014	0.00000	8.46779	0.0	-0.12520	1.00
4	0.97837	0.75123	-0.24333	0.30426	-0.48667	0.00000	1.0	-0.59272	0.06
				1.17691	0.00000	-0.48667	1.0	-2.29267	0.06
5	0.94133	0.60794	-0.35556	0.50796	-0.71112	0.00000	1.0	-1.75838	0.25
				1.20493	0.00000	-0.71112	1.0	-4.17104	0.25
6	0.50173	-0.43482	-0.26135	1.27054	-0.52271	0.00000	1.0	-2.66197	0.50
				0.22632	0.00000	-0.52271	1.0	-0.47418	0.50
7	-0.82925	-0.67191	0.26961	0.24512	0.53921	0.00000	0.0	-0.45458	0.50
				-0.52197	0.00000	0.53921	0.0	0.96802	0.50
8	-1.05655	-0.18790	0.45516	-0.22888	0.91032	0.00000	0.0	0.25143	1.00
				-0.38167	0.00000	0.91032	0.0	0.41927	1.00
9	-0.80511	0.23137	0.58156	-0.11295	1.16311	0.00000	0.0	0.09711	1.00
				-0.06252	0.00000	1.16311	0.0	0.05376	1.00
10	-0.70800	0.28512	0.69118	-1.1E-03	1.38235	0.00000	0.0	8.0E-04	1.00
				-1.1E-02	0.00000	1.38235	0.0	7.8E-03	1.00
11	-0.70721	0.29295	0.70699	-1.4E-04	1.41398	0.00000	0.0	1.0E-04	1.00
				8.0E-05	0.00000	1.41398	0.0	-5.7E-05	1.00
12	**-0.70711**	**0.29289**	**0.70711**	**1.2E-08**	**1.41421**	**0.00000**	**0.0**	**-8.4E-09**	**1.00**
				-2.5E-08	**0.00000**	**1.41421**	**0.0**	**1.8E-08**	**1.00**

Table 4-8: SQP iterations with globalization (starting point 3).

4.5 Interior point

Interior point methods are primal-dual methods applying a barrier penalization to the inequality constraints. The KKT conditions of the penalized problem are solved by Newton's method by gradually lowering the barrier. These methods are interesting for problems with a large number of inequality constraints. This section introduces some basic principles of interior point methods. More detailed explanations are given in chapter 5 on linear programming.

4.5.1 Barrier problem

The standard problem (4.1) is reformulated by transforming the inequalities into equalities with positive slack variables s.

$$\min_{x,s} f(x) \ \text{ s.t. } \begin{cases} c_E(x) = 0 \\ c_I(x) + s = 0 \\ s \geq 0 \end{cases} \tag{4.126}$$

We apply a penalty barrier B (4.42) to the constraints $s \geq 0$. The penalized problem has only equality constraints and it is formulated as

$$\min_{x,s} f_\rho(x,s) = f(x) - \frac{1}{\rho}\sum_{j=1}^{q} \ln s_j \ \text{ s.t. } \begin{cases} c_E(x) = 0 \\ c_I(x) + s = 0 \end{cases} \tag{4.127}$$

The inequality constraints $s \geq 0$ are ignored in this formulation, but they will be taken into account in the Newton iterations developed below.

The Lagrangian L_ρ of problem (4.127) is expressed with multipliers $\lambda \in \mathbb{R}^p$ associated with the p constraints: $c_E = 0$, and with multipliers $\mu \in \mathbb{R}^q$ associated with the q constraints: $c_I + s = 0$.

$$L_\rho(x,s,\lambda,\mu) = f(x) - \frac{1}{\rho}\sum_{j=1}^{q} \ln s_j + \lambda^T c_E(x) + \mu^T [c_I(x) + s] \tag{4.128}$$

The first-order KKT conditions form a system of nonlinear equations.

$$\begin{cases} \nabla_x L_\rho(x,s,\lambda,\mu) = \nabla f(x) + \nabla c_E(x)\lambda + \nabla c_I(x)\mu = 0 \\ \nabla_s L_\rho(x,s,\lambda,\mu) = -\dfrac{1}{\rho s} + \mu = 0 \\ \nabla_\lambda L_\rho(x,s,\lambda,\mu) = c_E(x) = 0 \\ \nabla_\mu L_\rho(x,s,\lambda,\mu) = c_I(x) + s = 0 \end{cases} \tag{4.129}$$

We notice that the derivatives of the Lagrangian L_ρ with respect to x,λ,μ do not depend on ρ and are identical to the derivatives of the Lagrangian L of problem (4.1). The penalty ρ appears only in the second equation which takes the form

$$-\frac{1}{\rho s}+\mu=0 \;\Leftrightarrow\; s_j\mu_j=\frac{1}{\rho} \;,\; j=1 \text{ to } q \;\underset{\text{noted}}{\Leftrightarrow}\; SMe=\frac{1}{\rho}e \tag{4.130}$$

where the diagonal matrices $S,M\in\mathbb{R}^{q\times q}$ and the vector $e\in\mathbb{R}^q$ are defined by

$$S=\begin{pmatrix} s_1 & 0 & \cdots & 0 & 0\\ 0 & s_2 & \cdots & 0 & 0\\ \vdots & \vdots & \ddots & \vdots & \vdots\\ 0 & 0 & \cdots & s_{q-1} & 0\\ 0 & 0 & \cdots & 0 & s_q \end{pmatrix} ,\; M=\begin{pmatrix} \mu_1 & 0 & \cdots & 0 & 0\\ 0 & \mu_2 & \cdots & 0 & 0\\ \vdots & \vdots & \ddots & \vdots & \vdots\\ 0 & 0 & \cdots & \mu_{q-1} & 0\\ 0 & 0 & \cdots & 0 & \mu_q \end{pmatrix} ,\; e=\begin{pmatrix}1\\1\\ \vdots\\1\\1\end{pmatrix} \tag{4.131}$$

When $\rho\to\infty$, equations (4.130) give the complementarity conditions $s_j\mu_j=0$ associated with the inequality constraints $s\geq 0$. By convention, we use the inverse of the penalty: $h=\rho^{-1}$ called the **barrier height**. Its initially large value decreases to zero to return to the initial problem (4.126).

Let us apply Newton's method to the system (4.129) put in the form

$$F(x,s,\lambda,\mu)=\begin{pmatrix}\nabla_x L(x,s,\lambda,\mu)\\ SMe-he\\ c_E(x)\\ c_I(x)+s\end{pmatrix}=0 \tag{4.132}$$

Newton's iteration is defined with a step α_s for the primal variables (x,s) and a step α_μ for the dual variables (λ,μ). These steps will be chosen so that the variables s and μ remain positive. Their setting is discussed below.

$$\begin{pmatrix}x_{k+1}\\ s_{k+1}\\ \lambda_{k+1}\\ \mu_{k+1}\end{pmatrix}=\begin{pmatrix}x_k\\ s_k\\ \lambda_k\\ \mu_k\end{pmatrix}+\begin{pmatrix}\alpha_s d_x\\ \alpha_s d_s\\ \alpha_\mu d_\lambda\\ \alpha_\mu d_\mu\end{pmatrix} \quad \text{with } \nabla F(x_k,s_k,\lambda_k,\mu_k)^T\begin{pmatrix}d_x\\ d_s\\ d_\lambda\\ d_\mu\end{pmatrix}=-F(x_k,s_k,\lambda_k,\mu_k) \tag{4.133}$$

The Jacobian matrix of F of dimension $n+q+p+q$ has the expression

$$\nabla F^T = \begin{pmatrix} \nabla^2_{xx}L & 0 & \nabla c_E & \nabla c_I \\ 0 & M & 0 & S \\ \nabla c_E^T & 0 & 0 & 0 \\ \nabla c_I^T & I & 0 & 0 \end{pmatrix} \tag{4.134}$$

where $\nabla^2_{xx}L$ is the Hessian of the Lagrangian of the initial problem (4.1).
The Newton system is made symmetric by premultiplying the second row of the matrix ∇F^T by S^{-1}. The diagonal matrix S is invertible, as the variables s are kept strictly positive during the iterations.

$$\begin{pmatrix} \nabla^2_{xx}L & 0 & \nabla c_E & \nabla c_I \\ 0 & S^{-1}M & 0 & I \\ \nabla c_E^T & 0 & 0 & 0 \\ \nabla c_I^T & I & 0 & 0 \end{pmatrix} \begin{pmatrix} d_x \\ d_s \\ d_\lambda \\ d_\mu \end{pmatrix} = - \begin{pmatrix} \nabla_x L \\ Me - hS^{-1}e \\ c_E \\ c_I + s \end{pmatrix} \tag{4.135}$$

d_s can be expressed from the second line.

$$d_s = -M^{-1}Sd_\mu - M^{-1}S(Me - hS^{-1}e) = -M^{-1}Sd_\mu - s + hM^{-1}e \tag{4.136}$$

because $M^{-1}S = SM^{-1}$ (diagonal matrices) and $Se = s$.

The system (4.135) reduces to

$$\begin{pmatrix} \nabla^2_{xx}L & \nabla c_E & \nabla c_I \\ \nabla c_E^T & 0 & 0 \\ \nabla c_I^T & 0 & -M^{-1}S \end{pmatrix} \begin{pmatrix} d_x \\ d_\lambda \\ d_\mu \end{pmatrix} = - \begin{pmatrix} \nabla_x L \\ c_E \\ c_I + hM^{-1}e \end{pmatrix} \tag{4.137}$$

We can then express d_μ from the third line:

$$d_\mu = S^{-1}M\nabla c_I^T d_x + S^{-1}M(c_I + hM^{-1}e) = S^{-1}M\nabla c_I^T d_x + S^{-1}(Mc_I + he) \tag{4.138}$$

The system (4.141) reduces to

$$\begin{pmatrix} \nabla^2_{xx}L + \nabla c_I S^{-1} M \nabla c_I^T & \nabla c_E \\ \nabla c_E^T & 0 \end{pmatrix} \begin{pmatrix} d_x \\ d_\lambda \end{pmatrix} = - \begin{pmatrix} \nabla_x L + \nabla c_I S^{-1}(Mc_I + he) \\ c_E \end{pmatrix} \tag{4.139}$$

The solution of the linear system can be done from (4.137) or (4.139). The most interesting form depends on the number of inequality constraints and the sparsity of the matrix $\nabla^2_{xx}L$. If this matrix is sparse, the form (4.137) may be faster to solve numerically than the denser form (4.139).

Solving the linear system yields the move $(d_x, d_s, d_\lambda, d_\mu)$. In addition to the usual precautions associated with Newton's method (globalization, approximation of the Hessian), this move must be restricted to remain in the domain $\{s > 0; \mu > 0\}$ and the barrier height h must be progressively lowered to return to the initial problem.

4.5.2 Globalization

To stay within the domain $\{s > 0 ; \mu > 0\}$, the move $(d_x, d_s, d_\lambda, d_\mu)$ is reduced by a step α_s on the primal variables (x, s) and a step α_μ on the dual variables (λ, μ). The steps are set to keep a margin with the domain boundary, for example

$$\begin{cases} s_k + \alpha_s d_s \geq 0{,}01 s_k \\ \mu_k + \alpha_\mu d_\mu \geq 0{,}01 \mu_k \end{cases} \tag{4.140}$$

Small values of s and μ should be avoided, otherwise subsequent moves may be blocked by the domain boundary or the matrix ∇F^T (4.134) may become ill-conditioned.

The move is evaluated by a function of the form

$$f_m(x,s) = f(x) - h\sum_{j=1}^{q} \ln s_j + \rho_m \left(\|c_E(x)\| + \|c_I(x) + s\| \right) \tag{4.141}$$

with a penalization on the constraint violation. The penalty coefficient ρ_m is to be set independently of the barrier height h.

The globalization is performed by a line search along the direction (d_x, d_s). The step α_s decreases from the maximum value (4.140) until an Armijo condition on the merit function is met. Alternatively, one can accept the move on the basis of a filter. An alternative approach is to solve the KKT system (4.129) by a SQP algorithm with a trust region. The implementation is more complex, but allows better control of the progress towards the boundary of the feasible domain.

The matrix $\nabla^2_{xx} L$ is approximated by a quasi-Newton method (section 3.2.2). To converge to a minimum, the matrix of the system (4.135) reduced on the tangent space to the constraints (defined by the Jacobian $(\nabla c_E, \nabla c_I)^T$) must be positive. This is achieved by adding a diagonal matrix: $\nabla^2_{xx} L + \tau I$ with τ large enough.

4.5.3 Barrier height

The solution of the barrier problem (4.129) is a function of the barrier height h. When h is zero, the KKT conditions of the original problem (4.1) are retrieved. The value of h should be reduced gradually, in order to avoid the cancelling or the premature reducing any of the variables s_i and μ_i. If the values of these variables are too small, there is a risk of blocking further moves.

A first strategy is to solve the system (4.129) completely before changing h. The system is solved with a high tolerance at the beginning and then lower as h decreases. For example, one can stop the resolution when

$$\|F(x,s,\lambda,\mu)\| = \left\| \begin{matrix} \nabla_x L(x,s,\lambda,\mu) \\ SMe - he \\ c_E(x) \\ c_I(x)+s \end{matrix} \right\| \le h \tag{4.142}$$

and then reduce the value of h by a fixed factor, for example $h' = h/10$.

A second more efficient strategy is to modify h at each iteration of Newton (4.133) according to the distance to the domain boundary $\{s>0\,;\,\mu>0\}$. This distance is measured by the average of the products $s_i\mu_i$ called the **duality measure** ν in the point (s,μ).

$$\nu = \frac{1}{q} s^T \mu = \frac{1}{q}\sum_{i=1}^{q} s_i\mu_i \tag{4.143}$$

The value of h is defined with a weighting factor σ between 0 and 1.

$$h = \sigma\nu = \frac{\sigma}{q}\sum_{i=1}^{q} s_i\mu_i \tag{4.144}$$

A small value of σ decreases the barrier height and allows for a faster approach to the boundary of the domain $\{s>0\,;\,\mu>0\}$. The two setting options for σ are similar to those applied for linear problems (section 5.2).
The first setting option is based on the distribution of products $s_i\mu_i$. If these products are close to each other, one can expect a rapid simultaneous decrease using a value of $\sigma \approx 0$. Otherwise, σ is set according to the smallest product $s_i\mu_i$ in order not to cancel it prematurely. For example, σ can be set to $\dfrac{\nu}{\min(s_i\mu_i)} - 1$.

The second setting option (of prediction-type) is to compute the Newton move (4.133) with $h = 0$. This move bounded by (4.140) gives a point $(\tilde{s}, \tilde{\mu})$ with a duality measure $\tilde{\nu}$. The value of σ is then set by the empirical formula used in linear programming (section 5.2.4):

$$\sigma = \left(\frac{\tilde{\nu}}{\nu}\right)^3 \tag{4.145}$$

Section 5.2 provides illustrative examples for linear problems.

4.6 Augmented Lagrangian

Augmented Lagrangian methods are dual methods applying a quadratic penalty to the constraints. Their principle is to solve the dual problem by updating the multipliers and the penalty after each minimization of the Lagrangian with respect to the primal variables. This leads to a sequence of unconstrained problems. The advantage of these methods is that the exact solution can be obtained without excessively increasing the penalty coefficient.

4.6.1 Dual problem

For the standard form optimization problem

$$\min_{x \in \mathbb{R}^n} f(x) \;\text{ s.t. }\; \begin{cases} c_E(x) = 0 \\ c_I(x) \le 0 \end{cases} \tag{4.146}$$

the dual problem consists in maximizing the dual function ψ:

$$\max_{\lambda, \mu \ge 0} \psi(\lambda, \mu) \quad \text{with} \quad \begin{cases} \psi(\lambda, \mu) = \min\limits_{x \in D} L(x, \lambda, \mu) \\ L(x, \lambda, \mu) = f(x) + \lambda^T c_E(x) + \mu^T c_I(x) \end{cases} \tag{4.147}$$

The dual function is defined on the domain: $D = \{\lambda, \mu \ge 0 \,/\, \psi(\lambda, \mu) > -\infty\}$.
By grouping the constraints and multipliers

$$c \underset{def}{=} \begin{pmatrix} c_E \\ c_I \end{pmatrix}, \quad \nu \underset{def}{=} \begin{pmatrix} \lambda \\ \mu \end{pmatrix} \tag{4.148}$$

the dual problem (4.147) can be written as

$$\max_{\nu} \psi(\nu) \quad \text{with} \quad \begin{cases} \psi(\nu) = \min\limits_{x \in D} L(x, \nu) \;,\; D = \{\nu \,/\, \psi(\nu) > -\infty\} \\ L(x, \nu) = f(x) + \nu^T c(x) \end{cases} \tag{4.149}$$

Let us first compute the gradient and the Hessian of the dual function ψ. The value of x minimizing the Lagrangian $L(x,\nu)$ for ν is denoted $x_L(\nu)$ and it satisfies

$$\nabla_x L\left[x_L(\nu),\nu\right]=0 \quad , \quad \forall\nu \tag{4.150}$$

Deriving with respect to ν

$$\begin{aligned}\frac{d}{d\nu}\nabla_x L\left[x_L(\nu),\nu\right] &= \frac{dx_L}{d\nu}\nabla^2_{xx}L(x_L,\nu)+\nabla^2_{\nu x}L(x_L,\nu)\\ &= \frac{dx_L}{d\nu}\nabla^2_{xx}L(x_L,\nu)+\nabla c(x_L)^T\end{aligned} \tag{4.151}$$

we obtain the derivative of $x_L(\nu)$

$$\frac{dx_L}{d\nu}=-\nabla c(x_L)^T\left(\nabla^2_{xx}L(x_L,\nu)\right)^{-1} \tag{4.152}$$

Let us now look for the maximum of the dual function ψ which is expressed as

$$\psi(\nu)=\min_x L(x,\nu)=L\left[x_L(\nu),\nu\right] \tag{4.153}$$

By deriving ψ with (4.150) and (4.152), we obtain the gradient and the Hessian.

$$\begin{cases}\nabla\psi=\dfrac{dx_L}{d\nu}\nabla_x L(x_L,\nu)+\nabla_\nu L(x_L,\nu)=\nabla_\nu L(x_L,\nu)=c(x_L)\\ \nabla^2\psi=\dfrac{dx_L}{d\nu}\nabla c(x_L)=-\nabla c(x_L)^T\left(\nabla^2_{xx}L(x_L,\nu)\right)^{-1}\nabla c(x_L)\end{cases} \tag{4.154}$$

The dual methods consist of solving the problem (4.149) which is a simple unconstrained maximization. They use the gradient and the Hessian above.

The **Uzawa method** consists in maximizing ψ by a simple gradient method with a step s to be adjusted. The iterations on ν are of the form

$$\nu_{k+1}=\nu_k+s\nabla\psi(\nu_k)=\nu_k+sc\left[x_L(\nu_k)\right] \tag{4.155}$$

The step s is fixed to avoid a costly line search. Indeed, each step trial requires a minimization of the Lagrangian (4.153) to evaluate the value of ψ. With a fixed step length, the iteration in ν (4.155) is immediate, but the convergence of this steepest descent method is generally slow (section 3.5.2).

An improvement consists in performing a **Newton iteration** to define ν_{k+1}.

$$\nu_{k+1}=\nu_k-\left(\nabla^2\psi(\nu_k)\right)^{-1}\nabla\psi(\nu_k) \tag{4.156}$$

The calculation of $\nabla^2\psi$ (4.154) uses the Hessian of the Lagrangian $\nabla^2_{xx}L(x_L,\nu)$

or an approximation of the latter. This approximation is available if the minimization with respect to x (4.153) is performed by a quasi-Newton method. After a potential modification to make the matrix $\nabla^2\psi$ negative (to maximize ψ), Newton's iteration with a step s is applied.

$$\nu_{k+1} = \nu_k + s\left[\nabla c(x_L)^T \left(\nabla^2_{xx} L(x_L,\nu)\right)^{-1} \nabla c(x_L)\right]^{-1} c(x_L) \tag{4.157}$$

The step s is to be adjusted by an Armijo condition for the solution to improve. The improvement is measured by merit function or by filter (section 4.1.3).

The following example compares the iterations of Uzawa and Newton methods.

Example 4-11: Uzawa and Newton methods

Consider the optimization problem of example 4-10 :

$\min\limits_x f(x) = x_1 + x_2 \;\text{ s.t. }\; c(x) = x_1^2 + (x_2-1)^2 - 2 = 0$

The Lagrangian $L(x,\lambda) = x_1 + x_2 + \lambda\left(x_1^2 + (x_2-1)^2 - 2\right)$ has its minimum at

$x_L(\lambda) = \left(-\dfrac{1}{2\lambda}, -\dfrac{1}{2\lambda} + 1\right)$

which yields the dual function: $\psi(\lambda) = \min\limits_x L(x,\lambda) = L[x_L(\lambda),\lambda] = 1 - 2\lambda - \dfrac{1}{2\lambda}$.

The gradient and Hessian are: $\nabla\psi(\lambda) = \dfrac{1}{2\lambda^2} - 2 \;,\; \nabla^2\psi(\lambda) = -\dfrac{1}{\lambda^3}$.

The solution to the dual problem is: $\max\limits_\lambda \psi(\lambda) \Rightarrow \lambda^* = \dfrac{1}{2} \rightarrow x^* = \begin{pmatrix}-1\\0\end{pmatrix}$.

This dual solution is identical to the primal solution. It is therefore a saddle point (section 1.4.2). Let us apply the methods of Uzawa and Newton to the problem:

- the Uzawa iteration is defined by: $\lambda_{k+1} = \lambda_k + s\left(\dfrac{1}{2\lambda_k^2} - 2\right)$;
- the Newton iteration is defined by: $\lambda_{k+1} = \lambda_k + s\lambda_k^3\left(\dfrac{1}{2\lambda_k^2} - 2\right)$.

Let us first examine the behavior of Uzawa method depending on the value of the step length s = 0.1 or 0.2. Table 4-9 shows the iterations for these two step values. The solution is reached in both cases, but with oscillations in the second case. It would then be preferable to reduce the step length to converge more quickly.

Uzawa method with s = 0.1

Iter	x_1	x_2	λ	s
1	-0.10000	1.00000	1.00000	0.1
2	-0.50000	0.50000	0.85000	0.1
3	-0.58824	0.41176	0.71920	0.1
4	-0.69521	0.30479	0.61587	0.1
5	-0.81186	0.18814	0.54769	0.1
6	-0.91292	0.08708	0.51438	0.1
7	-0.97205	0.02795	0.50335	0.1
8	-0.99334	0.00666	0.50070	0.1
9	-0.99861	0.00139	0.50014	0.1
10	-0.99972	0.00028	0.50003	0.1
11	-0.99994	5.63E-05	0.50001	0.1
12	-0.99999	1.13E-05	0.50000	0.1

Uzawa method with s = 0.2

Iter	x_1	x_2	λ	s
1	-0.10000	1.00000	1.00000	0.2
2	-0.50000	0.50000	0.70000	0.2
3	-0.71429	0.28571	0.50408	0.2
4	-0.99190	0.00810	0.49763	0.2
5	-1.00476	-0.00476	0.50145	0.2
6	-0.99711	0.00289	0.49914	0.2
7	-1.00172	-0.00172	0.50052	0.2
8	-0.99896	0.00104	0.49969	0.2
9	-1.00062	-0.00062	0.50019	0.2
10	-0.99963	0.00037	0.49989	0.2
11	-1.00022	-2.24E-04	0.50007	0.2
12	-0.99987	1.34E-04	0.49996	0.2

Table 4-9: Effect of the step length on Uzawa method.

Uzawa method with s = 0.1 and Newton method with an initial step s = 1 adjusted by an Armijo condition (with a reduction of a factor of 2 at each trial) are now compared. Table 4-10 shows the iterations of each method. Newton method is faster and more accurate. The value of the step s set for Uzawa method slows down the convergence.

Uzawa method

Iter	x_1	x_2	λ	s
1	-0.10000	1.00000	1.00000	0.1
2	-0.50000	0.50000	0.85000	0.1
3	-0.58824	0.41176	0.71920	0.1
4	-0.69521	0.30479	0.61587	0.1
5	-0.81186	0.18814	0.54769	0.1
6	-0.91292	0.08708	0.51438	0.1
7	-0.97205	0.02795	0.50335	0.1
8	-0.99334	0.00666	0.50070	0.1
9	-0.99861	0.00139	0.50014	0.1
10	-0.99972	0.00028	0.50003	0.1

Newton method

Iter	x_1	x_2	λ	s
1	-0.10000	1.00000	1.00000	0.25
2	-0.50000	0.50000	0.62500	0.50
3	-0.80000	0.20000	0.53711	1.00
4	-0.93091	0.06909	0.49577	1.00
5	-1.00854	-0.00854	0.49995	1.00
6	-1.00011	-0.00011	0.50000	1.00
7	-1.00000	-1.72E-08	0.50000	1.00
8	-1.00000	0.00E+00	0.50000	1.00

Table 4-10: Uzawa and Newton iterations.

The Uzawa and Newton methods do not take into account the following aspects:
- the domain of definition D of the dual function is not known in advance;
- if there is no saddle point, the dual solution may be different from the primal (theorem 1-7 and example 1-10).

When the multipliers are not in the domain D, the Lagrangian has no minimum. There is no way to detect this situation during the iterations of Uzawa (4.155) or Newton (4.157), which makes these methods not very robust.

4.6.2 Augmented dual problem

The difficulties related to the definition domain of the dual function can be circumvented by introducing a quadratic penalization.

Let us first consider a problem with m equality constraints. The treatment of inequality constraints is discussed later in section 4.6.3.

$$\min_x f(x) \quad \text{s.t.} \quad c(x)=0 \tag{4.158}$$

Its Lagrangian is expressed with multipliers $\nu \in \mathbb{R}^m$.

$$L(x,\nu)=f(x)+\nu^T c(x) \tag{4.159}$$

Note (x^*,ν^*) a solution of the KKT conditions of order 1 and 2.

$$\begin{cases} \nabla_x L(x^*,\nu^*)=\nabla f(x^*)+\nabla c(x^*)\nu^*=0 \\ d^T\nabla^2_{xx}L(x^*,\nu^*)d>0 \ , \ \forall d\in\mathbb{R}^n \ , d\neq 0 \ \text{ such that } \nabla c(x^*)d=0 \end{cases} \tag{4.160}$$

The augmented problem is defined from (4.158) by adding a quadratic penalization to the cost function. The equality constraints remain.

$$\min_x f_\rho(x)=f(x)+\frac{1}{2}\rho\|c(x)\|_2^2 \quad \text{s.t.} \quad c(x)=0 \tag{4.161}$$

The Lagrangian of this problem noted L_ρ is called the **augmented Lagrangian**.

Keeping the notation $\nu\in\mathbb{R}^m$ for multipliers, it is expressed as

$$L_\rho(x,\nu)=f_\rho(x)+\nu^T c(x)=L(x,\nu)+\frac{1}{2}\rho\|c(x)\|_2^2 \tag{4.162}$$

Let us calculate the gradient and Hessian of L_ρ with respect to the variables x.

$$\nabla_x L_\rho(x,\nu)=\nabla_x L(x,\nu)+\rho\nabla c(x)c(x) \tag{4.163}$$

To calculate the Hessian, we explicitly write the sum over the m constraints.

$$\nabla_x L_\rho(x,\nu) = \nabla_x L(x,\nu) + \rho\sum_{j=1}^{m} c_j(x)\nabla c_j(x) \tag{4.164}$$

Derivation with respect to x gives

$$\begin{aligned}\nabla^2_{xx} L_\rho(x,\nu) &= \nabla^2_{xx} L(x,\nu) + \rho\sum_{j=1}^{m} \nabla c_j(x)\nabla c_j(x)^T + \rho\sum_{j=1}^{m} c_j(x)\nabla^2 c_j(x) \\ &= \nabla^2_{xx} L(x,\nu) + \rho\nabla c(x)\nabla c(x)^T \quad + \rho\sum_{j=1}^{m} c_j(x)\nabla^2 c_j(x)\end{aligned} \tag{4.165}$$

At a feasible point: $c(x) = 0$, the gradient and Hessian simplify to

$$\begin{cases}\nabla_x L_\rho(x,\nu) &= \nabla_x L(x,\nu) \\ \nabla^2_{xx} L_\rho(x,\nu) &= \nabla^2_{xx} L(x,\nu) + \rho\nabla c(x)\nabla c(x)^T\end{cases} \tag{4.166}$$

These formulas apply in particular to the solution x* of problem (4.158).
They are used to demonstrate the following property.

Property 4-5: Saddle point of the augmented problem

For ρ large enough, the solution of the KKT conditions of problem (4.158) is a saddle point of the augmented problem (4.161).

Demonstration

We try to show that the point (x^*, ν^*) solution of the KKT conditions of problem (4.161) achieves also: $\max_\nu L_\rho(x^*,\nu) = L_\rho(x^*,\nu^*) = \min_x L_\rho(x,\nu^*)$.

The equality on the left can be deduced from (4.162) : $L_\rho(x^*,\nu) = f(x^*)$, which shows that $L_\rho(x^*,\nu)$ does not depend on ν.

For the equality on the right, we must check that x* is a minimum of $L_\rho(x,\nu^*)$.

The conditions for a minimum are: $\begin{cases}\nabla_x L_\rho(x^*,\nu^*) = 0 \\ \nabla^2_{xx} L_\rho(x^*,\nu^*) > 0\end{cases}$.

The condition on the gradient is satisfied using (4.166) and (4.160).
$\nabla_x L_\rho(x^*,\nu^*) = \nabla_x L(x^*,\nu^*) = 0$

It remains to show that the condition on the Hessian is met for ρ large enough.

Let us assume by contradiction that for any integer k, there exists a value $\rho_k > k$ such that the matrix $\nabla^2_{xx} L_{\rho_k}(x^*, v^*)$ is not positive definite.

We can then find a unit vector d_k such that: $d_k^T \nabla^2_{xx} L_{\rho_k}(x^*, v^*) d_k \le 0$.

Replacing in (4.166), we obtain the inequality denoted (I).

$d_k^T \nabla^2_{xx} L(x^*, v^*) d_k + \rho_k \|\nabla c(x^*) d_k\|^2 \le 0$ **(inequality I)**

The vectors d_k belong to the unit sphere which is compact.

The sequence (d_k) admits an accumulation point d on the unit sphere: $\|d\| = 1$.

Let us write inequality (I) as: $\|\nabla c(x^*) d_k\|^2 < -\frac{1}{\rho_k} d_k^T \nabla^2_{xx} L(x^*, v^*) d_k$

and take the limit $k \to \infty$: $\|\nabla c(x^*) d\| \le \lim_{k\to\infty} -\frac{1}{\rho_k} d_k^T \nabla^2_{xx} L(x^*, v^*) d_k = 0$

This limit is 0, because $\rho_k \to \infty$ and the numerator is bounded for $\|d_k\| = 1$.

The vector d belongs thus to the constraint tangent space in x*: $\nabla c(x^*) d = 0$

and it therefore satisfies the second KKT condition (4.160) : $d^T \nabla^2_{xx} L(x^*, v^*) d > 0$

Furthermore, inequality (I) gives: $d_k^T \nabla^2_{xx} L(x^*, v^*) d_k \le -\rho_k \|\nabla c(x^*) d_k\|^2 \le 0$

or by taking the limit $k \to \infty$: $d^T \nabla^2_{xx} L(x^*, v^*) d \le 0$

We arrive at a contradiction. The hypothesis $\nabla^2_{xx} L(x^*, v^*)$ not positive definite is therefore false. For ρ large enough $\nabla^2_{xx} L(x^*, v^*)$ is positive definite. The point x* is then a minimum of $L_\rho(x, v^*)$ and the saddle point conditions are satisfied.

The quadratic penalization creates a saddle point that does not necessarily exist in the original problem and this saddle point coincides with the KKT solution when the penalty is large enough. The example 4-12 illustrates this property.

Example 4-12: Creation of a saddle point by penalization

Consider the problem: $\min_x f(x) = -x_1 x_2$ s.t. $c(x) = x_1 + x_2 - 1 = 0$.

The Lagrangian is: $L(x, \lambda) = f(x) + \lambda c(x) = -x_1 x_2 + \lambda(x_1 + x_2 - 1)$.

The KKT solution is: $\lambda^* = \frac{1}{2}$, $x_1^* = x_2^* = \frac{1}{2}$ $\to$ $L(x^*, \lambda^*) = -\frac{1}{4}$.

Let us look at the "raw" dual problem and then at the penalized dual problem.

Raw dual problem

The dual function $\psi(\lambda) = \min_x L(x,\lambda)$ is not defined for any value of λ.

Indeed, taking $x_1 = -\lambda$, we have $L(x,\lambda) = 2\lambda x_2 - \lambda^2 - \lambda$ which is not bounded. This problem does not admit a saddle point and Uzawa method is inapplicable.

Penalized dual problem

Let us now apply a quadratic penalization on the cost function.
The penalized problem is

$$\min_x f_\rho(x) = -x_1x_2 + \frac{1}{2}\rho(x_1 + x_2 - 1)^2 \quad \text{s.t.} \quad c(x) = x_1 + x_2 - 1 = 0$$

The augmented Lagrangian is the Lagrangian of this penalized problem.

$$L_\rho(x,\lambda) = f_\rho(x) + \lambda c(x) = -x_1x_2 + \frac{1}{2}\rho(x_1 + x_2 - 1)^2 + \lambda(x_1 + x_2 - 1)$$

Let us find the domain of definition of the dual function: $\psi_\rho(\lambda) = \min_x L_\rho(x,\lambda)$.

The Lagrangian $L_\rho(x,\lambda)$ admits a minimum if:

- the gradient is zero

$$\nabla_x L_\rho = 0 \Rightarrow \begin{cases} -x_2 + \rho(x_1 + x_2 - 1) + \lambda = 0 \\ -x_1 + \rho(x_1 + x_2 - 1) + \lambda = 0 \end{cases} \Rightarrow x_1(\lambda) = x_2(\lambda) = \frac{\lambda - \rho}{1 - 2\rho};$$

- the Hessian is positive: $\nabla^2_{xx} L_\rho > 0 \Rightarrow \begin{pmatrix} \rho & \rho - 1 \\ \rho - 1 & \rho \end{pmatrix} > 0$

  The eigenvalues of the Hessian $\begin{cases} \sigma_1 = 1 \\ \sigma_2 = 2\rho - 1 \end{cases}$ are positive if $\rho > \frac{1}{2}$.

If $\rho > \frac{1}{2}$, the dual function is defined for all λ, and it has the expression

$$\psi_\rho(\lambda) = -\left(\frac{\lambda - \rho}{1 - 2\rho}\right)^2 + \frac{1}{2}\rho\left(\frac{2\lambda - 1}{1 - 2\rho}\right)^2 + \lambda\frac{2\lambda - 1}{1 - 2\rho}$$

Let us then solve the dual problem: $\max_\lambda \psi_\rho(\lambda)$. The maximum is obtained for

$$\frac{d}{d\lambda}\psi_\rho(\lambda) = 0 \Rightarrow -2(\lambda - \rho) + 2\rho(2\lambda - 1) + (1 - 2\rho)(4\lambda - 1) = 0$$

which gives the solution: $\lambda^* = \frac{1}{2} \Rightarrow \psi_\rho(\lambda^*) = -\frac{1}{4}$ and $x_1(\lambda^*) = x_2(\lambda^*) = \frac{1}{2}$
which is identical to the KKT solution given at the beginning of the example.
The duality gap is zero: $\psi_\rho(\lambda^*) = -\frac{1}{4} = L(x^*,\lambda^*)$.

The penalized problem thus admits a saddle point corresponding to the KKT solution, which allows us to apply a dual method. The saddle point is created by the quadratic penalization which makes the problem convex, as soon as $\rho > 0.5$. A dual method applied to the penalized problem thus allows the exact solution to be found without indefinitely increasing the penalty.

The augmented Lagrangian has its minimum in x* when the multiplier is ν^* and if the penalty is large enough. In practice, the values of ν^* and the minimum penalty are not known in advance. By minimizing the augmented Lagrangian for a multiplier $\nu = \nu^* + \delta\nu$ and a penalty ρ, we obtain a solution $x = x^* + \delta x$ which satisfies the minimum condition

$$\nabla_x L_\rho(x^* + \delta x, \nu^* + \delta\nu) = 0 \tag{4.167}$$

Let us perform an expansion to order 1 from formula (4.163) and using the relations satisfied by the KKT solution: $\nabla_x L(x^*, \nu^*) = 0$ and $c(x^*) = 0$.

$$\nabla_x L_\rho(x^* + \delta x, \nu^* + \delta\nu) = \nabla_x L(x^* + \delta x, \nu^* + \delta\nu) + \rho \nabla c(x^* + \delta x) c(x^* + \delta x)$$

$$\text{with} \begin{cases} \nabla_x L(x^* + \delta x, \nu^* + \delta\nu) & \approx \nabla^2_{xx} L(x^*, \nu^*)\delta x + \nabla^2_{x\nu} L(x^*, \nu^*)\delta\nu \\ & = \nabla^2_{xx} L(x^*, \nu^*)\delta x + \nabla c(x^*)\delta\nu \\ \nabla c(x^* + \delta x) c(x^* + \delta x) & \approx \nabla c(x^*) \nabla c(x^*)^T \delta x \end{cases} \tag{4.168}$$

Replacing into (4.167) and dividing by ρ, we obtain the relation

$$\left[\frac{1}{\rho} \nabla^2_{xx} L(x^*, \nu^*) + \nabla c(x^*) \nabla c(x^*)^T \right] \delta x = -\frac{1}{\rho} \nabla c(x^*) \delta\nu \tag{4.169}$$

This relation suggests two ways to reduce the deviation δx from the solution: decrease $\delta\nu$ or increase ρ. These two approaches are combined in the algorithm presented in section 4.6.4. The improvement of the multiplier is prioritized to avoid increasing prematurely the penalty and degrading the problem conditioning.

4.6.3 Inequality constraints

The inequality constraints are taken into account by reformulating the problem (4.146) with q positive slack variables s associated with the inequality constraints.

$$\min_{x, s \geq 0} f(x) \quad \text{s.t.} \begin{cases} c_E(x) = 0 \\ c_I(x) + s = 0 \end{cases} \tag{4.170}$$

By grouping the constraints into a single vector of dimension m

$$c(x,s)=\begin{pmatrix} c_E(x) \\ c_I(x)+s \end{pmatrix} \tag{4.171}$$

the problem takes the form

$$\min_{x,s\geq 0} f(x) \quad \text{s.t.} \quad c(x,s)=0 \tag{4.172}$$

This problem is analogous to problem (4.158) except that the variables s must be positive. Noting (ν_E,ν_I) the respective multipliers of the constraints (c_E, c_I+s), the augmented Lagrangian is expressed as

$$L_\rho(x,s,\nu)=f(x)+\frac{1}{2}\rho\|c_E(x)\|_2^2+\nu_E^T c_E(x)+\frac{1}{2}\rho\|c_I(x)+s\|_2^2+\nu_I^T\left[c_I(x)+s\right] \tag{4.173}$$

There are two possible approaches to minimizing the augmented Lagrangian.

The first approach consists in explicitly calculating the variables s. Indeed, the Lagrangian (4.173) is a quadratic function of s and one can explicitly determine the values of s minimizing $L_\rho(x,s,\nu)$ for a fixed x.

$$\min_{s\geq 0} L_\rho(x,s,\nu) \Leftrightarrow \min_{s\geq 0}\frac{1}{2}\rho\|c_I(x)+s\|_2^2+\nu_I^T\left[c_I(x)+s\right] \tag{4.174}$$

The minimum is obtained for

$$s=\max\left[0\,,\,-c_I(x)-\frac{\nu_I}{\rho}\right] \tag{4.175}$$

By replacing in (4.173) with the notation: $c_I^+(x)=\max\left[c_I(x)\,,\,-\frac{\nu_I}{\rho}\right]$, we obtain a function of x (denoted F_ρ) whose minimum with respect to x is sought.

$$\min_x F_\rho(x,\nu)=f(x)+\frac{1}{2}\rho\|c_E(x)\|_2^2+\nu_E^T c_E(x)+\frac{1}{2}\rho\|c_I^+(x)\|_2^2+\nu_I^T c_I^+(x) \tag{4.176}$$

This function is not derivable when $\rho c_I(x)=-\nu_I$, which raises difficulties.

A second approach is to perform a line search along a projected direction to stay within the domain $s\geq 0$.

Let us note (d_x,d_s) the components of the direction of descent in (x,s).

The d_x component can come from a quasi-Newton method and the d_s component can be the gradient of L_ρ (4.173) : $d_s = -\nabla_s L_\rho = -\left[\rho c_I(x) + \rho s + \nu_I\right]$.
A move of step length α along the direction (d_x, d_s) leads to the point

$$\begin{cases} x' = x + \alpha d_x \\ s' = s + \alpha d_s \end{cases} \tag{4.177}$$

Assume that the $d_{s,1}$ component is negative. The maximum associated step to stay within the domain $s \ge 0$ is: $\alpha_1 = -\dfrac{s_1}{d_{s,1}}$.

Let us also assume that the component $d_{s,1}$ gives the smallest step among all the negative components of d_s. We will then cancel the $d_{s,1}$ component for steps $\alpha \ge \alpha_1$ in order to stay on the bound $s_1 = 0$. The same process is applied to each negative component of d_s by treating them in increasing order of steps α_j. For the positive components of d_s, the step is not bounded $(\alpha_j = \infty)$.
The line search direction is thus defined by the function intervals of α.

$$\begin{array}{llllllll} d_s = (d_{s,1} & d_{s,2} & d_{s,3} & \cdots & d_{s,q-1} & d_{s,q}) & \text{if } 0 \le \alpha < \alpha_1 \\ d_s = (\;0 & d_{s,2} & d_{s,3} & \cdots & d_{s,q-1} & d_{s,q}) & \text{if } \alpha_1 \le \alpha < \alpha_2 \\ d_s = (\;0 & 0 & d_{s,3} & \cdots & d_{s,q-1} & d_{s,q}) & \text{if } \alpha_2 \le \alpha < \alpha_3 \\ \vdots \\ d_s = (\;0 & 0 & 0 & \cdots & 0 & d_{s,q}) & \text{if } \alpha_{q-1} \le \alpha < \alpha_q \end{array} \tag{4.178}$$

The values $s_1, s_2, \cdots$ cancel out together with the components $d_{s1}, d_{s2}, \cdots$ at each interval change. This projected direction makes it possible to perform the line search by staying in the domain $s \ge 0$, and by looking for a step that satisfies for example an Armijo condition.

4.6.4 Algorithm

Each iteration involves minimizing the augmented Lagrangian with respect to x and then updating the multipliers ν or the penalty ρ.
The minimization is performed either by line search or by trust region with the Hessian of the augmented Lagrangian estimated by a quasi-Newton method. This minimization does not need to be accurate during the first iterations.
Let us note (x_k, ν_k) the values of the variables and multipliers at iteration k with a penalty value ρ_k.

By comparing the minimum condition of the augmented Lagrangian (4.162) and the KKT condition (4.160) of the original problem

$$\begin{aligned} &\nabla_x L_\rho(x_k, \nu_k) = 0 \;\Rightarrow\; \nabla f(x_k) + \nabla c(x_k)\left[\nu_k + \rho_k c(x_k)\right] = 0 \\ &\nabla_x L(x^*, \nu^*) \;= 0 \;\Rightarrow\; \nabla f(x^*) + \nabla c(x^*)\nu^* \qquad\qquad = 0 \end{aligned} \tag{4.179}$$

we deduce the following simple strategy for updating the multipliers.

$$\nu_{k+1} = \nu_k + \rho_k c(x_k) \tag{4.180}$$

The decision to update the multipliers ν depends on the constraint values. If the constraints are poorly respected, it is better to increase the penalty coefficient.
In practice, a precision threshold ε_k is set for each iteration on the minimization of the Lagrangian and a tolerance threshold η_k on the respect of the constraints. The minimization of the Lagrangian is performed up to the stopping criterion

$$\left\|\nabla_x L_\rho(x_k, \nu_k)\right\| \le \varepsilon_k \tag{4.181}$$

If the constraints are met with the requested tolerance, the multipliers are updated and the thresholds are reduced without changing the penalty.

$$\left\|c(x_k)\right\| \le \eta_k \quad \rightarrow \quad \begin{cases} \nu_{k+1} = \nu_k + \rho_k c(x_k) \\ \rho_{k+1} = \rho_k \\ \varepsilon_{k+1} = \varepsilon_k / \rho_{k+1} \\ \eta_{k+1} = \eta_k / (\rho_{k+1})^{0,9} \end{cases} \tag{4.182}$$

Otherwise, the penalty coefficient and the thresholds are increased without changing the multipliers.

$$\left\|c(x_k)\right\| > \eta_k \quad \rightarrow \quad \begin{cases} \nu_{k+1} = \nu_k \\ \rho_{k+1} = 10\rho_k \\ \varepsilon_{k+1} = 1/\rho_{k+1} \\ \eta_{k+1} = 1/(\rho_{k+1})^{0,1} \end{cases} \tag{4.183}$$

The above setting coefficients are those of the Lancelot software.
The iterations stop when the thresholds on the constraints and the gradient reach sufficiently small values. The following example illustrates the iterations of an augmented Lagrangian algorithm.

Example 4-13: Augmented Lagrangian algorithm

Consider the optimization problem (presented in [R3]):
$\min_x f(x) = 2(x_1^2 + x_2^2 - 1) - x_1 \quad \text{s.t.} \quad c(x) = x_1^2 + x_2^2 - 1 = 0$

The solution is: $x_1 = 1,\ x_2 = 0,\ \lambda = -1.5$.

Table 4-11 shows the iterations of the augmented Lagrangian algorithm starting from the initial point: $x_1 = 0.5$, $x_2 = 1.3$, $\lambda = 0$. We observe the penalty increases (by a factor of 10), the convergence to 0 of the constraint and the norm of the Lagrangian, and the number of iterations for the minimization in x.

Iter	x_1	x_2	λ	ρ	c(x)	$\|\nabla L_\rho(x,\lambda)\|$	Newton
1	**0.50000**	**1.30000**	**0.00000**	1	-0.71238	0.90050	1
2	0.40707	0.34917	-0.71238	10	-0.05788	0.90016	1
3	0.73467	0.63433	-1.29122	10	-0.00905	0.50091	2
4	0.91556	0.39077	-1.38175	10	0.00635	0.41807	2
5	0.98869	0.16985	-1.38175	100	0.00188	0.62061	2
6	0.99953	0.04158	-1.30283	100	-0.00188	0.01728	2
7	0.99905	-0.00320	-1.49103	100	-0.00009	0.00172	1
8	0.99995	0.00171	-1.50003	100	2.06E-06	0.00057	3
9	**1.00000**	**0.00045**	**-1.50003**	100	1.85E-06	0.00031	

Table 4-11: Iterations of the augmented Lagrangian algorithm.

Figure 4-15 shows the iterations in the plane (x_1,x_2) with a right zoom on the final convergence. The zero-level line of the constraint is the solid circle. The level lines of the cost function are the dashed circles. We observe a first phase of convergence towards the constraint, followed by a progression along the zero-level line to the optimum in (1; 0).

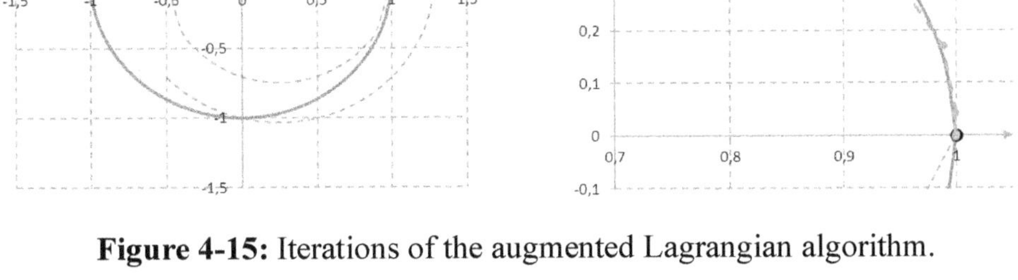

Figure 4-15: Iterations of the augmented Lagrangian algorithm.

For comparison, the SQP algorithm is applied to the same problem from the same starting point. Table 4-12 and figure 4-16 show the iterations of both algorithms. It can be seen that the SQP algorithm tries less to satisfy the constraint in the first iterations and therefore reaches the solution faster.

	SQP			Augmented Lagrangian		
Itér	x_1	x_2	λ	x_1	x_2	λ
1	**0.50000**	**1.30000**	**0.00000**	**0.50000**	**1.30000**	**0.00000**
2	0.59665	0.90129	-1.38660	0.40707	0.34917	-0.71238
3	1.12042	0.46118	-1.70047	0.73467	0.63433	-1.29122
4	1.18366	-0.19988	-1.57065	0.91556	0.39077	-1.38175
5	1.03482	0.02190	-1.52359	0.98869	0.16985	-1.38175
6	1.00084	-0.00103	-1.50118	0.99953	0.04158	-1.30283
7	**1.00000**	**0.00000**	**-1.50000**	0.99905	-0.00320	-1.49103
8				0.99995	0.00171	-1.50003
9				**1.00000**	**0.00045**	**-1.50003**

Table 4-12: Comparison SQP – Augmented Lagrangian.

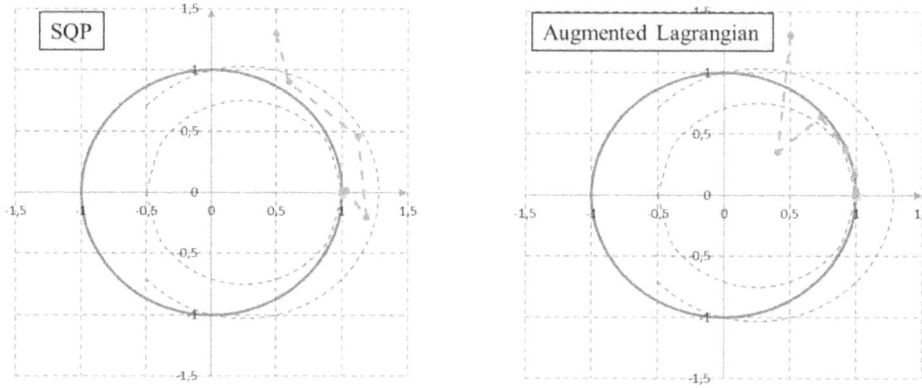

Figure 4-16: SQP – Augmented Lagrangian comparison.

4.7 Conclusion

4.7.1 The key points

- A merit function or a filter is used to compare solutions taking into account the minimization of the cost function and the respect of constraints;
- active constraint methods select saturated inequalities at each iteration and treat them as equalities;
- the penalization methods reduce to an unconstrained problem whose solution approaches (quadratic penalization) or coincides (L_1 penalization) with the exact solution if the penalty is large enough;
- constrained optimization methods approach the problem in its primal (reduced gradient), primal-dual (quadratic sequential, interior point) or dual (augmented Lagrangian) formulation;
- the reduced gradient method has the advantage of yielding feasible iterations. The SQP and interior point methods are faster and well suited to large problems. The augmented Lagrangian methods allow to reduce to a sequence of unconstrained problems.

4.7.2 To go further

- *Programmation mathématique (M. Minoux, Lavoisier 2008, 2e édition)*

  Chapter 5 presents the projected/reduced gradient method and the SQP method. Chapter 6 presents the penalization and augmented Lagrangian methods. Theoretical results are given on the convergence of the algorithms.

- *Introduction à l'optimisation différentiable (M. Bierlaire, Presses polytechniques et universitaires normandes 2006)*

  The projected gradient method is presented in chapter 18. Interior point methods are presented in chapter 19 for linear problems. Chapter 20 presents the augmented Lagrangian methods. Chapter 21 presents the SQP method. Numerous detailed examples illustrate how the algorithms work.

- *Numerical optimization (J. Nocedal, S.J. Wright, Springer 2006)*

  This book is one of the most comprehensive books on nonlinear programming. It details many algorithmic options with practical advices on implementation and settings. The available softwares are reviewed with their strengths and weaknesses and the prospects for the evolution of the various methods are discussed. Quadratic programming is covered in chapter 16, penalization and augmented Lagrangian methods in chapter 17, SQP methods in chapter 18 and interior point methods in chapter 19.

- *Practical optimization (P.E. Gill, W. Murray, M.H. Wright, Elsevier 2004)*

  Chapter 6 is devoted to nonlinear programming methods: penalization (section 6.2), projected gradient and reduced gradient (section 6.3), augmented Lagrangian (section 6.4), sequential quadratic programming (section 6.5). Chapter 21 presents the SQP method. A lot of practical advice is given in chapters 7 and 8 concerning the implementation of the algorithms and the numerical difficulties that can hamper convergence.

5. Linear programming

Linear programming concerns problems with linear cost function and constraints. These problems are of great practical importance, which justifies the development of specific algorithms.

Section 1 presents the simplex method. The theoretical study of a linear problem shows that the solution has at least a certain number of zero variables. The simplex algorithm exploits this property to perform an ordered search among all possible combinations. The solution is found after a finite number of trials, usually much less than the number of possible combinations. Although the number of trials can become exponential in the worst cases, the simplex method performs extremely well on most applications, even with large numbers of variables. The dual simplex method is based on the same algorithm applied to the dual linear problem. Its interest is to be able to start from a known solution when the problem is modified by the addition of a constraint. The complementary simplex method is an adaptation for solving quadratic-linear convex problems.

Section 2 presents the interior point methods. Contrary to the simplex method, which seeks the solution directly at the boundary of the feasible domain, these methods seek to approach the solution continuously from the interior of the domain. Their principle is to solve the optimality conditions of Karush, Kuhn and Tucker by a Newton method, taking care at each iteration not to approach the boundary of the feasible domain. These methods are more complex to manage than the simplex method, but they are competitive on large problems and have polynomial complexity (whereas the simplex has exponential complexity in the worst cases).

5.1 Simplex

The simplex method is specific to linear programming problems. This method developed by G. Dantzig in the late 1940s exploits the properties of the solution of a linear problem to reduce the search domain. It has proven to be extremely efficient on most applications. Improvements made since 1950 have extended its field of application to large problems and to mixed linear programming.

5.1.1 Standard form

The standard form of a linear programming problem is as follows:

$$\min_{x \in \mathbb{R}^n} z = c^T x \quad \text{s.t.} \begin{cases} Ax = b \\ x \geq 0 \end{cases} \quad \text{with } A \in \mathbb{R}^{m \times n}, b \in \mathbb{R}^m, c \in \mathbb{R}^n \tag{5.1}$$

The cost function and the constraints are linear. The variables x are positive.

The number m of constraints is less than the number n of variables and the matrix A is assumed to have full rank: $\text{rang}(A) = m \leq n$. These assumptions ensure the existence of feasible solutions. The problem (5.1) is denoted by (LP).
Any linear programming problem can be put into the standard form (5.1) using the following transformations:

- an inequality constraint is transformed into an equality by introducing a positive slack variable:

$$a^T x \leq b \Leftrightarrow \begin{cases} a^T x + x' = b \\ x' \geq 0 \end{cases} \Leftrightarrow \begin{cases} \begin{pmatrix} a^T & 1 \end{pmatrix} \begin{pmatrix} x \\ x' \end{pmatrix} = b \\ x' \geq 0 \end{cases} \tag{5.2}$$

- a lower bound x_l is treated by changing the variable, an upper bound x_u is treated by introducing a positive slack variable:

$$\begin{aligned} x_l \leq x \leq x_u \Leftrightarrow 0 \leq x - x_l \leq x_u - x_l \Leftrightarrow & \begin{cases} x' = x - x_l\,, & x' \geq 0 \\ x' \leq x_u - x_l \end{cases} \\ \Leftrightarrow & \begin{cases} x' = x - x_l\,, & x' \geq 0 \\ x' + x'' = x_u - x_l\,, & x'' \geq 0 \end{cases} \end{aligned} \tag{5.3}$$

- a free variable is expressed as the difference of two positive variables:

$$x \in \mathbb{R} \Leftrightarrow x = x'' - x' \quad \text{with } x' \geq 0\,, x'' \geq 0 \tag{5.4}$$

Example 5-1: Putting a linear problem into standard form

Consider the linear problem (LP)

$$\min_{x_1,x_2,x_3} x_1+2x_2+3x_3 \quad \text{s.t.} \begin{cases} -x_1+3x_2 = 5 \\ 2x_1 - x_2+3x_3 \geq 6 \\ x_1 \in \mathbb{R},\ x_2 \geq 1,\ x_3 \leq 4 \end{cases}$$

This problem is put into standard form in two stages:

- treatment of the bounds on the variables:

$$\begin{cases} x_1 \in \mathbb{R} \\ x_2 \geq 1 \\ x_3 \leq 4 \end{cases} \Leftrightarrow \begin{cases} x_1 = x_1" - x_1' & \to x_1', x_1" \geq 0 \\ x_2' = x_2 - 1 & \to x_2' \geq 0 \\ x_3' = 4 - x_3 & \to x_3' \geq 0 \end{cases}$$

- treatment of the inequality constraints:

$$2x_1 - x_2 + 3x_3 \geq 6 \Leftrightarrow 2x_1 - x_2 + 3x_3 - x_2" = 6 \quad \to x_2" \geq 0$$

This gives the standard form (LP') equivalent to the original problem (LP).

$$\min_{x_1',x_1",x_2",x_2',x_3'} x_1" - x_1' + 2x_2' - 3x_3' + 14 \quad \text{s.t.} \begin{cases} x_1' - x_1" + 3x_2' = 2 \\ -2x_1' + 2x_1" - x_2' - 3x_3' - x_2" = -5 \end{cases}$$

This problem has 5 positive variables and 2 equality constraints.

The feasible domain defined by the linear constraints of problem (LP) forms a **polytope** denoted P in $\mathbb{R}^n$. A polytope is the n-dimensional generalization of a polygon (n = 2) or a polyhedron (n = 3).

$$P = \left\{x \in \mathbb{R}^n \ / \ Ax = b,\ x \geq 0\right\} \tag{5.5}$$

The polytope P is a subspace of $\mathbb{R}^n$ of dimension n−m (n variables and m linear equations). We call a **vertex** of the polytope P is any point x of the polytope that cannot be expressed as a convex linear combination of two other points of P. In geometric terms, the point x does not lie on any segment joining two points of the polytope as illustrated in figure 5-1 in dimension 2.

$$x \text{ vertex of } P \Leftrightarrow \begin{cases} \forall y \in P, z \in P \\ \forall \alpha,\ 0 < \alpha < 1 \end{cases},\ x \neq \alpha y + (1-\alpha)z \tag{5.6}$$

The calculation of the vertices uses the notion of basis as will be explained in section 5.1.2

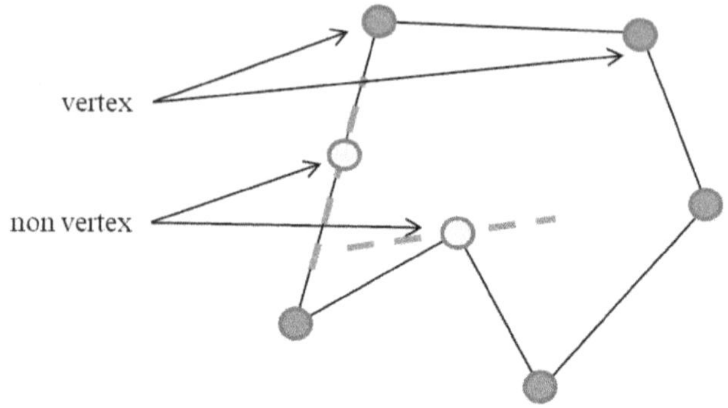

Figure 5-1:
Vertices of a polytope.

Since the cost function $z = c^T x$ is linear on $\mathbb{R}^n$, we can guess that it will reach its minimum in a vertex. Property 5-1 clarifies this idea.

Property 5-1: Optimal vertex

If problem (LP) has a solution, then there exists an optimal vertex.

Demonstration

Assume that problem (LP) has a finite cost solution f*.
Consider the set Q of points of the polytope P such that: $c^T x = f^*$.
Q is a sub-polytope of P defined by: $Q = \left\{ x \in \mathbb{R}^n \,/\, Ax = b \,,\, c^T x = f^* \,,\, x \geq 0 \right\}$.
Let x* be a vertex of Q: x* is then a solution of problem (LP).

Assume by contradiction that x* is not a vertex of P.
Then x* is a convex linear combination of 2 points y and z of P distinct from x*:
$x^* = \alpha y + (1-\alpha) z$ with $0 < \alpha < 1$ (**equation E**)
f* being by hypothesis the minimum of the cost function on P, we have

$$\begin{cases} c^T x^* \leq c^T y \Rightarrow \alpha c^T y + (1-\alpha) c^T z \leq c^T y \Rightarrow c^T z \leq c^T y \\ c^T x^* \leq c^T z \Rightarrow \alpha c^T y + (1-\alpha) c^T z \leq c^T z \Rightarrow c^T y \leq c^T z \end{cases} \quad \text{because } 0 < \alpha < 1$$

from which we can deduce: $c^T y = c^T z$.

By replacing in the equation E, we get: $c^T x^* = c^T y = c^T z = f^*$.
The points y and z distinct from x* therefore belong to Q. They satisfy the equation E, which contradicts the initial assumption that x* is a vertex of Q.
Therefore x* (which is a solution of the LP problem) is a vertex of P.

5.1.2 Basis

A **basis** of the matrix $A \in \mathbb{R}^{n \times m}$ is an invertible square submatrix $B \in \mathbb{R}^{m \times m}$ formed from m columns of the matrix A. Such a matrix exists because the matrix A is assumed to have full rank m.

By permuting the columns of A, it is assumed that the matrix B is formed from the first m columns of A. The remaining n−m columns form the **non-basis** matrix noted $N \in \mathbb{R}^{m \times (n-m)}$. The splitting of A into the basis matrix B and the non-basis matrix N leads to a partition of the variables $x \in \mathbb{R}^n$ into basic **variables** $x_B \in \mathbb{R}^m$ and **non-basic variables** $x_N \in \mathbb{R}^{n-m}$ as illustrated in figure 5-2.

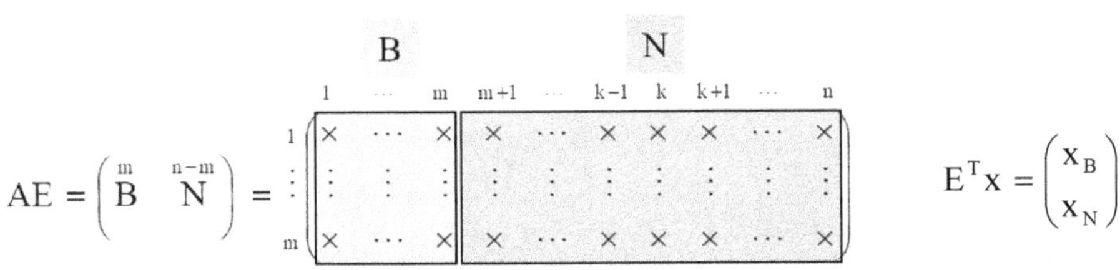

Figure 5-2: Matrix and basic/non-basic variables.

The matrix $E \in \mathbb{R}^{n \times n}$ represents the permutation to place the basis columns in the first position. A permutation of the columns j and k of the matrix A is obtained by post-multiplying A by the permutation matrix $E_{jk} \in \mathbb{R}^{n \times n}$ shown in figure 5-3. The matrix E is the product of matrices of the form E_{jk}.

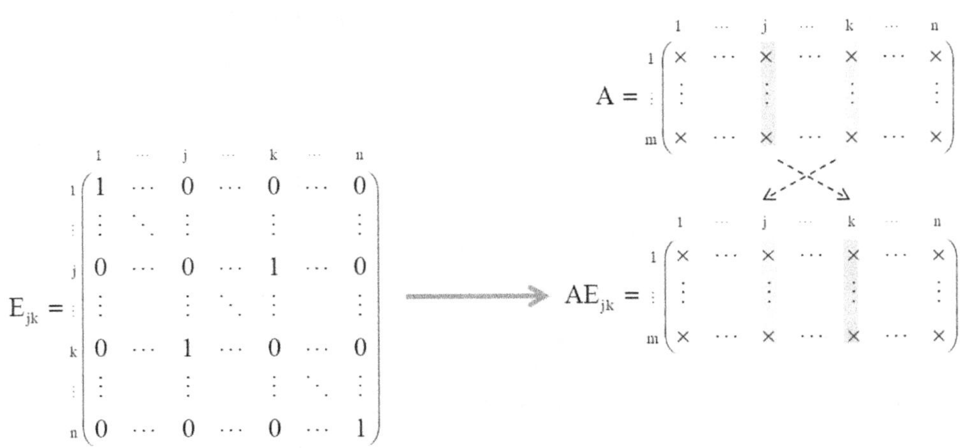

Figure 5-3: Column permutation matrix.

Let us now introduce the concept of a basis solution associated with the matrix B.

Basis solution

The basic variables are also called **linked** or **dependent** variables and the non-basic variables are also called **free** or **independent** variables. Indeed, the equality constraints give the relation (the matrix B being invertible)

$$Ax = b \;\Leftrightarrow\; Bx_B + Nx_N = b \;\Leftrightarrow\; x_B = B^{-1}(b - Nx_N) \tag{5.7}$$

We can therefore choose arbitrarily the values of x_N and deduce the values of x_B that allow the equality constraints to be satisfied. The point x obtained by setting $x_N = 0$ is called the **basis solution** associated with B, which gives

$$E^T x = \begin{pmatrix} B^{-1}b \\ 0 \end{pmatrix} \tag{5.8}$$

A basis B is said to be **feasible** if the corresponding basis solution satisfies the constraints $x \geq 0$ of the polytope P (5.5).

$$B \text{ feasible} \;\Leftrightarrow\; B^{-1}b \geq 0 \tag{5.9}$$

The property 5-2 relates the basis solutions and the vertices of the polytope.

Property 5-2: Vertices and basis solutions

x is a vertex of the polytope P $\Leftrightarrow$ x is a feasible basis solution.

Demonstration of the implication $\Leftarrow$

Let x be a feasible basis solution: $E^T x = \begin{pmatrix} x_B = B^{-1}b \geq 0 \\ x_N = 0 \end{pmatrix}$.

Let us assume by contradiction that x is not a vertex of P.
Then x is expressed as a convex linear combination of two points y and z of P distinct from x: $x = \alpha y + (1-\alpha)z$ with $0 < \alpha < 1$.

We express the basic and non-basic components: $\begin{cases} x_B = \alpha y_B + (1-\alpha)z_B \\ x_N = \alpha y_N + (1-\alpha)z_N \end{cases}$.

The equation: $x_N = \alpha y_N + (1-\alpha) z_N = 0$ with $x_N = 0$ (by assumption on x) and with $y_N, z_N \geq 0$ (because $y, z \in P$) leads to: $y_N = z_N = 0$.

The equation: $Ay = b$ (because $y \in P$) leads to: $By_B + Ny_N = b \Rightarrow y_B = B^{-1}b$ because $y_N = 0$.

The equation: $Az = b$ (because $z \in P$) leads to: $Bz_B + Nz_N = b \Rightarrow z_B = B^{-1}b$ because $z_N = 0$.

We obtain: $y = z = \begin{pmatrix} B^{-1}b \\ 0 \end{pmatrix} = x$, which contradicts the assumption (y and z distinct from x). x is therefore a vertex of P.

Demonstration of the implication $\Rightarrow$

Let x be a vertex of P with p non-zero components which are assumed to be the first p (by means of a permutation): $E^T x = \begin{pmatrix} \overset{1}{x_1} \cdots \overset{p}{x_p} & \overset{p+1}{0} \cdots\cdots \overset{n}{0} \end{pmatrix}^T$.

Let us assume by contradiction that the first p columns of A (denoted $A_{.1}, \cdots, A_{.p}$) are dependent. Then there exists a linear combination with non-zero coefficients $\lambda_1, \cdots, \lambda_p$ such that: $\lambda_1 A_{.1} + \cdots + \lambda_p A_{.p} = 0$.

Let us form the point $x' = x + \varepsilon d$, with d defined by: $E^T d = \begin{pmatrix} \overset{1}{\lambda_1} \cdots \overset{p}{\lambda_p} & \overset{p+1}{0} \cdots\cdots \overset{n}{0} \end{pmatrix}^T$.

This point satisfies: $Ax' = A(x + \varepsilon d) = Ax + \varepsilon Ad = b + \varepsilon(\lambda_1 A_{.1} + \cdots + \lambda_p A_{.p}) = b$.

For ε small, this point satisfies also: $x' \geq 0$ ($x_1, \cdots, x_p$ are strictly positive).

The point x' is therefore a point of the polytope. By choosing a step $\varepsilon_y < 0$, then a step $\varepsilon_z > 0$, we can thus construct two points y and z of P aligned with x and on either side of x, which is in contradiction with the assumption that x is a vertex.

Consequently, the columns $A_{.1}, \cdots, A_{.p}$ are independent and the number p of non-zero components of x is less than m (since the columns $A_{.k}$ have dimension m).

We can then form a basis B by completing the p columns $A_{.1}, \cdots, A_{.p}$ with $m-p$ columns chosen from the remaining columns of A (this is always possible, as the matrix A is of rank m).

The feasible point x (because $x \in P$) is then a basis solution associated to B (because its non-basic components are zero), which completes the demonstration.

Property 5-2 indicates that at least $n-m$ components (corresponding to non-basic variables) of a vertex are zero. If, on the other hand, the basic variables are zero, the basis is said to be **degenerate**.

The relation between vertex and feasible basis is not a bijection, as several feasible bases can define the same vertex as shown in the following example.

Example 5-2: Computing the vertices of a polytope (presented in [R3])

Consider the following polytope P in standard form in $\mathbb{R}^4$:

$$P = \{(x_1, x_2, x_3, x_4) \,/\, Ax = b,\ x \geq 0\} \text{ with } A = \begin{pmatrix} 1 & 1 & 1 & 0 \\ 1 & -1 & 0 & 1 \end{pmatrix},\ b = \begin{pmatrix} 1 \\ 1 \end{pmatrix}$$

In order to visualize the vertices of P, we express x_3 and x_4 in terms of x_1 and x_2.

$$Ax = b \Leftrightarrow \begin{cases} x_1 + x_2 + x_3 = 1 \\ x_1 - x_2 + x_4 = 1 \end{cases} \Leftrightarrow \begin{cases} x_3 = 1 - x_1 - x_2 \\ x_4 = 1 - x_1 + x_2 \end{cases}$$

$$x \geq 0 \Rightarrow \begin{cases} x_3 \geq 0 \\ x_4 \geq 0 \end{cases} \Rightarrow \begin{cases} x_1 + x_2 \leq 1 \\ x_1 - x_2 \leq 1 \end{cases}$$

A reduced polytope P' in $\mathbb{R}^2$ is thus defined.

$$P' = \left\{ \begin{pmatrix} x_1 \\ x_2 \end{pmatrix} / \begin{cases} x_1 + x_2 \leq 1 \\ x_1 - x_2 \leq 1 \end{cases}, \begin{cases} x_1 \geq 0 \\ x_2 \geq 0 \end{cases} \right\}$$

This polytope forms the triangle ABC shown in figure 5-4.

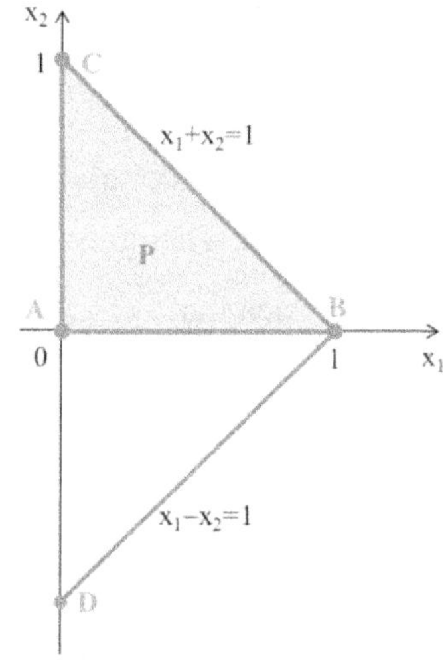

Figure 5-4: Reduced polytope in $\mathbb{R}^2$.

The polytope P is the standard form of the reduced polytope P'. Thanks to the reduction in $\mathbb{R}^2$, we can visualize the points (x_1, x_2) associated with the vertices of the polytope P. Let us now examine all the bases of P.

Choosing a basis consists in choosing two independent columns of the matrix A. There are at most six possible bases. The associated basis solution is obtained by setting the two non-basic variables to 0 and calculating the two basic variables to satisfy the equality constraints. The basis is feasible if positive basic variables are obtained. The point is then a vertex.

- Basis associated with (x_1, x_2)

  $$B = \begin{pmatrix} 1 & 1 \\ 1 & -1 \end{pmatrix}, B^{-1} = \begin{pmatrix} 0,5 & 0,5 \\ 0,5 & -0,5 \end{pmatrix}, x_B = \begin{pmatrix} 1 \\ 0 \end{pmatrix} \rightarrow x = \begin{pmatrix} 1 & 0 & 0 & 0 \end{pmatrix}^T$$

  $\rightarrow$ feasible basis (point B)

- Basis associated with (x_1, x_3)

  $$B = \begin{pmatrix} 1 & 1 \\ 1 & 0 \end{pmatrix}, \quad B^{-1} = \begin{pmatrix} 0 & 1 \\ 1 & -1 \end{pmatrix}, \quad x_B = \begin{pmatrix} 1 \\ 0 \end{pmatrix} \rightarrow x = \begin{pmatrix} 1 & 0 & 0 & 0 \end{pmatrix}^T$$

  $\rightarrow$ feasible basis (point B)

- Basis associated with (x_1, x_4)

  $$B = \begin{pmatrix} 1 & 0 \\ 1 & 1 \end{pmatrix}, \quad B^{-1} = \begin{pmatrix} 1 & 0 \\ -1 & 1 \end{pmatrix}, \quad x_B = \begin{pmatrix} 1 \\ 0 \end{pmatrix} \rightarrow x = \begin{pmatrix} 1 & 0 & 0 & 0 \end{pmatrix}^T$$

  $\rightarrow$ feasible basis (point B)

- Basis associated with (x_2, x_3)

  $$B = \begin{pmatrix} 1 & 1 \\ -1 & 0 \end{pmatrix}, B^{-1} = \begin{pmatrix} 0 & -1 \\ 1 & 1 \end{pmatrix}, \quad x_B = \begin{pmatrix} -1 \\ 2 \end{pmatrix} \rightarrow x = \begin{pmatrix} 0 & -1 & 2 & 0 \end{pmatrix}^T$$

  $\rightarrow$ non-feasible basis (point D)

- Basis associated with (x_2, x_4)

  $$B = \begin{pmatrix} 1 & 0 \\ -1 & 1 \end{pmatrix}, B^{-1} = \begin{pmatrix} 1 & 0 \\ 1 & 1 \end{pmatrix}, \quad x_B = \begin{pmatrix} 1 \\ 2 \end{pmatrix} \rightarrow x = \begin{pmatrix} 0 & 1 & 0 & 2 \end{pmatrix}^T$$

  $\rightarrow$ non-feasible basis (point C)

- Basis associated with (x_3, x_4)

  $$B = \begin{pmatrix} 1 & 0 \\ 0 & 1 \end{pmatrix}, \quad B^{-1} = \begin{pmatrix} 1 & 0 \\ 0 & 1 \end{pmatrix}, \quad x_B = \begin{pmatrix} 1 \\ 1 \end{pmatrix} \rightarrow x = \begin{pmatrix} 0 & 0 & 1 & 1 \end{pmatrix}^T$$

  $\rightarrow$ feasible basis (point A)

This yields a total of 5 feasible bases:
- 1 for the vertex A;
- 3 for the vertex B;
- 1 for the vertex C.

Property 5-1 states that there exists at least one optimal vertex and property 5-2 shows that we can determine all the vertices of the polytope by browsing the C_n^m possible bases. This systematic enumeration makes it possible to solve the problem (LP) with certainty, as in example 5-3 presented in [R3].

Example 5-3: Resolution by systematic enumeration of bases

Consider the linear problem in standard form

$$\min_{x_1,x_2,x_3,x_4} z=-x_1-2x_2 \text{ s.t. } \begin{cases} x_1+x_2+x_3=1 \\ x_1-x_2+x_4=1 \\ x_1,x_2,x_3,x_4 \geq 0 \end{cases}$$

The constraints (identical to those in example 5-2) are represented by the polytope P' in $\mathbb{R}^2$: $P' = \left\{ \begin{pmatrix} x_1 \\ x_2 \end{pmatrix} / \begin{cases} x_1+x_2 \leq 1 \\ x_1-x_2 \leq 1 \end{cases}, \begin{cases} x_1 \geq 0 \\ x_2 \geq 0 \end{cases} \right\}$.

Figure 5-5 shows the polytope P' and the level lines of the cost function.
Table 5-1 shows the enumeration of the six bases.
The best basis is the C vertex.

Basis	x	z
x_1,x_2	$(1\ 0\ 0\ 0) \rightarrow B$	−1
x_1,x_3	$(1\ 0\ 0\ 0) \rightarrow B$	−1
x_1,x_4	$(1\ 0\ 0\ 0) \rightarrow B$	−1
x_2,x_3	$(0\ -1\ 2\ 0) \rightarrow D$	Not feasible
x_2,x_4	**$(0\ 1\ 0\ 2) \rightarrow C$**	**−2**
x_3,x_4	$(0\ 0\ 1\ 1) \rightarrow A$	0

Table 5-1: List of bases.

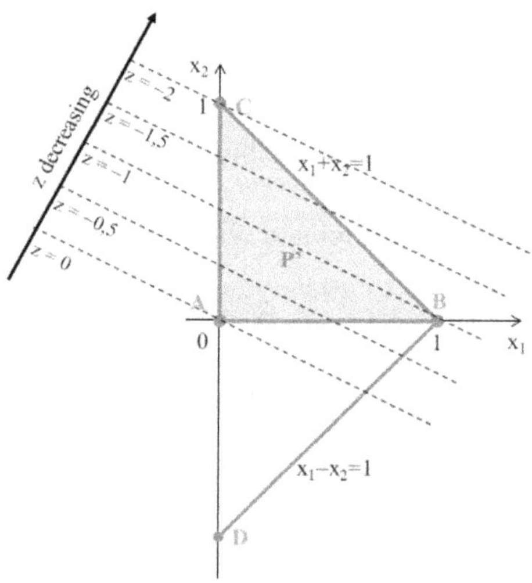

Figure 5-5: Graphical solution.

A systematic enumeration becomes unfeasible as soon as the dimension of the problem increases. A more efficient search method is to progress from vertex to vertex by choosing vertices of decreasing cost.
The move between vertices is based on the notion of basis direction.

Basis direction

Let us place ourselves in a vertex x associated with a feasible basis B.

$$E^T x = \begin{pmatrix} x_B \\ x_N \end{pmatrix} = \begin{pmatrix} B^{-1}b \\ 0 \end{pmatrix} \geq 0 \tag{5.10}$$

A move $d \in \mathbb{R}^n$ from x with components d_B (basis) and d_N (non-basis) is feasible if $x+d$ belongs to the polytope, which leads to the conditions

$$x+d \in P \quad \Leftrightarrow \quad \begin{cases} A(x+d)=b \\ x+d \geq 0 \end{cases} \quad \Leftrightarrow \quad \begin{cases} Ad=0 \\ x+d \geq 0 \end{cases} \quad \Leftrightarrow \quad \begin{cases} d_B = -B^{-1}Nd_N \\ x_B + d_B \geq 0 \\ d_N \geq 0 \end{cases} \tag{5.11}$$

The non-basic components d_N must be positive, as the non-basic variables x_N are zero. A component $d_{Nj} < 0$ would move outside the polytope.

Let us choose the non-basic components d_N all zero except the one associated with the k-th non-basic variable set to 1.

$$E^T \begin{pmatrix} 0 \\ d_N \end{pmatrix} = \begin{pmatrix} \overset{1}{0} & \cdots & \overset{k-1}{0} & \overset{k}{1} & \overset{k+1}{0} & \cdots & \overset{n}{0} \end{pmatrix}^T \underset{\text{noted}}{=} e_k \tag{5.12}$$

e_k denotes the k-th vector of the canonical basis of $\mathbb{R}^n$.

The basic components d_B are calculated by (5.11) to satisfy: $Ad = 0$.

Denoting $A_{.k}$ the k-th column of the matrix AE and d_j the components of d_N (all worth zero except d_k), we obtain

$$d_B = -B^{-1}Nd_N = -B^{-1} \sum_{j \text{ non-basic}} A_{.j} d_j = -B^{-1} A_{.k} \tag{5.13}$$

The k-th **basis direction** in x is the direction d^k defined by (5.12) and (5.13).

$$E^T d^k = \begin{pmatrix} d_B^k \\ d_N^k \end{pmatrix} = \begin{pmatrix} -B^{-1}A_{.,k} \\ 0 \end{pmatrix} + e_k = \begin{pmatrix} \overset{1}{d_{B1}} \cdots \overset{m}{d_{Bm}} & \overset{m+1}{0} \cdots \overset{k-1}{0} & \overset{k}{1} & \overset{k+1}{0} \cdots \overset{n}{0} \end{pmatrix}^T \tag{5.14}$$

Consider the point $x + sd^k$, where $s > 0$ is a step length along d^k.

The basis direction d^k is said to be feasible if the point $x + sd^k$ belongs to the polytope for s sufficiently small. In other words, a small move along d^k is possible without leaving the polytope of constraints.

This requires that the basic components remain positive.

$$x_B + sd_B \geq 0 \tag{5.15}$$

Two situations are possible depending on whether the basis is degenerate or not:
- if the basis is non-degenerate ($x_B > 0$), one can always find s small satisfying (5.15) and all basis directions are feasible;
- if the basis is degenerate, the components of d_B corresponding to the zero basic variables must be positive: $x_{Bj} = 0 \;\Rightarrow\; d_{Bj} \ge 0$. Non-feasible basis directions may then exist.

The example 5-4 illustrates the procedure for calculating basis directions.

Example 5-4: Calculation of basis directions (presented in [R3])

Let us retrieve the polytope P of example 5-2.

$$P = \left\{(x_1, x_2, x_3, x_4) \,/\, Ax = b,\; x \ge 0\right\} \;\text{ with } A = \begin{pmatrix} 1 & 1 & 1 & 0 \\ 1 & -1 & 0 & 1 \end{pmatrix},\; b = \begin{pmatrix} 1 \\ 1 \end{pmatrix}$$

To calculate the basis directions in a vertex associated with a basis B:
- we set one component of d_N to 1 (variable number k) and the others to 0;
- we calculate d_B by (5.13) : $d_B = -B^{-1}A_{.k}$;
- if the basis is degenerate, we check if the direction is feasible $x_{Bj} = 0 \Rightarrow d_{Bj} \ge 0$.

In the case of the polytope P, there are 2 basis directions in each vertex associated with the 2 non-basic variables.
Let us look at three feasible bases and their basis directions:

- the feasible basis associated with (x_2, x_4) corresponds to the vertex C.

$$B = \begin{pmatrix} 1 & 0 \\ -1 & 1 \end{pmatrix},\; B^{-1} = \begin{pmatrix} 1 & 0 \\ 1 & 1 \end{pmatrix},\quad x_B = \begin{pmatrix} 1 \\ 2 \end{pmatrix} \;\rightarrow\; x = \begin{pmatrix} 0 & 1 & 0 & 2 \end{pmatrix}^T$$

The 1st basis direction d^1 is associated with the non-basic variable x_1.

$$d_B = -B^{-1}A_{.1} = -\begin{pmatrix} 1 & 0 \\ 1 & 1 \end{pmatrix}\begin{pmatrix} 1 \\ 1 \end{pmatrix} = \begin{pmatrix} -1 \\ -2 \end{pmatrix} \;\rightarrow\; d^1 = \begin{pmatrix} 1 & -1 & 0 & -2 \end{pmatrix}^T$$

Since the basis is non-degenerate, the direction d^1 is feasible.

The 2nd basis direction d^3 is associated with the non-basic variable x_3.

$$d_B = -B^{-1}A_{.3} = -\begin{pmatrix} 1 & 0 \\ 1 & 1 \end{pmatrix}\begin{pmatrix} 1 \\ 0 \end{pmatrix} = \begin{pmatrix} -1 \\ -1 \end{pmatrix} \;\rightarrow\; d^3 = \begin{pmatrix} 0 & -1 & 1 & -1 \end{pmatrix}^T$$

Since the basis is non-degenerate, the direction d^3 is feasible;

- the feasible basis associated with (x_1, x_4) corresponds to the vertex B.

$$B = \begin{pmatrix} 1 & 0 \\ 1 & 1 \end{pmatrix}, \quad B^{-1} = \begin{pmatrix} 1 & 0 \\ -1 & 1 \end{pmatrix}, \quad x_B = \begin{pmatrix} 1 \\ 0 \end{pmatrix} \quad \rightarrow x = \begin{pmatrix} 1 & 0 & 0 & 0 \end{pmatrix}^T$$

<u>The 1st basis direction</u> d^2 is associated with the non-basic variable x_2.

$$d_B = -B^{-1}A_{.2} = -\begin{pmatrix} 1 & 0 \\ -1 & 1 \end{pmatrix}\begin{pmatrix} 1 \\ -1 \end{pmatrix} = \begin{pmatrix} -1 \\ 2 \end{pmatrix} \quad \rightarrow d^2 = \begin{pmatrix} -1 & 1 & 0 & 2 \end{pmatrix}^T$$

The basis is degenerate ($x_4 = 0$). We must have: $d_4^2 \geq 0$, which is the case.
The direction d^2 is feasible.

<u>The 2nd basis direction</u> d^3 is associated with the non-basic variable x_3.

$$d_B = -B^{-1}A_{.3} = -\begin{pmatrix} 1 & 0 \\ -1 & 1 \end{pmatrix}\begin{pmatrix} 1 \\ 0 \end{pmatrix} = \begin{pmatrix} -1 \\ 1 \end{pmatrix} \quad \rightarrow d^3 = \begin{pmatrix} -1 & 0 & 1 & 1 \end{pmatrix}^T$$

The basis is degenerate ($x_4 = 0$). We must have: $d_4^3 \geq 0$, which is the case.
The direction d^3 is feasible;

- the feasible basis associated with $(x_1 ; x_2)$ corresponds to the vertex B.

$$B = \begin{pmatrix} 1 & 1 \\ 1 & -1 \end{pmatrix}, B^{-1} = \begin{pmatrix} 0,5 & 0,5 \\ 0,5 & -0,5 \end{pmatrix}, x_B = \begin{pmatrix} 1 \\ 0 \end{pmatrix} \quad \rightarrow x = \begin{pmatrix} 1 & 0 & 0 & 0 \end{pmatrix}^T$$

<u>The 1st basis direction</u> d^3 is associated with the non-basic variable x_3.

$$d_B = -B^{-1}A_{.3} = -\begin{pmatrix} 0,5 & 0,5 \\ 0,5 & -0,5 \end{pmatrix}\begin{pmatrix} 1 \\ 0 \end{pmatrix} = \begin{pmatrix} -0,5 \\ -0,5 \end{pmatrix} \quad \rightarrow d^3 = \begin{pmatrix} -0,5 & -0,5 & 1 & 0 \end{pmatrix}^T$$

The basis is degenerate ($x_2 = 0$). We must have: $d_2^3 \geq 0$, which is not the case.
The direction d^3 is not feasible.

<u>The 2nd basis direction</u> d^4 is associated with the non-basic variable x_4.

$$d_B = -B^{-1}A_{.4} = -\begin{pmatrix} 0,5 & 0,5 \\ 0,5 & -0,5 \end{pmatrix}\begin{pmatrix} 0 \\ 1 \end{pmatrix} = \begin{pmatrix} -0,5 \\ 0,5 \end{pmatrix} \quad \rightarrow d^4 = \begin{pmatrix} -0,5 & 0,5 & 0 & 1 \end{pmatrix}^T$$

The basis is degenerate ($x_2 = 0$). We must have: $d_2^4 \geq 0$, which is the case.
The direction d^4 is feasible.

Figure 5-6 shows the basis directions in each vertex.
Feasible directions follow the edges of the polytope, while non feasible directions leave the polytope.

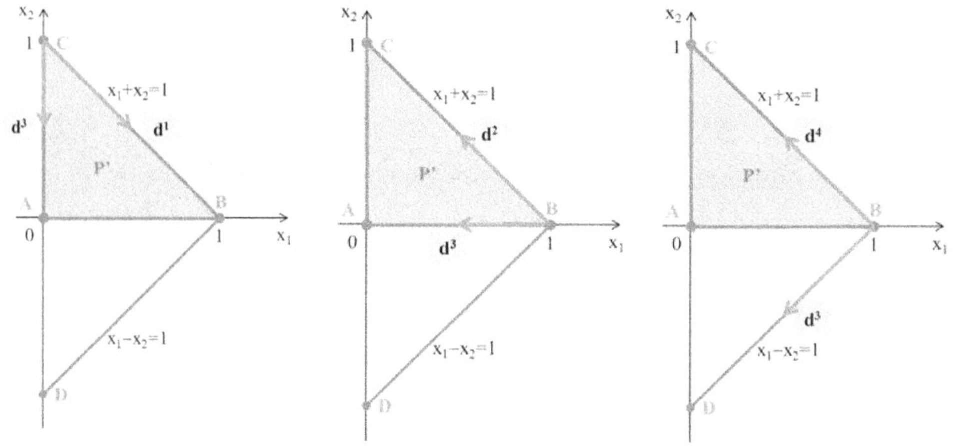

Figure 5-6: Basis directions for the bases (x_2 , x_4) , (x_1 , x_4) and (x_1 , x_2).

The basis directions have the property 5-3.

> **Property 5-3:** Feasible directions
>
> Any feasible direction from a vertex x can be expressed as a linear combination of the basis directions at x.

Demonstration

Let $d \in \mathbb{R}^n$ be a feasible direction of components d_B (basic) and d_N (non-basic).
Note α_j the components of d_N in the canonical basis of $\mathbb{R}^n$: $d_N = \sum\limits_{j\ \text{non-basic}} \alpha_j e_j$.

d being feasible, the components of d_B are: $d_B = -B^{-1} N d_N = B^{-1} \sum\limits_{j\ \text{non-basic}} \alpha_j A_{.j}$.

By grouping the components, we obtain: $d = \begin{pmatrix} d_B \\ d_N \end{pmatrix} = \sum\limits_{j\ \text{non-basic}} \alpha_j \left[\begin{pmatrix} B^{-1} A_{.j} \\ 0 \end{pmatrix} + e_j \right]$

which is a linear combination of the basis directions defined by (5.14).

Solving the linear problem (5.1) requires assessing a series of basis solutions with their associated basis directions. Putting the problem in canonical form allows to simplify significantly this sequence of calculations.

Canonical form

Let us retrieve the linear problem in standard form.

$$\min_{x \in \mathbb{R}^n} z = c^T x \quad \text{s.t.} \begin{cases} Ax = b \\ x \geq 0 \end{cases} \quad \text{with } A \in \mathbb{R}^{m \times n}, b \in \mathbb{R}^m, c \in \mathbb{R}^n \tag{5.16}$$

Take a basis $B \in \mathbb{R}^{m \times m}$ and split the variables into basic x_B and non-basic x_N.

$$AE = \begin{pmatrix} B & N \end{pmatrix}, \; x = E\begin{pmatrix} x_B \\ x_N \end{pmatrix} \quad \rightarrow \begin{cases} Ax = Bx_B + Nx_N \\ c^T x = c_B^T x_B + c_N^T x_N \end{cases} \tag{5.17}$$

Linear constraints allow for the elimination of basic variables

$$Ax = b \;\; \Rightarrow Bx_B + Nx_N = b \;\; \Rightarrow x_B = B^{-1}(b - Nx_N) \tag{5.18}$$

and lead to a problem with n – m variables and no equality constraints

$$\min_{x_N \in R^{n-m}} z = c_B^T B^{-1} b + (c_N^T - c_B^T B^{-1} N) x_N \quad \text{with} \begin{cases} x_B = B^{-1}(b - Nx_N) \geq 0 \\ x_N \geq 0 \end{cases} \tag{5.19}$$

This reduced problem is called the **canonical form of** the problem (LP).

To simplify its formulation, the terms $\bar{b}, \bar{z}, \bar{c}_N$ are defined.

$$\min_{\substack{x_N \in R^{n-m} \\ x_N \geq 0}} z = \bar{z} + \bar{c}_N^T x_N \quad \text{with } x_B = \bar{b} - B^{-1} N x_N \geq 0 \;\; \text{and} \begin{cases} \bar{b} = B^{-1} b \\ \bar{z} = c_B^T \bar{b} \\ \bar{c}_N^T = c_N^T - c_B^T B^{-1} N \end{cases} \tag{5.20}$$

The unknowns of the problem in canonical form are the non-basic variables x_N.

We know that the solution to the problem is in a vertex (property 5-1) and that the vertices correspond to the basis solutions defined by $x_N = 0$ (property 5-2).
The interest of the canonical form (5.20) is to show directly the quantities associated with the basis B:

- the vector $\bar{b} \in \mathbb{R}^m$ gives the value of the basic variables x_B;
- the real $\bar{z}$ gives the value of the cost function z;
- the vector $\bar{c}_N \in \mathbb{R}^{n-m}$ called **reduced cost** indicates if the solution is optimal.

Property 5-4: Reduced cost and optimality

If all reduced costs in a basis B are positive or zero, then the basis solution associated with B is optimal.

Demonstration

Let us apply a move (d_B, d_N) to the feasible basis solution (x_B, x_N).

The point $(x_B + d_B\,;\, x_N + d_N)$ is feasible if: $\begin{cases} d_N \geq 0 & \text{because } x_N = 0 \\ d_B = -B^{-1}Nd_N & \text{because } A(x+d) = b \\ d_{Bj} \geq 0 & \text{if } x_{Bj} = 0 \end{cases}$.

The change in cost is: $\delta z = c^T d = c_B^T d_B + c_N^T d_N = -c_B^T B^{-1} N d_N + c_N^T d_N = \bar{c}_N^T d_N$.

If $\bar{c}_N \geq 0$, no feasible move $(d_N \geq 0)$ can decrease the cost.

The reduced cost $\bar{c}_N$ is the gradient of the reduced cost function: $z(x_N) = \bar{z} + \bar{c}_N^T x_N$

It should strictly speaking be called the reduced gradient.

We observe that if a single non-basic component (associated with a variable x_k) is varied, then the direction of move $(d_B = -B^{-1}Nd_N\,,\, d_N)$ is the basis direction d^k defined in (5.14). The components of the vector $\bar{c}_N \in \mathbb{R}^{n-m}$ are therefore the **directional derivatives** along the n – m basis directions.

5.1.3 Pivoting

If the reduced cost associated with a non-basic variable x_k is negative, then the basis is not optimal and the basis direction d^k is a descent direction.

If we move along this direction:

- non-basic variables other than x_k remain zero (and thus remain non-basic);
- the variable x_k takes on a positive non-zero value (and thus enters the basis);
- the initially positive basic variables are modified according to d_B.

Pivoting consists in moving as far as possible along the direction d^k.
The move is bounded by the constraints: $x_B \geq 0$.
Two situations are possible:

- if no component of d_B is negative, then the move is not bounded and the linear problem has no solution;
- otherwise, the variables x_{Bj} for which $d_{Bj} < 0$ decrease. The move is bounded by the first basic variable cancelling along the direction d^k. This variable will be removed from the basis and replaced by the variable x_k.

Pivoting leads to a vertex **adjacent** to the starting vertex. These two adjacent vertices are linked by an edge of the polytope directed along d^k and their bases differ only by one variable (x_k).

Let us now look at the choice of the entering variable, the choice of the leaving variable and the basis change formulas (called pivoting formulas).

Choice of the entering variable

The first stage of pivoting is to select a non-basic variable that will enter the basis. The variable entering the basis is to be selected among the variables having a negative reduced cost. Several selection criteria are possible:

- choose the most negative reduced cost variable (**Dantzig's 1st rule**), which corresponds to the direction of steepest descent;
- choose the variable with the smallest index (**Bland's rule**, 1977), which avoids possible cycling (infinite loop) when the basis is degenerate;
- choose bases in ascending lexicographical order (according to their column numbers) in case of degeneracy;
- choose the variable that produces the greatest reduction in cost after pivoting, which requires evaluating all possibilities;
- choose the variable randomly with a probability proportional to the reduced cost.

Despite the risk of cycling in case of degeneracy, Dantzig 1st rule is generally the most effective. However, this is not always the case.

For example, consider the figure 5-7 with a polytope in $\mathbb{R}^2$ and the cost function defined by: $z(x_1,x_2)=-x_2$. The initial vertex is at the top, the solution is the whole horizontal segment at the bottom (degenerate solution).

Moving to the left gives the steepest descent (Dantzig's 1st rule) and also a greater reduction in cost (rule 4 above) than moving to the right. However, this initial choice will require 12 pivoting to arrive at the solution, whereas 2 pivoting are sufficient by going to the right. It is in fact impossible to determine the best choice for pivoting for sure.

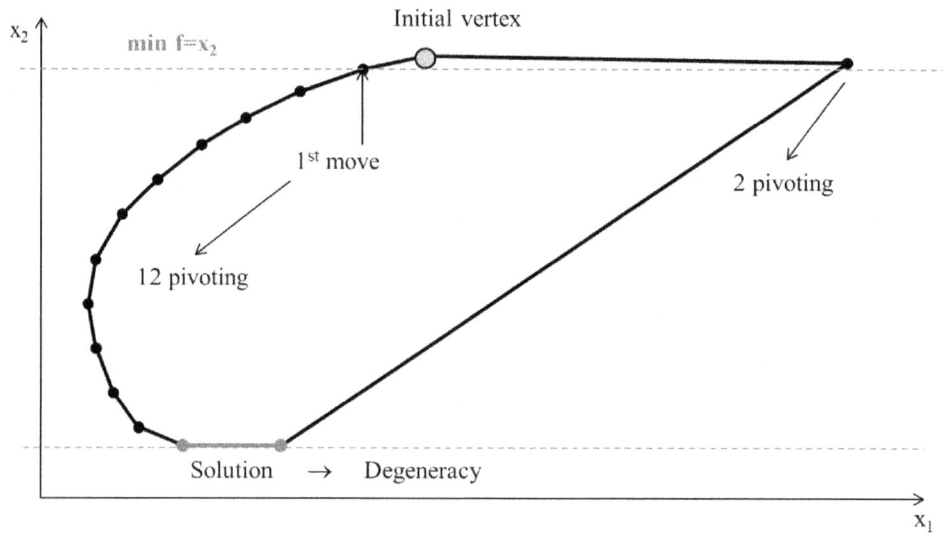

Figure 5-7: Pivoting sequence leading to the optimum.

Choice of the leaving variable

The second stage of pivoting is to select the basic variable that will leave the basis. With the entering variable x_e chosen, the leaving variable is necessarily the first one that cancels along the associated basis direction d^e.

This constitutes **Dantzig's 2nd rule**.

The canonical form (5.20) expresses the basic variables in terms of the non-basic variables. Noting $B^{-1}N \underset{\text{noted}}{=} (\bar{a}_{ij})_{i\in B, j\in N}$, we have

$$x_i = \bar{b}_i - \sum_{j\in N} \bar{a}_{ij} x_j \geq 0 \quad , \quad i \in B \tag{5.21}$$

In the basis direction d^e (5.14), all non-basic variables remain zero except for x_e which becomes positive. Expression (5.21) simplifies to

$$x_i = \bar{b}_i - \bar{a}_{ie} x_e \geq 0 \quad , \quad i \in B \tag{5.22}$$

Note s_{ie} the maximum value that the variable x_e can take while satisfying the constraint linked to any basic variable x_i :

- if $\bar{a}_{ie} > 0$, then: $s_{ie} = \dfrac{\bar{b}_i}{\bar{a}_{ie}}$;
- if $\bar{a}_{ie} \leq 0$, then: $s_{ie} = +\infty$ (unbounded move).

The maximum feasible step along the direction d^e is then: $s_e = \min\limits_{i\in B} s_{ie} = \min\limits_{\substack{i\in B \\ \bar{a}_{ie}>0}} \dfrac{\bar{b}_i}{\bar{a}_{ie}}$.

The leaving variable x_s is the one corresponding to the maximum step s_e.

<u>Pivoting formulas</u>

Let us take the canonical form (5.20) in the basis B, noting $B^{-1}N \underset{\text{noted}}{=} (\bar{a}_{ij})_{i\in B, j\in N}$.

$$\min_{x_N \geq 0} z = \bar{z} + \sum_{j\in N} \bar{c}_j x_j \quad \text{s.t.} \quad x_i = \bar{b}_i - \sum_{j\in N} \bar{a}_{ij} x_j \geq 0 \quad , \quad i \in B \tag{5.23}$$

By abuse of notation, B and N (basis and non-basis matrices) also refer to the basis and non-basis index sets.
Pivoting exchanges a basic variable x_s and a non-basic variable x_e.
This change of basis is written as

$$\begin{aligned} B' &= B - \{s\} + \{e\} \\ N' &= N - \{e\} + \{s\} \end{aligned} \tag{5.24}$$

After pivoting, the canonical form in the new basis B' becomes

$$\min_{x_N \geq 0} z = \bar{z}' + \sum_{j\in N'} \bar{c}_j' x_j \quad \text{s.t.} \quad x_i = \bar{b}_i' - \sum_{j\in N'} \bar{a}_{ij}' x_j \geq 0 \quad , \quad i \in B' \tag{5.25}$$

The pivoting formulas express the values $\bar{b}', \bar{a}', \bar{z}', \bar{c}_N'$ in the new basis B' in terms of the values $\bar{b}, \bar{a}, \bar{z}, \bar{c}_N$ in the old basis B. They are as follows:

- terms $\bar{b}', \bar{a}'$ associated with the new basic variable x_e

$$i = e \rightarrow \begin{cases} \bar{b}_e' = \dfrac{\bar{b}_s}{\bar{a}_{se}} \\ \bar{a}_{es}' = \dfrac{1}{\bar{a}_{se}} & \rightarrow j = s \\ \bar{a}_{ej}' = \dfrac{\bar{a}_{sj}}{\bar{a}_{se}} & \rightarrow j \in N - \{e\} \end{cases} \tag{5.26}$$

- terms $\bar{b}', \bar{a}'$ associated with the other basic variables x_i, $i \in B - \{s\}$

$$i \in B - \{s\} \rightarrow \begin{cases} \bar{b}_i' = \bar{b}_i - \dfrac{\bar{a}_{ie}}{\bar{a}_{se}} \bar{b}_s \\ \bar{a}_{is}' = -\dfrac{\bar{a}_{ie}}{\bar{a}_{se}} & \rightarrow j = s \\ \bar{a}_{ij}' = \bar{a}_{ij} - \dfrac{\bar{a}_{ie}}{\bar{a}_{se}} \bar{a}_{sj} & \rightarrow j \in N - \{e\} \end{cases} \tag{5.27}$$

- terms $\bar{z}', \bar{c}_N'$

$$\begin{cases} \bar{z}' = \bar{z} + \bar{c}_e \dfrac{\bar{b}_s}{\bar{a}_{se}} \\ \bar{c}_s' = -\dfrac{\bar{c}_e}{\bar{a}_{se}} & \rightarrow j = s \\ \bar{c}_j' = \bar{c}_j - \bar{c}_e \dfrac{\bar{a}_{sj}}{\bar{a}_{se}} & \rightarrow j \in N - \{e\} \end{cases} \tag{5.28}$$

Demonstration

Formulas (5.26)

Let us start with the expression of the selected basic variable x_s in terms of the initial non-basic variables N.

$x_s = \bar{b}_s - \sum_{j \in N} \bar{a}_{sj} x_j = \bar{b}_s - \bar{a}_{se} x_e - \sum_{j \in N - \{e\}} \bar{a}_{sj} x_j$ from which we deduce

$$x_e = \frac{\bar{b}_s}{\bar{a}_{se}} - \frac{1}{\bar{a}_{se}} x_s - \sum_{j \in N - \{e\}} \frac{\bar{a}_{sj}}{\bar{a}_{se}} x_j$$

We express x_e in terms of the final non-basic variables N': $x_e = \bar{b}_e' - \sum_{j \in N'} \bar{a}_{ej}' x_j$.

By identification, we obtain the formulas (5.26).

Formulas (5.27)

Let us similarly express the other basic variables x_i in terms of the initial non-basic variables N.

$$x_i = \bar{b}_i - \sum_{j \in N} \bar{a}_{ij} x_j = \bar{b}_s - \bar{a}_{ie} x_e - \sum_{j \in N-\{e\}} \bar{a}_{ij} x_j$$

We replace x_e using (5.26)

$$x_e = \bar{b}_e' - \sum_{j \in N'} \bar{a}_{ej}' x_j = \frac{1}{\bar{a}_{se}} \bar{b}_s - \frac{1}{\bar{a}_{se}} x_s - \sum_{j \in N-\{e\}} \frac{\bar{a}_{sj}}{\bar{a}_{se}} x_j$$

which yields for x_i

$$x_i = \bar{b}_i - \frac{\bar{a}_{ie}}{\bar{a}_{se}} \bar{b}_s + \frac{\bar{a}_{ie}}{\bar{a}_{se}} \bar{b}_s x_s - \sum_{j \in N-\{e\}} \left(\bar{a}_{ij} - \frac{\bar{a}_{ie}}{\bar{a}_{se}} \bar{a}_{sj} \right) x_j$$

We express x_i in terms of the final non-basic variables N': $x_i = \bar{b}_i' - \sum_{j \in N'} \bar{a}_{ij}' x_j$.

By identification, we obtain the formulas (5.27).

Formulas (5.28)

Let us express the cost in terms of the initial non-basic variables N, using (5.26) to replace x_e .

$$z = \bar{z} + \sum_{j \in N} \bar{c}_j x_j = \bar{z} + \bar{c}_e x_e + \sum_{j \in N-\{e\}} \bar{c}_j x_j = \bar{z} + \bar{c}_e \left(\frac{1}{\bar{a}_{se}} \bar{b}_s - \frac{1}{\bar{a}_{se}} x_s - \sum_{j \in N-\{e\}} \frac{\bar{a}_{sj}}{\bar{a}_{se}} x_j \right) + \sum_{j \in N-\{e\}} \bar{c}_j x_j$$

We express the cost in terms of the final non-basic variables N': $z = \bar{z}' + \sum_{j \in N'} \bar{c}_j' x_j$

By identification, we obtain the formulas (5.28).

The pivoting formulas allow us to go from the canonical form in the basis B to the canonical form in the basis $B' = B - \{s\} + \{e\}$. To apply them in a simple way, the calculations are arranged in a table called the simplex array.

5.1.4 Simplex array

The canonical form of the linear problem in the basis B

$$\min_{x_N \geq 0} z = \bar{z} + \bar{c}_N^T x_N \quad \text{s.t.} \quad x_B + B^{-1} N x_N = \bar{b} \geq 0 \tag{5.29}$$

is summarized in an array called the **simplex array** (or **simplex tableau**).

$$T_E = \begin{array}{c} \begin{array}{cc} x_B & x_N \end{array} \\ \begin{array}{|cc|c|} \hline I & B^{-1}N & \bar{b} \\ \hline 0 & \bar{c}_N^T & -\bar{z} \\ \hline \end{array} \end{array} \quad \begin{array}{l} \to x_B + B^{-1} N x_N = \bar{b} \\ \to \bar{c}_N^T x_N = z - \bar{z} \end{array} \tag{5.30}$$

The notation T_E indicates that a permutation of columns has been performed in order to place the basic variables in the first columns.

This array has m + 1 rows and n + 1 columns:
- the first m rows express the constraints with their matrix and second member;
- the row m + 1 expresses the cost function: $z - \bar{z}$;
- the columns 1 to m correspond to the basic variables x_B;
- the columns m + 1 to n correspond to the non-basic variables x_N.

The important elements of the array are the last column and the last row.
The last column shows the values of the basic variables $(x_B = \bar{b})$ and the opposite of the cost value $(z = -\bar{z})$ associated with the basis solution $(x_N = 0)$.
The last row shows the reduced costs, which allow to know whether the solution is optimal (positive or zero costs) or to select the entering variable for the next pivoting (negative cost).

In the array T_E (5.30), the basic variables correspond to the first m columns. This permutation is not useful in practice and the initial order of the variables of the linear problem is kept. The array T is then presented in the form

$$T = \begin{array}{|c|c|} \hline B^{-1}A & \bar{b} \\ \hline \bar{c}^T & -\bar{z} \\ \hline \end{array} = \begin{array}{c} \begin{array}{ccccc} x_1 & & x_j & & x_n \end{array} \\ \begin{array}{|ccccc|c|} \hline B^{-1}A_{.1} & \cdots & B^{-1}A_{.j} & \cdots & B^{-1}A_{.n} & \bar{b} \\ \hline \bar{c}_1 & \cdots & \bar{c}_j & \cdots & \bar{c}_n & -\bar{z} \\ \hline \end{array} \end{array} \tag{5.31}$$

where $A_{.k}$ is the j-th column of the matrix A.

Let us now detail the simplex algorithm on the T array.

Stage 0: create the initial array

This initialization assumes that a feasible initial basis is known and that the problem is put in canonical form (5.29) in this basis. Section 5.1.5 presents the method for determining a feasible initial basis.
The canonical form allows the array T_E to be filled in directly. By resetting the basic and non-basic variables to their initial indexes, we obtain the array T.

Stage 1: recognize basic and non-basic variables

In the array T, the basic variables are easily recognized, as their columns are those of the identity matrix and their reduced costs are zero. Their values $\bar{b}$ can be read in the last column.

Stage 2: select the entering non-basic variable x_e

The variable is selected from among those with negative reduced cost (last line). In general, the variable with the most negative cost is chosen (Dantzig's 1st rule) or the first in their natural order (Bland's rule).

Stage 3: calculate the maximum feasible step

We place ourselves on the column of the entering variable x_e and calculate for each row the associated maximum move: $s_{ie} = \dfrac{\bar{b}_i}{\bar{a}_{ie}}$ if $\bar{a}_{ie} > 0$.

Stage 4: select the leaving basic variable x_s

This variable is the one giving the smallest feasible step: $s_e = \min_{i \in B} s_{ie}$.

The value $\bar{a}_{se}$ is called the pivot (by analogy with Gauss pivot method for solving a linear system).

Stage 5: update the array associated with the new basis

The row s (associated with the variable x_s) is divided by the pivot $\bar{a}_{se}$ to make the value 1 appear instead of the pivot. Each row (including the last row of the reduced costs) is combined with the row of the pivot to make zeros appear in column e. As a result of these operations, column e is a column of the identity matrix, which reflects the entry of the variable x_e into the basis. Its reduced cost is zero and its value appears in the last column. This stage is illustrated below.

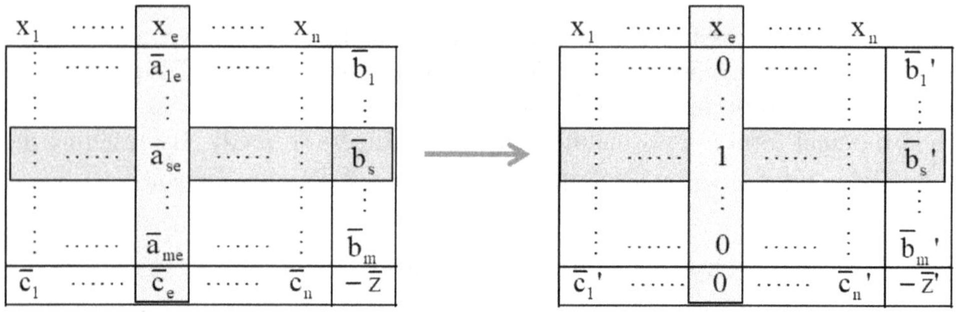

Figure 5-8: Pivoting on the simplex array.

Stages 1 to 5 should be repeated as long as there are negative reduced costs. The last table is the optimal basis. It gives the values of the basic variables in the last column and the opposite of the cost at the bottom of the last column

The example 5-5 details the iterations of the simplex algorithm with the updating of the array at each pivoting.

Example 5-5: Simplex array

Consider the linear problem with 3 variables (x_1, x_2, x_3)

$$\min_{x_1,x_2,x_3} -10x_1 - 12x_2 - 12x_3 \quad \text{s.t.} \begin{cases} x_1 + 2x_2 + 2x_3 \le 20 \\ 2x_1 + x_2 + 2x_3 \le 20 \\ 2x_1 + 2x_2 + x_3 \le 20 \\ x_1, x_2, x_3 \ge 0 \end{cases}$$

The problem is put into standard form with 3 slack variables (x_4, x_5, x_6).

$$\min_{x_1,x_2,x_3,x_4,x_5,x_6} -10x_1 - 12x_2 - 12x_3 \quad \text{s.t.} \begin{cases} x_1 + 2x_2 + 2x_3 + x_4 = 20 \\ 2x_1 + x_2 + 2x_3 + x_5 = 20 \\ 2x_1 + 2x_2 + x_3 + x_6 = 20 \\ x_1, x_2, x_3, x_4, x_5, x_6 \ge 0 \end{cases}$$

This problem is in canonical form in the basis of the variables (x_4, x_5, x_6).

Indeed, these variables (x_4, x_5, x_6) appear in the constraint rows with an identity matrix and they do not appear in the cost function. Moreover, the basis solution $(x_4, x_5, x_6) = (20, 20, 20)$ is feasible, because the variables are positive.

We thus have the double "chance" of having directly a canonical form with a feasible basis.
The initial array is filled with the 3×6 matrix of constraints, their second members in the last column, the reduced costs in the last row, and the cost function cell initialized to 0 (corresponding to the basis solution $x_4 = x_5 = x_6 = 0$).

Initial array

Basis: (x_4, x_5, x_6)

x_1	x_2	x_3	x_4	x_5	x_6		
1	2	2	1	0	0	20	x_4
2	1	2	0	1	0	20	x_5
2	2	1	0	0	1	20	x_6
-10	-12	-12	0	0	0	0	-z

The basis is not optimal, as there exists negative reduced costs. Pivoting is applied with Bland's rule (first variable of negative reduced cost).

Pivoting 1

Current basis: (x_4, x_5, x_6) → Entering variable: x_1

Limit step on (x_4, x_5, x_6) : $s_{15} = 10$ → Leaving variable: x_5

x_1	x_2	x_3	x_4	x_5	x_6			Step
1	2	2	1	0	0	20	x_4	→ s_{14}=20
2	1	2	0	1	0	20	x_5	→ s_{15}=10
2	2	1	0	0	1	20	x_6	→ s_{16}=10
-10	-12	-12	0	0	0	0	z	

Pivot: $\bar{a}_{51} = 2$ → Division of row 2 by $\bar{a}_{51}$

Elimination in column 1 → New basis (x_4, x_1, x_6) / cost $z = -100$

x_1	x_2	x_3	x_4	x_5	x_6		
0	1.5	1	1	-0.5	0	10	x_4
1	0.5	1	0	0.5	0	10	x_1
0	1	-1	0	-1	1	0	x_6
0	-7	-2	0	5	0	100	z

<u>Pivoting 2</u>

Current basis: (x_4, x_1, x_6) → Entering variable: x_2

Limit step on (x_4, x_1, x_6) : $s_{26} = 0$ → Leaving variable: x_6

x_1	x_2	x_3	x_4	x_5	x_6			Step
0	1.5	1	1	-0.5	0	10	x_4	→ s_{24}=20/3
1	0.5	1	0	0.5	0	10	x_1	→ s_{21}=20
0	1	-1	0	-1	1	0	x_6	→ s_{26}=0
0	-7	-2	0	5	0	100	z	

Pivot: $\bar{a}_{62} = 1$ → Division of row 3 by $\bar{a}_{62}$

Elimination in column 2 → New basis (x_4, x_1, x_2) / cost $z = -100$

x_1	x_2	x_3	x_4	x_5	x_6		
0	0	2.5	1	1	-1.5	10	x_4
1	0	1.5	0	1	-0.5	10	x_1
0	1	-1	0	-1	1	0	x_2
0	0	-9	0	-2	7	100	z

<u>Pivoting 3</u>

Current basis: (x_4, x_1, x_2) → Entering variable: x_3

Limit step on (x_4, x_1, x_2) : $s_{34} = 4$ → Leaving variable: x_4

x_1	x_2	x_3	x_4	x_5	x_6			Step
0	0	2.5	1	1	-1.5	10	x_4	→ s_{34}=4
1	0	1.5	0	1	-0.5	10	x_1	→ s_{31}=20/3
0	1	-1	0	-1	1	0	x_2	→ s_{32}=+∞
0	0	-9	0	-2	7	100	z	

Pivot: $\bar{a}_{43} = 2.5$ → Division of row 1 by $\bar{a}_{43}$

Elimination in column 3 → New basis (x_3, x_1, x_2) / cost $z = -136$

x_1	x_2	x_3	x_4	x_5	x_6		
0	0	1	0.4	0.4	-0.6	4	x_3
1	0	0	-0.6	0.4	0.4	4	x_1
0	1	0	0.4	-0.6	0.4	4	x_2
0	0	0	3.6	1.6	1.6	136	z

All reduced costs are positive or zero: the basis (x_3, x_1, x_2) is optimal.
The resulting solution is: $(x_1, x_2, x_3, x_4, x_5, x_6) = (4,4,4,0,0,0)$ of cost $z = -136$.
It can be observed that the optimal basis (x_3, x_1, x_2) only contains the original variables of the problem (no slack variables). If this is not the case, additional pivoting can be performed (without cost modification) to obtain a basis with only the original variables.
We also notice that the basis at the beginning of the 2nd pivoting is degenerate $(x_6 = 0)$, which leads to a zero step. A possible cycling (infinite loop) may occur in this situation depending on the entering selection rule. Bland's rule based on the variable index avoids any cycling.

Table 5-2 summarizes the iterations with the successive pivoting.

k	B	c			x_1	x_2	x_3	x_4	x_5	x_6	d_B			s_{max}	e	s	z
0	4 5 6	-10	-12	-12	0	0	0	20	20	20	-1	-2	-2	10	1	5	0
1	1 4 6	-7	-2	5	10	0	0	10	0	0	-1.5	-0.5	-1	0	2	6	-100
2	1 2 4	-9	-2	7	10	0	0	10	0	0	-2.5	-1.5	1	4	3	4	-100
3	1 2 3	3.6	31.6	1.6	4	4	4	0	0	0							-136

Table 5-2: Summary of simplex iterations.

In example 5-5, we had the double "chance" of having a problem already in canonical form associated to a positive basis solution. In practice, there is no reason for the linear problem to be naturally in this form. Section 5.1.5 presents the method of constructing a feasible basis.

5.1.5 Auxiliary problem

Let us retrieve the linear problem in standard form.

$$\min_{x\in\mathbb{R}^n} z = c^T x \quad \text{s.t.} \begin{cases} Ax = b \\ x \geq 0 \end{cases} \quad \text{with} \quad A \in \mathbb{R}^{m\times n}, b \in \mathbb{R}^m, c \in \mathbb{R}^n \tag{5.32}$$

The simplex method requires to know a feasible basis to fill the initial array. To construct such a basis, we consider the following linear problem called **auxiliary problem**.

$$\min_{\substack{x\in\mathbb{R}^n \\ y\in\mathbb{R}^m}} 0^T x + e^T y \quad \text{s.t.} \begin{cases} Ax + y = b \\ x, y \geq 0 \end{cases} \tag{5.33}$$

The m new variables $y \in \mathbb{R}^m$ are positive and associated with the m constraints.

The cost function is the sum of the new variables y: $e^T = (1,1,\cdots,1) \in \mathbb{R}^m$.

Assume that x_0 is a feasible solution of the linear problem (5.32).

Then $(x = x_0 \,;\, y = 0)$ is a feasible solution of the auxiliary problem (5.33). This zero-cost solution is optimal for problem (5.33), because the cost function of (5.33) is zero (which is its minimum achievable value). Solving the auxiliary problem (5.33) thus provides a feasible solution of problem (5.32) if the optimal cost is zero, or indicates that there is no feasible solution if the cost is non-zero.

Let us now try to put the problem (5.33) into canonical form. Using the constraints to express the cost function

$$Ax + y = b \quad \Rightarrow \quad e^T y = e^T b - e^T Ax \tag{5.34}$$

the auxiliary problem can be reformulated as

$$\min_{\substack{x\in\mathbb{R}^n \\ y\in\mathbb{R}^m}} e^T b - e^T Ax \quad \text{s.t.} \begin{cases} Ax + y = b \\ x, y \geq 0 \end{cases} \tag{5.35}$$

We observe that this problem is already in canonical form in the basis associated with the y variables. Indeed, these y variables appear with an identity matrix in the constraints and they do not appear in the cost function. For this basis to be feasible, the solution of the associated basis $y = b$ must be positive.

We can always return to the case $b \geq 0$. Indeed, if the component b_i is negative, we can rewrite the constraint i by changing the sign of the two members.

$$\sum_{j=1}^{n} a_{ij} x_j = b_i \quad \Leftrightarrow \quad \sum_{j=1}^{n} (-a_{ij}) x_j = -b_i \tag{5.36}$$

Since the auxiliary problem (5.35) is in canonical form in the feasible basis formed by the variables y, we can apply the simplex algorithm directly to it starting with the initial array

$$T = \begin{array}{c} \quad x \qquad\quad y \qquad\qquad \\ \begin{array}{|cc|c|} \hline A & I & b \\ \hline -e^T A & 0 & -e^T b \\ \hline \end{array} \end{array} \tag{5.37}$$

The following example shows the process of determining a feasible initial basis by solving the auxiliary problem.

Example 5-6: Solving the auxiliary problem using the simplex algorithm

Consider the linear problem with 3 variables (x_1, x_2, x_3):

$$\min_{x_1,x_2,x_3,x_4} \; x_1 + x_2 + x_3 \;\text{ s.t. } \begin{cases} x_1 + 2x_2 + 3x_3 & = 3 \\ -x_1 + 2x_2 + 6x_3 & = 2 \\ \quad 4x_2 + 9x_3 & = 5 \\ \qquad 3x_3 + x_4 & = 1 \\ x_1, x_2, x_3, x_4 \geq 0 \end{cases}$$

The auxiliary problem is formulated with 4 additional variables (y_1, y_2, y_3, y_4).

$$\min_{\substack{x_1,x_2,x_3,x_4 \\ y_1,y_2,y_3,y_4}} \; y_1 + y_2 + y_3 + y_4 \;\text{ s.t. } \begin{cases} x_1 + 2x_2 + 3x_3 + y_1 & = 3 \\ -x_1 + 2x_2 + 6x_3 + y_2 & = 2 \\ \quad 4x_2 + 9x_3 + y_3 & = 5 \\ \qquad 3x_3 + x_4 + y_4 & = 1 \\ x_1, x_2, x_3, x_4, y_1, y_2, y_3, y_4 \geq 0 \end{cases}$$

This problem is reformulated by expressing $y_1 + y_2 + y_3 + y_4$ from the constraints.

$$\min_{\substack{x_1,x_2,x_3,x_4 \\ y_1,y_2,y_3,y_4}} \; -8x_2 - 21x_3 - x_4 + 11 \;\text{ s.t. } \begin{cases} x_1 + 2x_2 + 3x_3 + y_1 & = 3 \\ -x_1 + 2x_2 + 6x_3 + y_2 & = 2 \\ \quad 4x_2 + 9x_3 + y_3 & = 5 \\ \qquad 3x_3 + x_4 + y_4 & = 1 \\ x_1, x_2, x_3, x_4, y_1, y_2, y_3, y_4 \geq 0 \end{cases}$$

This gives a canonical form in the basis (y_1, y_2, y_3, y_4).

The basis solution $(y_1, y_2, y_3, y_4) = (3, 2, 5, 1) \geq 0$ is feasible.

This allows the initial simplex array to be filled in directly.

Initial array

Basis: (y_1, y_2, y_3, y_4)

x_1	x_2	x_3	x_4	y_1	y_2	y_3	y_4		
1	2	3	0	1	0	0	0	3	y_1
-1	2	6	0	0	1	0	0	2	y_2
0	4	9	0	0	0	1	0	5	y_3
0	0	3	1	0	0	0	1	1	y_4
0	-8	-21	-1	0	0	0	0	-11	-z

Pivoting is applied with Bland's rule (first variable of negative reduced cost).

Pivoting 1

Current basis: (y_1, y_2, y_3, y_4) / Entering variable: x_2 / Leaving variable: y_2

x_1	x_2	x_3	x_4	y_1	y_2	y_3	y_4		
1	2	3	0	1	0	0	0	3	y_1
-1	2	6	0	0	1	0	0	2	y_2
0	4	9	0	0	0	1	0	5	y_3
0	0	3	1	0	0	0	1	1	y_4
0	-8	-21	-1	0	0	0	0	-11	-z

x_1	x_2	x_3	x_4	y_1	y_2	y_3	y_4		
2	0	-3	0	1	-1	0	0	1	y_1
-1/2	1	3	0	0	1/2	0	0	1	x_2
2	0	-3	0	0	-2	1	0	1	y_3
0	0	3	1	0	0	0	1	1	y_4
-4	0	3	-1	0	4	0	0	-3	-z

Pivoting 2

Current basis: (y_1, x_2, y_3, y_4) / Entering variable: x_1 / Leaving variable: y_1

x_1	x_2	x_3	x_4	y_1	y_2	y_3	y_4		
2	0	-3	0	1	-1	0	0	1	y_1
-1/2	1	3	0	0	1/2	0	0	1	x_2
2	0	-3	0	0	-2	1	0	1	y_3
0	0	3	1	0	0	0	1	1	y_4
-4	0	3	-1	0	4	0	0	-3	-z

x_1	x_2	x_3	x_4	y_1	y_2	y_3	y_4		
1	0	-3/2	0	1/2	-1/2	0	0	1/2	x_1
0	1	9/4	0	1/4	1/4	0	0	5/4	x_2
0	0	0	0	-1	-1	1	0	0	y_3
0	0	3	1	0	0	0	1	1	y_4
0	0	-3	-1	2	2	0	0	-1	-z

Pivoting 3

Current basis: (x_1, x_2, y_3, y_4) / Entering variable: x_3 / Leaving variable: y_4

x_1	x_2	x_3	x_4	y_1	y_2	y_3	y_4		
1	0	-3/2	0	1/2	-1/2	0	0	1/2	x_1
0	1	9/4	0	1/4	1/4	0	0	5/4	x_2
0	0	0	0	-1	-1	1	0	0	y_3
0	0	3	1	0	0	0	1	1	y_4
0	0	-3	-1	2	2	0	0	-1	-z

x_1	x_2	x_3	x_4	y_1	y_2	y_3	y_4		
1	0	0	1/2	1/2	-1/2	0	1/2	1	x_1
0	1	0	-3/4	1/4	1/4	0	-3/4	1/2	x_2
0	0	0	0	-1	-1	1	0	0	y_3
0	0	1	1/3	0	0	0	1/3	1/3	x_3
0	0	0	0	2	2	0	1	0	-z

The reduced costs are non-negative: the basis (x_1, x_2, y_3, x_3) is optimal for the auxiliary problem. The solution $(x_1, x_2, x_3, x_4, y_1, y_2, y_3, y_4) = \left(1, \frac{1}{2}, \frac{1}{3}, 0, 0, 0, 0, 0\right)$ of cost $z = 0$ corresponds to a feasible solution of the original problem.
We see in the third row that the coefficients of the x variables are all zero. The third constraint present in the initial problem is thus redundant (it is indeed the sum of the first two constraints) and the matrix of constraints is not of full rank. The resolution of the auxiliary problem has revealed this redundancy.
By removing this constraint and the column associated with the additional variable y_3, we obtain the solution array of the auxiliary problem in the basis (x_1, x_2, x_3). This basis is feasible for the original linear problem.

Solution array

Current basis: (x_1, x_2, x_3)

x_1	x_2	x_3	x_4	y_1	y_2	y_4		
1	0	0	1/2	1/2	-1/2	1/2	1	x_1
0	1	0	-3/4	1/4	1/4	-3/4	1/2	x_2
0	0	1	1/3	0	0	1/3	1/3	x_3
0	0	0	0	2	2	1	0	-z

In the case where the constraints are not redundant, additional pivoting is applied to exchange the remaining y variables with non-basic x variables. Such pivoting corresponds to zero moves which do not change the obtained solution and allow to obtain a feasible basis of the initial linear problem. The complete process is described in section 5.1.6.

5.1.6 Two-phase method

Solving the auxiliary problem gives a feasible solution of the original problem. The next stage is to move from the solution array of the auxiliary problem to the array of the original problem.

The solution array is not directly usable for two reasons:

- the solution basis of the auxiliary problem may contain variables y that are not part of the original problem;
- the cost row of the auxiliary problem is not that of the original problem.

The process of shifting from the solution array of the auxiliary problem to the initial array of the original problem is shown in figure 5-9. The crosses show the parts of the array to be deleted (figure 5-9a) or completed (figure 5-9b).

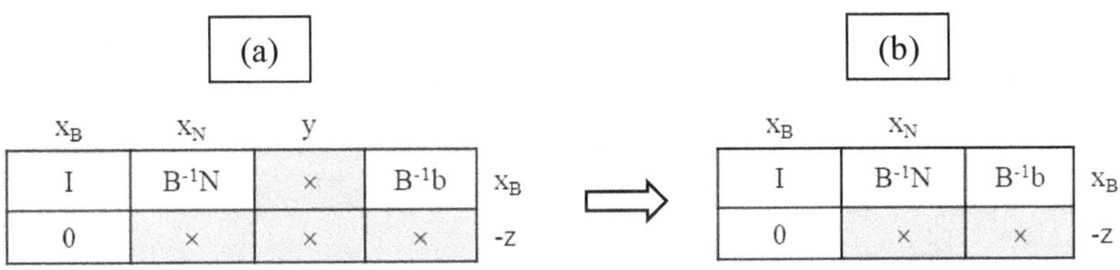

Figure 5-9: Shifting from the auxiliary solution array to the initial array.

Let us present the two successive operations that allow us to construct the initial array from the auxiliary array.

<u>Switching to a basis of x variables</u>

In order to bring out the variables y that may be present in the solution basis, it is sufficient to exchange them by pivoting with non-basic variables x having a non-zero pivot. If the matrix A is of full rank, there exists necessarily such a pivot.

Since the variable y leaving the basis has the value 0 (because the optimum of the auxiliary problem is $y = 0$), the corresponding move is necessarily zero. The only effect of the pivoting is to change the basis without changing the solution. When the basis only contains x variables, the columns associated with the y variables are removed from the array (crosses in figure 5-9a)

Passing to the costs of the original problem

Since the feasible basis B is known, it is necessary to calculate the reduced costs in this basis and the cost function associated with the basis solution. Let us retrieve the formulas (5.20).

$$\begin{cases} \bar{c}_N^T = c_N^T - c_B^T B^{-1} N \\ \bar{z} \ = c_B^T B^{-1} b \end{cases} \tag{5.38}$$

To perform these matrix calculations in a simple way, it can be noted that the matrices I and $B^{-1}N$ associated with the costs c_N and c_B respectively are already present in the resulting array (figure 5-9). Indeed, these matrices represent the canonical form of the constraints in the basis B. We can therefore directly apply the formulas (5.38) from the solution array in figure 5-9.

In practice, by arranging the costs c_B^T and c_N^T on a row above the array as shown figure 5-10, a matrix multiplication can be performed between the vectors c_B^T, c_N^T and the matrices I, $B^{-1}N$ to obtain the costs to be placed in the last row. The array of the initial problem is thus filled in the basis B (shaded boxes in figure 5-10).

We can then apply the simplex algorithm from this array.

c_B^T	c_N^T		
I	$B^{-1}N$	$B^{-1}b$	x_B
0	$c_N^T - c_B^T B^{-1} N$	$-c_B^T B^{-1} b$	$-z$

⟹

x_B	x_N		
I	$B^{-1}N$	$B^{-1}b$	x_B
0	$\bar{c}_N$	$-z$	$-z$

Figure 5-10: Calculation of the costs of the initial array.

The **two-phase method** consists of first solving the auxiliary problem by the simplex algorithm (first phase), transforming the solution array into the original problem array, and then solving the original problem by the simplex algorithm (second phase). The example 5-7 details these successive phases.

Example 5-7: Two-phase method

Consider the linear problem with 5 variables $(x_1, x_2, x_3, x_4, x_5)$

$$\min_{x_1,x_2,x_3,x_4,x_5} \; 2x_1+3x_2+3x_3+x_4-2x_5 \;\text{ s.t. } \begin{cases} x_1+3x_2 \quad +4x_4+x_5=2 \\ x_1+2x_2 \quad -3x_4+x_5=2 \\ -x_1-4x_2+3x_3 \quad =1 \\ x_1, x_2, x_3, x_4, x_5 \ge 0 \end{cases}$$

First phase: building a feasible basis

The auxiliary problem is formulated with 3 additional variables (y_1, y_2, y_3).

$$\min_{\substack{x_1,x_2,x_3,x_4,x_5\\ y_1,y_2,y_3}} \; y_1+y_2+y_3 \;\text{ s.t. } \begin{cases} x_1+3x_2 \quad +4x_4+x_5+y_1 \quad =2 \\ x_1+2x_2 \quad -3x_4+x_5 \quad +y_2 \quad =2 \\ -x_1-4x_2+3x_3 \quad +y_3=1 \\ x_1, x_2, x_3, x_4, x_5, y_1, y_2, y_3 \ge 0 \end{cases}$$

This problem is reformulated by expressing $y_1+y_2+y_3$ from the constraints.

$$\min_{\substack{x_1,x_2,x_3,x_4\\ y_1,y_2,y_3,y_4}} \; -x_1-x_2-3x_3-x_4-2x_5+5 \;\text{ s.t. } \begin{cases} x_1+3x_2 \quad +4x_4+x_5+y_1 \quad =2 \\ x_1+2x_2 \quad -3x_4+x_5 \quad +y_2 \quad =2 \\ -x_1-4x_2+3x_3 \quad +y_3=1 \\ x_1, x_2, x_3, x_4, x_5, y_1, y_2, y_3 \ge 0 \end{cases}$$

This gives a canonical form in the basis (y_1, y_2, y_3).
The basis solution $(y_1, y_2, y_3)=(2,2,1) \ge 0$ is feasible.
This allows the initial simplex array to be filled in directly.

Initial array

Basis: (y_1, y_2, y_3)

x_1	x_2	x_3	x_4	x_5	y_1	y_2	y_3		
1	3	0	4	1	1	0	0	2	y_1
1	2	0	-3	1	0	1	0	2	y_2
-1	-4	3	0	0	0	0	1	1	y_3
-1	-1	-3	-1	-2	0	0	0	-5	$-z$

Pivoting is applied with Bland's rule.

Pivoting 1

Current basis: (y_1, y_2, y_3) / Entering variable: x_1 / Leaving variable: y_1

x_1	x_2	x_3	x_4	x_5	y_1	y_2	y_3		
1	3	0	4	1	1	0	0	2	y_1
1	2	0	-3	1	0	1	0	2	y_2
-1	-4	3	0	0	0	0	1	1	y_3
-1	-1	-3	-1	-2	0	0	0	-5	-z

x_1	x_2	x_3	x_4	x_5	y_1	y_2	y_3		
1	3	0	4	1	1	0	0	2	x_1
0	-1	0	-7	0	-1	1	0	0	y_2
0	-1	3	4	1	1	0	1	3	y_3
0	2	-3	3	-1	1	0	0	-3	-z

Pivoting 2

Current basis: (x_1, y_2, y_3) / Entering variable: x_3 / Leaving variable: y_3

x_1	x_2	x_3	x_4	x_5	y_1	y_2	y_3		
1	3	0	4	1	1	0	0	2	x_1
0	-1	0	-7	0	-1	1	0	0	y_2
0	-1	3	4	1	1	0	1	3	y_3
0	2	-3	3	-1	1	0	0	-3	-z

x_1	x_2	x_3	x_4	x_5	y_1	y_2	y_3		
1	3	0	4	1	1	0	0	2	x_1
0	-1	0	-7	0	-1	1	0	0	y_2
0	-1/3	1	4/3	1/3	1/3	0	1/3	1	x_3
0	1	0	7	0	2	0	1	0	-z

All reduced costs are positive or zero: the basis (x_1, y_2, x_3) is optimal.
The solution is: $(x_1,x_2,x_3,x_4,x_5,y_1,y_2,y_3)=(2,0,1,0,0,0,0,0)$ of cost $z=0$.
This basis includes the variable y_2 which is not present in the initial problem.
To reduce to a basis with only the initial variables, we choose a non-basic variable x having a non-zero pivot on the row of y_2.
Here, we can choose x_2 (pivot $= -1$) or x_4 (pivot $= -7$), but not x_5 (pivot $= 0$).

Let us choose x_2 to perform a pivoting with y_2. The basic variable y_2 has the value 0 (because the optimal solution of the auxiliary problem is $y=0$) and it keeps the value 0 when leaving the basis.
Pivoting therefore corresponds to a zero move which does not change the solution.

Pivoting 3

Current basis: (x_1, y_2, x_3) / Entering variable: x_2 / Leaving variable: y_2

x_1	x_2	x_3	x_4	x_5	y_1	y_2	y_3		
1	3	0	4	1	1	0	0	2	x_1
0	-1	0	-7	0	-1	1	0	0	y_2
0	-1/3	1	4/3	1/3	1/3	0	1/3	1	x_3
0	1	0	7	0	2	0	1	0	-z

x_1	x_2	x_3	x_4	x_5	y_1	y_2	y_3		
1	0	0	-17	1	2	3	0	2	x_1
0	1	0	7	0	-1	-1	0	0	x_2
0	0	1	3.67	1/3	2/3	-1/3	1/3	1	x_3
0	0	0	0	0	1	1	1	0	-z

We now have the feasible basis (x_1, x_2, x_3) for the original problem.

Second phase: Solving the linear problem

To construct the initial array in the feasible basis (x_1, x_2, x_3), we first delete the columns of the additional variables (y_1, y_2, y_3) introduced for the auxiliary problem.

Initial array (before calculating the costs)

Basis: (x_1, x_2, x_3)

x_1	x_2	x_3	x_4	x_5		
1	0	0	-17	1	2	x_1
0	1	0	7	0	0	x_2
0	0	1	3.67	1/3	1	x_3
0	0	0	0	0	0	-z

The last row of the array should then be filled in with: $\begin{cases} \overline{c}_N^T = c_N^T - c_B^T B^{-1} N \\ \overline{z} = c_B^T B^{-1} b \end{cases}$

(shaded boxes above). The useful matrices are already present in the array.

$$B^{-1}N = \begin{pmatrix} -17 & 1 \\ 7 & 0 \\ 3{,}67 & 1/3 \end{pmatrix} , \quad B^{-1}b = \begin{pmatrix} 2 \\ 0 \\ 1 \end{pmatrix}$$

The cost function of the initial problem is: $z = 2x_1 + 3x_2 + 3x_3 + x_4 - 2x_5$.

The vector $c^T = (2,3,3,1,-2)$ is copied in a row above the array to facilitate these calculations. This gives the following initial array.

Initial array (with costs)

Basis: (x_1, x_2, x_3) Reduced costs for basis: $(x_1, x_2, x_3) \rightarrow (0,0,0)$

for non-basis: $(x_4, x_5) \rightarrow (3,-5)$

	x_1	x_2	x_3	x_4	x_5		
c =	2	3	3	1	-2		
	1	0	0	-17	1	2	x_1
	0	1	0	7	0	0	x_2
	0	0	1	3.67	1/3	1	x_3
	0	0	0	3	-5	-7	-z

From this array, we apply the simplex algorithm (with Bland's rule).

Pivoting 1

Current basis: (x_1, x_2, x_3) / Entering variable: x_5 / Leaving variable: x_1

x_1	x_2	x_3	x_4	x_5		
1	0	0	-17	1	2	x_1
0	1	0	7	0	0	x_2
0	0	1	3.67	1/3	1	x_3
0	0	0	3	-5	-7	-z

x_1	x_2	x_3	x_4	x_5		
1	0	0	-17	1	2	x_5
0	1	0	7	0	0	x_2
-1/3	0	1	9.33	0	1/3	x_3
5	0	0	-82	0	3	-z

Pivoting 2

Current basis: (x_5, x_2, x_3) / Entering variable: x_4 / Leaving variable: x_2

x_1	x_2	x_3	x_4	x_5		
1	0	0	-17	1	2	x_5
0	1	0	7	0	0	x_2
-1/3	0	1	9.33	0	1/3	x_3
5	0	0	-82	0	3	-z

x_1	x_2	x_3	x_4	x_5		
1	2.43	0	0	1	2	x_5
0	0.14	0	1	0	0	x_4
-1/3	-1.33	1	0	0	1/3	x_3
5	11.71	0	0	0	3	-z

All reduced costs are positive or zero: the basis (x_3, x_4, x_5) is optimal.

The resulting solution is: $(x_1, x_2, x_3, x_4, x_5) = \left(0, 0, \frac{1}{3}, 0, 2\right)$ of cost $z = -3$.

The simplex method can solve most linear problems very efficiently.
The following sections present the main improvements and extensions to the original Dantzig algorithm.

5.1.7 Revised simplex

The array method presented in section 5.1.4 modifies the entire constraint matrix at each pivoting. For a large problem, the succession of many pivoting leads to a progressive loss of numerical accuracy (due to the finite machine precision) and may prevent the solution from being achieved.
The **revised simplex** method presented below systematically starts from the initial matrix at each pivoting in order to avoid this loss of precision. It also aims to minimize the memory storage and the amount of computation.

Let us retrieve the linear problem in standard form.

$$\min_{x\in\mathbb{R}^n} z = c^T x \quad \text{s.t.} \begin{cases} Ax = b \\ x \geq 0 \end{cases} \quad \text{with } A \in \mathbb{R}^{m\times n}, b \in \mathbb{R}^m, c \in \mathbb{R}^n \tag{5.39}$$

Suppose we have a basis $B \in \mathbb{R}^{m\times m}$ which is a submatrix of A.
The variables $x \in \mathbb{R}^n$ are partitioned into (x_B, x_N) and the cost vector $c \in \mathbb{R}^n$ is partitioned into (c_B, c_N). The elements needed for pivoting are:
- the choice of the entering variable x_e according to the reduced costs $\bar{c}_N$;
- the calculation of the associated basis direction d^e;
- the choice of the leaving variable x_s according to the feasible steps along d^e.
Let us look at how to determine these elements from the initial matrices.

The reduced costs are calculated by

$$\bar{c}_N = c_N - N^T B^{-T} c_B = c_N - N^T \lambda \quad \text{with } B^T \lambda = c_B \tag{5.40}$$

The basis direction is defined by

$$d^e = \begin{pmatrix} d_B^e \\ d_N^e \end{pmatrix} \quad \text{with} \begin{cases} B d_B^e = -A_{.e} \\ d_N^e = \begin{pmatrix} \overset{1}{0} \cdots \overset{e-1}{0} & \overset{e}{1} & \overset{e+1}{0} \cdots \overset{n-m}{0} \end{pmatrix}^T \end{cases} \tag{5.41}$$

where $A_{.e}$ denotes the column number e of the matrix A.
The feasible steps with respect to the basic variables are calculated by

$$s_{ie} = \frac{\bar{b}_i}{\bar{a}_{ie}} \quad \text{with} \begin{cases} B\bar{b} = b \\ B^{-1}N = (\bar{a}_{ij})_{i\in B, j\in N} \rightarrow \bar{a}_{ie} = (d_B^e)_i \end{cases} \tag{5.42}$$

We observe that it is sufficient to solve three linear systems of matrix B.

$$\begin{cases} B^T\lambda = c_B \\ Bd_B^e = -A_{.e} \\ B\bar{b} = b \end{cases} \tag{5.43}$$

The first system gives the reduced costs (5.40) and allows the choice of x_e.
The second system gives the components of the direction d_B^e (5.41).
The third system calculates the steps (5.42) and allows the choice of x_s.

These three systems are solved at each iteration to perform the basis change. They use the matrices A, b, c of the initial problem, which maintains the same numerical accuracy at each iteration. The information to be stored in memory is reduced to the matrices A, b, c and the column numbers forming the matrix B.

The basis matrix B is only changed by one column at each iteration. To save computational time when solving linear systems, one can store its LU factorization in memory and update it when changing the basis. The revised simplex method based on the systems (5.43) allows for an economical and accurate implementation of the simplex algorithm.

5.1.8 Dual simplex

For a linear problem, it is equivalent to solve the primal problem or the dual problem (section 1.4.3). Table 5-3 shows the correspondence between the variables and constraints of these two problems (see also table 1-4).

Primal (P) $\min_{x\in\mathbb{R}^n} c^Tx$	dimension	Dual (D) $\max_{y\in\mathbb{R}^m} b^Ty$
$Ax = b$	m	$y \in \mathbb{R}^m$
$Ax \le b$	m	$y \le 0$
$x \ge 0$	n	$A^Ty \le c$
$x \in \mathbb{R}^n$	n	$A^Ty = c$

Table 5-3: Correspondence primal problem – dual problem.

In some situations, solving the dual problem is simpler. This is particularly the case in mixed linear programming when a feasible basis of the dual problem is already known.

To apply the simplex method to the dual problem, we must first write its canonical form. Let us show how to pass from the canonical form of the primal (P) to the canonical form of the dual (D) in a given basis.

Canonical form of the dual

Let us start from the canonical form of the primal (P) in a basis B.

$$(P) \quad \min_{x_B, x_N} \bar{c}_N^T x_N \quad \text{s.t.} \begin{cases} x_B + B^{-1}Nx_N = \bar{b} & \rightarrow \text{m constraints} \\ x_B, x_N \geq 0 & \rightarrow \text{n variables} \end{cases} \tag{5.44}$$

The positive variables x_B play the role of slack variables. By removing them, we reformulate an equivalent problem (P') with inequality constraints.

$$(P') \quad \min_{x_N} \bar{c}_N^T x_N \quad \text{s.t.} \begin{cases} B^{-1}Nx_N \leq \bar{b} & \rightarrow \text{m constraints} \\ x_N \geq 0 & \rightarrow \text{n} - \text{m variables} \end{cases} \tag{5.45}$$

We pass to the dual (D') of (P') using the correspondences in table 5-3.

$$(D') \quad \max_{y} \bar{b}^T y \quad \text{s.t.} \begin{cases} \left(B^{-1}N\right)^T y \leq \bar{c}_N & \rightarrow \text{n} - \text{m constraints} \\ y \leq 0 & \rightarrow \text{m variables} \end{cases} \tag{5.46}$$

The problem is reformulated as a minimization with the variable change $y_B = -y$

$$(D') \quad \min_{y_B} \bar{b}^T y_B \quad \text{s.t.} \begin{cases} -\left(B^{-1}N\right)^T y_B \leq \bar{c}_N & \rightarrow \text{n} - \text{m constraints} \\ y_B \geq 0 & \rightarrow \text{m variables} \end{cases} \tag{5.47}$$

We put the dual in standard form (D) with positive slack variables y_N.

$$(D) \quad \min_{y_B, y_N} \bar{b}^T y_B \quad \text{s.t.} \begin{cases} y_N - \left(B^{-1}N\right)^T y_B = \bar{c}_N & \rightarrow \text{n} - \text{m constraints} \\ y_B, y_N \geq 0 & \rightarrow \text{n variables} \end{cases} \tag{5.48}$$

We observe that the problem (D) is in **canonical form** in the basis B, provided we consider y_N as the basic variables and y_B as the non-basic variables. This linear problem can be solved by the simplex method based on the array.

Dual problem array

Let us form the simplex array of the dual in the basis B from its canonical form (5.48). This array is denoted T_D.

$$T_D = \begin{array}{c|c|c} y_N & y_B & \\ \hline I & -(B^{-1}N)^T & \bar{c}_N \\ \hline 0 & \bar{b}^T & -\bar{z} \end{array} \quad \begin{array}{l} \rightarrow y_N - \left(B^{-1}N\right)^T y_B = \bar{c}_N \\ \rightarrow \bar{b}^T y_B = z - \bar{z} \end{array} \tag{5.49}$$

The interpretation of this dual array is similar to that of a primal array, with a reversal of the roles of B (non-basis columns) and N (basis columns):

- the basic variables are y_N and have the values $\bar{c}_N$;
- the non-basic variables are y_B and have reduced costs $\bar{b}$;
- the constraint matrix of the dual A^T (Table 5-3) yields the reduced matrix $(B^{-1}N)^T$ in the array T_D;
- the basis solution associated with B is: $\begin{cases} y_N = \bar{c}_N \\ y_B = 0 \end{cases}$;
- the basis B is said to be **dual-feasible** if: $y_N = \bar{c}_N \geq 0$.

The simplex pivoting rules apply as before:

- entering of a non-basic variable y_B of negative reduced cost;
- leaving of the first basic variable y_N which cancels.

Note y_e, $e \in B$ the entering variable of reduced cost $\bar{b}_e < 0$.
As in (5.21) and (5.22), the variables y_B remain zero during the move, except for y_e. The positivity constraints on the basic variables y_N give

$$y_i = \bar{c}_{Ni} + \bar{a}_{ei} y_e \geq 0 \quad , \quad i \in N \tag{5.50}$$

keeping the notation already used in (5.21) : $B^{-1}N \underset{\text{noted}}{=} (\bar{a}_{ij})_{i \in B, j \in N}$.

Noting s_{ie} the maximum value that the variable y_e can take, while respecting the constraint linked to any basic variable y_i, we have:

- if $\bar{a}_{ei} < 0$, then: $s_{ie} = \dfrac{\bar{c}_{Ni}}{-\bar{a}_{ie}}$;
- if $\bar{a}_{ei} \geq 0$, then: $s_{ie} = +\infty$ (unbounded move).

The maximum feasible step along the basis direction associated with y_e is

$$s_e = \min_{i \in N} s_{ie} = \min_{\substack{i \in N \\ \bar{a}_{ei} < 0}} \frac{\bar{c}_{Ni}}{-\bar{a}_{ie}} \tag{5.51}$$

The leaving variable y_s , $s \in N$ is the one corresponding to the minimum step s_e. The pivoting is done with the pivot $\bar{a}_{se} < 0$ and makes a column of the identity matrix appear in column e.

<u>Dual simplex algorithm</u>

We note that the previous rotation can be carried out <u>on the primal array</u> T_P without explicitly forming the dual array T_D. In fact, these two arrays contain the same elements arranged differently. In particular, the reduced costs $\bar{c}_N$ and the basic values $\bar{b}$ are transposed whereas the reduced constraint matrix is transposed and <u>changed in sign</u>.

$$T_D = \begin{array}{c|c|c} y_N & y_B & \\ \hline I & -(B^{-1}N)^T & \bar{c}_N \\ \hline 0 & \bar{b}^T & -\bar{z} \end{array} \qquad T_P = \begin{array}{c|c|c} x_B & x_N & \\ \hline I & (B^{-1}N) & \bar{b} \\ \hline 0 & \bar{c}_N^T & -\bar{z} \end{array} \tag{5.52}$$

Let us take the pivoting rules stated above for the dual array T_D and transcribe them into the primal array T_P:

- entering variable for the dual:

  Initial rule in T_D: select y_B with negative reduced cost $(\bar{b}_e < 0)$
  Transcribed rule in T_P: select x_B with negative value $(\bar{b}_e < 0)$;

- leaving variable for the dual:

  Initial rule in T_D: select y_N from minimum step $\min\limits_{\substack{i \in N \\ \bar{a}_{ei} < 0}} \frac{\bar{c}_{Ni}}{-\bar{a}_{ie}}$ (pivot < 0)

  Transcribed rule in T_P: select x_N from maximum step $\max\limits_{\substack{i \in N \\ \bar{a}_{ei} < 0}} \frac{\bar{c}_{Ni}}{\bar{a}_{ie}}$ (pivot < 0)

Since the constraint matrix $B^{-1}N$ has the opposite sign in the dual array, it is equivalent to looking for the maximum step by directly taking the values $\bar{a}_{ie}$ present in the primal array;

- pivoting application:

  Initial rule in T_D: make an identity column appear for y_{Be}
  Transcribed rule in T_P: make an identity column appear for x_{Ns}.

Pivoting according to these rules is called **dual pivoting**, to remind us that it deals with the resolution of the dual problem, although we are working on the primal array T_P. These rules define the dual simplex algorithm.

Vocabulary notes:

- *although the variable* x_B *is an entering variable for the dual, it is convenient to call it a leaving variable in order to preserve the nomenclature of the primal array.*
  Similarly, the leaving variable x_N *for the dual is called an entering variable;*
- *a basis is said primal-feasible if the basic variables are positive* $(\bar{b} \geq 0)$ *and it is said dual-feasible if the reduced costs are positive* $(\bar{c}_N \geq 0)$.

The primal simplex algorithm starts from a primal-feasible basis $(\bar{b} \geq 0)$ and seeks to achieve positive reduced costs $(\bar{c}_N \geq 0)$ by primal pivoting.
The dual simplex algorithm starts from a dual-feasible basis $(\bar{c}_N \geq 0)$ and seeks to achieve positive variables $(\bar{b} \geq 0)$ by dual pivoting.

Depending on the initial basis available, it is more advantageous to use the primal or the dual simplex. The dual simplex algorithm is very interesting when a problem has already been solved and one wants to solve it again either by adding constraints, or by changing the constraint thresholds, or by fixing the value of some variables. These operations modify the constraints part of the simplex array, but they do not change the reduced costs which remain positive. The basis of the already solved problem thus remains dual-feasible for the new problem. This property is used especially in mixed linear programming.

The example 5-8 illustrates the solution of a linear problem by the dual simplex algorithm.

Example 5-8: Dual simplex method

Consider the linear problem with 5 variables $(x_1, x_2, x_3, x_4, x_5)$

$$\min_{x_1,x_2,x_3,x_4,x_5} \; x_1 + 2x_2 + 2x_3 + 3x_4 + x_5 \;\text{ s.t. }\begin{cases} x_1 + x_2 = 1 \\ -x_2 - x_3 + x_5 = 0 \\ -x_1 + x_3 + x_4 = 0 \\ x_1, x_2, x_3, x_4, x_5 \geq 0 \end{cases}$$

The array is filled with the matrices

$$A = \begin{pmatrix} 1 & 1 & 0 & 0 & 0 \\ 0 & -1 & -1 & 0 & 1 \\ -1 & 0 & 1 & 1 & 0 \end{pmatrix} \;,\; b = \begin{pmatrix} 1 \\ 0 \\ 0 \end{pmatrix} \;,\; c^T = \begin{pmatrix} 1 & 2 & 2 & 3 & 1 \end{pmatrix}$$

Initial array

(non-canonical)

x_1	x_2	x_3	x_4	x_5	
1	1	0	0	0	1
0	-1	-1	0	1	0
-1	0	1	1	0	0
1	2	2	3	1	0

Let us put the problem in canonical form in the basis (x_2, x_3, x_4) by row combinations, in order to have the identity matrix on the columns (x_2, x_3, x_4) and zero costs on the last row.

x_1	x_2	x_3	x_4	x_5	
1	1	0	0	0	1
0	-1	-1	0	1	0
-1	0	1	1	0	0
1	2	2	3	1	0

Elimination x_2 →

x_1	x_2	x_3	x_4	x_5	
1	1	0	0	0	1
1	0	-1	0	1	1
-1	0	1	1	0	0
-1	0	2	3	1	-2

x_1	x_2	x_3	x_4	x_5	
1	1	0	0	0	1
1	0	-1	0	1	1
-1	0	1	1	0	0
-1	0	2	3	1	-2

Elimination x_3 →

x_1	x_2	x_3	x_4	x_5	
1	1	0	0	0	1
-1	0	1	0	-1	-1
0	0	0	1	1	1
1	0	0	3	3	0

x_1	x_2	x_3	x_4	x_5	
1	1	0	0	0	1
-1	0	1	0	-1	-1
0	0	0	1	1	1
1	0	0	3	3	0

Elimination x_4 →

x_1	x_2	x_3	x_4	x_5	
1	1	0	0	0	1
-1	0	1	0	-1	-1
0	0	0	1	1	1
1	0	0	0	0	-3

This yields the simplex array in the basis (x_2, x_3, x_4).

Initial array (canonical)

Basis: (x_2, x_3, x_4)

x_1	x_2	x_3	x_4	x_5		
1	1	0	0	0	1	x_2
-1	0	1	0	-1	-1	x_3
0	0	0	1	1	1	x_4
1	0	0	0	0	-3	-z

This array shows that the basis (x_2, x_3, x_4) is not primal-feasible, as $x_3 < 0$. This basis is on the other hand dual-feasible, because the reduced costs in the last row are positive. We can therefore apply the dual simplex algorithm from this array. The basis is not dual-optimal, because some variables are negative. We apply a dual pivoting by choosing the first negative variable (here x_3).

Pivoting 1

Current basis: (x_2, x_3, x_4) → Leaving variable: x_3

Limit step on x_1, x_5 : $\max(s_{31} = -1 , s_{35} = 0)$ → Entering variable: x_5

x_1	x_2	x_3	x_4	x_5		
1	1	0	0	0	1	x_2
-1	0	1	0	-1	-1	x_3
0	0	0	1	1	1	x_4
1	0	0	0	0	-3	-z

Pivot: $\bar{a}_{35} = -1$ → Division of row 2 by $\bar{a}_{35}$

Elimination in column 5 → New basis (x_2, x_5, x_4) / cost $z = 3$

x_1	x_2	x_3	x_4	x_5		
1	1	0	0	0	1	x_2
1	0	-1	0	1	1	x_5
-1	0	1	1	0	0	x_4
1	0	0	0	0	-3	-z

All basic variables are positive or zero: the basis (x_2, x_4, x_5) is optimal.

The resulting solution is: $(x_1, x_2, x_3, x_4, x_5) = (0,1,0,0,1)$ of cost $z = 3$.

The final array is optimal for both the primal and the dual problem.

5.1.9 Complementary simplex

The simplex method can be adapted to solve a convex quadratic-linear problem of the form

$$\min_{x\in\mathbb{R}^n} \frac{1}{2}x^TQx+c^Tx \quad \text{s.t.} \begin{cases} Ax \geq b \\ x \geq 0 \end{cases}, Q\in\mathbb{R}^{n\times n}, A\in\mathbb{R}^{m\times n}, b\in\mathbb{R}^m, c\in\mathbb{R}^n \quad (5.53)$$

The cost function is assumed to be convex with a matrix A symmetric positive definite. The Lagrangian is expressed with multipliers μ and σ associated with the inequality constraints $-(Ax-b)\leq 0$ and $-x\leq 0$ respectively. The inequality constraints are expressed as $\leq$ (standard form) and their multipliers are positive.

$$L(x,\mu,\sigma)=\frac{1}{2}x^TQx+c^Tx-\mu^T(Ax-b)-\sigma^Tx \quad (5.54)$$

The KKT conditions of order 1 give the system

$$\begin{cases} Qx+c-A^T\mu-\sigma=0 \\ Ax-b\geq 0 \\ x\geq 0, \mu\geq 0, \sigma\geq 0 \\ \sigma_i x_i=0 , \quad i=1 \text{ to } n \\ \mu_j(Ax-b)_j=0 , \quad j=1 \text{ to } m \end{cases} \quad (5.55)$$

The KKT conditions of order 2 are satisfied, as the Q matrix is positive definite. Let us rewrite the first-order conditions by introducing m positive slack variables noted r associated with the m constraints $Ax-b\geq 0$.

$$\begin{cases} Qx+c-A^T\mu-\sigma=0 \\ Ax-b-r=0 \\ x\geq 0, r\geq 0, \mu\geq 0, \sigma\geq 0 \\ \sigma_i x_i=0 , \quad i=1 \text{ to } n \\ \mu_j(Ax-b)_j=0 , \quad j=1 \text{ to } m \end{cases} \quad (5.56)$$

We group the variables (x,μ) on the one hand, the variables (σ,r) on the other.

By introducing the notations: $y=\begin{pmatrix} x \\ \mu \end{pmatrix}, \rho=\begin{pmatrix} \sigma \\ r \end{pmatrix}, G=\begin{pmatrix} Q & -A^T \\ A & 0 \end{pmatrix}, \gamma=\begin{pmatrix} c \\ -b \end{pmatrix}$

the system (5.56) takes the compact form

$$\begin{cases} I\rho-Gy=\gamma \\ y\geq 0, \rho\geq 0 \\ y_i\rho_i=0 , \quad i=1 \text{ to } n+m \end{cases} \quad (5.57)$$

By analogy with the simplex method, we observe that the equation $I\rho - Gy = \gamma$ is in canonical form if we consider:

- $\rho \in \mathbb{R}^{n+m}$ as the basic variables (matrix I);
- $y \in \mathbb{R}^{n+m}$ as the non-basic variables (matrix $-G$).

The last row adds complementarity conditions. The variables y_i and ρ_i are called complementary variables.

The basis $\begin{cases} \rho = \gamma \\ y = 0 \end{cases}$ satisfies the complementarity conditions, but not necessarily the positivity conditions $\rho \geq 0$.

Let us note $z = (y, \rho)$ the set of $2(n+m)$ variables of the problem (5.57). Half of the variables are in basis, the other half are out of basis.

The purpose of the complementary simplex method is to perform basis changes maintaining the complementarity conditions until the positivity conditions are satisfied. Each basis change consists of a standard pivoting followed if necessary by complementary pivoting to restore complementarity.

A **standard pivoting** consists in choosing a negative basic variable and entering its complementary variable (which is necessarily non-basic) into the basis.

Note z_e the leaving variable:

- $z_e = \rho_e$ if the chosen basic variable is y_e;
- $z_e = y_e$ if the chosen basic variable is ρ_e.

The constraints on the z variables are of the form

$$z_i + \sum_{j \in N} \alpha_{ij} z_j = \beta_i \ , \ i \in B \tag{5.58}$$

When z_e increases, with the other non-basic variables $(z_j)_{j \in N, j \neq e}$ staying zero, these constraints reduce to

$$z_i + \alpha_{ie} z_e = \beta_i \ , \ i \in B \tag{5.59}$$

The maximum value that z_e can take is defined by the first basic variable that cancels. If $\beta_i = 0$ and $a_{ie} > 0$, the maximum value is zero: $z_e = 0$.

Otherwise, if β_i and a_{ie} have the same sign, the maximum value is given by

$$z_e = \min \left\{ \frac{\beta_i}{\alpha_{ie}} \ , \ \frac{\beta_i}{\alpha_{ie}} > 0 \right\}_{i \in B} \tag{5.60}$$

This rule is similar to the usual simplex rule, except that the second member β_i can be negative.

Let us note z_s the leaving basic variable. This variable is not necessarily the complementary of z_e, which destroys the basis complementarity.
Complementary pivoting is intended to restore complementarity.

Complementary pivoting consists in choosing the complementary variable of z_s as the entering variable. In fact, the variable z_s having just left the basis, we try to bring in its complement. Let us note z'_e this entering variable.

The same pivoting rule (5.60) as above is then applied. The leaving variable z'_s is the first basic variable that cancels as z'_e increases. To avoid cycling, the previous variable z_e must be excluded from this search. Complementary pivoting is to be repeated as long as the basis is not complementary, taking care not to return to a basis already obtained. When a complementary basis is restored, a standard pivoting can be applied again if there are still negative basic variables.

The example 5-9 illustrates the solution of a quadratic-linear problem by the complementary simplex algorithm.

Example 5-9: Complementary simplex method

Consider the problem: $\min\limits_{x_1,x_2} -6x_1+2x_1^2-2x_1x_2+2x_2^2 \quad \text{s.t.} \begin{cases} x_1+x_2 \le 2 \\ x_1,x_2 \ge 0 \end{cases}$.

We introduce a multiplier $\mu_1 \ge 0$ for the constraint: $x_1+x_2-2 \le 0$ and multipliers $\sigma_1,\sigma_2 \ge 0$ for the constraints: $x_1,x_2 \ge 0$. The Lagrangian is expressed as
$L(x_1,x_2,\mu_1,\sigma_1,\sigma_2) = -6x_1+2x_1^2-2x_1x_2+2x_2^2+\mu_1(x_1+x_2-2)-\sigma_1x_1-\sigma_2x_2$.

The KKT conditions are formulated by introducing a slack variable $r_1 \ge 0$.

$$\begin{cases} -6+4x_1-2x_2+\mu_1-\sigma_1=0 \\ -2x_1+4x_2+\mu_1-\sigma_2=0 \\ x_1+x_2 \le 2 \\ \sigma_1x_1=0 \\ \sigma_2x_2=0 \\ \mu_1(x_1+x_2-2)=0 \\ x_1,x_2,\mu_1,\sigma_1,\sigma_2 \ge 0 \end{cases} \longrightarrow \begin{cases} -4x_1+2x_2-\mu_1+\sigma_1=-6 \\ 2x_1-4x_2-\mu_1+\sigma_2=0 \\ x_1+x_2+r_1=2 \\ \sigma_1x_1=0 \\ \sigma_2x_2=0 \\ \mu_1 r_1=0 \\ x_1,x_2,\mu_1,\sigma_1,\sigma_2,r_1 \ge 0 \end{cases}$$

The KKT conditions of order 2 are on the matrix: $\nabla^2_{xx}L = \begin{pmatrix} 4 & -2 \\ -2 & 4 \end{pmatrix}$.

Its eigenvalues are 2 and 6. It is positive definite and the problem is convex.
To solve the KKT system, we form the complementary simplex array.
The pairs of complementary variables are: (x_1, σ_1), (x_2, σ_2), (μ_1, r_1).

The initial problem is in canonical form in the basis $(\sigma_1, \sigma_2, r_1)$.

Initial array

Basis: $(\sigma_1, \sigma_2, r_1)$

x_1	x_2	μ_1	σ_1	σ_2	r_1		
-4	2	-1	1	0	0	-6	σ_1
2	-4	-1	0	1	0	0	σ_2
1	1	0	0	0	1	2	r_1

The basis solution is non-feasible, because the variable σ_1 is negative $(\sigma_1 = -6)$. The first pivoting brings the variable x_1 (complementary of σ_1) into the basis. The purpose of this standard pivoting is to bring σ_1 out of the basis to cancel it. The increase of x_1 is limited by the first basic variable which cancels. The maximum moves with respect to the basic variables $(\sigma_1, \sigma_2, r_1)$ are respectively $(1{,}5 , 0 , 2)$. The first variable cancelling out when x_1 increases is therefore σ_2 with a zero move.

Pivoting 1: standard

Current basis: $(\sigma_1, \sigma_2, r_1)$ → Entering variable: x_1

Step on σ_1, σ_2, r_1 : $\min(s_{\sigma 1} = 1{,}5 , s_{\sigma 2} = 0 , s_{r1} = 2)$ → Leaving variable: σ_2

x_1	x_2	μ_1	σ_1	σ_2	r_1			Step
-4	2	-1	1	0	0	-6	σ_1	→ s = 3/2
2	-4	-1	0	1	0	0	σ_2	→ s = 0
1	1	0	0	0	1	2	r_1	→ s = 2

Pivot: $\alpha_{\sigma_2 x_1} = 2$ → Division of row 2 by $\alpha_{\sigma_2 x_1}$

Elimination in column 1 → New basis (σ_1, x_1, r_1)

x_1	x_2	μ_1	σ_1	σ_2	r_1		
0	-6	-3	1	2	0	-6	σ_1
1	-2	-1/2	0	1/2	0	0	x_1
0	3	1/2	0	-1/2	1	2	r_1

The basis (σ_1, x_1, r_1) is no longer complementary.

The variable σ_2 having been taken out of the basis, we try to bring back in the basis its complementary variable x_2 by a complementary pivoting. The maximum moves with respect to the basic variables (σ_1, x_1, r_1) are respectively $(1, \infty, 2/3)$. The step on x_1 is not bounded, because the second member is zero (line 2) and the coefficient of x_2 is negative. The variable x_2 can therefore increase indefinitely with respect to this constraint. The first variable that cancels when x_2 increases is therefore r_1.

Pivoting 2: complementary

Current basis: (σ_1, x_1, r_1) → Entering variable: x_2

Step on σ_1, x_1, r_1 : $\min(s_{\sigma 1} = 1, s_{x1} = \infty, s_{r1} = 2/3)$ → Leaving variable: r_1

x_1	x_2	μ_1	σ_1	σ_2	r_1			Step
0	-6	-3	1	2	0	-6	σ_1	→ s = 1
1	-2	-1/2	0	1/2	0	0	x_1	→ unbounded
0	3	1/2	0	-1/2	1	2	r_1	→ s = 2/3

Pivot: $\alpha_{r_1 x_2} = 3$ → Division of row 3 by $\alpha_{r_1 x_2}$

Elimination in column 2 → New basis (σ_1, x_1, x_2)

x_1	x_2	μ_1	σ_1	σ_2	r_1		
0	0	-2	1	1	2	-2	σ_1
1	0	-1/6	0	1/6	2/3	4/3	x_1
0	1	1/6	0	-1/6	1/3	2/3	x_2

The (σ_1, x_1, x_2) basis is still not complementary.
The variable r_1 having been taken out of the basis, we try to bring back in the basis its complementary variable μ_1 by a complementary pivoting. The maximum moves with respect to the basic variables (σ_1, x_1, x_2) are respectively $(1, \infty, 4)$. The step on x_1 is not bounded, because the second member and the coefficient of μ_1 are of opposite signs. The first variable that cancels when μ_1 increases is therefore σ_1.

<u>Pivoting 3: complementary</u>

Current basis: (σ_1, x_1, x_2) → Entering variable: μ_1

Step on σ_1, x_1, x_2 : $\min(s_{\sigma 1} = 1, s_{x1} = \infty, s_{x2} = 4)$ → Leaving variable: σ_1

x_1	x_2	μ_1	σ_1	σ_2	r_1			Step
0	0	-2	1	1	2	-2	σ_1	→ s = 1
1	0	-1/6	0	1/6	2/3	4/3	x_1	→ unbounded
0	1	1/6	0	-1/6	1/3	2/3	x_2	→ s = 4

Pivot: $\alpha_{\sigma_1\mu_2} = -2$ → Division of row 1 by $\alpha_{\sigma_1\mu_2}$

Elimination in column 3 → New basis (μ_1, x_1, x_2)

x_1	x_2	μ_1	σ_1	σ_2	r_1		
0	0	1	-1/2	-1/2	-1	1	μ_1
1	0	0	-1/12	1/12	1/2	3/2	x_1
0	1	0	1/12	-1/12	1/2	1/2	x_2

The basis (μ_1, x_1, x_2) is complementary and the basic variables are all positive. This basis is therefore optimal. The associated basis solution is

$$x_1 = \frac{3}{2},\ x_2 = \frac{1}{2},\ \mu_1 = 1,\ \sigma_1 = 0,\ \sigma_2 = 0,\ r_1 = 0$$

This solution satisfies all the first-order KKT conditions written at the beginning of this example. The table 5-4 summarizes the pivoting performed.

k	Basis			x_1	x_2	μ_1	σ_1	σ_2	r_1	Pivoting	Step	e	s
0	σ_1	σ_2	r_1	0	0	0	-6	0	2	standard	0	x_1	σ_2
1	σ_1	x_1	r_1	0	0	0	-6	0	2	complementary	2/3	x_2	r_1
2	σ_1	x_1	x_2	4/3	2/3	0	-2	0	0	complementary	1	μ_1	σ_1
3	μ_1	x_1	x_2	3/2	1/2	1	0	0	0	complementary			

Table 5-4: Summary of the iterations of the complementary simplex.

The complementary simplex method can only be applied if the quadratic-linear problem is convex. As with the usual simplex, there are different ways of performing the standard and complementary pivoting (choice of entering variable). It can be shown that the solution is obtained in a finite number of iterations, provided that one guards against possible cycling, for example by choosing entering variables of increasing numbers.

5.2 Interior point

The simplex method solves a linear problem by exploring the polytope vertices, but this can lead in some cases to a large number of iterations, as in figure 5-7.

In worse cases, all C_n^m vertices may be visited before arriving at the solution (as illustrated by Klee and Minty problem, 1972).

Interior point methods use a different strategy. Rather than moving along the edges of the polytope, one starts from the inside and moves directly towards the solution vertex. These methods have been developed since the 1980s and they are proving to be competitive especially for large problems.

5.2.1 Central path

Let us retrieve the linear problem in standard form.

$$\min_{x\in\mathbb{R}^n} c^T x \quad \text{s.t.} \begin{cases} Ax = b \\ x \geq 0 \end{cases} \quad \text{with } A \in \mathbb{R}^{m\times n}, b \in \mathbb{R}^m, c \in \mathbb{R}^n \tag{5.61}$$

The Lagrangian is expressed with multipliers $\lambda \in \mathbb{R}^m$ for the equality constraints and multipliers $s \in \mathbb{R}^n$ for the inequality constraints.

$$L(x,\lambda,s) = c^T x + \lambda^T(Ax - b) - s^T x \tag{5.62}$$

The KKT conditions of order 1 give a system in variables (x,λ,s).

$$\begin{cases} Ax-b=0 \\ c-A^T\lambda-s=0 \\ x_i s_i=0 \ , \ i=1 \text{ to } n \\ x,s\ge 0 \end{cases} \tag{5.63}$$

This nonlinear system has a combinatorial aspect because of the complementarity conditions $x_i s_i = 0$. Furthermore, we know that the solution of the linear problem (5.61) is a vertex of the constraint polytope (property 5-1) and that the associated basis solution (property 5-2) has at least $n-m$ zero components (non-basic variables).

A natural idea is to solve the system (5.63) by Newton's method, taking care not to leave the domain $\{x\ge 0\,;\,s\ge 0\}$. If Newton's iteration goes outside this domain, it can be corrected either by projection or by step reduction.

This approach is very sensitive to the initial point. Indeed, when approaching the boundary of the domain $\{x\ge 0\,;\,s\ge 0\}$, the "projected" Newton iterations tend to follow the edge and the progress becomes very slow. In order to converge towards the solution, a premature cancelling of the variables x_i or s_i must be avoided before arriving in the vicinity of the optimal vertex.

This brings us to consider a problem of the form

$$\min_{x\in\mathbb{R}^n} f_h(x)=c^Tx-h\sum_{i=1}^{n}\ln x_i \quad \text{s.t.} \quad \begin{cases} Ax=b \\ x\ge 0 \end{cases} \tag{5.64}$$

The cost function is penalized by a barrier function B introduced in section 4.2.5.

$$B(x)=-\sum_{i=1}^{n}\ln x_i \tag{5.65}$$

The barrier function B has a push-back effect with respect to the boundary of the domain $x\ge 0$. The parameter $h\ge 0$ is called the height of the barrier. The problem (5.64) is called **barrier problem** and its solution is denoted x_h:

- when $h=0$, the point $x_h = x^*$ is the solution of the linear problem (5.61);
- when $h=\infty$, the point $x_h = x_\infty$ is the solution to the problem

$$\min_{x\in\mathbb{R}^n} -\sum_{i=1}^{n}\ln x_i \quad \text{s.t.} \quad \begin{cases} Ax=b \\ x\ge 0 \end{cases} \tag{5.66}$$

  This point is called the center of the polytope. It is the "farthest" point from the bounds $x_i = 0$.

The set of points $\{x_h \ ; h \geq 0\}$ forms the **central primal path** which goes from the polytope center (for $h = \infty$) to the solution of the linear problem (for $h = 0$). This path is called primal, because it deals only with the primal variables x, as opposed to the primal-dual path defined below.

Example 5-10: Analytical center of the polytope (example presented in [R3])

Consider the polytope in $\mathbb{R}^3$ defined by: $\begin{cases} x_1 + x_2 + x_3 = 1 \\ x_1, x_2, x_3 \geq 0 \end{cases}$.

This polytope shown in figure 5-11 forms an equilateral triangle (ABC).

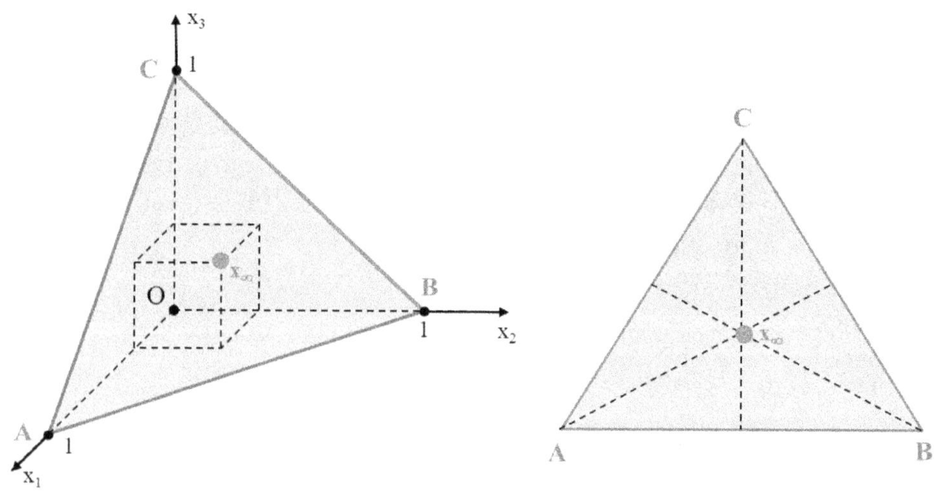

Figure 5-11: Analytical center of the polytope.

The coordinates of the polytope center are obtained by solving the problem

$$\min_{x_1, x_2, x_3} \ -(\ln x_1 + \ln x_2 + \ln x_3) \quad \text{s.t.} \quad \begin{cases} x_1 + x_2 + x_3 = 1 \\ x_1, x_2, x_3 \geq 0 \end{cases}$$

Observing that the 3 coordinates play symmetrical roles, we obtain the solution directly from the constraint: $x_1 = x_2 = x_3 = \dfrac{1}{3}$.

The center of the polytope noted x_∞ is the center of the equilateral triangle (ABC). This point is equidistant from the 3 vertices A, B and C of the polytope.

Let us form the Lagrangian of the barrier problem (5.64) by temporarily noting $\mu \in \mathbb{R}^n$ the multipliers of the inequality constraints

$$L_h(x,\lambda,\mu) = c^T x - h\sum_{i=1}^{n} \ln x_i + \lambda^T(Ax-b) - \mu^T x \tag{5.67}$$

and write the first-order KKT conditions with the notation: $\frac{1}{x} = \left(\frac{1}{x_i}\right)_{i=1 \text{ to } n} \in \mathbb{R}^n$.

$$\begin{cases} Ax - b = 0 \\ c - \frac{h}{x} - A^T\lambda - \mu = 0 \\ x_i\mu_i = 0 \quad , \quad i = 1 \text{ to } n \\ x \geq 0 \, , \, \mu \geq 0 \end{cases} \tag{5.68}$$

By changing the variable: $s = \mu + \frac{h}{x}$, these equations become

$$\begin{cases} Ax - b = 0 \\ c - A^T\lambda - s = 0 \\ x_i s_i = h \quad , \quad i = 1 \text{ to } n \\ x \geq 0 \, , \, s \geq 0 \end{cases} \tag{5.69}$$

We note that this system (5.69) is identical to the KKT system of the original linear problem (5.63), except for the complementarity conditions taking the value h instead of 0.

The set of solutions $\{(x_h, \lambda_h, s_h) \, ; \, h \geq 0\}$ of the KKT system (5.69) forms the **primal-dual central path**, which includes the primal x and dual variables λ, s. For $h = 0$, we obtain the point $(x^*; \lambda^*; s^*)$ solution of the initial KKT system.

The example 5-11 illustrates the geometrical construction of the central path.

Example 5-11: Central primal-dual path (example presented in [R3])

Consider the problem in $\mathbb{R}^3$: $\min\limits_{x_1,x_2,x_3} x_1 + 2x_2 + 3x_3$ s.t. $\begin{cases} x_1 + x_2 + x_3 = 1 \\ x_1, x_2, x_3 \geq 0 \end{cases}$.

The constraints correspond to the polytope of example 5-10. To illustrate geometrically the previous calculations, the linear problem and the barrier problem are solved successively.

Resolution of the linear problem

The KKT conditions of the linear problem (5.63) are treated in the following order.

$$x_i s_i = 0 \quad \Rightarrow \begin{cases} x_1 s_1 = 0 \\ x_2 s_2 = 0 \\ x_3 s_3 = 0 \end{cases} \quad \Rightarrow \begin{cases} x_1 = 0 \text{ or } s_1 = 0 \\ x_2 = 0 \text{ or } s_2 = 0 \\ x_3 = 0 \text{ or } s_3 = 0 \end{cases}$$

→ 6 possible combinations

$$A^T\lambda + s = c \quad \Rightarrow \begin{cases} \lambda + s_1 = 1 \\ \lambda + s_2 = 2 \\ \lambda + s_3 = 3 \end{cases} \quad \Rightarrow \begin{cases} s_1 = 1-\lambda \\ s_2 = 2-\lambda \\ s_3 = 3-\lambda \end{cases} \Rightarrow \begin{cases} s_1 = s_2 - 1 \\ s_2 = s_3 - 1 \end{cases}$$

$$s \geq 0 \quad \Rightarrow \begin{cases} s_1 \geq 0 \\ s_2 = 1 + s_1 \geq 1 \\ s_3 = 1 + s_2 \geq 2 \end{cases} \quad \Rightarrow \begin{cases} x_1 = 0 \text{ or } s_1 = 0 \\ x_2 = 0 \\ x_3 = 0 \end{cases}$$

→ 2 possible combinations remain

$$Ax - b = 0 \quad \Rightarrow x_1 + x_2 + x_3 = 1 \quad \Rightarrow x_1 = 1$$

The solution: $\begin{cases} x_1 = 1 \\ x_2 = 0 \\ x_3 = 0 \end{cases}$, $\begin{cases} s_1 = 0 \\ s_2 = 1 \\ s_3 = 2 \end{cases}$, $\lambda = 1$ corresponds to vertex A (figure 5-12).

Resolution of the barrier problem

The KKT conditions of the barrier problem (5.69) are treated in the following order.

$$A^T\lambda + s = c \quad \Rightarrow \begin{cases} \lambda + s_1 = 1 \\ \lambda + s_2 = 2 \\ \lambda + s_3 = 3 \end{cases} \quad \Rightarrow \begin{cases} s_1 = 1-\lambda \\ s_2 = 2-\lambda \\ s_3 = 3-\lambda \end{cases}$$

$$x_i s_i = h \quad \Rightarrow \begin{cases} x_1 s_1 = h \\ x_2 s_2 = h \\ x_3 s_3 = h \end{cases} \quad \Rightarrow \begin{cases} x_1 = \dfrac{h}{1-\lambda} = \dfrac{h}{\mu - 1} \\ x_2 = \dfrac{h}{2-\lambda} = \dfrac{h}{\mu} \\ x_3 = \dfrac{h}{3-\lambda} = \dfrac{h}{\mu + 1} \end{cases} \quad \text{by noting } \mu = 2 - \lambda$$

$$Ax - b = 0 \quad \Rightarrow x_1 + x_2 + x_3 = 1 \quad \Rightarrow \frac{h}{\mu - 1} + \frac{h}{\mu} + \frac{h}{\mu + 1} = 1$$

$$\Rightarrow \mu^3 - 3h\mu^2 - \mu + h = 0$$

We obtain a 3rd degree equation in μ. This equation can have 1, 2 or 3 roots depending on the barrier height h. The solution of the KKT equations is then a function of μ given by

$$\lambda_h = 2 - \mu \quad , \quad \begin{cases} s_h = \begin{pmatrix} 1-\lambda_h & 2-\lambda_h & 3-\lambda_h \end{pmatrix} \\ x_h = h\begin{pmatrix} \dfrac{1}{1-\lambda_h} & \dfrac{1}{2-\lambda_h} & \dfrac{1}{3-\lambda_h} \end{pmatrix} \end{cases}$$

The conditions: $\begin{cases} x_h \geq 0 \\ s_h \geq 0 \end{cases}$ allow to choose μ among the 3 possible roots.

These parametric equations are used to draw the central path contained in the triangle (ABC). It starts from the center of the triangle $x_\infty\left(\frac{1}{3};\frac{1}{3};\frac{1}{3}\right)$ and ends in the vertex $A(1;0;0)$, which is the solution of the linear problem (for $h=0$).

For this plot, a coordinate system is defined in the triangle (ABC) with origin A and axes AB and CD (height from vertex C).
This system of axes is shown in figure 5-12.

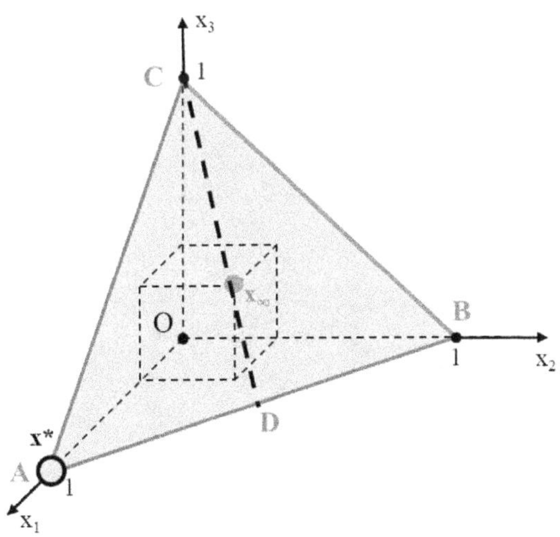

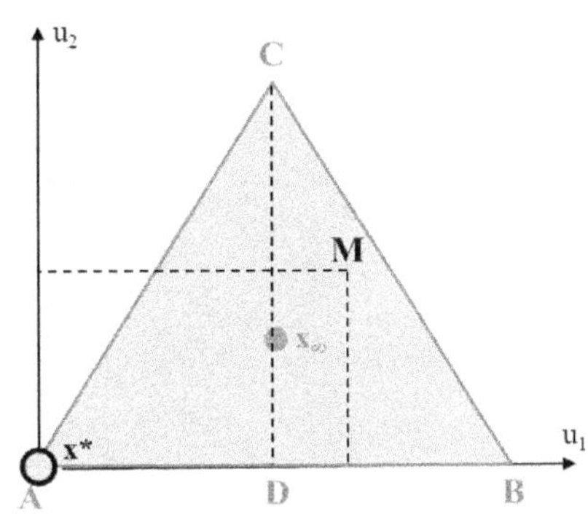

Figure 5-12: System of axes of the polytope.

In these axes, the coordinates $(u_1;u_2)$ of a point M on the central path are

$$\overrightarrow{AM} = u_1\overrightarrow{AB} + u_2\overrightarrow{DC} \Leftrightarrow \begin{pmatrix} x_1-1 \\ x_2 \\ x_3 \end{pmatrix} = u_1\begin{pmatrix} -1 \\ 1 \\ 0 \end{pmatrix} + u_2\begin{pmatrix} -1/2 \\ -1/2 \\ 1 \end{pmatrix} \Rightarrow \begin{cases} u_1 = x_2 + \dfrac{1}{2}x_3 \\ u_2 = x_3 \end{cases}$$

The equation in μ is solved for different values of h, from which x_h and $(u_1;u_2)$ are derived. Table 5-5 shows the solutions obtained for a barrier height decreasing from 10 000 to 0. Figure 5-13 shows the central path in the axes $(u_1;u_2)$ of the triangle (ABC). This path appears almost straight.

h	x_1	x_2	x_3	s_1	s_2	s_3	λ
10000	0.33335	0.33334	0.33333	29998.5	29999.5	30000.5	-29997.5
1000	0.33342	0.33331	0.33320	2999.2	3000.2	3001.2	-2998.2
100	0.33444	0.33332	0.33222	299.0	300.0	301.0	-298.0
10	0.34457	0.33309	0.32236	29.0	30.0	31.0	-28.0
1	0.45162	0.31112	0.23729	2.2142	3.2142	4.2142	-1.2142
0.1000	0.86308	0.08962	0.04726	0.1159	1.1159	2.1159	0.8841
0.0100	0.98507	0.00990	0.00497	0.0102	1.0102	2.0102	0.9898
0.0010	0.99863	0.00100	0.00050	0.0010	1.0010	2.0010	0.9990
0.0001	0.99970	0.00010	0.00005	0.0001	1.0001	2.0001	0.9999
0.0000	**1.00000**	**0.00000**	**0.00000**	**0.0000**	**1.0000**	**2.0000**	**1.0000**

Table 5-5: Central path points as a function of barrier height.

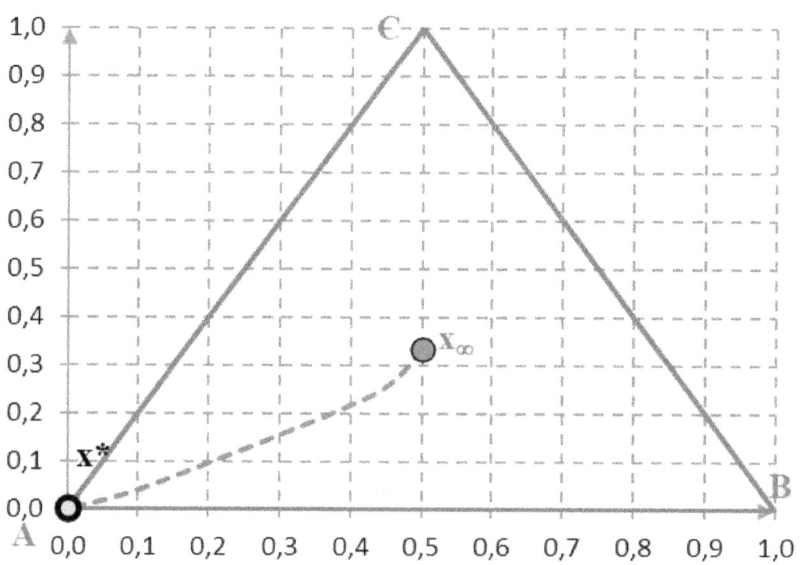

Figure 5-13: Central path in the polytope in axes (u_1 ; u_2).

5.2.2 Direction of move

Let us try to solve the KKT system (5.69) by Newton's method.

By defining the diagonal matrices $X, S \in \mathbb{R}^{n\times n}$ and the vector $e \in \mathbb{R}^n$

$$X=\begin{pmatrix} x_1 & 0 & \cdots & 0 & 0 \\ 0 & x_2 & \cdots & 0 & 0 \\ \vdots & \vdots & \ddots & \vdots & \vdots \\ 0 & 0 & \cdots & x_{n-1} & 0 \\ 0 & 0 & \cdots & 0 & x_n \end{pmatrix} , \quad S=\begin{pmatrix} s_1 & 0 & \cdots & 0 & 0 \\ 0 & s_2 & \cdots & 0 & 0 \\ \vdots & \vdots & \ddots & \vdots & \vdots \\ 0 & 0 & \cdots & s_{n-1} & 0 \\ 0 & 0 & \cdots & 0 & s_n \end{pmatrix} , \quad e=\begin{pmatrix} 1 \\ 1 \\ \vdots \\ 1 \\ 1 \end{pmatrix} \tag{5.70}$$

the equations (5.69) are in the form

$$F(x,\lambda,s)=\begin{pmatrix} Ax-b \\ A^T\lambda+s-c \\ XSe-he \end{pmatrix}=0 \tag{5.71}$$

Newton's iteration is defined with a step α to be set so as not to go outside the domain $\{x \ge 0 \,;\, s \ge 0\}$ (this setting is discussed in section 5.2.3).

$$\begin{pmatrix} x_{k+1} \\ \lambda_{k+1} \\ s_{k+1} \end{pmatrix}=\begin{pmatrix} x_k \\ \lambda_k \\ s_k \end{pmatrix}+\alpha\begin{pmatrix} d_x \\ d_\lambda \\ d_s \end{pmatrix} \quad \text{with} \quad \nabla F(x_k,\lambda_k,s_k)^T\begin{pmatrix} d_x \\ d_\lambda \\ d_s \end{pmatrix}=-F(x_k,\lambda_k,s_k) \tag{5.72}$$

By detailing the function F and its gradient, we obtain the system giving the direction of move (d_x, d_λ, d_s).

$$\begin{pmatrix} A & 0 & 0 \\ 0 & A^T & I \\ S_k & 0 & X_k \end{pmatrix}\begin{pmatrix} d_x \\ d_\lambda \\ d_s \end{pmatrix}=-\begin{pmatrix} Ax_k-b \\ A^T\lambda_k+s_k-c \\ X_kS_ke-he \end{pmatrix} \tag{5.73}$$

Note that the first two KKT equations (5.71) are linear. Therefore, if the point $(x_k \,;\, \lambda_k \,;\, s_k)$ satisfies these equations, the point $(x_{k+1} \,;\, \lambda_{k+1} \,;\, s_{k+1})$ resulting from Newton's iteration will also satisfy them.

Such points belong to the set of **interior points** X_{int} defined by

$$(x,\lambda,s)\in X_{int} \iff \begin{cases} Ax-b=0 \\ A^T\lambda+s-c=0 \\ x>0 \,,\, s>0 \end{cases} \tag{5.74}$$

Starting from an initial interior point, the second member of (5.73) is simplified.

$$\begin{pmatrix} A & 0 & 0 \\ 0 & A^T & I \\ S_k & 0 & X_k \end{pmatrix}\begin{pmatrix} d_x \\ d_\lambda \\ d_s \end{pmatrix} = -\begin{pmatrix} 0 \\ 0 \\ X_k S_k e - he \end{pmatrix} \tag{5.75}$$

Solving this linear system gives the direction (d_x, d_λ, d_s). This direction depends on the barrier height h which defines a central path point, between the center of the polytope (for $h = \infty$) and the solution of the linear problem (for $h = 0$).

Recall that the central path points (5.69) satisfy: $(x_i s_i)_{i=1 \text{ to } n} = h$.

Suppose we are in the point $(x;s)$, and we seek to return to the central path to favor the progression of Newton's method. A natural idea for targeting the point on the central path "closest" to $(x;s)$ is to set h as the average of the products $x_i s_i$. This average is called the **duality measure** ν in the point $(x;s)$.

$$\nu = \frac{1}{n} x^T s = \frac{1}{n}\sum_{i=1}^{n} x_i s_i \tag{5.76}$$

In practice, this average is weighted by the **centering parameter** σ, comprised between 0 and 1. The value of h for calculating the direction (d_x, d_λ, d_s) is thus

$$h = \sigma\nu = \frac{\sigma}{n}\sum_{i=1}^{n} x_i s_i \quad \text{with} \quad 0 \le \sigma \le 1 \tag{5.77}$$

- for $\sigma = 1$, the direction will be towards the central path;
- for $\sigma = 0$, the direction will be towards the edge of the domain $\{x \ge 0 \,;\, s \ge 0\}$.

The parameter σ corrects Newton's direction in order to avoid approaching the boundary of the feasible domain too quickly.

The example 5-12 illustrates the directions obtained depending on the value of σ.

Example 5-12: Effect of the centering parameter (example presented in [R3])

Let us retrieve to the problem of example 5-11.

$$\min_{x_1, x_2, x_3} x_1 + 2x_2 + 3x_3 \quad \text{s.t.} \begin{cases} x_1 + x_2 + x_3 = 1 \\ x_1, x_2, x_3 \ge 0 \end{cases}$$

The matrices of this linear problem are: $A = \begin{pmatrix} 1 & 1 & 1 \end{pmatrix}$, b=1, $c^T = \begin{pmatrix} 1 & 2 & 3 \end{pmatrix}$.

The interior points satisfy: $\begin{cases} Ax-b=0 \\ A^T\lambda+s-c=0 \\ x,s>0 \end{cases} \Rightarrow \begin{cases} x_1+x_2+x_3=1 \\ \lambda+s_1=1 \\ \lambda+s_2=2 \\ \lambda+s_3=3 \end{cases}$, $x_i>0$, $s_i>0$.

To illustrate geometrically the effect of the centering parameter σ, we choose three interior points and calculate the move in these points for $\sigma=0$ or 1.

Coordinates of the three interior points

The selected interior points have the same values of λ and s: $\lambda=0 \Rightarrow \begin{cases} s_1=1 \\ s_2=2 \\ s_3=3 \end{cases}$

and for respective values of x: $(x_1\,;\,x_2\,;\,x_3)=\begin{cases} (0{,}6\,;\,0{,}2\,;\,0{,}2) \rightarrow \text{point n°1} \\ (0{,}2\,;\,0{,}6\,;\,0{,}2) \rightarrow \text{point n°2} \\ (0{,}2\,;\,0{,}2\,;\,0{,}6) \rightarrow \text{point n°3} \end{cases}$

These points are equidistant from the center of the polytope in $x_\infty=\left(\frac{1}{3}\,;\,\frac{1}{3}\,;\,\frac{1}{3}\right)$.

Newton's move

Equation (5.75) gives:
$$\begin{pmatrix} 1 & 1 & 1 & 0 & 0 & 0 & 0 \\ 0 & 0 & 0 & 1 & 1 & 0 & 0 \\ 0 & 0 & 0 & 1 & 0 & 1 & 0 \\ 0 & 0 & 0 & 1 & 0 & 0 & 1 \\ s_1 & 0 & 0 & 0 & x_1 & 0 & 0 \\ 0 & s_2 & 0 & 0 & 0 & x_2 & 0 \\ 0 & 0 & s_3 & 0 & 0 & 0 & x_3 \end{pmatrix}\begin{pmatrix} d_{x1} \\ d_{x2} \\ d_{x3} \\ d_\lambda \\ d_{s1} \\ d_{s2} \\ d_{s3} \end{pmatrix}=\begin{pmatrix} 0 \\ 0 \\ 0 \\ 0 \\ -x_1s_1+h \\ -x_2s_2+h \\ -x_3s_3+h \end{pmatrix}$$

from which we deduce
$$\begin{cases} d_\lambda\left(\dfrac{x_1}{s_1}+\dfrac{x_2}{s_2}+\dfrac{x_3}{s_3}\right)=x_1+x_2+x_3-h\left(\dfrac{1}{s_1}+\dfrac{1}{s_2}+\dfrac{1}{s_3}\right) \\ d_{s1}=d_{s2}=d_{s3}=-d_\lambda \\ d_{x1}=\dfrac{h+x_1d_\lambda}{s_1}-x_1 \ , \ d_{x2}=\dfrac{h+x_2d_\lambda}{s_2}-x_2 \ , \ d_{x3}=\dfrac{h+x_3d_\lambda}{s_3}-x_3 \end{cases}$$

The barrier height h is set by $h=\sigma\nu$ where ν is the duality measure.
$$\nu=\frac{1}{n}x^Ts=\frac{1}{3}(x_1s_1+x_2s_2+x_3s_3)$$

Newton's move is fully determined by choosing a value of the centering parameter σ between 0 and 1.

The moves obtained for $\sigma = 0$ (Newton's direction) or $\sigma = 1$ (central direction) are calculated from the three chosen interior points. Tables 5.6, 5.7 and 5.8 give the move components $(dx_1, dx_2, dx_3, ds_1, ds_2, ds_3, d\lambda)$. Figures 5.14, 5.15 and 5.16 show the moves in the plane of the polytope ABC (figure 5-12):

- directions for $\sigma = 0$ or $\sigma = 1$ in the point $(x_1 ; x_2 ; x_3) = (0.6 ; 0.2 ; 0.2)$:

	v	x_1	x_2	x_3	s_1	s_2	s_3	λ
	0.53333	0.6	0.2	0.2	1.0	2.0	3.0	0.0
σ	h	dx_1	dx_2	dx_3	ds_1	ds_2	ds_3	$d\lambda$
0	0.00000	0.18261	-0.06957	-0.11304	-1.30435	-1.30435	-1.30435	1.30435
1	0.53333	-0.04928	0.06957	-0.02029	-0.02899	-0.02899	-0.02899	0.02899
Central path	0.53333	0.54971	0.27070	0.17956	0.97020	1.97020	2.97020	0.02980

Table 5-6: Central and Newton's direction in point (0.6; 0.2; 0.2).

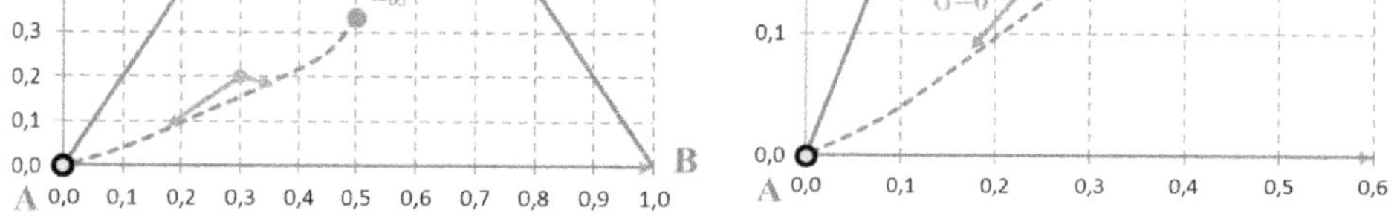

Figure 5-14: Central and Newton's direction in point (0.6; 0.2; 0.2).

The initial point is very close to the central path. It is observed that:
- the move for $\sigma = 0$ (Newton's direction) progresses clearly towards the solution vertex A;
- the move for $\sigma = 1$ (central direction) returns to the central path, but moving away from the solution vertex A.

- directions for $\sigma = 0$ or $\sigma = 1$ in the point $(x_1 ; x_2 ; x_3) = (0.2 ; 0.6 ; 0.2)$:

	v	x_1	x_2	x_3	s_1	s_2	s_3	λ
	0.66667	0.2	0.6	0.2	1.0	2.0	3.0	0.0
σ	h	dx_1	dx_2	dx_3	ds_1	ds_2	ds_3	$d\lambda$
0	0.00000	0.15294	-0.07059	-0.08235	-1.76471	-1.76471	-1.76471	1.76471
1	0.66667	0.38824	-0.38431	-0.00392	0.39216	0.39216	0.39216	-0.39216
Central path	0.66667	0.50965	0.28884	0.20153	1.30808	2.30808	3.30808	-0.30808

Table 5-7: Central and Newton's direction in point (0.2; 0.6; 0.2).

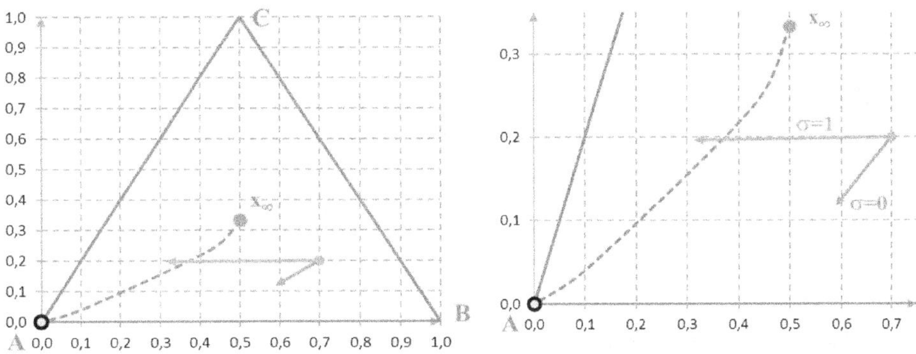

Figure 5-15: Central and Newton's direction in point (0.2; 0.6; 0.2).

The initial point is far from the central path. It is observed that:
- the move for $\sigma = 0$ (Newton's direction) is closer to the edge (AB) with a risk of subsequent blockage;
- the move for $\sigma = 1$ (central direction) returns to the central path while progressing towards the solution vertex A.

- directions for $\sigma = 0$ or $\sigma = 1$ in the point $(x_1 ; x_2 ; x_3) = (0.2 ; 0.2 ; 0.6)$:

	v	x_1	x_2	x_3	s_1	s_2	s_3	λ
	0.80000	0.2	0.2	0.6	1.0	2.0	3.0	0.0
σ	h	dx_1	dx_2	dx_3	ds_1	ds_2	ds_3	$d\lambda$
0	0.00000	0.20000	0.00000	-0.20000	-2.00000	-2.00000	-2.00000	2.00000
1	0.80000	0.41333	0.10667	-0.52000	0.93333	0.93333	0.93333	-0.93333
Central path	0.80000	0.48130	0.30051	0.21845	1.66217	2.66217	3.66217	-0.66217

Table 5-8: Central and Newton's direction in point (0.2; 0.2; 0.6).

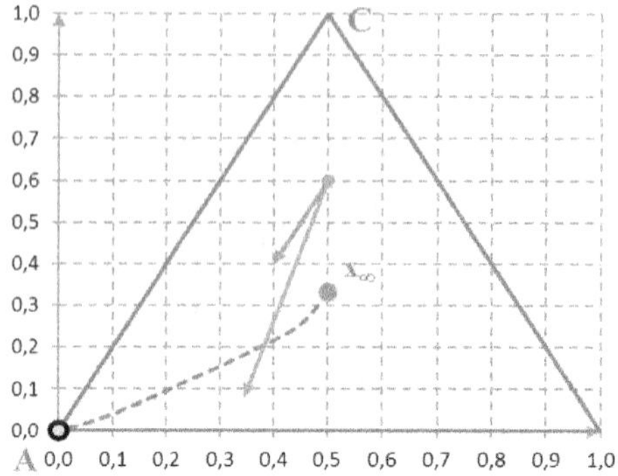

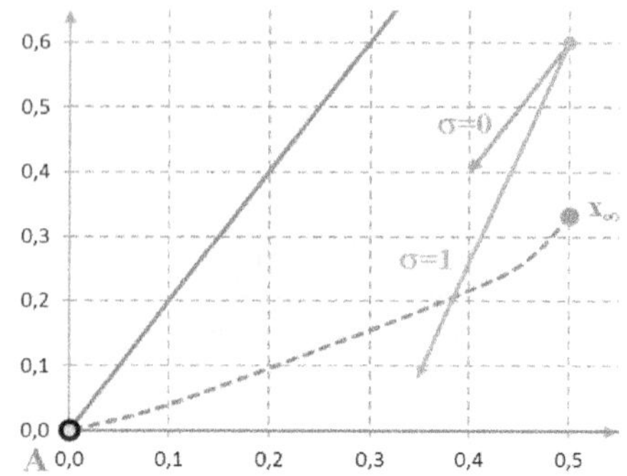

Figure 5-16: Central and Newton's direction in point (0.2; 0.2; 0.6).

The initial point is far from the central path. It is observed that:
- the move for $\sigma = 0$ (Newton's direction) progresses slightly towards the solution vertex A;
- the move for $\sigma = 1$ (central direction) returns to the central path and progressing further towards the solution vertex A.

In the end, it can be seen that the value $\sigma = 0$ is only interesting if one is already close to the central path (point number 1). Otherwise, it can lead to a blockage by approaching an edge too quickly (point number 2). It is therefore important to be close to the central path to enable Newton's method to progress quickly.

5.2.3 Step length

Each Newton iteration involves a direction calculation (d_x, d_λ, d_s) by equations (5.75) with the barrier height set by (5.77)

$$\begin{pmatrix} A & 0 & 0 \\ 0 & A^T & I \\ S_k & 0 & X_k \end{pmatrix} \begin{pmatrix} d_x \\ d_\lambda \\ d_s \end{pmatrix} = -\begin{pmatrix} 0 \\ 0 \\ X_k S_k e - \sigma \nu_k e \end{pmatrix} \quad \text{with } \nu_k = \frac{1}{n} x_k^T s_k \tag{5.78}$$

followed by a move of step length α along that direction

$$\begin{pmatrix} x_{k+1} \\ \lambda_{k+1} \\ s_{k+1} \end{pmatrix} = \begin{pmatrix} x_k \\ \lambda_k \\ s_k \end{pmatrix} + \alpha \begin{pmatrix} d_x \\ d_\lambda \\ d_s \end{pmatrix} \tag{5.79}$$

The direction depends on the centering parameter σ (between 0 and 1) and the move depends on the step length α. The objective is to set the values of σ and α to progress close to the central path and avoid premature approach to the boundary of the domain $\{x \ge 0 \,;\, s \ge 0\}$.

Let us first define the neighborhood of the central path.

<u>Neighborhood of the central path</u>

Consider a point $(x;s)$ with duality measure: $\nu = \frac{1}{n}x^T s = \frac{1}{n}\sum_{i=1}^{n} x_i s_i$.

The point on the central path "closest" to $(x;s)$ is the one with the same duality measure ν. The distance from the point $(x;s)$ to the central path is defined by

$$\delta = \frac{1}{\nu}\left\| \begin{pmatrix} x_1 s_1 \\ \vdots \\ x_n s_n \end{pmatrix} - \begin{pmatrix} \nu \\ \vdots \\ \nu \end{pmatrix} \right\| = \frac{1}{\nu}\|XSe - \nu e\| \tag{5.80}$$

This normalized distance is:

- $\delta = 0$ if (x,s) is on the central path, because $(x_i s_i)_{i=1 \text{ to } n} = \nu$;
- $\delta = \|e\|$ if (x,s) is the solution to the linear problem, because $(x_i s_i)_{i=1 \text{ to } n} = 0$.

The neighborhood of the central path is the set of interior points such that $\delta \le \delta_m$ where the neighborhood width δ_m is chosen between 0 and 1. Recall that the interior points also satisfy equations (5.74).

Two main types of neighborhoods are considered, depending on the norm chosen.

By using the L_2 norm

$$\frac{1}{\nu}\|XSe - \nu e\|_2 \le \delta_m \;\Rightarrow\; \sum_{i=1}^{n}\left(\frac{x_i s_i}{\nu} - 1\right)^2 \le \delta_m^2 \tag{5.81}$$

we define the **restricted neighborhood** noted $V_2(\delta_m)$

$$V_2(\delta_m) = \left\{ (x,\lambda,s) \in X_{int} \;/\; \sum_{i=1}^{n}\left(\frac{x_i s_i}{\nu} - 1\right)^2 \le \delta_m^2 \right\} \tag{5.82}$$

By using the L∞ norm

$$\frac{1}{\nu}\|XSe-\nu e\|_\infty \le \delta_m \Rightarrow -\nu\delta_m \le x_i s_i - \nu \le \nu\delta_m$$
$$\Rightarrow \nu(1-\delta_m) \le x_i s_i \le \nu(1+\delta_m) \tag{5.83}$$

and changing δ_m to $1-\delta_m$, we define the **large neighborhood** noted $V_{-\infty}(\delta_m)$

$$V_{-\infty}(\delta_m) = \{(x,\lambda,s) \in X_{int} \;/\; x_i s_i \ge \nu\delta_m\} \tag{5.84}$$

This large neighborhood only consider the lower value of the products $x_i s_i$, as the aim is to avoid approaching too soon the boundary of the domain $\{x \ge 0 \,;\, s \ge 0\}$.

Let us now look at two simple strategies based on these neighborhoods.

<u>Restricted-step algorithm</u>

This algorithm tries to stay in the restricted neighborhood by setting only the parameter σ (which determines the direction) and by fixing the step $\alpha = 1$.
The new point $(x_{k+1}\,;\,\lambda_{k+1}\,;\,s_{k+1})$ must satisfy

$$\frac{1}{\nu_k}\|X_{k+1}S_{k+1}e-\nu_k e\|_2 \le \delta_m \tag{5.85}$$

where ν_k is the duality measure in the initial point $(x_k\,;\,\lambda_k\,;\,s_k)$.

To calculate the product $X_{k+1}S_{k+1}e$, we introduce as in (5.70) the diagonal matrices D_x, D_s associated respectively with the vectors d_x, d_s.

$$D_x = \begin{pmatrix} d_{x,1} & 0 & \cdots & 0 & 0 \\ 0 & d_{x,2} & \cdots & 0 & 0 \\ \vdots & \vdots & \ddots & \vdots & \vdots \\ 0 & 0 & \cdots & d_{x,n-1} & 0 \\ 0 & 0 & \cdots & 0 & d_{x,n} \end{pmatrix}, \quad D_s = \begin{pmatrix} d_{s,1} & 0 & \cdots & 0 & 0 \\ 0 & d_{s,2} & \cdots & 0 & 0 \\ \vdots & \vdots & \ddots & \vdots & \vdots \\ 0 & 0 & \cdots & d_{s,n-1} & 0 \\ 0 & 0 & \cdots & 0 & d_{s,n} \end{pmatrix} \tag{5.86}$$

Let us develop the product $X_{k+1}S_{k+1}e$.

$$X_{k+1}S_{k+1}e = (X_k + D_x)(S_k + D_s)e = X_kS_ke + X_kD_se + D_xS_ke + D_xD_se \tag{5.87}$$

Observing that $D_x S_k = S_k D_x$ (the matrices are diagonal), $D_x e = d_x$, $D_s e = d_s$, and neglecting the $D_x D_s e$ term of order 2 (for small moves), we obtain

$$X_{k+1} S_{k+1} e = X_k S_k e + X_k d_s + S_k d_x \tag{5.88}$$

The direction (5.78) satisfies: $S_k d_x + X_k d_s = -(X_k S_k e - \sigma \nu_k e)$, which leads to

$$X_{k+1} S_{k+1} e = \sigma \nu_k e \tag{5.89}$$

Let us replace in the inequality (5.85) with $\|e\|_2 = \sqrt{n}$ and $0 \le \sigma \le 1$.

$$\frac{1}{\nu_k}\|X_{k+1} S_{k+1} e - \nu_k e\|_2 = \frac{1}{\nu_k}\|(\sigma - 1)\nu_k e\|_2 = |\sigma - 1|\sqrt{n} = (1-\sigma)\sqrt{n} \le \delta_m \tag{5.90}$$

We obtain the minimum value of the centering parameter σ so that the new point $(x_{k+1}; \lambda_{k+1}; s_{k+1})$ remains in the neighborhood $V_2(\delta_m)$.

$$\sigma \ge 1 - \frac{\delta_m}{\sqrt{n}} \tag{5.91}$$

Since this inequality was obtained by neglecting the second-order terms in (5.87), it is necessary to check that the new point remains actually in $V_2(\delta_m)$ and possibly increase the value of σ to get closer to the central path.
A neighborhood width δ_m about 0.5 is used to initialize the setting of σ.

Long-step algorithm

This algorithm tries to stay in the large neighborhood by fixing the parameter σ (which determines the direction) and by setting only the step length α. This strategy is the "reverse" of the restricted step algorithm.
The new point $(x_{k+1}; \lambda_{k+1}; s_{k+1})$ must satisfy

$$\min_{i=1 \text{ to } n} x_i s_i \ge \nu_k \delta_m \tag{5.92}$$

We first calculate the direction (d_x, d_λ, d_s) by (5.78) with the chosen value of σ.

The step length is then set by dichotomy starting from $\alpha = 1$ and halving until the condition (5.92) is satisfied.
The large neighborhood $V_{-\infty}(\delta_m)$ being not very restrictive, we can fix a small width: $\delta_m = 0.001$ and a small centering parameter: $\sigma = 0.1$.

Figure 5-17 illustrates the progression of the long-step algorithm. In the primal-dual space (products $x_i s_i$), the neighborhood $V_{-\infty}(\delta_m)$ is a cone from the origin. Each iteration moves away from the bound before returning to it.

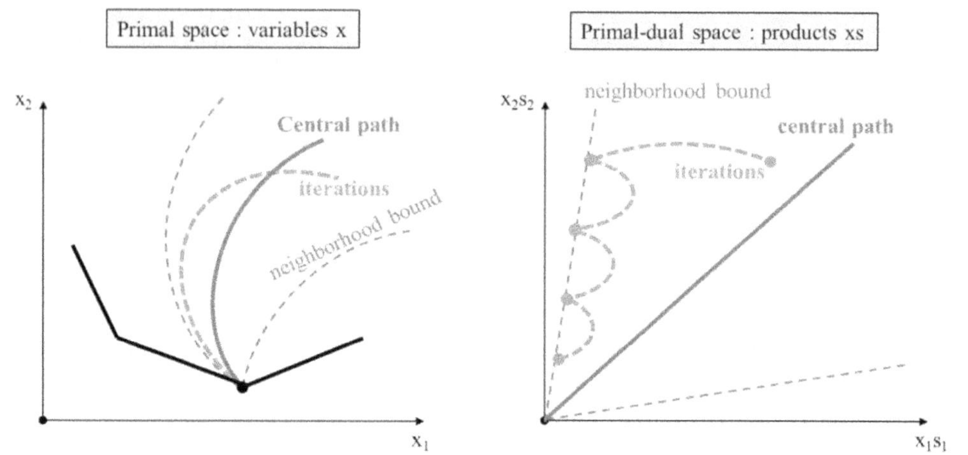

Figure 5-17: Iterations of the long-step algorithm.

The example 5-13 compares the performance of the restricted-step algorithm and the long-step algorithm.

Example 5-13: Restricted-step and long-step algorithms (see [R3])

Let us compare the restricted-step and long-step algorithms on the problem of example 5-12.

$$\min_{x_1,x_2,x_3} x_1 + 2x_2 + 3x_3 \text{ s.t. } \begin{cases} x_1 + x_2 + x_3 = 1 \\ x_1, x_2, x_3 \geq 0 \end{cases}$$

The starting points are the 3 interior points of example 5-12, whose respective coordinates are

$$\lambda = 0 \Rightarrow \begin{cases} s_1 = 1 \\ s_2 = 2 \\ s_3 = 3 \end{cases} \text{ and } (x_1 ; x_2 ; x_3) = \begin{cases} (0{,}6 \;;\; 0{,}2 \;;\; 0{,}2) \rightarrow \text{point n°1} \\ (0{,}2 \;;\; 0{,}6 \;;\; 0{,}2) \rightarrow \text{point n°2} \\ (0{,}2 \;;\; 0{,}2 \;;\; 0{,}6) \rightarrow \text{point n°3} \end{cases}$$

The table 5-9, figures 5-18 and 5-19 show the decrease of the duality measure ν for both algorithms.

Restricted-step algorithm ($\delta_m = 0.4$)

x_1	0.6	0.2	0.2
x_2	0.2	0.6	0.2
x_3	0.2	0.2	0.6
Iteration	v	v	v
0	**0.53333**	**0.66667**	**0.80000**
1	0.41017	0.51271	0.61525
2	0.31544	0.39430	0.47316
3	0.24259	0.30324	0.36389
4	0.18657	0.23321	0.27985
5	0.14348	0.17935	0.21522
6	0.11035	0.13793	0.16552
7	0.08486	0.10608	0.12729
8	0.06526	0.08158	0.09790
9	0.05019	0.06274	0.07529
10	0.03860	0.04825	0.05790
20	0.00279	0.00349	0.00419
30	0.00020	0.00025	0.00030
40	0.00001	0.00002	0.00002
45	**0.00000**	**0.00000**	**0.00001**

Long-step algorithm ($\delta_m = 10^{-3}$, $\sigma = 0.1$)

x_1	0.6	0.2	0.2
x_2	0.2	0.6	0.2
x_3	0.2	0.2	0.6
Iteration	v	v	v
0	**0.53333**	**0.66667**	**0.80000**
1	0.29333	0.36667	0.44000
2	0.16133	0.20167	0.24200
3	0.08873	0.02017	0.02420
4	0.00887	0.00202	0.00242
5	0.00089	0.00020	0.00024
6	0.00009	0.00002	0.00002
7	0.00001	0.00000	0.00000
8	0.00000	0.00000	0.00000

Restricted-step

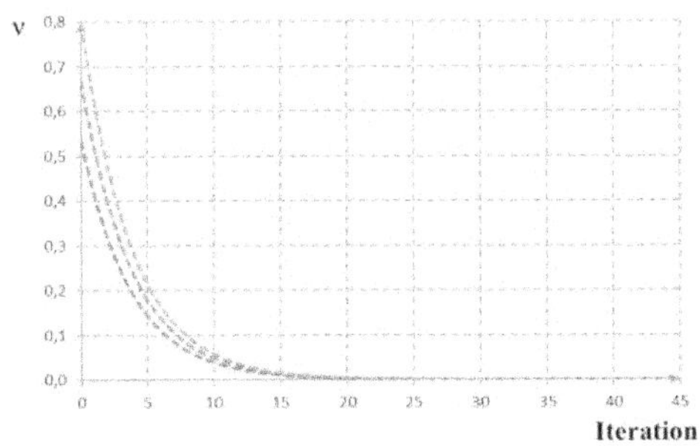

Long-step

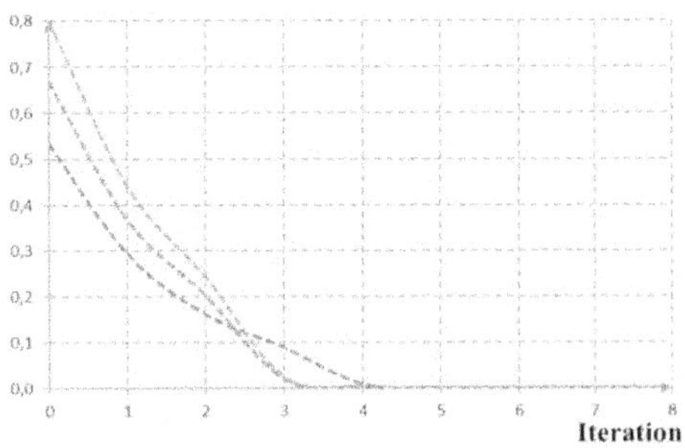

Table 5-9: Restricted and long-step algorithms: duality measure.

The figures 5-18 and 5-19 show the progression towards the solution vertex A in the plane of the polytope.

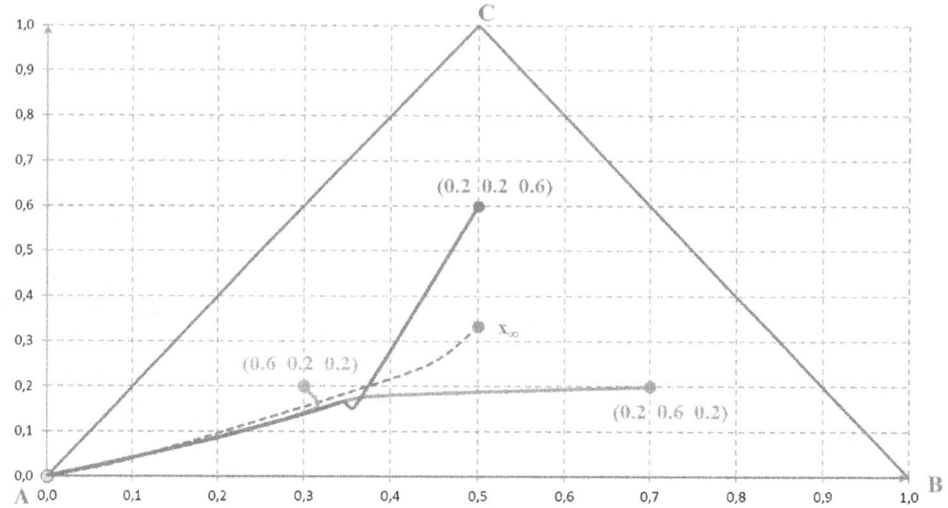

Figure 5-18: Restricted-step algorithm: progression in the polytope plane.

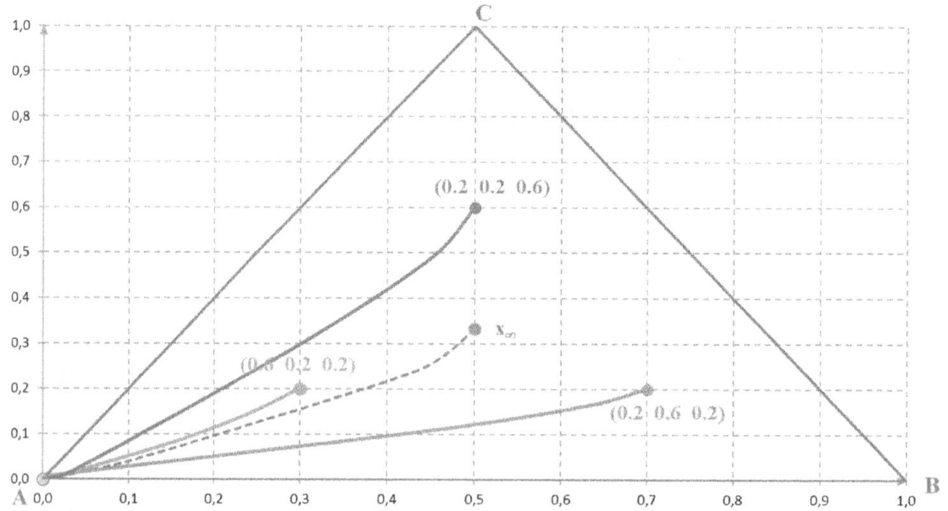

Figure 5-19: Long-step algorithm: progression in the polytope plane.

We observe that the long-step algorithm is more efficient: the constraint of staying in the neighborhood of the central path is less strong, which allows a more direct progression towards the solution.

5.2.4 Prediction-correction algorithm

Restricted-step or long-step strategies are simple, but they do not fully exploit the possibilities of combined setting of the parameters σ and α.
The prediction-correction algorithm implemented in most software combines these settings in three stages at each iteration: prediction, correction and recentering.

The **prediction** stage directly aims at the solution of the linear problem with a zero centering parameter: $\sigma = 0$. The associated move noted $(d_{px}, d_{p\lambda}, d_{ps})$ is obtained by solving the linear system

$$\begin{pmatrix} A & 0 & 0 \\ 0 & A^T & I \\ S_k & 0 & X_k \end{pmatrix} \begin{pmatrix} d_{px} \\ d_{p\lambda} \\ d_{ps} \end{pmatrix} = -\begin{pmatrix} 0 \\ 0 \\ X_k S_k e \end{pmatrix} \tag{5.93}$$

Applying this move, we would get

$$X_{k+1} S_{k+1} e = (X_k + D_{px})(S_k + D_{ps}) e = X_k S_k e + X_k D_{ps} e + D_{px} S_k e + D_{px} D_{ps} e \tag{5.94}$$

where D_{px}, D_{ps} are the diagonal matrices associated respectively with d_{px}, d_{ps}.

This expression is simplified by observing that $D_x e = d_x$, $D_s e = d_s$ and using the last equation of (5.93), which gives: $S_k d_{px} + X_k d_{ps} + X_k S_k e = 0$.
This leaves only the second-order term, which was expected because Newton's iteration only considers first-order terms.

$$X_{k+1} S_{k+1} e = D_{px} D_{ps} e \tag{5.95}$$

The **correction** stage aims to cancel this second-order term. The associated move noted $(d_{cx}, d_{c\lambda}, d_{cs})$ is obtained by solving the linear system

$$\begin{pmatrix} A & 0 & 0 \\ 0 & A^T & I \\ S_k & 0 & X_k \end{pmatrix} \begin{pmatrix} d_{cx} \\ d_{c\lambda} \\ d_{cs} \end{pmatrix} = -\begin{pmatrix} 0 \\ 0 \\ D_{px} D_{ps} e \end{pmatrix} \tag{5.96}$$

The prediction $(d_{px}, d_{p\lambda}, d_{ps})$ and correction $(d_{cx}, d_{c\lambda}, d_{cs})$ moves do not take into account the domain boundaries $\{x \geq 0 \,;\, s \geq 0\}$.

The **recentering** stage introduces the centering parameter σ. To estimate this parameter, the maximum step achievable along the direction $(d_{px}, d_{p\lambda}, d_{ps})$ without leaving the domain $\{x \geq 0 ; s \geq 0\}$ is calculated. The move is limited by the components of d_{px} and d_{ps} becoming negative. The maximum step α_{px} along d_{px} and the maximum step α_{ps} along d_{ps} are given by

$$\alpha_{px} = \min_{i, d_{px,i}<0}\left(-\frac{x_{k,i}}{d_{px,i}}, 1\right) \quad , \quad \alpha_{ps} = \min_{i, d_{ps,i}<0}\left(-\frac{s_{k,i}}{d_{ps,i}}, 1\right) \tag{5.97}$$

Applying these steps α_{px} and α_{ps} from the point $(x_k ; s_k)$, we would obtain the duality measure $\tilde{\nu}$

$$\tilde{\nu} = \frac{1}{n}(x_k + \alpha_{px} d_{px})^T (s_k + \alpha_{ps} d_{ps}) \tag{5.98}$$

The centering parameter σ is set by comparing $\tilde{\nu}$ and the initial value ν_k.

If $\tilde{\nu} < \nu_k$, the progress in the prediction direction is good and does not require recentering. If $\tilde{\nu} \approx \nu_k$, the progress along the prediction direction is limited by the step, and recentering is required.

In practice, the following empirical setting gives good results.

$$\sigma = \left(\frac{\tilde{\nu}}{\nu_k}\right)^3 \tag{5.99}$$

At the end of the three stages of prediction, correction and recentering, the direction of move (d_x, d_λ, d_s) is obtained by solving the linear system

$$\begin{pmatrix} A & 0 & 0 \\ 0 & A^T & I \\ S_k & 0 & X_k \end{pmatrix} \begin{pmatrix} d_x \\ d_\lambda \\ d_s \end{pmatrix} = -\begin{pmatrix} 0 \\ 0 \\ X_k S_k e + D_{px} D_{ps} e - \sigma \nu_k e \end{pmatrix} \tag{5.100}$$

The new point is defined by applying a step α_x along d_x (primal step) and a step α_s along d_λ and d_s (dual step).

$$\begin{pmatrix} x_{k+1} \\ \lambda_{k+1} \\ s_{k+1} \end{pmatrix} = \begin{pmatrix} x_k \\ \lambda_k \\ s_k \end{pmatrix} + \begin{pmatrix} \alpha_x d_x \\ \alpha_s d_\lambda \\ \alpha_s d_s \end{pmatrix} \tag{5.101}$$

The steps α_x and α_s are set so that the new point $(x_{k+1}; \lambda_{k+1}; s_{k+1})$ remains in a large neighborhood $V_{-\infty}(\delta_m)$ of the central path. The width δ_m of this neighborhood may initially be small $(\delta_m = 0.1)$, and then grow as the solution vertex is approached.

The steps α_x and α_s can also be set directly with respect to the domain boundary $\{x \geq 0 \,;\, s \geq 0\}$, without considering any notion of neighborhood.

$$\alpha_x = \min_{i, d_{px,i} < 0} \left(-\frac{x_{k,i}}{d_{x,i}} , 1 \right) \quad , \quad \alpha_s = \min_{i, d_{ps,i} < 0} \left(-\frac{s_{k,i}}{d_{s,i}} , 1 \right) \tag{5.102}$$

These steps are applied in (5.101) with a reduction coefficient η slightly inferior to 1, so as to remain strictly within the feasible domain.

Solving the linear system

The prediction-correction algorithm requires solving three linear systems having the same matrix at each iteration. These systems are of the form

$$\begin{pmatrix} A & 0 & 0 \\ 0 & A^T & I \\ S_k & 0 & X_k \end{pmatrix} \begin{pmatrix} d_x \\ d_\lambda \\ d_s \end{pmatrix} = \begin{pmatrix} \beta_x \\ \beta_\lambda \\ \beta_s \end{pmatrix} \tag{5.103}$$

The matrices X_k and S_k are non-singular diagonal (because $x_k > 0$, $s_k > 0$) and have inverses X_k^{-1} and S_k^{-1} respectively. By expressing d_s from the last equation, and permuting the first two equations, we obtain a symmetric reduced system.

$$\begin{pmatrix} -X_k^{-1}S_k & A^T \\ A & 0 \end{pmatrix} \begin{pmatrix} d_x \\ d_\lambda \end{pmatrix} = \begin{pmatrix} \beta_\lambda - X_k^{-1}\beta_s \\ \beta_x \end{pmatrix} \tag{5.104}$$

with $d_s = X_k^{-1}\beta_s - X_k^{-1}S_k d_x$

This formulation is well suited to large problems when the matrix A is sparse.

The solution can be further developed by expressing d_x from the first equation.

There remains a linear system in d_λ from which d_s and d_x can be calculated.

$$\begin{cases} AS_k^{-1}X_kA^Td_\lambda = \beta_x + AS_k^{-1}X_k\beta_\lambda - AS_k^{-1}\beta_s \\ d_s = \beta_\lambda - A^Td_\lambda \\ d_x = S_k^{-1}\beta_s - S_k^{-1}X_kd_s \end{cases} \tag{5.105}$$

Depending on the structure of the matrix A, it is simpler to use (5.104) or (5.105).

Initialization

The behavior of the algorithm is quite sensitive to the initial point. Ideally, an interior point not too close to the boundary of the domain $\{x \geq 0 \,;\, s \geq 0\}$ should be chosen.

Recall that the set of **interior points** X_{int} is defined by (5.74).

$$(x, \lambda, s) \in X_{int} \iff \begin{cases} Ax - b = 0 \\ A^T\lambda + s - c = 0 \\ x, s > 0 \end{cases} \tag{5.106}$$

A good initialization can be achieved by solving the following two problems

$$\begin{aligned} &\min_x \frac{1}{2} x^T x \quad \text{s.t.} \quad Ax = b \\ &\min_{\lambda,s} \frac{1}{2} s^T s \quad \text{s.t.} \quad A^T\lambda + s = c \end{aligned} \tag{5.107}$$

whose respective solutions are

$$\begin{cases} x = A^T(AA^T)^{-1}b \\ \lambda = (AA^T)^{-1}Ac \\ s = c - A^T\lambda \end{cases} \tag{5.108}$$

The negative components of the vectors x and s are then arbitrarily corrected to come back into the domain $\{x \geq 0 \,;\, s \geq 0\}$. This initial point does not satisfy the equalities (5.106), but the second members of these linear equations will be immediately cancelled by the first Newton iteration (5.73).

5.2.5 Extensions

The formulation developed in section 5.2.1 and leading to the KKT system (5.69) can be extended in a similar way to nonlinear problems. In particular, let us consider the case of a quadratic-linear problem and the case of a nonlinear problem with equality constraints.

Quadratic problem

A quadratic-linear problem has the standard form

$$\min_{x \in \mathbb{R}^n} \frac{1}{2} x^T Q x + c^T x \quad \text{s.t.} \quad \begin{cases} Ax = b \\ x \geq 0 \end{cases} \tag{5.109}$$

The Lagrangian is expressed with multipliers $\lambda \in \mathbb{R}^m$ for the equality constraints and multipliers $s \in \mathbb{R}^n$ for the inequality constraints.

$$L(x,\lambda,s) = \frac{1}{2}x^TQx + c^Tx + \lambda^T(Ax-b) - s^Tx \tag{5.110}$$

The KKT conditions of order 1 give a system in variables (x,λ,s)

$$\begin{cases} Ax-b=0 \\ Qx+c-A^T\lambda-s=0 \\ x_is_i=0 \ , \ x_i \ge 0 \, , s_i \ge 0 \ , \ i=1 \text{ to } n \end{cases} \tag{5.111}$$

which is transformed into a barrier problem with a height parameter h.

$$F(x,\lambda,s) = \begin{pmatrix} Ax-b \\ Qx+c-A^T\lambda-s \\ XSe-he \end{pmatrix} = 0 \tag{5.112}$$

Newton's method gives the linear system

$$\begin{pmatrix} A & 0 & 0 \\ Q & A^T & I \\ S_k & 0 & X_k \end{pmatrix}\begin{pmatrix} d_x \\ d_\lambda \\ d_s \end{pmatrix} = -\begin{pmatrix} Ax_k-b \\ Q_kx_k+A^T\lambda_k+s_k+c \\ X_kS_ke-he \end{pmatrix} \tag{5.113}$$

very similar to system (5.73) except for the matrix Q.

Nonlinear problem

A nonlinear problem with equality constraints is of the form

$$\min_{x\in\mathbb{R}^n} f(x) \quad \text{s.t.} \quad \begin{cases} c(x)=0 \\ x \ge 0 \end{cases} \tag{5.114}$$

The Lagrangian is expressed with multipliers $\lambda \in \mathbb{R}^m$ for the equality constraints and multipliers $s \in \mathbb{R}^n$ for the bound constraints on x.

$$L(x,\lambda,s) = f(x) + \lambda^Tc(x) - s^Tx \tag{5.115}$$

The KKT conditions of order 1 give a system in variables (x,λ,s)

$$\begin{cases} c(x)=0 \\ \nabla_x L(x,\lambda,s)=0 \\ x_is_i=0 \ , \ x_i \ge 0 \, , s_i \ge 0 \ , \ i=1 \text{ to } n \end{cases} \tag{5.116}$$

which is transformed into a barrier problem with a height parameter h.

$$F(x,\lambda,s)=\begin{pmatrix} c(x) \\ \nabla_x L(x,\lambda,s) \\ XSe-he \end{pmatrix}=0 \tag{5.117}$$

Newton's method gives the linear system

$$\begin{pmatrix} \nabla c(x_k)^T & 0 & 0 \\ \nabla^2_{xx} L(x_k,\lambda_k,s_k) & \nabla c(x_k) & I \\ S_k & 0 & X_k \end{pmatrix}\begin{pmatrix} d_x \\ d_\lambda \\ d_s \end{pmatrix}=-\begin{pmatrix} c(x_k) \\ \nabla_x L(x_k,\lambda_k,s_k) \\ X_k S_k e-he \end{pmatrix} \tag{5.118}$$

This system is more complex than the system (5.73), as it requires an estimation of the Hessian of the Lagrangian. Moreover, taking into account inequality constraints requires algorithmic improvements described in section 4.5.

5.3 Conclusion

5.3.1 The key points

- The solution of a linear problem corresponds to a vertex of the polytope of constraints. It has at least n–m zero variables, where n is the number of variables and m is the number of equality constraints;
- the simplex method browses the vertices in an ordered fashion to arrive at the solution. This is the most commonly used method and it performs very well on the vast majority of applications, although the computation time can be exponential in the worst cases;
- interior point methods seek to solve the optimality conditions by a Newton method by progressing within the polytope of constraints. They are more difficult to tune than the simplex, but they guarantee polynomial computation time and perform well on large problems.

5.3.2 To go further

- *Programmation mathématique (M. Minoux, Lavoisier 2008, 2e édition)*

  Chapter 2 presents in detail the theoretical results of linear programming, the simplex method and the dual simplex method. The explanations are continued in chapter 7 to address integer linear programming.

- *Introduction à l'optimisation différentiable (M. Bierlaire, Presses polytechniques et universitaires normandes 2006)*

  The simplex method is presented in chapter 17 and the interior point methods in chapter 19. The explanations are accompanied by detailed examples of the application of the algorithms.

- *Numerical optimization (J. Nocedal, S.J. Wright, Springer 2006)*

  The simplex method is presented in chapter 13 and the interior point methods in chapter 14. Theoretical convergence results are given with their demonstrations. Much practical advice is also given on the implementation of the algorithms.

- *Practical methods of optimization (R. Fletcher, Wiley 1987, 2nd edition)*

  Chapter 8 presents the simplex method. Chapter 10 on quadratic programming describes the simplex technique with complementary pivoting. Many practical tips are given on how to use the algorithms.

Index

Short bibliography

[R1] Introduction to numerical continuation methods
(E.L. Allgower, K. Georg, Siam 2003)

[R2] Analyse numérique
(Collectif direction J. Baranger, Hermann 1991)

[R3] Introduction à l'optimisation différentiable
(M. Bierlaire, Presses polytechniques et universitaires normandes 2006)

[R4] Optimisation continue
(F.J. Bonnans, Dunod 2006)

[R5] Numerical optimization
(J.F. Bonnans, J.C. Gilbert, C. Lemaréchal, C.A. Sagastizabal, Springer 2003)

[R6] Numerical methods for unconstrained optimization and nonlinear equations
(J.E. Dennis, R.B. Schnabel, Siam 1996)

[R7] Métaheuristiques pour l'optimisation difficile
(J. Dréo, A. Pétrowski, P. Siarry, E. Taillard, Eyrolles 2003)

[R8] Practical methods of optimization
(R. Fletcher, Wiley 1987, 2nd edition)

[R9] Practical optimization
(P.E. Gill, W. Murray, M.H. Wright, Elsevier 2004)

[R10] Les mathématiques du mieux-faire
(J.B. Hiriart-Urruty, Ellipses 2008)

[R11] Programmation mathématique
(M. Minoux, Lavoisier 2008, 2e édition)

[R12] Numerical optimization
(J. Nocedal, S.J. Wright, Springer 2006)

[R13] Proximal algorithms
(N. Parikh, Foundations and Trends in Optimization, Vol 1, No 3, 2013)

[R14] Ant colony optimization
(M. Rodrigo, T. Stützle, MIT Press 2004)

[R15] Stochastic optimization
(J.J. Schneider, S. Kirkpatrick, Springer 2006)

[R16] Practical mathematical optimization
(J.A. Snyman, Springer 2005)

www.ingramcontent.com/pod-product-compliance
Lightning Source LLC
Chambersburg PA
CBHW071409230926
51846CB00001B/37
9782759831623